AF294693

Springer Series on Fluorescence
Methods and Applications

O. Wolfbeis
Editor-in-Chief

Springer

*Berlin
Heidelberg
New York
Barcelona
Hong Kong
London
Milan
Paris
Tokyo*

New Trends in Fluorescence Spectroscopy

Bernard Valeur · Jean-Claude Brochon (Eds.)

Applications to Chemical and Life Sciences

With 187 Figures and 39 Tables

Springer

Fluorescence spectroscopy, fluorescence imaging and fluorescent probes are indispensible tools in numerous fields of modern medicine and science, including molecular biology, biophysics, biochemistry, clinical diagnosis and analytical and environmental chemistry. Applications stretch from spectroscopy and sensor technology to microscopy and imaging, to single molecule detection, to the development of novel fluorescent probes, and to proteomics and genomics. The *Springer Series on Fluorescence* aims at publishing state-of-the-art articles that can serve as invaluable tools for both practitioners and researchers being active in this highly interdisciplinary field. The carefully edited collection of papers in each volume will give continuous inspiration for new research and will point to exciting new trends.

– Springer WWW home page: http://www.springer.de

ISSN 1617-1306
ISBN 978-3-642-63214-3   ISBN 978-3-642-56853-4 (eBook)
DOI 10.1007/978-3-642-56853-4

Library of Congress Cataloging-in-Publication Data

New trends in fluorescence spectroscopy : applications to chemical and life science / Bernard Valeur, Jean-Claude Brochon (eds.).
p. cm. – (Springer series on fluorescence ; 1), Includes bibliographical references and index.
ISBN 978-3-642-63214-3
1. Fluorescence spectroscopy. I. Valeur, Bernard, 1944- II. Brochon, Jean-Claude, 1944- III. Series.

QD96.F56 N48 2001
543′.08584–dc21

http://www.springer.de
© Springer-Verlag Berlin Heidelberg 2001
Originally published by Springer-Verlag Berlin Heidelberg New York in 2001
Softcover reprint of the hardcover 1st edtion 2001

Typesetting: Fotosatz-Service Köhler GmbH, Würzburg
Production editor: Christiane Messerschmidt, Rheinau
Cover: design & production, Heidelberg
Printed on acid-free paper   SPIN: 10993446   52/3111 – 5 4 3 2

Editor-in-Chief

Professor Dr. Otto Wolfbeis
University of Regensburg
Institute of Analytical Chemistry, Chemo- and Biosensors
93040 Regensburg
Germany

Volume Editor

Professor Bernard Valeur
Conservatoire National des Arts et Métiers
Laboratoire de Chimie Général
292 rue Saint-Martin
75141 Paris Cedex 03
France
e-mail: valeur@cnam.fr

Dr. Jean-Claude Brochon
Ecole Normale Supérieure de Cachan
Photobiologie Moléculaire L.B.P.A.
61 Av. du Président Wilson
94235 Cachan Cedex
France
e-mail: Brochon@lbpa.ens-cachan.fr

Preface

Fluorescence is more and more widely used as a tool of investigation, analysis, control and diagnosis in many fields relevant to physical, chemical, biological and medical sciences. New technologies continuously emerge thanks to the progress in the design of light sources (e.g. laser diodes), detectors (3D, 4D) and compact ultrafast electronic devices. In particular, much progress has been made in time-resolved fluorescence microscopy (FLIM: Fluorescence Lifetime Imaging Microscopy; FCS: Fluorescence Correlation Spectroscopy). Furthermore, the sensitivity now allows one to detect a single molecule in the restricted field of a confocal microscope, which actually offers the possibility to study phenomena at a molecular level.

The development of new fluorescent probes is still a necessity. In particular, the growing use of lasers implies high resistance to photodegradation. Fluorescence emission at long wavelengths is also a distinct advantage. Furthermore, in vivo inclusion of new fluorescent aromatic residues in proteins offer new potentialities in biology.

Fluorescence-based selective detection of ions and molecules is still the object of special attention. Considerable effort is being made in the design of supramolecular systems in which the recognition event is converted into a fluorescence signal easily detected. New fluorescent sensors for clinical diagnosis and detection of pollutants in atmosphere and water are extensively developed.

All these developments justify the regular publication of books giving the state-of-the-art of the methods and applications of fluorescence spectroscopy.

The present book collects articles written by invited speakers (and other participants upon invitation) at the 6th International Conference on Methods and Applications of Fluorescence Spectroscopy (MAFS-6), held in Paris, France, in September 1999. It is the aim of this series of conferences to bring together researchers working in various fields which employ fluorescence as a tool of investigation for both fundamental and applied purposes. Applications in chemistry, physics, biology and medicine were covered. The presentations as well as the number of participants (more than 300 from 32 countries) reflected the ever-growing interest in fluorescence and showed the rise of new methodologies.

Attention was paid in this meeting – and in this book as well – to retain the balance between basic science and applied science, and between physicochemical sciences and life sciences. An original aspect of this meeting was the short opening session devoted to great characters in the history of fluorescence. We have considered that it is worth including the relevant articles in the present book. It is indeed important to know not only where we are going to but also where we are coming from.

The book is divided into the following sections:

I   Historical Aspects of Fluorescence
II   Fluorescence of Molecular and Supramolecular Systems
III   Fluorescence in Sensing Applications
IV   New Techniques of Fluorescence Microscopy in Biology
V   Proteins and Their Interactions as Studied by Fluorescence Methods

We hope that this book will provide a useful tool for scientists working in academic institutions and industry. The relevance to very different areas should inspire researchers to use technologies and/or methodologies that are already employed in fields other than theirs.

Paris, January 2001   Bernard Valeur,
                                                   Jean-Claude Brochon

Contents

Part 2 Fluorescence of Molecular and Supramolecular Systems

4 Investigation of Femtosecond Chemical Reactivity by Means of Fluorescence Up-Conversion 61
J.-C. Mialocq, T. Gustavsson

8 Phototunable Metal Cation Binding Ability of Some Fluorescent Macrocyclic Ditopic Receptors . . . 157

J.-P. Desvergne, E. Perez-Inestrosa, H. Bouas-Laurent, G. Jonusauskas, J. Oberlé, C. Rullière

Part 3 Fluorescence in Sensing Applications

9 The Design of Molecular Artificial Sugar Sensing Systems   173

S. Shinkai, A. Robertson

10 PCT (Photoinduced Charge Transfer) Fluorescent Molecular Sensors for Cation Recognition 187

B. Valeur, I. Leray

11   Fluorometric Detection of Anion Activity and Temperature Changes

L. Fabbrizzi, M. Licchelli, A. Poggi, G. Rabaioli,
A. Taglietti

12   Oxygen Diffusion in Polymer Films for Luminescence Barometry Applications

X. Lu, I. Manners, M.A. Winnik

Part 4 New Techniques of Fluorescence Microscopy in Biology

Part 5 Proteins and Their Interactions as Studied by Fluorescence Methods

19 About the Prediction of Tryptophan Fluorescence Lifetimes and the Analysis of Fluorescence Changes in Multi-Tryptophan Proteins

A. SILLEN, Y. ENGELBORGHS

Contributors

H. Alpha-Bazin
CIS bio international, BP 175 – 30203
Bagnols sur Cèze Cedex, France

M. Auer
Novartis Forschungsinstitut GmbH,
Dermatology, A-1235 Vienna, Austria
(e-mail: manfred.auer@pharma.-
novartis.com)

P. Ballet
Laboratory for Molecular Dynamics
and Spectroscopy, Chemistry
Department K. U. Leuven,
Celestijnenlaan 200F, 3001 Heverlee,
Belgium

P. I. H. Bastiaens
Cell Biophysics Laboratory, Imperial
Cancer Research Fund, 44 Lincoln's
Inn Fields, WC2A 3PX London, UK
(e-mail: p.bastiaens@icrf.icnet.uk)

H. Bazin
CIS bio international, BP 175-30203
Bagnols sur Cèze Cedex, France

M. N. Berberan-Santos
Centro de Química-Física Molecular,
Instituto Superior Técnico,
1049 – 001 Lisboa, Portugal
(e-mail: berberan@ist.utl.pt)

H. Bouas-Laurent
Laboratoire de Chimie Organique et
Organométallique LCOO, CNRS UMR
5802, Université Bordeaux 1,
351 Cours de la Libération,
33405 Talence Cedex, France

F. V. Bright
Department of Chemistry, Natural
Sciences Complex, University at
Buffalo, State University of New York,
Buffalo, New York 14260 – 3000, U.S.A.
(e-mail: chefvb@acsu.buffalo.edu)

C. Buelher
Paul Scherrer Institute, PSI,
WMHA/C24, 5232 Villingen PSI,
Switzerland

Y. Chen
Laboratory for Fluorescence
Dynamics, University of Illinois
at Urbana-Champaign, Urbana, IL
81801, USA

S. N. Daniel
Department of Chemistry, Natural
Sciences Complex, University at
Buffalo, State University of New York,
Buffalo, New York 14260 – 3000, USA

J.-P. Desvergne
Laboratoire de Chimie Organique et
Organométallique LCOO, CNRS UMR
5802, Université Bordeaux 1,
351 Cours de la Libération,
33405 Talence Cedex, France
(e-mail: jp.desvergne@lcoo.u-bor-
deaux.fr)

J. S. Eid
Laboratory for Fluorescence
Dynamics, University of Illinois
at Urbana-Champaign, Urbana,
IL 81801, USA

Y. ENGELBORGHS
Laboratory of Biomolecular
Dynamics, K. U. Leven,
Celestijnenlaan, 200 D,
3001 Leuven, Belgium
(e-mail: yves.engelborghs@fys.kuleu-
ven.ac.br)

L. FABBRIZZI
Dipartimento di Chimica Generale,
Università di Pavia, viale Taramelli
12, 27100 Pavia, Italy
(e-mail: fabbrizz@unipv.it)

E. GRATTON
Laboratory for Fluorescence
Dynamics, University of Illinois
at Urbana-Champaign, Urbana,
IL 81801, USA
(e-mail: enrico@aries.scs.uiuc.edu)

T. GUSTAVSSON
CEA/Saclay,
DSM/DRECAM/SCM/URA 331
CNRS, 91191 Gif-sur-Yvette, France

D.-P. HERTEN
Physikalisch-Chemisches Institut,
Universität Heidelberg,
Im Neuenheimer Feld 253,
69124 Heidelberg, Germany

C. HUBER
University of Regensburg, Institute of
Analytical Chemistry, Chemo-
and Biosensors, 93040 Regensburg,
Germany

D. M. JAMESON
Department of Genetics and
Molecular Biology, University of
Hawaii, 1960 East-West road,
Honolulu, Hawaii 96822, USA
(e-mail: djameson@hawaii.edu)

K. JEURIS
Laboratory for Molecular Dynamics
and Spectroscopy, Chemistry
Department K. U. Leuven,
Celestijnenlaan 200F, 3001 Herverlee,
Belgium

G. JONUSAUSKAS
Centre de Physique Moléculaire
Optique et Hertzienne CPMOH,
CNRS UMR 5798, Université
Bordeaux 1, 351 Cours de la
Libération, 33405 Talence Cedex,
France

M. A. KANE
Department of Chemistry, Natural
Sciences Complex, University at
Buffalo, State University of New York,
Buffalo, New York 14260–3000, USA

K. KEMNITZ
EuroPhoton GmbH, Mozartstraße 27,
12247 Berlin, Germany
(e-mail: kkemnitz@t-online.de;
www.europhoton.de)

I. KLIMANT
University of Regensburg, Institute of
Analytical Chemistry, Chemo-
and Biosensors, 93040 Regensburg,
Germany

H. LAGUITTON-PASQUIER
Laboratoire d'Electrochimie
Organique et de Photochimie Redox,
Université J. Fourier de Grenoble,
UMR CNRS 5360, B. P. 38041
Grenoble, France
(e-mail: helene.laguitton-
pasquier@ujfgrenoble.fr)

I. LERAY
Laboratoire PPSM (CNRS UMR
8531), Département de Chimie, ENS-
Cachan, 61 Av. du Président Wilson,
F-94235 Cachan cedex,
and Laboratoire de Chimie
Générale, Conservatoire National des
Arts et Métiers, 292 rue St Martin,
75141 Paris Cedex 03, France

G. LIEBSCH
University of Regensburg, Institute of
Analytical Chemistry, Chemo-
and Biosensors, 93040 Regensburg,
Germany

H. Lemmetyinen
Institute of Materials Chemistry,
Tampere University of Technology,
P.O. Box 541, 33101 Tampere, Finland

M. Licchelli
Dipartimento di Chimica Generale,
Università di Pavia, viale Taramelli
12, 27100 Pavia, Italy

X. Lu
Department of Chemistry, University
of Toronto, 80 St. George St., Toronto
ON M5S 3H6, Canada
(e-mail: xlu@chem.utoronto.ca)

I. Manners
Department of Chemistry, University
of Toronto, 80 St. George St., Toronto
ON M5S 3H6, Canada
(e-mail: imanners@chem.utoronto.ca)

G. Mathis
CIS bio international, BP 175, 30203
Bagnols sur Cèze Cedex, France
(e-mail: gmathis@compuserve.com)

J.-C. Mialocq
CEA/Saclay,
DSM/DRECAM/SCM/URA 331
CNRS, 91191 Gif-sur-Yvette, France
(e-mail: mialocq@drecam.saclay.-
cea.fr)

D. P. Millar
Department of Molecular Biology
The Scripps Research Institute,
10550 N. Torrey Pines Rd., La Jolla,
CA 92037, USA
(e-mail: millar@scripps.edu)

J. D. Müller
Laboratory for Fluorescence
Dynamics, University of Illinois
at Urbana-Champaign, Urbana, IL
81801, USA

M. Neumann
Physikalisch-Chemisches Institut,
Universität Heidelberg, Im
Neuenheimer Feld 253, 69124
Heidelberg, Germany

G. Neurauter
University of Regensburg, Institute of
Analytical Chemistry, Chemo-
and Biosensors, 93040 Regensburg,
Germany

E. D. Niemeyer
Department of Chemistry, Natural
Sciences Complex, University at
Buffalo, State University of New York,
Buffalo, New York 14260–3000, USA

J. Oberlé
Centre de Physique Moléculaire
Optique et Hertzienne CPMOH,
CNRS UMR 5798, Université
Bordeaux 1, 351 Cours de la
Libération, 33405 Talence Cedex,
France

E. Perez-Inestrosa
Laboratoire de Chimie Organique et
Organométallique LCOO, CNRS UMR
5802, Université Bordeaux 1,
351 Cours de la Libération,
33405 Talence Cedex, France

D. Pevenage
Laboratory for Molecular Dynamics
and Spectroscopy, Chemistry
Department K. U. Leuven,
Celestijnenlaan 200F, 3001 Heverlee,
Belgium

A. Poggi
Dipartimento di Chimica Generale,
Università di Pavia, viale Taramelli
12, 27100 Pavia, Italy

M. Préaudat
CIS bio international – BP 175, 30203
Bagnols sur Cèze Cedex, France

G. Rabaioli
Dipartimento di Chimica Generale,
Università di Pavia, viale Taramelli
12, 27100 Pavia, Italy

W. Rettig
Institut für Physikalische und
Theoretische Chemie, Humboldt
Universität zu Berlin, Bunsenstr. 1,
10117 Berlin, Germany
(e-mail: rettig@chemie.hu-berlin.de)

A. ROBERTSON
Department of Chemistry and
Biochemistry, Graduate School of
Engineering, Kyushu University,
Fukuoka, 812–8581, Japan

C. RULLIÈRE
Centre de Physique Moléculaire
Optique et Hertzienne CPMOH,
CNRS UMR 5798, Université
Bordeaux 1, 351 Cours de la
Libération, 33405 Talence Cedex,
France

K. RURACK
Institut für Physikalische und
Theoretische Chemie, Humboldt
Universität zu Berlin, Bunsenstr. 1,
10117 Berlin, Germany

M. SAUER
Physikalisch-Chemisches Institut,
Universität Heidelberg,
Im Neuenheimer Feld 253,
69124 Heidelberg, Germany
(e-mail: sauer@urz.uni-heidelberg.de)

M. SCZEPAN
Institut für Physikalische und
Theorische Chemie, Humboldt
Universität zu Berlin, Bunsenstr. 1,
10117 Berlin, Germany

S. SHINKAI
Department of Chemistry and
Biochemistry, Graduate School
of Enginering, Kyushu University,
Fukuoka, 812–8581, Japan
(e-mail: seijitcm@mbox.nc.kyushu-
u.ac.jp)

A. SILLEN
Laboratory of Biomolecular
Dynamics, K. U. Leuven,
Celestijnenlaan, 200D, 3001 Leuven,
Belgium

A. SQUIRE
Cell Biophysics Laboratory, Imperial
Cancer Research Fund,
44 Lincoln's Inn Fields, WC2A 3PX,
London, UK

A. STANGELMAYER
University of Regensburg, Institute of
Analytical Chemistry, Chemo-
and Biosensors, 93040 Regensburg,
Germany

K. STOECKLI
Novartis Pharma Research,
CTA/LFU/NAT, 4002 Basel,
Switzerland
(e-mail: kurt.stoeckli@pharma.novar-
tis.com)

A. TAGLIETTI
Dipartimento di Chimica Generale,
Università di Pavia, viale Taramelli
12, 27100 Pavia, Italy

E. TRINQUET
Cis bio international – BP 175, 30203
Bagnols sur Cèze Cedex, France

B. VALEUR
Laboratoire de Chimie Générale,
Conservatoire National des Arts et
Métiers, 75141 Paris Cedex 03, France
and Laboratoire PPSM (CNRS UMR
8531), ENS-Cachan, 94235 Cachan
Cedex, France
(e-mail: valeur@cnam.fr)

P. A. W. VAN DEN BERG
Laboratory of Biochemistry,
Wageningen University,
P. O. Box 8128, 6700 ET Wageningen,
The Netherlands

M. VAN DER AUWERAER
Laboratory for Molecular Dynamics
and Spectroscopy, Chemistry
Department K. U. Leuven,
Celestijnenlaan 200F, 3001 Heverlee,
Belgium
(e-mail: mark.vanderauweraer-
@chem.kuleuven.ac.be)

P. J. VERVEER
Cell Biophysics Laboratory, Imperial
Cancer Research Fund,
44 Lincoln's Inn Fields, WC2A 3PX,
London, UK

A. J. W. G. Visser
Laboratory of Biochemistry,
Wageningen University,
P. O. Box 8128, 6700 ET Wageningen,
The Netherlands
(e-mail: Ton.Visser@laser.bc.wau.nl)

E. Vuorimaa
Institute of Materials Chemistry,
Tampere University of Technology,
P. O. Box 541, 33101 Tampere, Finland
(e-mail: vuorimaa@butler.cc.tut.fi)

M. A. Winnik
Department of Chemistry, University
of Toronto, 80 St. George St., Toronto
ON M5S 3H6, Canada
(e-mail: mwinnik@chem.utoronto.ca)

O. S. Wolfbeis
University of Regensburg, Institute of
Analytical Chemistry, Chemo-
and Biosensors, 93040 Regensburg,
Germany
(e-mail: otto.wolfbeis@chemie.uni-
regensburg.de)

F. Wouters
Cell Biophysics Laboratory, Imperial
Cancer Research Fund,
44 Lincoln's Inn Fields, WC2A 3PX,
London, UK

Introduction: On the Origin of the Terms Fluorescence, Phosphorescence, and Luminescence

B. VALEUR

As an introduction to the historical part of this book, it is worth recalling the origin of the terms *fluorescence, phosphorescence, luminescence*. When and how these phenomena were discovered are of course the basic questions [1], but the first step of a historical research is the understanding of the etymology of a word invented for designating a phenomenon. *Fluorescence* is a beautiful example of a term whose etymology is not obvious at all; in particular, it is strange, at first sight, that it contains *fluor* which is not a fluorescent element! In contrast, the etymologies of *phosphorescence* and *luminescence* are straightforward: both these terms contain light ($\phi\omega\varsigma$ in Greek and *lumen* in Latin, respectively). Let us examine first these two terms.

The term *phosphorescence* comes from the Greek: $\phi\omega\varsigma$ = light (genitive case: $\phi o\tau o\varsigma \rightarrow$ photon) and $\phi o\varrho\varepsilon\iota\nu$ = to bear. Therefore, *phosphor* means "which bears light". The term *phosphor* has indeed been assigned since the Middle Ages to materials that glow in the dark after exposure to light. There are many examples of minerals reported a long time ago that exhibit this property, and the most famous of them (but not the first one) was the *Bolognian phosphor* discovered by a cobbler of Bologna in 1602, Vincenzo Cascariolo, whose hobby was alchemy. One day he went for a walk in the Monte Paterno area and he picked up some strange heavy stones. After calcination with coal, he observed that these stones glowed in the dark after exposure to light. It was recognized later that the stones contained barium sulfate which, by calcination with coal, led to barium sulfide, a phosphorescent compound. Later, the same name *phosphor* was assigned to the element isolated by Brandt in 1677 (despite the fact that it is chemically very different) because, when exposed to air, it burns and emits vapors that glow in the dark.

The term *luminescence* comes from the Latin (*lumen* = light) and was first introduced as *luminescenz* by the physicist and historian of science, Eilhardt Wiedemann in 1888 for "all those phenomena of light which are not solely conditioned by the rise in temperature", as opposed to incandescence. Luminescence is *cold light* whereas incandescence is *hot light*.

We now turn our attention to the term *fluorescence* which was introduced by Sir George Gabriel Stokes, a physicist and professor of mathematics at Cambridge in the middle of the last century. Before describing Terms Stokes' work, it should be recalled that the first reported observation of fluorescence was made by a Spanish physician, Nicolas Monardes, in 1565. He described the wonderful peculiar blue color of an infusion of a wood called *Lignum nephriticum*. This wood was further investigated by Boyle, Newton and others, but the phenomenon was not understood.

In 1833, David Brewster, a Scottish preacher, reported [2] that a beam of white light passing through an alcoholic extract of leaves (chlorophyll) appears to be red when observed from the side, and he pointed out the similarity with the blue light coming from a light beam passing through fluorspar crystals. In 1845, John Herschel, the famous astronomer, considered that the blue color at the surface of solutions of quinine sulphate and *lignum nephriticum* was "a case of superficial colour presented by a homogeneous liquid, internally colourless" and he called this phenomenon *epipolic dispersion*, from the Greek *επιπολη* = surface [3]. The solutions observed by Herschel were very concentrated so that the major part of the incident light was absorbed and all the blue color appeared to be only at the surface. Herschel used a prism to show that the *epipolic dispersion* could be observed only upon illumination by the blue end of the spectrum, and not the red end. The crude spectral analysis with the prism revealed blue, green and a small amount of yellow light, but Herschel did not realize that the superficial light was of longer wavelength than the incident light.

The phenomena were reinvestigated by Stokes who published a famous paper entitled "On the refrangibility of light" in 1852 [4]. He demonstrated that the phenomenon was an emission of light following absorption of light. It is worth describing one of the Stokes experiments which is spectacular and remarkable by its simplicity. Stokes formed the solar spectrum by means of a prism. When he moved a tube filled with a solution of quinine sulfate through the visible part of the spectrum, nothing happened: the solution simply remained transparent. But beyond the violet portion of the spectrum, i.e., in the non-visible zone corresponding to ultraviolet radiations, the solution glowed with a blue light. Stokes wrote: "It was certainly a curious sight to see the tube instantaneously light up when plunged into the invisible rays; it was literally *darkness visible.*" This experiment provided compelling evidence that there was absorption of light followed by emission of light. Stokes stated that the emitted light is always of longer wavelength than the exciting light. This statement later became Stokes' law.

Stokes' paper led Edmond Becquerel, a French physicist [5], to "réclamation de priorité" for this kind of experiment [6]. In fact, Becquerel published an outstanding paper in 1842 [7] in which he described the light emitted by calcium sulfide deposited on paper when exposed to solar light beyond the violet part of the spectrum. He was the first to state that the emitted light is of longer wavelength than the incident light.

In his first paper [4], Stokes called the observed phenomenon *dispersive reflexion* but in a footnote he wrote "I confess I do not like this term. I am almost inclined to coin a word, and call the appearance *fluorescence*, from fluorspar, as the analogous term *opalescence* is derived from the name of a mineral." Most of the varieties of fluorspar or fluospath (minerals containing calcium fluoride (fluorite)) indeed exhibit the above-mentioned property. In his second paper [8], Stokes definitely resolved to use the word *fluorescence*.

We understand now why *fluorescence* contains the term *fluor*, but what is the origin of *fluorspar* or *fluorspath* and why are these materials fluorescent? *Spar* (in English) and *spath* (in German) were the names given in the eighteenth century [9] to "stones" that are more or less transparent and crystallized with a lamellar texture. Since these materials can be easily melted, and some of them can help to melt other materials, the word *fluor* [10] was often employed in order to express some fluidity (fluere = to flow in Latin; this is the origin of the name of the element *fluor*, isolated by Moissan in 1886). Many spaths are known to be colored owing to the presence of small amounts of impurities which explains the fluorescence properties since fluorite itself is not fluorescent. These fluorogenic impurities are difficult to characterize. The blue and red fluorescences are due to divalent and trivalent europium ions, respectively. Yttrium and dysprosium can also be present and yield a yellow fluorescence [11].

The difference between the Stokes and Becquerel experiments described above is that quinine sulfate is fluorescent whereas calcium sulfide is phosphorescent, but they are both relevant to photoluminescence. In the nineteenth century, the distinction between fluorescence and phosphorescence was made on an experimental basis: fluorescence was considered as an emission of light that disappears simultaneously with the end of excitation, whereas in phosphorescence the emitted light persists after the end of excitation. Now we know that in both cases emission lasts longer than excitation and a distinction based on the duration of emission is not sound because there are long-lived fluorescences (e.g., uranyl salts) and short-lived phosphorescences (e.g., violet luminescence of zinc sulfide). The first theoretical distinction between fluorescence and phosphores-

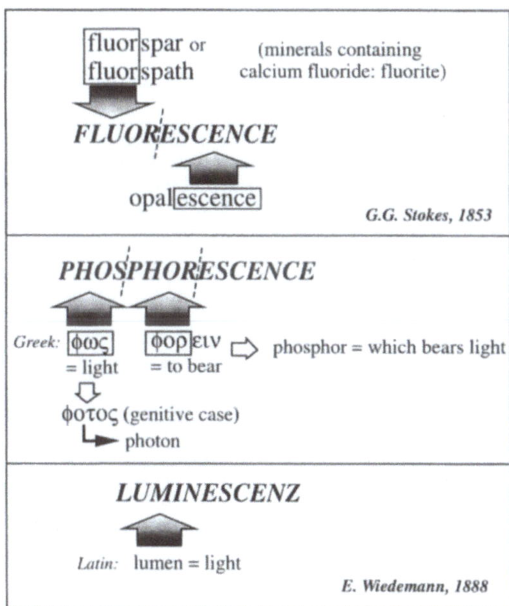

Fig. 1.1. Illustration of the origin of the terms fluorescence, phosphorescence, and luminescence

cence was provided by Francis Perrin [12]: "If the molecules pass, between absorption and emission, through a stable or unstable intermediate state and are thus no longer able to reach the emission state without receiving from the medium a certain amount of energy, there is phosphorescence". Further works clarified the distinction between fluorescence, delayed-fluorescence and phosphorescence [13].

It is to be hoped that this note will have given readers a clear understanding of the origin of the terms *fluorescence, phosphorescence,* and *luminescence,* as summarized in Fig. 1.1, and that it will be useful as an etymological introduction to courses on fluorescence.

Acknowledgements. I am grateful to Profs. Elisabeth Bardez and Mario Nuno Berberan-Santos for very helpful suggestions.

References

1. Harvey EN (1957) History of luminescence. The American Philosophical Society, Philadelphia
2. Brewster D (1833) Trans Roy Soc Edinburgh 12:538–545
3. Herschel JFW (1945) Phil Trans 143–145; (1945) Phil Trans 147–153
4. Stokes GG (1852) Phil Trans 142:463–562
5. Edmond Becquerel is the father of Henri Becquerel who discovered radioactivity, Edmond Becquerel invented the famous phosphoroscope which bears his name
6. Cosmos (1854) 3:509–510
7. Becquerel E (1842) Annales de Chimie et Physique (3) 9:257–322
8. Stokes GG (1853) Phil Trans 143:385–396
9. Macquer PJ (1779) Dictionnaire de Chymie, p 462
10. Macquer PJ (1779) Dictionnaire de Chymie, p 464
11. Robbins M (1994) Fluorescence. Gems and minerals under ultraviolet light. Geoscience Press
12. Perrin F (1929) Doctoral thesis, Paris; Annales de Physique 12:2252–2254
13. Nickel B (1996) Pioneers in photochemistry. From the Perrin diagram to the Jablonski diagram. EPA Newsletter 58:9–38

CHAPTER 2

Pioneering Contributions of Jean and Francis Perrin to Molecular Luminescence

M. N. BERBERAN-SANTOS

> Le but suprême que doit se donner tout homme de Science doit être
> de travailler pour la plus grande gloire de l'esprit humain.
>
> (The supreme goal of a scientist must be to work for the greater
> glory of the human spirit.)
>
> Francis Perrin, *Leçon terminale au Collège de France* (1972)

2.1 Introduction

> … in science, what X misses today Y will surely hit upon tomorrow (or maybe the day after tomorrow) (…) Scientists are entitled to be proud of their accomplishments, and what accomplishments can they call "theirs" except the things they have done or thought of first?
>
> P. Medawar (1964) [1]

> *Timon of Athens* could not have been written,
> *Les demoiselles d'Avignon* not have been painted,
> Had Shakespeare and Picasso not existed.
> But of how many scientific achievements can this be claimed?
> One could almost say that, with very few exceptions,
> It is not the men that make science;
> It is science that makes the men.
> What A does today, B or C or D could surely do tomorrow.
>
> E. Chargaff (1968) [2].

We know this to be very true. In science, priority is a key issue, and independent multiple discoveries are not uncommon. In art, the probability that two different creators will produce the same work is close to zero, and Pierre Menard's *Don Quixote* [3] stands as an isolated (and imaginary) exception.

On the other hand, the uniqueness of every scientist is undeniable. Remember Lagrange's famous comment about Lavoisier's death [4]:

> Il ne leur a fallu qu'un moment
> Pour faire tomber cette tête.
> Cent années, peut-être, ne suffiront pas
> Pour en reproduire une semblable.
>
> (It took them only a brief moment to chop off this head. Yet, a hundred years
> will not be perhaps sufficient to produce a similar one.)

In the same vein, C.P. Snow wrote of Einstein's "prodigious year" (1905): It is pretty safe to say that, so long as physics lasts, no one will again hack out three major breakthroughs in one year [5].

How can these two apparently opposite views be reconciled?

A work of art is always the combined result of two distinct aspects: first, skills; and second, the artist's culture, imagination, and emotions. Furthermore, it relies weakly on previous works: progress does not exist in art.

To accomplish a scientific work, however remarkable, skills and creativity are also decisive, but the outcome is silent about the scientist's personality, and only the skills are apparent in it (even the reporting style is not free, following strict rules [6]). This is so because science refers to those parts of reality for which a general rational consensus can exist [7]. This essential difference between art and science was aptly summarized long ago: l'Art, c'est *moi*; la Science, c'est *nous* (Claude Bernard) [8]. Subjectivity, the very essence of Art, is not welcomed in science. Scientists (including mathematicians) believe in an independent and pre-existing reality that they progressively unravel (see Fig. 2.12). They are like the Renaissance navigators that discovered new worlds or established new routes across the oceans, connecting civilizations. Vasco da Gama or Magellan (had he survived) could claim no more (and no less!) than having been the first. Even the serendipitous voyage of Columbus finds many a parallel in science [9–11].

Of course, aesthetics is not foreign to science. The study of nature is a source of intellectual joy, *la joie de connaître*. After surmounting many difficulties, each newly found harmony, however small, is rewarding and beautiful. And a sense of enchantment comes from being the first and only one to know. The subsequent communication of the findings can convey and revive in the audience the same sense of harmony. But that is neither the ultimate purpose, nor the main quality of a scientific work.

Another distinctive aspect of scientific works, as opposed to artistic ones, is the cooperative nature of the scientific endeavor:

1. In general, it builds upon previous contributions; given a common starting point, not all persons can accomplish the same, but a few always will.
2. It evolves within a given intellectual atmosphere and common view (the scientific community). The collective nature reaches extremes in some collaborative works, whose lists of authors extend for many lines. But papers with a few authors are very common.

Paradoxically, it is the collective nature of the scientific work that gives science its human dignity. While artistic expression is individual, and is mostly meaningful within a given culture, science cannot even be said to have a nationality: La science n'a pas de patrie, ou plutôt la patrie de la science embrasse l'humanité tout entière (Pasteur) (Science has no fatherland or rather, the fatherland of science is the whole mankind) [12].

The uniqueness of each scientist comes from the ensemble of his/her work, that reflects talents, tastes, motivations, and a personal evolution, and also results from a culture and an epoch.

It is with all these aspects in mind that the pioneering contributions of Jean and Francis Perrin to molecular luminescence will be discussed. In this chapter, after brief biographical sketches of Jean Perrin and Francis Perrin, their contributions to three different subjects will be described (Fig. 2.1).

Jean Perrin (1870-1942)   **Francis Perrin (1901-1992)**

Fig. 2.1. Jean Perrin and Francis Perrin (J. Perrin, Copyright Palais de la Découverte; F. Perrin, Copyright Robert Cohen, Reportages Photographiques, 3 rue Fontaine, Paris)

2.2
Biographical Sketches of Jean Perrin and Francis Perrin

The great French physicist Jean Perrin (1870–1942) is mostly remembered for the elegant experiments carried out before the First World War [13], that established the reality of atoms and molecules (Fig. 2.2), and for which he received the 1926 Nobel Prize in Physics. These studies are described in detail in his book (still in print) *Les Atomes* [14]. His earlier work on the nature of cathode rays (electrons) and X-rays [13], and his later work on molecular luminescence and photochemistry [15–32], are also remarkable. He was one of the first to apply the ideas of quantum theory to the absorption and emission of radiation by molecules. In particular, he was probably the first to have presented a molecular energy diagram with transitions between states [17, 18, 23]. In 1918 he proposed the mechanism of resonance energy transfer, refined in subsequent years (*induction moléculaire par résonance* [15, 23, 26, 27]). He also provided the correct model for thermally activated delayed fluorescence (E-type delayed fluorescence), based on the concept of metastable state, that he nevertheless supposed could also account for the phenomenon of phosphorescence [23].

Extensive biographical information on Jean Perrin is readily available, and here only a synoptic table (Table 2.1) and some selected topics are addressed. Relevant bibliography is discussed in Sect. 2.7.

Fig. 2.2. First Conseil de Physique Solvay (Brussels, 1911), where Jean Perrin presented his studies of the Brownian motion, finally demonstrating the existence of atoms and molecules. (Institut International de Physique Solvay, courtesy AIP Emilio Segrè Visual Archives)

Francis Perrin (1901–1992) was raised in a very rich cultural and scientific atmosphere. The inner circle of relations of his parents included Pierre Curie (deceased in 1906) and Marie Curie, the physicist Paul Langevin, the chemist and artist Georges Urbain, and the mathematician Émile Borel. It was only at the age of 11, already conversant with calculus, that Francis went to school. Before that age, he was taught by his parents and parents' friends. Following his father's footsteps, he became a *normalien* (graduate from the École Normale Supérieure). In the *post-scriptum* of a letter to Einstein (November 1919), Jean Perrin proudly announces the result of the admission examination: Francis a été reçu le premier (à seize ans et demi) à l'École Normale (Sciences). Il sera un meilleur physicien que son père (Francis was admitted to the École Normale (Sciences) in the top position. He will be a better physicist than his father) [33].

Working in his father's laboratory from 1924, Francis Perrin continued and extended the luminescence studies, with important personal contributions [30, 34–56]. His *Docteur ès Sciences physiques* thesis (1929) [46] deserves to be considered a classic for both style and scientific content. In the previous year he had already obtained the equivalent academic degree in mathematics, with a thesis on the mathematics of rotational Brownian motion [45], under the supervision of Borel.

Table 2.1. Jean-Baptiste Perrin (biographical synopsis)

Year	Major event	Biographical	Achievements
1870	Franco-Prussian war	Birth (Lille)	
1891–1894		Undergraduate studies at École Normale Supérieure (section sciences)	
1895–1898		ENS (agrégé préparateur)	
1897	Electron (Thomson)	Thèse de Doctorat: Rayons cathodiques et rayons de Roentgen. Marriage	Cathode rays are negatively charged particles (1895)
1898		Reader (chargé de cours) of the newly created physical chemistry course (Sorbonne)	
1900	Quantum (Planck)		
1901		Birth of Francis	First planetary model of the atom
1905	Theories of special relativity, of the photoelectric effect, and of the Brownian motion (Einstein)		
1908–1911		Chair of Physical Chemistry (Sorbonne, University of Paris, 1910)	Studies of the Brownian motion: first direct proofs of molecular reality
1911	Rutherford atomic model	First Solvay Council	
1913	Bohr atomic model	Book: Les Atomes	
1914		First studies of luminescence	
1914–1918	First World War	Military activities	Myriaphone (1916)
1916	Absorption and emission of radiation (Einstein)		
1918		Fluorescence studies	Radiative theory of chemical reactions. Induction par resonance
1919			Origin of stellar radiation
1922–1926		Laboratoire de Chimie Physique	
1923		Académicien	
1924	Matter waves (de Broglie)		

Table 2.1 (continued)

Year	Major event	Biographical	Achievements
1925	Quantum Mechanics (Heisenberg, Born)		Delayed fluorescence. Induction par resonance
1926	Wave Mechanics (Schrödinger) Photon (Lewis)	Nobel Prize in Physics	
1927		Institut de Biologie Physico-Chimique	
1929		Last fluorescence studies	
1935	Jablonski's diagram		
1936–1938		Sous-secrétariat d'État à la Recherche	Creation of astronomical observatories, laboratories. Palais de la Découverte
1939			Creation of the CNRS
1939–1945	Second World War		
1941		Departure to the USA	
1942		Death (NY)	
1948		Funérailles nationales (Panthéon)	

Fig. 2.3. Francis Perrin, Jean Perrin, and Otto Stern (of Stern-Volmer and Stern-Gerlach fame) in 1928. (AIP Emilio Segrè Visual Archives, Segrè Collection)

The main experimental and theoretical achievements of F. Perrin in the field of molecular luminescence include the following:

1. The active sphere model for quenching (1924) [34].
2. The relation between quantum yield and lifetime (1926) [38, 39, 41].
3. The theory of fluorescence polarization, in relation to rotational Brownian motion (1926) [36, 38, 46, 55]. In particular, from the key (Perrin) equation, he was able to determine accurately, for the first time, the fluorescence lifetime of dyes in solution.
4. The first qualitative theory of depolarization by resonance energy transfer (*transfert d'activation*) (1929) [46].

From the 1930s onwards, F. Perrin's scientific and professional activity was mainly devoted to nuclear physics [57, 58]. In this field, he was again the author of several important contributions. For example, he introduced the concept of critical mass (*dimension critique*) for a nuclear chain reaction, and provided the first estimate of its value for uranium (1939).

The life and scientific achievements of Francis Perrin are summarized in Table 2.2.

It is interesting to note that Jean Perrin and Francis Perrin devoted their attention to luminescence for approximately the same amount of time: a decade (Jean Perrin started his studies shortly before the outbreak of the First World War, but he resumed them only after its end). As can be seen from the tables, the two contributed to it in more than one important way. Unlike another contemporary father-son combination (William Henry and William Lawrence Bragg),

Fig. 2.4. The team of the Laboratoire de Chimie Physique. First row, from left to right: Pierre Girard, Jean Perrin, Francis Perrin, Costa; second row, standing: Nine Choucroun, André Marcelin, Audubert, Hibben, Pierre Auger, and Platard. The bust of Jean Perrin, as Dionysos, sculpted by Urbain, can be seen at top right. (Copyright Palais de la Découverte)

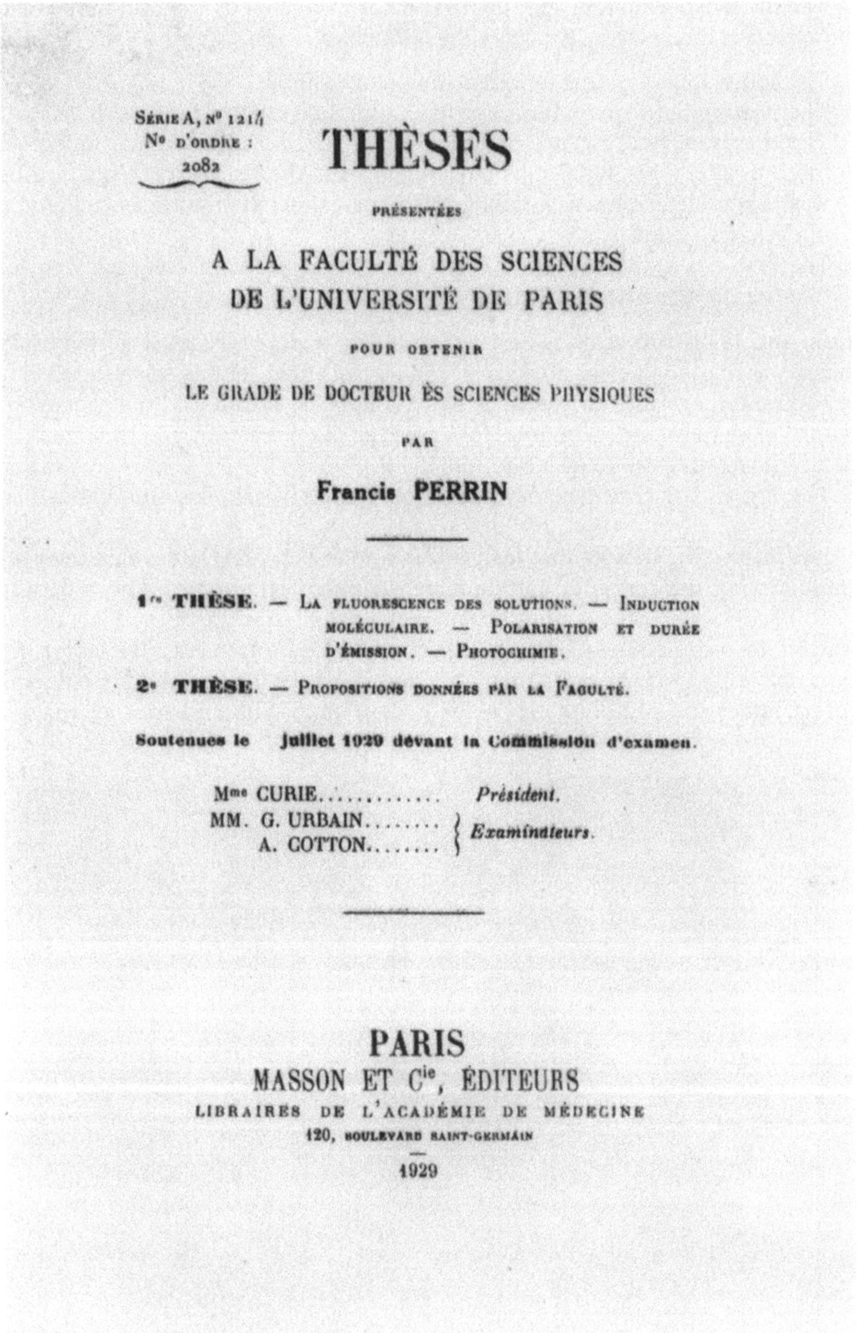

Série A, Nº 1214
Nº d'ordre : 2082

THÈSES

PRÉSENTÉES

A LA FACULTÉ DES SCIENCES
DE L'UNIVERSITÉ DE PARIS

POUR OBTENIR

LE GRADE DE DOCTEUR ÈS SCIENCES PHYSIQUES

PAR

Francis PERRIN

1ʳᵉ THÈSE. — LA FLUORESCENCE DES SOLUTIONS. — INDUCTION MOLÉCULAIRE. — POLARISATION ET DURÉE D'ÉMISSION. — PHOTOCHIMIE.

2ᵉ THÈSE. — PROPOSITIONS DONNÉES PAR LA FACULTÉ.

Soutenues le   juillet 1929 devant la Commission d'examen.

Mᵐᵉ CURIE.............   *Président.*
MM. G. URBAIN........ } *Examinateurs.*
     A. COTTON........ }

PARIS
MASSON ET Cⁱᵉ, ÉDITEURS
LIBRAIRES DE L'ACADÉMIE DE MÉDECINE
120, BOULEVARD SAINT-GERMAIN

1929

Fig. 2.5. Frontispiece of Francis Perrin Ph.D. thesis in physics. The first part of the thesis was published in the Annales de Physique [46]

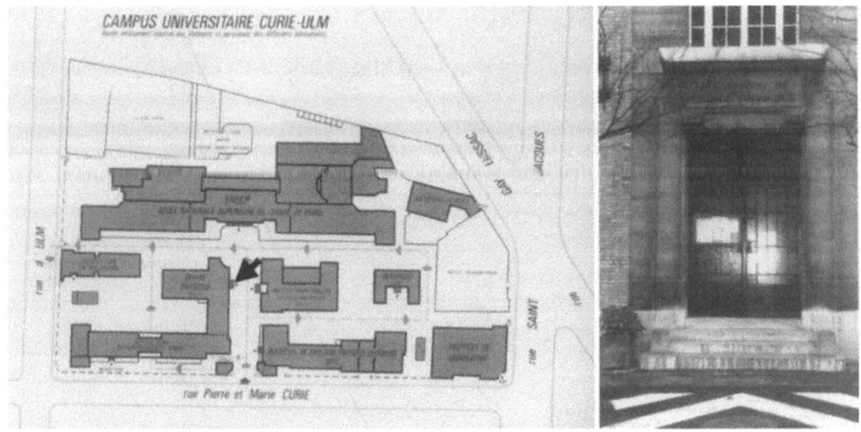

Fig. 2.6. Location and façade of the Laboratoire de Chimie Physique (f. 1922–6; present situation of the campus), where the Perrins did most of their luminescence studies. Before, Jean Perrin's laboratory was located in the old buildings of the Sorbonne

Table 2.2. Francis-Henri-Jean-Siegfried Perrin (biographical synopsis)

Year	Major event	Biographical	Achievements
1901		Birth (Paris)	
1918–1922		Studies at École Normale Supérieure (section sciences)	
1919	Stern-Volmer relation		
1920	Polarization of solutions (Weigert) Bohr-Grotrian diagram	Licencié	
1922	Sensitized fluorescence (Cario and Franck)	Agrégé	
1923	Polarization (Vavilov and Levshin)	Military service	
1924		Assistant (lab. de chimie physique, Sorbonne)	Active sphere model
1925	Approximate theory of polarization (Levshin)		Approximate theory of polarization
1926	Direct determination of nanosecond lifetimes (Gaviola)		Quantum yield and lifetime theory of fluorescence polarization (sphere) Experimental determination of lifetimes in solution. Comparison with radiative lifetimes.
1928	Quantum theory of molecular interaction (Kallmann and London)	Ph.D. thesis in Mathematics: Étude mathématique du mouvement brownien de rotation	

Table 2.2 (continued)

Year	Major event	Biographical	Achievements
1929		Ph.D. thesis in Physics: La fluorescence des solutions	Concentration depolarization
1932			Smallness of neutrino's mass. Quantum theory of resonance transfer between atoms
1933		Maître de Conférences (Sorbonne)	Materialization of energy (production of particle + antiparticle)
1935	Jablonski's diagram	Professeur (Sorbonne)	
1936			Theory of fluorescence polarization (ellipsoid)
1936–1938		Political activity	
1939		Book: Mécanique statistique quantique	Critical mass of natural uranium nuclear chain reaction
1939–1945	Second World War Triplet state (Lewis and Kasha, 1944)	Military mobilisation (1939–40). Stay in the USA (Columbia University: 1941–43) France Forever movement	
1946		Commissariat à l'Energie Atomique. Collège de France	
1948	Quantum theory of RET (Förster)		
1950–1972	Additivity of anisotropies (Weber, 1952)	Haut Commissaire du CEA	French nuclear programme
1953		Académicien	Co-foundation of CERN
1957	Anisotropy (Jablonski)		
1972		Retirement	
1973			Proposal of a new periodic table
1992		Death	

most of the work was carried out in relative independence – only one publication [30] is co-signed. This, in spite of great affection and frequent scientific discussions, as is evident from the following two extracts, written by Francis Perrin and Jean Perrin, respectively:

> J'ai eu le très rare bonheur de faire ce travail au côté de mon père dans son laboratoire; en tant que disciple je tiens à exprimer ici ma profonde reconnaissance pour tout ce que je lui dois. (I had the very rare pleasure of doing this work beside my father, in his laboratory; As his pupil, I must express here my deep gratitude for all that I owe him.) [46]

> J'ai tenté, aux pages précédentes, d'esquisser le progrès, sans cesse plus rapide, des recherches tendues vers l'essence des Choses. Dans la mesure où j'ai pu réussir, j'ai une grande joie à dire l'aide précieuse que j'ai trouvée dans mes longues conversations avec mon fils Francis Perrin. (In the preceding pages I tried to sketch the increasingly fast progress of the research on the essence of all Things. I gladly declare that my possible success owes much to the precious help of my son Francis Perrin, with whom I had many long conversations.) [59]

If one compares the works of Jean and Francis Perrin on luminescence, it is apparent that Jean Perrin has a great physical intuition, chooses new and worthwhile subjects of research, and puts forward new concepts and views, establishing a general, qualitative framework, while Francis Perrin is particularly skilled in the precise physico-mathematical description and analysis of the phenomena. There is thus an effective scientific complementarity, even if it did not materialize in joint papers.

Jean and Francis Perrin held similar political and philosophical ideas. Both were socialists and atheists. Like many nineteenth century French men of science, Jean Perrin viewed science almost as a religion. In the words of Raspail, inscribed in the surviving pedestal of his statue, located near Francis Perrin's residence: …la Science, l'unique religion de l'avenir (Science, the only Religion of the future). Francis Perrin publicly denied all religions and gods, and was the President of Honour of the *Union des Athées*. He donated his body to science.

Fig. 2.7. Francis Perrin in the US, during the Second World War. (Copyright Fred Stein, 11525–25 Metropolitan Ave., Kew Gardens, NY 11418)

Fig. 2.8. Louis de Broglie, Maurice de Broglie, and Francis Perrin in 1951. (AIP Emilio Segrè Visual Archives)

Jean and Francis Perrin were, in their own ways and in the best French spirit, defenders of the *Droits de l'Homme*, and rejected all totalitarianisms. In the difficult period between wars, both actively opposed the rising fascism in Europe. The post-war positions of Francis Perrin on communism, at a time where it was fashionable among intellectuals, are worthy of mention.

Both actively labored in favor of science. Jean Perrin, in particular, was the founder of the CNRS, and of several important scientific institutions. He was also greatly concerned with the dissemination of scientific culture. One of his permanent concerns was the motivation for science of the greatest number of young people. For these purposes, he founded in Paris the Palais de la Découverte (1937), and wrote several books of popularization (e.g., [14, 59]), among other activities.

2.3
The Perrin-Jablonski Diagram

The history of the "Jablonski diagram" has been the subject of a recent and thorough investigation [60–62]. The views and reminiscences of the only living direct participant Michael Kasha, are presented in [63–65]. The name "scheme of Jablonski" was used for the first time by Lewis, Lipkin, and Magel (1941), in a paper on the phosphorescent state [66]. In a following paper by Lewis and Kasha (1944), where the phosphorescence state was identified with the lowest triplet state, the designation "Jablonski diagram" makes its first appearance [67]. Interestingly, in later papers by the same authors [68, 69], the diagram is simply referred to as "energy diagram." This usage remained common up to the 1970s. The name "Jablonski diagram" nevertheless made its way into the recent photochemical vocabulary. It is defined by the IUPAC as detailed below.

2.3.1
Jablonski Diagram

Originally, a diagram showing that the fluorescent state of a molecular entity is the lowest excited state from which the transition to the ground state is allowed, whereas the phosphorescent state is a metastable state below the fluorescent state, which is reached by radiationless transition. In the most typical cases the fluorescent state is the lowest singlet excited state and the phosphorescent state the lowest triplet state, the ground state being a singlet. Presently, modified Jablonski diagrams are frequently used and are actually state diagrams in which molecular electronic states, represented by horizontal lines displaced vertically to indicate relative energies, are grouped according to multiplicity into horizontally displaced columns. Excitation and relaxation processes that interconvert states are indicated in the diagrams by arrows. Radiative transitions are generally indicated with straight arrows, while radiationless transitions are generally indicated with wavy arrows [70].

As pointed out by B. Nickel [62], G.N. Lewis created a misnomer when he decided to refer to his own diagram (the original form of the diagram, according to the IUPAC definition above) as the "Jablonski diagram". In fact, most of the characteristics of the diagram, as presently perceived, are not due to A. Jablonski. This point is clearly recognized by Kasha: later use of the "Jablonski diagram" fails to recognize the limitations of his interpretations [63]. However, like Lewis, he fails to give appropriate credit to the contributions of Jean and Francis Perrin to the early stages of this diagram [62, 65].

Schematically, it may be said that:

1. The type of diagram, with electronic states represented by horizontal segments, arranged vertically according to relative energies, and arranged in columns according to the spin multiplicities, and including lines connecting states corresponding to electronic transitions, turns out to be an extension of the Bohr-Grotrian diagram for atoms (1920) [71–73]. This diagram, originally due to Bohr [71], and used at length by Grotrian (but not introduced by him, as stated in [64]), replaced the Fowler representation, where the energies plotted were those of the atomic lines, and not those of the terms.
2. The first use of an energy level diagram for molecules in connection with the absorption and emission of light is probably due to Jean Perrin [17, 18, 23].
3. Using his diagram, that contains a metastable state (Fig. 2.9), Jean Perrin correctly explains, long before Jablonski, the phenomenon of thermally activated delayed fluorescence (presently sometimes called E-type delayed fluores-

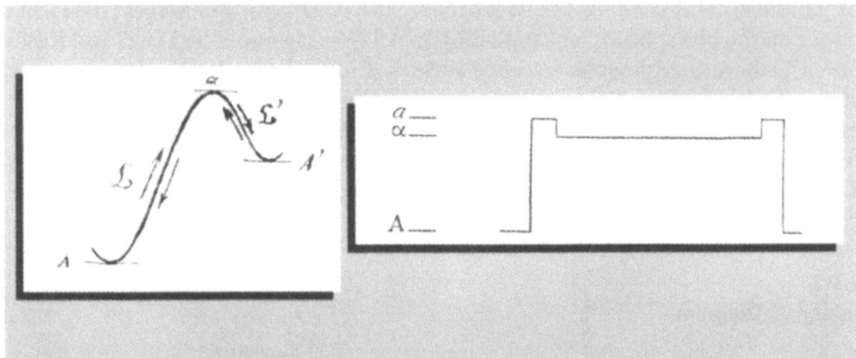

Fig. 2.9. Two forms of the diagram of Jean Perrin: *left* 1920 [17, 18], *right*, since 1922 [18, 23, 32]. In both diagrams, energy is the ordinate. In the second representation, time is the abscissa

cence), including its temperature dependence. Jean Perrin even uses the words "delayed fluorescence" (*fluorescence retardée*) [23, 32]. The model of Jean Perrin is discussed in detail by Francis Perrin in his Ph.D. thesis [46]. For unknown reasons, Lewis et al. [66] attributed the interpretation of thermally activated delayed fluorescence (called by him alpha process) to Jablonski and not to Perrin.

4. It was recently argued that the metastable state implied by Jean and Francis Perrin was a trap state (only possible in a solid matrix), generated by photoinduced charge separation [65]. In this way, the actual resemblance of their diagram to the "Jablonski diagram" would be superficial. Jean Perrin indeed consistently explained thermoluminescence on the basis of this diagram [14 (1936 edn.), 23]. However, in several publications [30, 41, 46], including the one invoked in [65] (see the excerpt given below, in point 5, and also [60–62]), both Perrins explicitly distinguish between the two cases: charge separation and metastable states analogous to those of atoms. The above argument is therefore invalid [62].

5. The diagram of Jean and Francis Perrin is, of course, still incomplete: the metastable state cannot revert radiatively or otherwise to the ground state, and has therefore an essentially infinite lifetime at low temperature. This is the only (and crucial) point where a difference exists between the scheme of Perrin and the scheme of Jablonski. By allowing such a transition, a second emission at longer wavelengths becomes possible (true phosphorescence), as well as a weak absorption. That is the merit of Jablonski's work [74]. It is, nevertheless, interesting to quote the two passages of the Perrins given below.

2.3.2
États Métastables – Phosphorescence

Lorsqu'un état activé peut être atteint directement, à partir de l'état normal, par absorption de lumière, réciproquement le retour à l'état normal est possible par émission spontanée de lumière. Mais il existe des états activés, dit *métastables,* pour lesquels aucune transformation de ce genre n'est possible; le retour à l'état normal ne peut alors se produire que par interaction avec d'autres molécules (désactivation induite) ou par passage intermédiaire par un autre état activé d'énergie interne plus grande. L'existence de ces états métastables, prévues théoriquement par Bohr, comme conséquence du principe de correspondance, a été démontrée expérimentalement dans plusieurs cas (hélium, mercure,…). Tous les principes de sélection en spectroscopie expriment d'ailleurs l'impossibilité de certaines transformations quantiques par rayonnement [46].

(Whenever an activated state can be attained directly from the ground state, by light absorption, then the return to the ground state by spontaneous emission of light is also possible. Nevertheless, there are some activated states, called *metastable*, for which this kind of transition is not possible. The return to the ground state can only occur by interaction with other molecules (induced deactivation) or via a further intermediate state of higher energy. The existence of metastable states, theoretically predicted by Bohr, as a consequence of the correspondence principle, has been experimentally demonstrated in several cases (helium, mercury, etc.). Furthermore, certain quantum transitions with the emission of radiation are not allowed by spectroscopic selection rules).

Il semble bien d'ailleurs que cette probabilité de désactivation par émission lumineuse spontanée ne soit jamais rigoureusement nulle, c'est-à-dire qu'il n'existe pas d'état activé absolument métastable. C'est ainsi que certains états activés des atomes d'azote et d'oxygène ionisés, considérés comme tout à fait métastables, donnent pourtant lieu à des émissions lumineuses lorsqu'ils sont soustraits à toute cause de désactivation induite pendant un temps suffisamment long, de l'ordre de la seconde (…) [30].

(It also appears that this probability of radiative decay is never strictly zero, that is, the absolutely metastable activated state does not exist. It is for this reason that certain activated states of ionized nitrogen and oxygen atoms, long considered as strictly metastable, are now known to slowly deactivate with the emission of light, if they are isolated from all causes of induced deactivation for sufficiently long times, of the order of one second (…)).

Other relevant citations are given by Nickel [60–62], which also discusses at length several shortcomings in Jablonski's 1935 paper [74]. In particular, the kinetic model and analysis are incorrect. Also, the proposed mechanism by which the weak long wavelength emission is possible is solvent perturbation (forced dipole) or quadrupole emission [62], and not spin-orbit coupling.

Following namely Mulliken's results, and Kasha's experiments, the role of the triplet state was finally advanced (and demonstrated) by Lewis and Kasha in 1944 [65, 67], and in subsequent works. Only then did a close parallel with the Bohr-Grotrian diagram become possible.

It is interesting to note that Pringsheim, in his 1943 book with Vogel [75], still uses the model of Perrin (the authors make a clear distinction between Perrin's monomolecular delayed fluorescence, then called phosphorescence, and bimolecular recombination afterglow), not mentioning Jablonski's 1933 and 1935 works, of which he was nevertheless aware. Only in his 1949 book, Fluorescence and Phosphorescence [76], after Lewis' studies, does he write the following (repeated in [77]): A theoretical explanation of the coexistence of the two phenomena [*in modern terminology: fluorescence and phosphorescence*] has been given by Jablonski in complete analogy to the energy-level scheme of the mercury atom (...) (p. 435).

What is meant by Pringsheim is best understood if another extract of his book [76] is also quoted: An instance in which all possible luminescence processes [*in modern terminology: prompt and delayed fluorescence, phosphorescence; these terms do not strictly apply to the mercury transitions mentioned by Pringsheim*] can be observed (...) is provided by mercury vapor (...) (p. 290).

A (modern) reader of the first statement above could be led to believe that Jablonski had already reasoned as if something analogous to the Bohr-Grotrian scheme should exist for molecules, while Pringsheim only implies that an energy level scheme accounting for the three types of processes was already known.

In fact, Jablonski (as the Perrins before him) does not discuss his diagram nor the nature of the metastable state in terms of spin multiplicity [62]. He even opposed until very late the triplet hypothesis [62].

The contribution of Jean and Francis Perrin to the formation of the "Jablonski diagram" was important. A better name, correct from the photokinetic point of view, would therefore be "Perrin-Jablonski diagram," although it would still leave aside the final contribution of Lewis and co-workers.

It should be remembered at this point that incorrect attributions are legion in science. In a very interesting article, Laidler [78] gives several examples pertaining to physical chemistry: Boyle's law, Le Chatelier principle, Arrhenius equation, etc.

It is appropriate to close this section with a quotation of G. N. Lewis, that he wrote in 1906: Perfection is rare in the science of chemistry. Our theories do not spring full-armed from the brow of the creator. They are subject of gradual growth ... [78].

2.4
Resonance Energy Transfer

The pioneering role of Jean and Francis Perrin in the field of molecular resonance energy transfer is generally acknowledged [79]. In the paper by Theodor Förster that laid down the standard quantitative theory of molecular resonance energy transfer [80], it is written that: J. Perrin was the first to note [23, 26] that in addition to radiation and reabsorption, a transfer of energy (*transfert d'activation*) could also take place through direct electrodynamic interaction between the primarily excited molecule and its neighbors. He presented a theory of such processes based on classical physics, and F. Perrin [52] later gave a corresponding quantum mechanical theory, the latter leaning on Kallmann and London's theory [81] of excitation energy transfer between various atoms in the gas phase [82].

In order to understand fully the early stage of molecular resonance energy transfer, which is not without its twists and turns, mention must be made of the studies of atoms in the gas phase.

In a famous experiment (1913), Franck and Hertz [83] showed that the collision of fast electrons with slow-moving atoms could result in the production of excited atoms and slow electrons. These collisions, where a transfer of energy takes place, with the conversion of translational energy into electronic energy, were called *collisions of the first kind*.

The reverse process, collision of slow electrons with excited atoms, resulting in the production of ground state atoms and fast electrons, was postulated by Klein and Rosseland [84] in 1921. These collisions, where a radiationless transfer of energy again takes place, with the conversion of electronic energy into translational energy, were called *collisions of the second kind*.

In 1922, Franck [85] extended this last mechanism to include collisions between atoms or molecules. In this case, the collision of an excited atom A* with an unexcited atom B can produce an unexcited atom A and an excited atom B*. Cario [86] and Cario and Franck [87] experimentally demonstrated in 1922 the existence of collisions of the second kind between atoms. A mixture of mercury and thallium atomic vapors, in conditions of selective photoexcitation of mercury (mercury resonance line at 254 nm), also displayed thallium (sensitized) green emission (535 nm). A more familiar situation where this process also occurs is the helium-neon laser.

It was later demonstrated by Beutler and Josephy [88] that the transfer occurs with high probability only when there is a near match of electronic transition energies between A and B. One thus speaks of *resonance energy transfer*. The need for near resonance is classically explained by the Franck principle [76], more familiar in its application to molecular spectroscopic transitions [89–91]. Kallmann and London developed the quantum theory of resonance energy transfer between atoms in 1928 [81]. The dipole-dipole interaction and the parameter R_0 are used for the first time in this work.

Jean Perrin's concept of molecular transfer of energy (*transfert d'activation*) evolved in the context of his first studies of fluorescence [15] and of his theory of unimolecular chemical reactions [16–18, 23]. In [15], based on the observation of photobleaching of several dyes and other organic compounds (and of the corresponding fluorescence recovery by molecular diffusion, also noted by him), he (incorrectly) concludes that the fluorescence emission implies the destruction of the emitting molecule. The (complex) effect of concentration on the photobleaching rate led Perrin to propose that the neighboring molecules had a protecting role on the molecule directly excited. He advanced the following explanation.

D'autres molécules, même très voisines, ne seraient pas directement absorbantes. Mais la masse mise en vibration dans la molécule sensible exercerait sur les masses semblables des molécules voisines une "induction" qui les ferait à leur tour, par une résonance secondaire, entrer en vibration, induction d'autant plus active que les molécules sont proches, et qui, en définitive, aurait encore pour effet de partager entre plusieurs molécules qui resteraient intactes, l'énergie qui aurait été nécessaire pour en remanier irréversiblement une. [15]

(The other molecules, even if very close to the excited one, would not absorb the exciting radiation. But the vibrating part of the excited molecule would exert over similar parts of the neighboring molecules an "induction" that would put them in vibration by means of a secondary resonance. This induction would be the stronger the closer the molecules, and would result in the sharing of the energy absorbed. This sharing would render harmless the absorbed energy, only destructive if concentrated in a single molecule.)

This induction was supposed to be of electromagnetic origin. The model used by Jean Perrin was that of a common transformer, with two synchronous circuits [22, 23, 26, 29–32]. Since these circuits, when in phase, repel each other, this provided him with the explanation for the conversion of electronic excitation energy into heat: the repulsion between the two molecules results in the production of kinetic energy [18, 22, 23]. This also accounted for the mechanism of heating by the absorption of radiation, a question that had been raised by Pierre Curie [18].

The proposal that fluorescence emission implies the destruction of the emitting molecule was afterwards withdrawn, on the basis of new experimental evidence (namely the effect of oxygen) [20]. Notwithstanding, Jean Perrin had measured in the meanwhile a continuous decrease of the intrinsic fluorescence yield of dyes in solution with an increase in concentration [19]. He explained it on the basis of the same mechanism of induction. The neighboring molecules, instead of helping to preserve the excited one, were acting as quenchers! [20].

It is interesting to note that the well-known exponential dependence on quencher concentration, for active-sphere quenching in rigid medium, known as the Perrin equation [92], was derived by Francis Perrin [34] precisely to account for the hypothetical self-quenching by resonance energy transfer previously postulated by Jean Perrin [20]. Only much later was it was demonstrated that the true cause of self-quenching of dyes is the formation of nonfluorescent aggregates that, apart from reducing the number of luminescent molecules, may also act as traps by means of nonradiative and radiative transfer [93]. Additionally, in some experimental conditions, radiative transport also decreases the macroscopic fluorescence yield [94].

It is not known when Jean Perrin became aware of the results for energy transfer between atoms in the gas phase. He cites them in 1929 [29, 30], the year of Francis Perrin's Ph.D. thesis [46], where the subject is also raised, and where the 3rd edition of Pringsheim's book *Fluorescenz und Phosphorescenz* (1928) is indicated as a comprehensive bibliographical source. In the same year, in a work with Nine Choucroun, probably motivated by Cario and Franck's experiments, it is conceded that: Si, par suite de l'induction, il peut se produire à distance une transformation intégrale du quantum d'activation en énergie cinétique, il doit également pouvoir se produire, sans répulsion (et plus particulièrement en milieu visqueux ou rigide), un passage intégral de ce quantum du premier circuit au second. Ce sera le *transfert d'activation* qui substitue, *à distance*, sans changer les vitesses, une molécule activée à une autre molécule activée [29]. (If, by induction, the quantum of activation can be totally converted in kinetic energy at distance, then it must also be possible to produce, in the absence of repulsion (particularly in a viscous or rigid medium) the total transfer of this quantum from the first to the second circuit. This would be a *transfer of activation* at a distance, replacing one activated molecule for another one, without any change of speeds.)

They then give experimental evidence for sensitized fluorescence of molecules in solution, and establish a parallel with the Cario and Franck studies cited above. A similar discussion appears in [30]. It was suggested [95] that the experimental results of Choucroun and Perrin could be at least in part due to radiative transfer.

Jean Perrin did not altogether abandon his original view of deactivation by direct electronic-to-translation transfer, as is clearly stated in [30, 32]. The same explanation is found in Francis Perrin's thesis [46].

There, when discussing the case of identical molecules, Francis Perrin makes the important remark that the transfer of energy is observable by the depolarization of fluorescence. In fact, it had been previously reported by several researchers that an increase in the concentration of dyes in viscous solvents was accompanied by a progressive depolarization [96–98], even for concentrations where quenching was still negligible, but no explanation had been given. To Francis Perrin, the cause can only be energy transfer: Il suffit qu'un transfert d'activation puisse se produire entre deux molécules voisines d'orientations différentes, c'est-à-dire portant des oscillateurs non parallèles, pour qu'il en résulte en moyenne une diminution de l'anisotropie de distribution des oscillateurs excités et par suite de la polarisation de la lumière emise [46]. (It suffices that a transfer of activation can occur between two neighboring molecules with different orientations, that is, with non-parallel oscillators, in order to have, on the average, a decrease in the anisotropy of the distribution of excited oscillators, and therefore a decrease of the polarization of the emitted light.)

Francis Perrin does not develop in his thesis the theory of the corresponding kinetics, that he considers quite difficult. He nevertheless estimated the distance at which the transfer may take place. From the concentration at which the depolarization becomes important for fluorescein in glycerol (10^{-3} g/cm^3), the average distance is found to be 80 Å (the calculation presupposes a cubic lattice; the value for a random distribution is 45 Å). It is thus concluded that the probability for transfer at this distance is reasonable within the excited state lifetime, that is, transfer takes place at distances much larger than molecular dimensions [46].

Francis Perrin latter developed a quantum theory of resonance energy transfer between atoms [52, 53], based on Kallman and London results [81]. He concluded that the transfer would be probable for distances of the order of one-quarter of the wavelength, in agreement with his previous estimates for molecules. He also qualitatively discussed the effect of the spectral overlap between the emission spectrum of the donor and the absorption spectrum of the acceptor, on the efficiency of transfer [52, 53], an effect that Jean Perrin had already recognized [26]. The development of the first quantitative theory of molecular resonance energy transfer, and corresponding kinetics were left to Theodor Förster [80, 99, 100].

2.5
Fluorescence Polarization

We saw in the previous section that Francis Perrin provided the correct interpretation of concentration depolarization in solution. He was particularly well placed to understand the problem, as he had been studying for some time the effect of molecular rotational motion on fluorescence polarization. This subject forms the core of his Ph.D. thesis in physics [46], while the mathematics of rotational Brownian motion was the subject of his Ph.D. thesis in mathematics [45].

Weigert discovered the polarization of the fluorescence of solutions in 1920, and already noted the effect of molecular size, solvent viscosity and temperature [101]. As a result of their own experiments, Vavilov and Levshin proposed in 1923 that the origin of depolarization was molecular rotation [102]. A first quantitative treatment was attempted by Levshin [98, 103], but the approximations made were too unrealistic. Francis Perrin published in 1925 his first results [36], not entirely correct because he supposed a circular oscillator and used a wrong definition of polarization. These two errors compensated in part, so that a good agreement with Levshin's experimental results [98] was obtained. In the following year, Francis Perrin publishes one of his best works on luminescence [41], where he corrects the problems mentioned and gives the equation that bears his name:

$$p = p_0 \frac{1}{1 + \left(1 - \frac{1}{3} p_0\right) \frac{RT}{V\eta} \tau} \tag{2.1}$$

where p is the polarization, p_0 is the polarization in the absence of rotation (limiting polarization), R is the gas constant, T is the temperature, V is the solute's effective molar volume, η is the viscosity of the solvent and τ is the lifetime of fluorescence. The modern, but equivalent form of the equation is deceptively simple:

$$\frac{r_0}{r} = 1 + \frac{\tau}{\tau_r} \tag{2.2}$$

where r is the anisotropy, r_0 is the fundamental anisotropy, τ_r is the rotational correlation time, $\tau_r = \eta V/(RT)$, and τ is the lifetime of fluorescence. This equation, valid for spheres, was later generalized to ellipsoids [55].

Francis Perrin applied the equation to experimental polarizations obtained by him for fluorescein in water-glycerol mixtures at 20 °C. A plot of $1/p$ vs $RT/(V\eta)$ gave a straight line, as predicted, and yielded $p_0 = 0.44$ and $\tau = 4.3$ ns [41]. This last value is in good agreement with the presently accepted value, 4.1 ns [104]. The Perrin equation thus allowed the determination of fluorescence lifetimes [38, 39]. This was particularly important, since no other reliable method was then known. More experimental lifetimes are given in Francis Perrin's thesis [46]. One, particularly interesting, refers to erythrosin (tetraiodofluorescein) in water, for which a lifetime of ca. 80 ps was found. Again, this value is in very good agreement with more recent determinations, $\tau = 75 \pm 5$ ps [105].

In 1926, Enrique Gaviola, working in Pringsheim's laboratory in Berlin, built the first phase-shift fluorimeter (based on the Kerr effect), for the direct measurement of short lifetimes [106, 107]. The values reported for uranin (fluorescein sodium salt) in water and in glycerol (4.4–4.5 ns) [107] agree with those of Perrin. But the value measured for erythrosin (1–2 ns) [107] is much higher than the correct one, showing the limitations of the apparatus for the measurement of picosecond lifetimes. Curiously, this last value still appears in a 1967 compilation of lifetimes [108], although it had been previously criticized by Förster [109]. The use of Perrin's equation for the determination of lifetimes is also of course subject to limitations [76].

In the cited 1926 paper of Francis Perrin [41], he also obtains for the first time the relation between lifetime and quantum yield. Computing the radiative lifetime of fluorescein from its absorption spectrum by means of a relation that was already known, he could show that the estimated quantum yield was close to unity, in good agreement with direct determinations by Vavilov (Fig. 2.10).

Francis Perrin also obtained the relation between p_0 and the angle made by the absorption and emission dipoles [46], predicting a variation between $-1/3$ and $1/2$ (the same result, in a less clear context [62], had been obtained before by Levshin [103]). The existence of negative values of polarization, experimentally established by Vavilov in 1929 [110], but incorrectly interpreted (attributed to an interaction with the magnetic field of the radiation), was also discussed by F. Perrin [51].

The last published work of Francis Perrin on molecular luminescence [56], results from his lecture in the first international luminescence conference, held in 1936 in Warsaw, where Jablonski and Pringsheim played major roles [61]. In

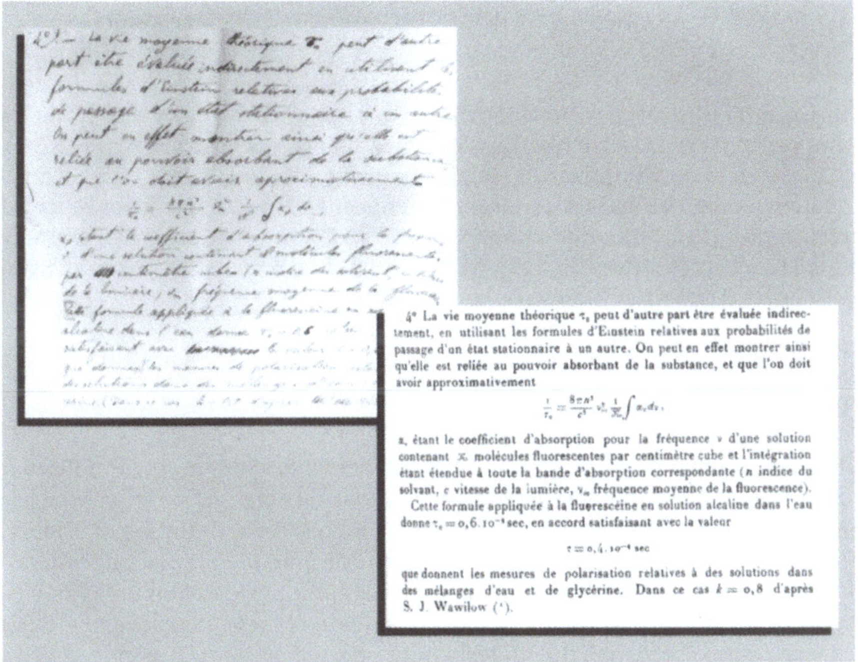

Fig. 2.10. Original manuscript and corresponding printed text of part of the article of Francis Perrin "Détermination de la vie moyenne dans l'état activé des molécules fluorescentes", Comptes Rendus 182 (1926) 219–221. (Archives de l'Académie des Sciences, Paris)

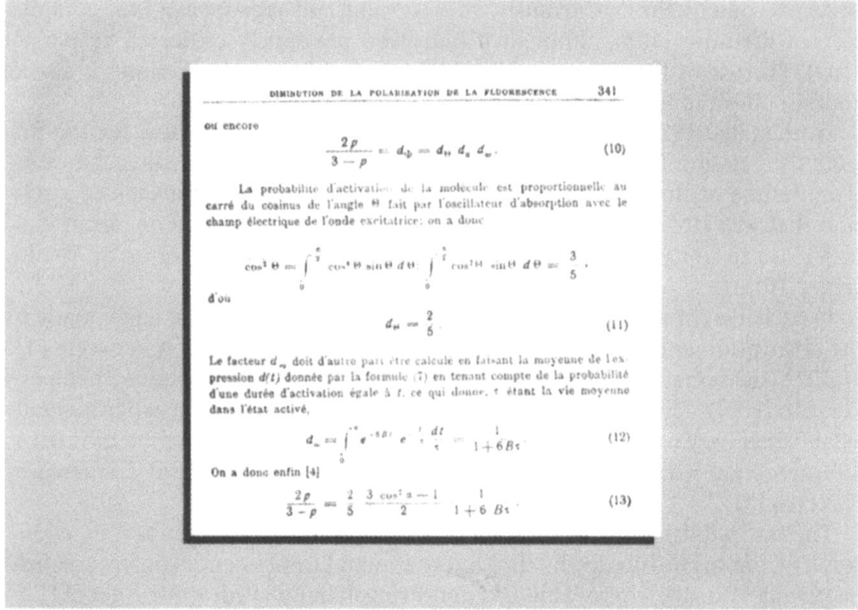

Fig. 2.11. Facsimile of part of the article of Francis Perrin "Diminution de la polarisation de la fluorescence des solutions résultant du mouvement brownien de rotation", Acta Phys. Polon. 5 (1936) 335–347. The quantity 2p/(3-p) was later named anisotropy by Jablonski

this work (Fig. 2.11), he reviews the theory of rotational depolarization, giving a new presentation. In particular, he uses a new quantity, denoted d_Φ (presently r), which renders equations much simpler (see, e.g., the two forms of the Perrin equation above). Jablonski later used the same quantity [111–115] that he called anisotropy, r [112], but only acknowledged Perrin's prior use once, in a footnote (at the request of a referee?) [114]. As a consequence, Francis Perrin's priority in this matter is not widely known.

2.6
Concluding Remarks

The main contributions of Jean and Francis Perrin to three areas of molecular luminescence have been reviewed. An aspect that becomes patent from what has been presented is the complex, nonlinear character of the evolution of science. To a student acquainted with theories and models (and the general understanding of a subject), only from their logical and smooth presentation in textbooks, it may come as a surprise that the birth and growth of these theories and models is much more disordered, full of hesitations and wrong steps. It is indeed like a succession of multi-authored drafts converging to the final, collective text. Yet that is the true way of science, in an ever-greater understanding of nature (Fig. 2.12):

Fig. 2.12. La Nature se dévoilant à la Science (E. Barrias, 1899). The original, initially intended for the Conservatoire National des Arts et Métiers, is now at the Musée d'Orsay. Similar statues by Barrias can be found in the Ancienne Faculté de Médecine in Paris, in the Pena palace (Sintra, Portugal), and in the Smart Museum (University of Chicago, USA)

La Nature déploie la même splendeur sans limites
dans l'Atome ou dans la Nébuleuse,
et tout moyen nouveau de connaissance
la montre plus vaste et diverse, plus féconde,
plus imprévue, plus belle,
plus riche d'insondable immensité.
Jean Perrin, Les Atomes [14]
(Nature displays the same boundless splendor
In the Atom and in the Nebula,
And every new way of knowledge
Shows her more vast and diverse, more fecund,
More unpredictable, more beautiful,
And more full of inscrutable immensity.)

2.7
Bibliographical Notes

The works of Jean Perrin are collected in two posthumous books, *La Science et l'Espérance* [116], containing his philosophical and social views, along with some general writings on science, and *Oeuvres Scientifiques* [13], edited by Francis Perrin, containing a selection of scientific papers and other writings. It also comprises the full list of Jean Perrin's publications, including books. Jean Perrin was recently the subject of a very readable but biased biography [117]. A much briefer but more balanced account of his scientific works and their significance, bearing especially on the experimental proofs of molecular reality, was delivered by Louis de Broglie at the Académie des Sciences in 1945, and is reproduced in *La Science et l'Espérance* [116]. A short biography of Jean Perrin can be found in the Nobel Foundation web page.

The works of Francis Perrin are collected in a recent book, *Écrits de Francis Perrin* [57]. It specifically contains a selection of scientific papers and a short biography. A complete list of his scientific publications, along with illustrated biographical notes and interesting recollections, can be found in *Hommage à Francis Perrin* [58]. Francis Perrin wrote in 1951 a brochure describing his scientific works [118] (reproduced in part in [57]).

All publications of Jean Perrin and Francis Perrin directly related to luminescence, sometimes incorrectly cited, are given in the reference list: [15–32] (Jean Perrin) and [30, 34–56] (Francis Perrin).

Acknowledgements. I am most grateful to Prof. Bernard Valeur (Conservatoire National des Arts et Métiers, Paris, and ENS Cachan) for having invited me to present the lecture from which the present text evolved, and for all the support and encouragement. I am grateful to Mme Demeulanaere-Douyère, Directrice of the Archives de l'Académie des Sciences (Paris) for all the help concerning the files of Jean Perrin and Francis Perrin. I also thank Dr. Bernhard Nickel (MPI für Biophysikalische Chemie, Göttingen) for reprints of his important articles, Prof. Piotr Targowski (University N. Copernicus, Torun) for copies of some of the earliest Jablonski's articles on anisotropy and of a brochure on A. Jablonski, Prof. Evgeny Bodunov (Russian Hydrometeorology University, St. Petersburg) for biographical information on S. Vavilov, and Dr. Jean-Claude Mialocq (CEA, Saclay) for an article on F. Perrin's periodical table. I am also grateful to the Center for the History of Physics of the American Institute of Physics for access to documents, recordings, and photographs, some of which are reproduced here, and to the Cultural Services of the German Embassy in Lisbon, and the Archives of the Humboldt University (Berlin) for biographical information on Walter Grotrian. Financial support from the ICCTI (Portugal)-CNRS (France) joint program, from the FCT (Portugal), and from Instituto Superior Técnico is gratefully acknowledged.

References

1. Medawar P (1982) Pluto's republic. Oxford UP, Oxford
2. Chargaff E (1968) Science 159:1448–1449 (see also Heraclitean Fire (1978) Rockefeller UP, New York)
3. Borges JL (1944) Ficciones. Sur, Buenos Aires
4. Poirier J-P (1993) Lavoisier. Pygmalion, Paris
5. Snow CP (1967) Variety of men. Scribner's, New York
6. Hoffmann R (1995) The same and not the same. Columbia UP, New York

7. Ziman J (1968) Public knowledge. Cambridge UP, Cambridge
8. Bernard C (1865) Introduction à l'étude de la médecine expérimentale, Paris
9. Kohn A (1989) Fortune or failure. Missed opportunities and chance discoveries. Blackwell, Oxford
10. Roberts RM (1989) Serendipity. Wiley, New York
11. Jacques J (1990) L'imprévu ou la science des objects trouvés. Odile Jacob, Paris
12. Vallery-Radot P (1943) Les plus belles pages de Pasteur. Flammarion, Paris
13. (1950) Oeuvres scientifiques de Jean Perrin. CNRS, Paris,
14. Perrin J (1913) Les atomes. Alcan, Paris; (1936) Rédaction nouvelle. Alcan, Paris
15. Perrin J (1918) La fluorescence. Ann Phys (Paris) 10:133–159
16. Perrin J (1919) Matière et lumière. Ann Phys (Paris) 11:5–108
17. Perrin J (1920) Atomes et lumière. Revue du Mois 21:113–166
18. Perrin J (1922) Radiation and chemistry. Trans Faraday Soc 17:546–572
19. Perrin J (1923) Observations sur la fluorescence. Comptes Rendus 177:469–475
20. Perrin J (1923) Radiochimie de la fluorescence. Comptes Rendus 177:612–618
21. Perrin J (1923) Radiochimie de la fluorescence (suite). Comptes Rendus 177:665–666
22. Perrin J, Choucroun N (1924) Fluorescence, et lois générales relatives aux vitesses de réaction. Comptes Rendus 178:1401–1406
23. Perrin J (1925) Lumière et réactions chimiques, rapport au 2e Conseil de Chimie Solvay, Bruxelles (pp 322–398 of Structure et activité chimique). Gauthier-Villars, Paris, 1926
24. Perrin J, Choucroun N (1926) Parallélisme entre le pouvoir fluorescent et la vitesse de réaction. Comptes Rendus 183:329–331
25. Perrin J, Choucroun N (1927) Rôle de l'induction moléculaire dans l'activation par choc. Comptes Rendus 184:985–987
26. Perrin J (1927) Fluorescence et induction moléculaire par résonance. Comptes Rendus 184:1097–1100
27. Perrin J, Choucroun N (1928) Vitesse des réactions photochimiques. Comptes Rendus 187:697–698
28. Perrin J (1928) Détermination du rôle de la lumière dans les réactions chimiques thermiques. Comptes Rendus 187:913–916
29. Perrin J, Choucroun N (1929) Fluorescence sensibilisée en milieu liquide (transfert d'activation par induction moléculaire). Comptes Rendus 189:1213–1216
30. Perrin F, Perrin J (1929) Activation et désactivation par induction moléculaire. In: Activation et structure des molécules. P.U.F., Paris, pp 354–382
31. Perrin J (1935) L'induction moléculaire. In: Jubilé de M. Marcel Brillouin, Gauthier-Villars, Paris, pp 349–359
32. Perrin J (1936) L'induction moléculaire, Acta Phys. Polon. 5:319–332
33. Balibar F (ed) (1989) Correspondances françaises d'Einstein. Seuil, Paris
34. Perrin F (1924) Loi de décroissance du pouvoir fluorescent en fonction de la concentration. Comptes Rendus 178:1978–1980
35. Perrin F (1924) Rôle de la viscosité dans les phénomènes de fluorescence. Comptes Rendus 178:2252–2254
36. Perrin F (1925) Théorie de la fluorescence polarisée (Influence de la viscosité). Comptes Rendus 180:581–583
37. Perrin F (1925) Sur le mouvement brownien de rotation. Comptes Rendus 181:514–516
38. Perrin F (1926) Détermination de la vie moyenne à l'état excité des molécules fluorescentes. J Physique 7:13S–15S
39. Perrin F (1926) Détermination de la vie moyenne dans l'état activé des molécules fluorescentes. Comptes Rendus 182:219–221
40. Perrin F (1926) Fluorescence à longue durée des sels d'urane solides et dissous. Comptes Rendus 182:929–931
41. Perrin F (1926) Polarisation de la lumière de fluorescence. Vie moyenne des molécules dans l'état excité. J Physique 7:390–401

42. Perrin F (1927) La désactivation induite des molécules et la théorie des antioxygènes. Comptes Rendus 184:1121–1124
43. Perrin F, Delorme R (1928) Mesure des durées de fluorescence des sels d'uranyle solides et de leurs solutions. Comptes Rendus 186:428–430
44. Perrin F (1928) La désactivation induite des molécules et la théorie des antioxygènes. J Chim Phys 25:531–534
45. Perrin F (1928) Étude mathématique du mouvement brownien de rotation. Ann Sc de l'École normale supérieure 45:1–51
46. Perrin F (1929) La fluorescence des solutions. Induction moléculaire – Polarisation et durée d'émission – Photochimie. Ann Phys (Paris) 12:169–275
47. Delorme R, Perrin F (1929) Durées de fluorescence des sels d'uranyle solides et de leurs solutions. J Physique 10:177–186
48. Perrin F (1930) Activation by light and by collisions in thermal equilibrium. Chemical Reviews 7:231–237
49. Perrin F (1931) L'association moléculaire et l'optimum de la fluorescence des solutions. Influence des sels. Comptes Rendus 192:1727–1729
50. Perrin F (1931) Fluorescence. Durée élémentaire d'émission lumineuse. Hermann, Paris
51. Perrin F (1931) Remarques et expériences sur la polarisation de fluorescence des solutions. J Physique 2:163S–166S
52. Perrin F (1932) Théorie quantique des transferts d'activation entre molécules de même espèce. Cas des solutions fluorescentes. Ann Phys(Paris) 17:283–314
53. Perrin F (1933) Interaction entre atomes normal et activé. Transferts d'activation. Formation d'une molécule activée. Ann Institut Poincaré 3:279–318
54. Perrin F (1934) Mouvement brownien d'un ellipsoïde (I). Dispersion diélectrique pour des molécules ellipsoïdales. J Physique 5:497–511
55. Perrin F (1936) Mouvement brownien d'un ellipsoïde (II). Rotation libre et dépolarisation des fluorescences. Translation et diffusion de molécules ellipsoïdales. J Physique 7:1–11
56. Perrin F (1936) Diminution de la polarisation de la fluorescence des solutions résultant du mouvement brownien de rotation. Acta Phys Polon 5:335–347
57. Neveu M, Baton J-P (eds) (1998) Écrits de Francis Perrin. CEA
58. (1994) Hommage à Francis Perrin. CEA
59. Perrin J (1935) Grains de matière et de lumière. Hermann, Paris
60. Nickel B (1996) EPA Newsletter 58:9–38
61. Nickel B (1997) EPA Newsletter 61:27–60
62. Nickel B (1998) EPA Newsletter 64:19–72
63. Kasha M (1984) J Chem Educ 61:204–215
64. Kasha M (1987) Acta Phys Pol A71:661–670
65. Kasha M (1999) Acta Phys Pol 95:15–36
66. Lewis GN, Lipkin D, Magel T (1941) J Am Chem Soc 63:3005–3018
67. Lewis GN, Kasha M (1944) J Am Chem Soc 66:2100–2116
68. Lewis GN, Kasha M (1945) J Am Chem Soc 67:994–1003
69. Kasha M (1947) Chem Rev 41:401–419
70. IUPAC Commission on Photochemistry (1996) Pure Appl Chem 68:2223–2286
71. Bohr N (1924) The theory of spectra and atomic constitution, 2nd edn. Cambridge UP, Cambridge
72. Andrade EN da C (1927) The structure of the atom, 3rd edn. Bell and Sons, London
73. Grotrian W (1928) Graphische darstellung der spektren von atomen und ionen mit ein, zwei und drei valenzelektronen. Springer, Berlin Heidelberg New York
74. Jablonski A (1935) Z Physik 94:38–46
75. Pringsheim P, Vogel M (1943) Luminescence of liquids and solids and its practical applications. Interscience, New York
76. Pringsheim P (1949) Fluorescence and phosphorescence. Interscience, New York
77. Kawski A (1997) EPA Newsletter 61:17–26

78. Laidler K (1995) Acc Chem Res 28:187–192
79. Juzeliunas G, Andrews DL (1999) In: Andrews DL, Demidov AA (eds) Resonance energy transfer. Wiley, Chichester, pp 65–107
80. Förster T (1948) Ann Phys 2:55–75 (1948)
81. Kallmann H, London F (1928) Z Physik Chem B2:207–243
82. Translation by RS Knox (1984) Univ of Rochester, New York
83. Franck J, Hertz G (1914) Vehr d Deutsch Phys Ges 16.12–19
84. Klein O, Rosseland S (1921) Z Physik 4:46–51
85. Franck J (1922) Z Physik 9:259–266
86. Cario G (1922) Z Physik 10:185–199
87. Cario G, Franck J (1922) Z Physik 11:161–166
88. Beutler H, Josephy B (1927) Naturwiss 15:540
89. Murata S, Tachiya M (1992) Chem Phys Lett 194:347–350
90. Franck J (1925) Trans Faraday Soc 21:536–542
91. Condon EU (1926) Phys Rev 28:1182–1201
92. Condon EU (1928) Phys Rev 32:858–872
93. Agranovich VM, Galanin MD (1982) Electronic excitation energy transfer in condensed matter. Elsevier, Amsterdam
94. Berberan-Santos MN, Nunes Pereira EJN, Martinho JMG (1999) In: Andrews DL, Demidov AA (eds) Resonance energy transfer. Wiley, Chichester, pp 108–149
95. Bowen EJ (1966) In: Phillips GO (ed) Energy transfer in radiation processes. Elsevier, Amsterdam
96. Gaviola E, Pringsheim P (1924) Z Physik 24:24–36
97. Weigert F, Käppler G (1924) Z Physik 25:99–117
98. Lewschin WL (1924) Z Physik 26:274–284
99. Förster T (1946) Naturwiss 33:166–175
100. Förster T (1949) Z Naturforsch 4a:321–327
101. Weigert F, (1920) Vehr d Deutsch Phys Ges (3) 1:100–102
102. Vavilov SI, Lewschin WL (1923) Z Physik 16:135–154
103. Lewschin WL (1925) Z Physik 32:307–326
104. Sjöback R, Nygren J, Kubista M (1995) Spectrochim Acta A51:L7–L21
105. Yu W, Pellegrino F, Grant M, Alfano RR (1977) J Chem Phys 67:1766–1773
106. Gaviola E (1926) Z Physik 35:748–756
107. Gaviola E (1927) Z Physik 42:853–861
108. Birks JB, Munro IH (1967) Prog React Kinet 4:239–303
109. Förster T (1951) Fluoreszenz Organischer Verbindungen, Vandenhoeck & Ruprecht, Göttingen
110. Wawilow SJ (1929) Z Physik 55:690–700
111. Jablonski A (1955) Acta Phys Pol 14:295–307
112. Jablonski A (1957) Acta Phys Pol 17:471–479
113. Jablonski A (1960) Bull Acad Pol Sci, sér Sci, Math, Astr et Phys 8:259–263
114. Jablonski A (1961) Z Naturforsch 16a:1–4
115. Jablonski A (1962) Bull Acad Pol Sci, sér Sci, Math, Astr et Phys 10:555–556
116. Perrin J (1948) La Science et l'Espérance. PUF, Paris
117. Charpentier-Morize M (1997) Perrin, savant et homme politique. Belin, Paris
118. Perrin F (1951) Notice sur les travaux scientifiques de Francis Perrin. Paris

CHAPTER 3

The Seminal Contributions of Gregorio Weber to Modern Fluorescence Spectroscopy

D. M. Jameson

Gregorio Weber is acknowledged to be the person responsible for many of the more important theoretical and experimental developments in modern fluorescence spectroscopy. In particular, Weber pioneered the application of fluorescence spectroscopy to the biological sciences. His list of achievements includes:

- The synthesis and use of dansyl chloride as a probe of protein hydrodynamics
- The extension of Perrin's theory of fluorescence polarization to fluorophores associated with random orientations with ellipsoids of revolution and to mixtures of fluorophores
- The first spectral resolution of the fluorescence of the aromatic amino acids and of intrinsic fluorescence of proteins
- The first demonstration that both FAD and NADH make internal complexes
- The first report on aromatic secondary amines, which are strongly fluorescent in apolar solvents, but hardly in water, the most spectacular case being the anilino-naphthalene sulfonates (ANS)
- The first description of the use of the fluorescence of small molecules as probes for the viscosity of micelles, with implications for membrane systems
- A general formulation of depolarization by energy transfer
- The discovery of the "red-edge" effect in homo-energy transfer
- The development of modern cross-correlation phase fluorometry
- The development of the excitation-emission matrix method for resolving contributions from multiple fluorophores
- The synthesis of several novel fluorophores, including pyrenebutyric acid, IAEDANS, bis-ANS, PRODAN, and LAURDAN, designed to probe dynamic aspects of biomolecules

In addition to these seminal contributions, Gregorio Weber also trained and inspired generations of spectroscopists and biophysicists who went on to make important contributions in their fields, including both basic research as well as the commercialization of fluorescence methodologies and their extension into the clinical and biomedical disciplines.

3.1
Overview

During the last few decades, fluorescence spectroscopy has evolved from a narrow, highly specialized technique into an important discipline widely utilized in the biological, chemical, and physical sciences. Fluorescence methodologies have also assumed an increasingly important role in the clinical and medical sciences. There are now world-renowned centers for fluorescence spectroscopy, highly successful commercial enterprises specializing in fluorescence instrumentation, fluorescence based clinical instruments in virtually all hospital laboratories, and thousands of practitioners worldwide. As in all scientific disciplines, the development of modern fluorescence spectroscopy has benefited from the contributions of many individuals from many countries. However, one individual, Gregorio Weber, can be singled out for his outstanding and far-reaching contributions to this field.

3.2
Early Years

Born in Buenos Aires, Argentina, on July 4, 1916, Weber demonstrated an early aptitude for science, mathematics, and linguistics. He was greatly impressed by his high school teacher of geology and mineralogy – not just by his teaching skills, but also by his broad knowledge of science. Weber told this teacher that he would like to have a career in science and asked which area – chemistry, physics, etc. – he would recommend. The teacher replied that it was very difficult to find steady employment in Argentina as a pure scientist at that time and he recommended that Weber go to Medical School, since that would afford him the opportunity to study diverse areas of science and, if all else failed, provide him with a profession with which to support himself. Taking this advice, Weber enrolled in the University of Buenos Aires and completed his M. D. degree in 1943. While a medical student, from 1939 to 1943, he worked in the Department of Physiology and Biochemistry as a teaching assistant to Bernardo Alberto Houssay who had already achieved fame as a physiologist for his work on the endocrine system and in particular the pituitary gland (Houssay shared the 1947 Nobel Prize for Physiology and Medicine with Carl and Gerty Cori). Houssay recognized the ability of his young protegé and suggested that Weber apply for a prestigious British Council Fellowship to support graduate studies toward a Ph.D. at Cambridge University. Although in 1943 the war was still raging in Europe, and London was being subjected to V-1 attacks, Weber enthusiastically embraced the opportunity to pursue his love of science.

3.3
Cambridge

Travel to England during the war years was an adventure and Weber's voyage took 44 days in a convoy, which languished off the coast of Africa for weeks and endured occasional U-boat attacks. In Cambridge, Weber initially thought to study colloid and surface chemistry and joined the laboratory of Eric Riddeal. Soon, however, he became enamored of proteins and went to talk to Malcolm Dixon, the well-known enzymologist, about applying techniques of physical chemistry to the study of proteins. Dixon suggested that Weber consider applying fluorescence techniques to the study of the naturally fluorescent flavin and flavoprotein systems. At that time, Weber knew little about fluorescence but soon learned that there were a number of low molecular weight flavin compounds, such as riboflavin and FAD, that differed greatly in fluorescence intensity. Weber was thus given the task of "sorting out" this area.

When Gregorio Weber began his graduate studies, the fluorescence of substances extracted from organisms had already attracted the attention of biologists and biochemists. In these early studies, however, fluorescence was used as an aid in isolation and purification and also in the quantitative determination of fluorescent substances such as riboflavin, porphyrins, and pterines. The relationship between the fluorescence from chlorophylls and biochemical aspects of the photosynthetic process seemed certain but was far from clear. Almost all of the work done on fluorescence substances had been descriptive in nature, i.e., concerned mainly with the conditions under which the fluorescence could be most readily observed as well as with the color and intensity of the fluorescent emission. The situation was different in the physical sciences, though. Physicists, such as Enrique Gaviola (a fellow Argentinean), Jean and Francis Perrin, S. I. Vavilov, F. Weigert, P. Pringsheim, and others had already introduced important concepts such as the excited state lifetime, the polarization of fluorescence, and the quantum yield of the emission process, yet these ideas had not penetrated into the chemical or biological fields (for a comprehensive, lucid, and exceptionally well-documented overview of the history of photoluminescence in the first half of the twentieth century, the reader is referred to the series of articles by Bernhard Nickel [1–3]). Weber recognized the need for a quantitative understanding of fluorescence phenomena and to this end he spared himself no pains. He wrote in his Ph.D. thesis [4] "I feel that a knowledge, as deep as possible, of the physical principles concerned is indispensable. Even close collaboration with a physicist cannot spare this task to the biochemist. I am tempted to believe that a biologist having n ideas related to the biological side of the problem and a physicist possessing another n relating to the physical side would result in some $2n$ useful combinations whereas the same ideas collected in one brain would lead to a number of combinations more like $n!$".

3.4
Francis Perrin's Influence

In later years, Weber would say that the work that most influenced him and which he liked the best was that of Francis Perrin. Weber's introduction to polarization started when he read Perrin's classic paper of 1926 in the Journal de Physique on the depolarization of fluorescence by Brownian rotations [5]. In Weber's own words [6] "I remember that Malcolm Dixon came to me one day, handed me a little piece of paper, and said that somebody at King's College – I wish I could remember his name – had said that there was a paper on fluorescence that I should read. The little piece of paper had written on it: F. Perrin, *J. de Physique*, 1926. So I went to the Cambridge library, which I positively thought of as a temple of learning and looks indeed like one, and I read the famous paper of Perrin on depolarization of the fluorescence by Brownian rotations, not one but many times. Argentine secondary education in the first half of the century included French language and literature so that I could not only understand the scientific content, but also enjoy the literary quality of the writing. It was written in that transparent, terse style of XVIII century France, which I have tried, perhaps unsuccessfully, to imitate from then onwards. The clarity of Perrin's thought and his ability to do the right experiment were really remarkable." Weber went on to note [6] "It was from reading Perrin's papers that I conceived three ideas on the use of polarization: determination of the change in the fluorescence lifetime as one quenches the fluorescence by addition of an appropriate chemical, determination of the molecular volume of proteins by fluorescent conjugates with known dyes and determination of the viscosity of a medium through the polarization of the emission from a known fluorescent probe."

3.5
Ph.D. Thesis

It goes without saying that all fluorescence instrumentation had to be home built at that time. The original apparatus built by Weber is shown in Fig. 3.1, which is reproduced from his thesis. The light source (L) was a carbon arc, originally developed for use in searchlights during the war. The exciting light was first filtered through a layer of concentrated $NaNO_2$ (U) to remove UV light (< 420 nm) and then polarized by a Nicol prism (N_1). Additional glass filters (F_1 and F_2) were used to delimit further the exciting light and to isolate the emission. The actual measurement of the polarization of the fluorescence was realized using visual compensation techniques involving observation of interference patterns as a "pile-of-plates" polarizer P (the compensator of Arago) was rotated. Using these simple methods and only his eye as the detector, Weber was able to quantify levels of polarized light reaching only 1% or 2%. There was a price to be paid, however, for these visual compensation methods. Like many of the pioneering spectroscopists, Weber suffered acute eye aliments in latter years as

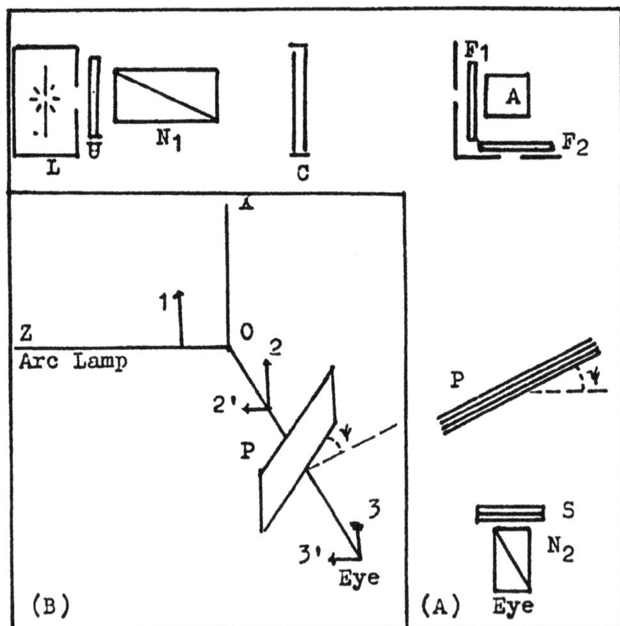

Fig. 3.1. Original drawing from Gregorio Weber's Ph.D. thesis showing the optical arrangement of the instrument he constructed for polarization measurements

Fig. 3.2. Gregorio Weber graduating from St. John's College, Cambridge University, 1947

a result of excessive exposure to infrared and ultraviolet light, which led to removal of his lenses, detached retinas, and eventually cornea transplants[1]. These difficulties resulted in a marked photophobia, which required Gregorio Weber to wear sunglasses most of the time – the sunglasses became almost a trademark for "The Professor" as he was known to his students.

A large portion of Weber's thesis was devoted to measurements on the quenching of fluorescence of riboflavin and on development of a general theory of quenching by complex formation. This led to his first publication entitled *The quenching of fluorescence in liquids by complex formation. Determination of the mean life of the complex* [7]. This paper was the first to demonstrate that fluorescence quenching can take place after formation of molecular complexes of finite duration rather than collisions. His second publication entitled *Fluorescence of riboflavin and flavin–adenine dinucleotide* [8], was the first demonstration of an internal complex in FAD. Years later he was to follow up this work with the first demonstration that NADH also formed an internal complex [9] and with more complete characterizations of the excited state properties of FAD and NADH [10–13]. Weber completed his doctoral thesis entitled *Fluorescence of riboflavin,*

---

[1] An insight into the rationality characteristic of Gregorio Weber is his remark, in the author's presence, to the physician who removed the bandages from the second eye, which had received the new cornea – "I have a homogeneous, clear, binocular visual field" – a statement conveying the maximum of information with the minimum of words.

diaphorase and related substances in 1947 (Fig. 3.2). Interestingly, the final chapter of his thesis was devoted to the application of polarization measurements to the determination of the microscopic viscosity of gels. This work followed on the original, seminal observations of F. Weigert [14] in 1920 that the percentage of polarized light from a dye in solution increased rapidly with the viscosity of the solution, and the more detailed studies of Vavilov and Levshin in 1923 who reported on the polarization of the fluorescence of 26 dyes in water and glycerol and also on dyes in colloidal solutions [15] – observations which were, of course, important for F. Perrin's experimental and theoretical studies. Weber observed that the polarization of fluorescein in gels such as agar, gelatin, and silicic acid did not increase as the gels solidified. These observations were important in deciding between different theories on the nature of colloidal suspensions, namely supporting the theory that the colloidal particles built a continuous, fibrillar structure, which would leave pockets of solvent in between the gel structures. These pockets could then accommodate the fluorescein molecules, which would rotate freely in the embedded solvent. Weber clearly articulated the concept of microviscosity as opposed to macroviscosity or "bulk" viscosity and suggested that fluorescence polarization might be fruitfully applied to the study of cell protoplasm. This prescient observation anticipated the work he would publish 24 years later which first delineated the application of fluorescence probes to study the physical state of lipid systems [16, 17]. In fact, Weber never lost his fascination with the rotation of small molecules. In later years he developed the theory of differential phase fluorometry [18] and applied the method to demonstrate the anisotropic rotation of small molecules in isotropic solvents [19, 20].

3.6
Postdoctoral

From 1948 to 1952 Weber carried out independent investigations at the Sir William Dunn Institute of Biochemistry at Cambridge, supported by a British Beit Memorial Fellowship. At that time he began to delve more deeply into the theory of fluorescence polarization and also began to develop methods which would allow him to study proteins which did not contain an intrinsic fluorophore such as FAD or NADH (the fluorescence of the aromatic amino acids had not yet been discovered). To this end, he invested considerable time and effort in synthesizing a fluorescent probe which could be covalently attached to proteins and which possessed absorption and emission characteristics appropriate for the instrumentation available in post-war England. The result of two years of effort was the still popular probe dimethylaminonaphthalene sulfonyl chloride or dansyl chloride. With this tool in hand and with new instrumentation he began to investigate several protein systems, publishing his theory and experimental results in two classic papers published in 1952, namely *Polarization of the fluorescence of macromolecules. I. Theory and experimental method* [21] and *Polarization of the fluorescence of macromolecules. II. Fluorescent conjugates of ovalbumin and bovine serum albumin* [22]. The theory paper (which interestingly contains an acknowledgement to F. Perrin for his suggestions) includes an extension of Perrin's theory of depolarization due to rotation of ellipsoidal molecules. Specifically, Weber showed that Perrin's complex equations, which required a knowledge of the orientation of the fluorophore's absorption and emission oscillators with respect to the axis of rotation of the ellipsoid, could be considerably simplified if the fluorophores carrying the oscillators were assumed to be randomly oriented on the macromolecule. This paper also contained a formulation of the law of additivity of polarizations, namely.

$$\left(\frac{1}{P_{obs}} - \frac{1}{3} \right)^{-1} = \Sigma f_i \left(\frac{1}{P_i} - \frac{1}{3} \right)^{-1}$$

where P_{obs} is the actual polarization observed arising from i-components, f_i represents the fractional contribution of the i-th component to the total emission intensity, and P_i is the polarization of the i-th component. The motivation for considering this additivity function arose from Weber's realization that a population of fluorescent molecules differing in their size or excited state lifetime (for example free and protein-bound probe) would contribute separately to the observed polarization. Similarly, a population of nonspherical proteins, such as prolate or oblate ellipsoids, could give rise to a distribution of rotational rates depending upon the orientation of the probe along the respective rotational axes. Hence, a clear understanding of the ways in which the individual contributions sum to the total signal was important. These considerations led immediately (in fact in the paper directly following Weber's two articles in *Biochemical Journal*) to the work of Laurence who first described the application of polarization methods to follow the binding of various small fluorescent ligands, such as fluorescein, eosin, acridine, and others, to bovine serum albumin [23]. Dandliker and co-workers later applied these principles explicitly to the study of antibody-antigen [24, 25] and hormone-binding site interaction [26] and these methodologies are still widely used in the biological sciences [27]. We may also note that 8 years after Weber's demonstration of the additivity principle of polarization, Jablonski [28] drew attention to the additive nature of the anisotropy function, defined as:

$$r = \frac{I_{\parallel} - I_{\perp}}{I_{\parallel} + 2I_{\perp}}$$

Since the relationship between anisotropy and polarization is given by:

$$r = \frac{2}{3} \left(\frac{1}{P} - \frac{1}{3} \right)^{-1}$$

the additivity property of anisotropy follows directly from Weber's earlier work[1]. Interestingly, at the same time that Weber was carrying out his theoretical and experimental studies in England, Singleterry and Weinberger, at the US

---

[1] Interestingly, in his article on the additivity of anisotropy [28], which appeared in 1960, Jablonski did not refer to any of Weber's articles on polarization – not even the 1952 article, which explicitly presented the formula for the additivity of polarization. Jablonski did, however, point out that the use of anisotropy leads to simplifications of many relevant equations.

Naval Research Laboratory, were independently applying fluorescence polarization methods to study the size of oil soluble soap micelles in non-polar solvents [29]. Steiner and McAlister [30] soon followed Weber's methods and published an elegant study of conjugates of dansyl chloride, anthracene sulfonyl chloride, and FITC with several proteins – fluorescence polarization measurements were made as well as phase-shift lifetime determinations.

3.7
Sheffield

Weber stayed at Cambridge as an independent researcher until 1953 when Hans Krebs recruited him for the new Biochemistry Department at Sheffield University. During his years at Sheffield, Weber continued to lay the foundations of modern fluorescence spectroscopy developing both fluorescence theory [31, 32] and instrumentation [33]. His pioneering contributions during these early years included his report with Laurence [34] of aromatic secondary amines, which were strongly fluorescent in apolar solvents but very weakly fluorescent in water, the most spectacular case being the anilino-naphthalene sulfonates (ANS). It is interesting to note that, even today, more than 50 years after that first report, ANS is still being used in protein studies, quite often as an indicator of the "molten globule state."

3.8
Intrinsic Protein Fluorescence

During those early years at Sheffield, Weber and his postdoctoral fellow, F. W. John Teale, began their studies on intrinsic protein fluorescence. At that time, compounds resembling the aromatic amino acids had been shown to possess appreciable fluorescence in the near ultraviolet but the fluorescence of the aromatic amino acids themselves had not yet been unequivocally characterized (although Debye and Edwards [35] had made observations on the phosphorescence of the aromatic amino acids and the position of these phosphorescence bands indicated to McClure [36] the probable existence of fluorescence bands in the near ultraviolet). In 1953, Weber hypothesized that emission bands for tyrosine and tryptophan should exist with maxima in the region 3000–4000 Å [31]. At about the same time Weber and Teale were carrying out their studies, Shore and Pardee adapted a Beckman DU spectrophotometer to view the ultraviolet fluorescence of tyrosine, tryptophan, and a number of proteins through a filter that passed wavelengths greater than about 300 nm [37]. Shore and Pardee could not record emission spectra with their apparatus, however, and the excitation spectra obtained were very approximate. In 1957, Weber and Teale published the first emission spectra of the aromatic amino acids, and the first accurate excitation spectra [38] (Fig. 7 from this paper has been reproduced many times and is reproduced here in Fig. 3.3). In the late 1950s and early 1960s, Weber and Teale published a series of important papers and communications on intrinsic protein fluorescence and the determination of absolute quantum yields [39–43]. Interestingly, the quantum yield Weber and Teale reported for tryptophan, 0.20, was

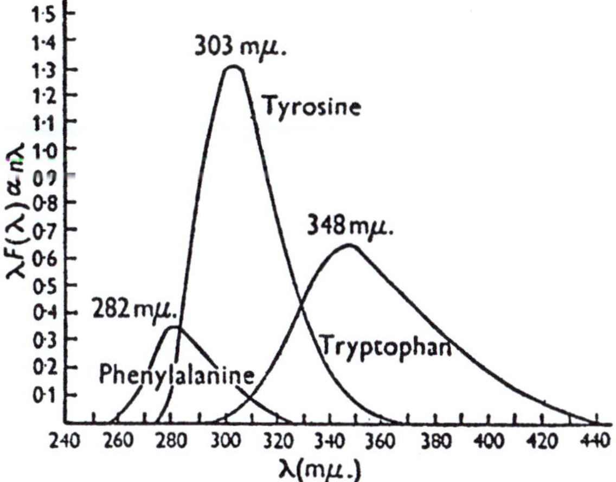

Fig. 3.3. Figure 7 from [38] giving the first emission spectra of the aromatic amino acids

later found to be somewhat higher than the currently accepted value near 0.14. At the time Weber and Teale carried out their experiments, however, the large temperature effect on tryptophan's lifetime and quantum yield was not appreciated and their work, reported as done at "room temperature," was in fact carried out in the winter in a Quonset hut without heating, which caused a marked increase in their tryptophan quantum yield relative to that expected for 25 °C. In 1960, Weber published the excitation polarization spectra of the aromatic amino acids and numerous proteins and also gave the first demonstration of electronic energy transfer among tyrosines and tryptophans and the critical transfer distances from tyrosine to tryptophan and among tyrosine or tryptophan residues [44, 45]. In 1959, Weber and Teale also demonstrated the first use of electronic energy transfer in the study of hemeproteins by comparing the fluorescence of hemoglobin and horseradish peroxidase before and after removal of the heme [41].

Weber's work on intrinsic tryptophan fluorescence inspired Velick to study the binding of NADH to dehydrogenases by following the quenching of the tryptophan fluorescence due to energy transfer [46]. Soon afterwards, Velick et al. [47] then applied this method to study the binding of aromatic ligands (in particular 2,4 dinitrophenol) to specific antibodies – work which led to the significant finding of affinity maturation (increase in the affinity of the antibodies produced with time after induction of antibody production), which was to have a major impact on the field of immunology. Weber followed up his early interest in heme protein fluorescence years later with the first report of the emission spectrum of hemoglobin [48]. During the last 20 years, numerous papers have appeared from many laboratories reporting studies on the intrinsic fluorescence of hemeproteins such as hemoglobin, myoglobin, and horseradish peroxidase. Weber's interest in the photophysics of tryptophan also resulted in a publication with Bernard Valeur, in 1977, of an important and often quoted paper [49] on the ^1L_a and ^1L_b excitation bands of indole and tryptophan. During the four decades since the first description of protein fluorescence, thousands of papers have been written on the fluorescence of tryptophan, tyrosine, or phenylalanine, or some aspect of intrinsic protein fluorescence. The study of intrinsic protein fluorescence has, in fact, become one of the most important techniques used in protein research and has been of great importance in establishing the dynamic nature of proteins. This potential was certainly not lost on Weber who presented a classic paper at the "Light and Life" conference held in 1960 and, in a true understatement, summarized his presentation by saying "There are many ways in which the properties of the excited state can be utilized to study points of ignorance of the structure and function of proteins" [50]. In fact, in an earlier communication [43] (presented at the annual meeting of the British Biochemical Society on April 3, 1959) Weber estimated that the excited state lifetime of tryptophan in proteins was of the order of 4 ns and commented "These values are too short to permit measurements of fluorescence polarization to be of value in the determination of the rotational relaxation times of proteins in solution, but can give useful information on local conditions about the tryptophan or tyrosine residues." Now that present day methods of site-directed mutagenesis permit the facile removal and/or addition of tryptophan residues to allow the creation of novel single-tryptophan containing proteins, Weber's vision of the utility of intrinsic protein fluorescence is being fully realized.

3.9
Red-Edge Effects

In 1924, Gaviola and Pringsheim first observed decreases in the polarization of solutions of some fluorophores in glycerol as the fluorophore concentration increased but the explanation of this phenomenon as being due to dipole-dipole energy transfer over distances larger than the contact distance was due to Jean Perrin [51], the father of Francis Perrin. One of the first quantitative treatments of concentration-dependent energy transfer was due to Weber [52]. In his original observations on energy transfer among the aromatic amino acids [44] Weber also pointed out that homotransfer, i.e., indole to indole or tryptophan to tryptophan, became much less effective upon excitation near the red-edge of the absorption. In 1970, Weber and Shinitsky [53] published a more comprehensive study and showed that this "red-edge effect" was a very general phenomenon of aromatic fluorophores. Homotransfer and the failure of such transfer upon red-edge excitation have been used to study dynamic aspects of numerous macromolecular systems (see, for example, [54, 55]). Weber's interest in red-edge effects persisted and in the late 1970s he published two articles with Bernard Valeur, which described a new red-edge effect producing apparent rotational anomalies of fluorophores [56, 57].

3.10
EEM

Always mindful of the effect of heterogeneity on observed signals, during his stay in Sheffield, Weber conceived of a method to elucidate the number of fluorescing compounds in mixtures of fluorophores by variation of the excitation and emission wavelength and construction of a matrix of the resulting intensities [58]. Years later, with the advent of computer controlled instrumentation and data analysis, Weber's matrix approach would become widely utilized in analytical chemistry and would be known as the EEM (Excitation-Emission Matrix) technique [59, 60].

3.11
Brandeis

In 1960, Weber was a Visiting Professor at Brandeis University. While there he gave a series of lectures in fluorescence and inspired a number of students and postdoctoral fellows with the potential of fluorescence methods. Among those were two individuals, Lubert Stryer and Ludwig Brand, who went on to establish themselves as leading researchers and who made many important contributions in the biological applications of fluorescence spectroscopy.

3.12
University of Illinois

At around this time, I. C. "Gunny" Gunsalus, then the head of the Biochemistry Division of the Department of Chemistry at the University of Illinois at Urbana-Champaign, recruited Weber. Gunny relates the story that while he was convincing his colleagues that Gregorio Weber was an exceptional scientist, someone commented that Weber didn't have as many publications as one might expect from a senior professor. Gunny explained that while this was true, *Weber's ratio of outstanding papers to total papers was unity* and that this ratio – known thereafter as the Weber ratio – was certainly the more important consideration. The reader should pause at this point to estimate his/her own "Weber ratio!"

Gregorio Weber joined the University of Illinois in 1962 and built a research program that continued actively until his death from leukemia on July 17, 1997. During the early years in Urbana, Weber continued to develop novel fluorescence instrumentation and probes and extended his studies of protein systems. In the mid-1960s, Philippe Wahl visited Weber's laboratory. Building on previous studies by Gottlieb and Wahl on dansyl labeled polymer systems [61], Wahl and Weber published one of the first reports delineating the effects of thermally activated local probe mobility on fluorescence depolarization studies of protein systems (specifically, dansyl conjugates of gamma globulins) [62]. One of Weber's lasting contributions to the biological fluorescence field, in fact, was his approach of deciding what questions he wanted to ask about a biomolecular system, and then designing and synthesizing a fluorophore with the optimum spectroscopic properties to achieve the goal. Among the fluorescence probes Weber developed in Urbana were pyrenebutyric acid [63] (which had a lifetime of 100–150 ns and thus extended the polarization method to proteins with molecular weights of 10^6), bis-ANS [64] (which binds to many proteins with much higher affinity than ANS and which also binds to many nucleotide binding sites), IAEDANS [65] (the first sulfhydryl specific fluorescence probe), and PRO-

DAN [66] (2-dimethylamino-6-propionyl-naphthalene, a probe designed by Weber to have an exceptionally large excited state dipole moment and hence to possess an extreme environmental sensitivity). Weber also made derivatives of PRODAN such as LAURDAN, which included a lauric acid tail to render the probe lipid soluble (LAURDAN has been very extensively used in recent years as a probe of membrane dynamics – see, for example, [67 – 70]), and DANCA, which had a cyclohexanoic group attached that increased the affinity of the probe for heme binding sites [71, 72]. Soon after PRODAN appeared another group synthesized ACRYLODAN, a sulfhydryl specific derivative of PRODAN [73].

One of Weber's colleagues at the University of Illinois was Nelson Leonard, a world-renowned organic chemist. Leonard once audited a course on fluorescence given by Weber (I can testify to the fact that Gregorio Weber's fluorescence courses were usually so packed with auditors that the poor graduate students actually taking the course for credit were hard-pressed to find seats!) and was inspired by these lectures to develop fluorescent analogs of nucleosides and coenzymes, which would clearly be of great value in investigations of coenzyme-enzyme and nucleic acid-protein interactions [74]. Leonard and his research group synthesized a series of analogs, such as $\in$ATP, $\in$CTP, cyclic $\in$AMP, and others, rendered fluorescent by reaction of the nucleoside with chloroacetaldehyde. Most of the early characterizations of the fluorescent properties of these analogs were made in Weber's laboratory [75 – 77] (Fig. 3.4).

Fig. 3.4. Gregorio Weber in his laboratory in the Roger Adams Laboratory building at the University of Illinois – ca. 1970

3.13
Phase Fluorometry

During his days at Sheffield, Weber began to work on designing a fluorescence lifetime instrument. Influenced perhaps by the work of fellow Argentinean Enrique Gaviola [78], who built the first phase fluorometer in 1926, Weber focused on phase fluorometry (one should note though that J.B. Birks in Manchester, England and others were also working on phase fluorometry in the late 1950s and early 1960s). It was only in Urbana, however, in the mid-1960s that Weber, together with his graduate student Richard Spencer, succeeded in constructing a highly versatile phase and modulation fluorometer utilizing the principle of cross-correlation [79]. (For an excellent overview of the early history of phase fluorometry the reader is referred to an article by F. W. John Teale [80].) Their cross-correlation method proved to be the key to modern phase fluorometry and is still used universally today. In addition to its use in commercial phase fluorometers, the cross-correlation method is also used in time-resolved fluorescence microscopy [81], and in clinical and biomedical instrumentation which apply frequency domain measurements of photon migration through thick tissues to study problems as diverse as blood oxygenation levels [82] and mammography [83]. Weber also extended the theory of phase fluorometry. For example, he and Spencer described the effect of Brownian rotation and energy transfer on phase lifetime determinations [84]. Weber also solved the daunting problem of deriving the analytical solution to resolving multiple lifetimes from multifrequency phase and modulation data [85]. While working on this problem, Weber developed a mathematical technique he had not seen before. He discussed this technique with his mathematician friends at the University of Illinois and one of them told him that he had seen this approach before and eventually found a reference. I remember going with Gregorio Weber to the Math Library on the University of Illinois campus and finding the article by R. de Prony in Volume 1 of the 1795 issue of J. Ecole Polytech. When Weber wrote his article on this topic, one of the section headings was entitled *Computation of the component lifetimes from the moments by Prony's method.* I asked Weber why he referenced de Prony's article – almost two centuries old – rather than simply state that he had developed the method himself. Weber replied that since de Prony had found the method first he must receive the credit! In the days when only two or three light modulation frequencies were readily available, Weber's algorithm was useful for resolving heterogeneous lifetimes (see, for example, [86]). However, as continuously variable frequency instrumentation developed (see below) it was found that Weber's algorithm was not generally applicable since the precision required in the phase and modulation lifetime values became impossibly high as the number of frequencies being utilized increased [87]. Hence, the phase and modulation field adopted nonlinear least-squares data fitting routines [87, 88]. In recent years, however, with the advent of time-resolved microscopy utilizing phase and modulation methodologies, and typically one or two light modulation frequencies, Weber's algorithm is again proving useful since it allows for an extremely rapid analysis of an unknown lifetime component if the second component is known [89].

While Enrico Gratton was a postdoctoral fellow in Gregorio Weber's laboratory from 1975 to 1976, he worked, at the suggestion of Weber, on development of a continuously variable frequency phase and modulation fluorometer. At that time the phase and modulation instrument used a Debye-Sears ultrasonic tank to achieve the light modulation and only two or three frequencies were readily available from each radio crystal utilized – changing crystals and extending the accessible frequency range was a time-consuming enterprise. Enrico returned to Urbana in 1978 as an Assistant Professor in the Physics Department (Fig. 3.5). By this time he had built the first true multifrequency phase and modulation instrument, utilizing a Pockels cell as the light modulator [90], thus completing Weber's vision. Gregorio Weber still had more contributions to make to the development of phase fluorometry, though, as he helped with the establishment of a multifrequency phase and modulation instrument at the Frascati ADONE Synchrotron Radiation Source which utilized the harmonic content of the light pulses to generate the modulation frequencies [91] – a method which is now widely utilized with pulsed laser sources. Still later Weber was involved with the setup

Fig. 3.5. Gregorio Weber and Enrico Gratton on the University of Illinois campus – ca. 1985

of another phase fluorometer at the Wisconsin Aladdin Synchrotron Radiation Center [92].

3.14
Polarization Revisited

During his years in Urbana, Weber continued to extend the theory of polarization. In 1971 he published an article describing a "phenomenological" treatment of depolarization due to rotational diffusion [93] and in 1972, with G. G. and R. L. Belford (Professors of Mathematics and Physical Chemistry, respectively, at the University of Illinois) published a re-examination of treatments by several groups (including Weber's own work which he critically re-examined) of the theory of fluorescence depolarization [94]. This article presents the generally accepted master equation – with five exponential terms – for the time-dependence of fluorescence depolarization owing to rotational diffusion of fluorophores attached to rigid macromolecules (this equation was also derived independently at the same time by Ehrenberg and Rigler [95]). Years later, in 1989, Weber showed that he still thought deeply about polarization and rotational diffusion as he published a paper entitled *Perrin revisited: Parametric theory of the motional depolarization of fluorescence* [96]. This article presented a formulation in which depolarization results from exchanges between a fixed number of oscillator orientations in thermodynamic equilibrium – a treatment which Weber considered would be appropriate for many cases of biological systems wherein the fluorophores may occupy only a finite number of positions.

3.15
Students, Postdocs and Visitors

During his career, Gregorio Weber trained a significant fraction of the people who later became leading fluorescence researchers. His students and postdoctoral fellows during his years in England included D. J. R. Laurence, James Longworth, Audrey White, and F.W. John Teale. Postdoctoral fellows and students during the early years in Illinois included Terry Pasby, Sonia Anderson, K. Rosenheck, Carl Rosen, John Brewer, Ezra Daniels, Jim Knopp, Allen Rawitch, Earl Hudson, Bill Vaughan, and Ana Jonas.

During the time that I was a graduate student in Weber's laboratory (1971–1977), I overlapped with graduate students David Kolb, Jim Stewart, Moraima Winkler, Kathy Gibbons, Joe Lakowicz, Alex Paladini Jr, J. Fenton Williams, John Wehrly, Bob Hall, Wayne Richards, and Tom Li, and with postdoctoral fellows Francisco Barrantes, Roberto Morero, Fumio Tanaka, I. Iweibo, Yuehhsiu Chien, Louise Slade, Bob Mustacich, Richard Spencer, George Mitchell, Bernard Valeur, Antoine Visser, Bill Mantulin, and Enrico Gratton. Other individuals who spent formative periods in Weber's laboratory include Philippe Wahl, Meir Shinitzky, John Olson, Ken Jacobson, Bob Clegg, Greg Reinhart, and George Fortes. In the 1980s and 1990s, Weber's students included Parkson Chong, Lan King, Cathy Royer, Sue Scarlata, Chris Luddington, Rob Macgregor, Peter Torgerson, and Gerard Marriott, and postdoctoral fellows included Maite Coppey,

Frank Kaufman, Mohamed Rholam, Dave Edmundson, Kancheng Ruan, Andre Kasprzak, Gen-Jun Xu, Larry Morrison, Edith Miles, Don Nealon, Leonardo Erijman, Patricio Rodriguez, Susana Sanchez, Jerson Silva, and Debora Foguel. During my years in Gregorio Weber's laboratory (as a student and later as a postdoc), visitors who came to carry out experiments included Nicole Cittanova, Bill Cramer, Andy Cossins, Pierre Sebban, Serge Pin, Bernard Alpert, Christian Zentz, Patrick Tauc, Maurice Eftink, Tiziana Parasassi, and José Delfino. No doubt I am missing some names and I apologize for my failing memory. I should mention that during most of Gregorio Weber's years in Urbana, his technician Fay Farris served as his hands and eyes in almost all of the chemical syntheses he undertook to design new fluorescence probes.

3.16
Commercialization of Fluorescence

In the late 1960s and early 1970s, Weber had three people working with him who would go on to make an important contribution to the commercialization of fluorescence. Richard Spencer was first a graduate student and then a postdoctoral fellow, George Mitchell was a postdoctoral fellow, and Dave Laker was a machinist in the Chemistry Department at the University of Illinois. Spencer and Mitchell were largely responsible for the development of a new generation of photon-counting instrumentation in Weber's laboratory [97, 98] – at a time when photon counting was still a novelty outside of physics and astronomy. In the early 1970s, Spencer, Laker and Mitchell formed the company SLM. In the beginning this company (originally Spencer and Laker Instruments, Inc.) worked out of a garage until they were able to lease some space, and Weber contributed fatherly support, in terms of advice and finances, to the fledgling enterprise. SLM went on to become one of the most innovative and important developers of research quality fluorescence instrumentation, which helped push the entire field forward.

Weber's influence in the commercialization of fluorescence extended to other companies as well. In the late 1960s and early 1970s he was a consultant for Hitachi-Perkin Elmer, which at that time was producing the MPF-2 and MPF-3 series of spectrofluorimeters. In 1975, David Kolb received his Ph.D. degree with Weber and immediately joined SPEX Industries where he went on to become Product Manager and helped to design new instrumentation. In the early 1980s, Enrico Gratton, consulting with an Italian Industrialist, helped to start Instrumenzione Scientificia Sperimentale – I.S.S. Eventually I.S.S., under the leadership of Beniamino Barbieri, became located in Urbana, Illinois which, due to Gregorio Weber's presence, had become the Mecca of fluorescence. In fact, the first commercial, continuously variable frequency, phase and modulation fluorometer, delivered by I.S.S. in 1984, was named the Greg 200 in honor of Gregorio Weber. In the mid-1970s, Abbott Laboratories consulted Weber about the development of a polarization instrument for clinical assays. The result was the Abbott TDx instrument, which has become one of the leading clinical instruments for analysis of a wide variety of biomolecules – tens of thousands of TDx instruments are currently in use.

3.17
National Laboratories

In 1986, the Laboratory for Fluorescence Dynamics was formed at the University of Illinois at Urbana-Champaign. The LFD, a National Research Resource supported by the National Institutes of Health, was started by Enrico Gratton and William Mantulin, both of whom had spent a postdoctoral period with Gregorio Weber in the mid-1970s. In the 1990s, the Center for Fluorescence Spectroscopy, supported by the National Institutes of Health, was started at the University of Maryland by Joseph Lakowicz, who had been a graduate student with Gregorio Weber in the early 1970s. More recently, the Gregorio Weber Laboratory for Protein Association and Virus Assembly was initiated at the Universidad Federal de Rio de Janeiro by Jerson Silva, who was a postdoctoral fellow with Gregorio Weber in the early 1980s.

3.18
Honors

Gregorio Weber's scientific achievements were recognized by many honors and awards. These include election to the US National Academy of Sciences, election to the American Academy of Arts and Sciences, election as a corresponding member to the National Academy of Exact Sciences of Argentina, the first National Lecturer of the Biophysical Society, the Rumford Premium of the Ameri-

Fig. 3.6. Gregorio Weber receiving the Rumford Premium. Also receiving awards are Robert L. Mills, and Chen Ning Yang

can Academy of Arts and Sciences, the ISCO Award for Excellence in Biochemical Instrumentation, the first Repligen Award for the Chemistry of Biological Processes (awarded by the American Chemical Society), and the first International Jablonski Award for Fluorescence Spectroscopy. It is worth noting that the Rumford Premium is one of the oldest scientific awards given in the United States. It was created by a bequest to the Academy from Benjamin Thompson, Count Rumford, in 1796 – previous awardees include J. Willard Gibbs, A. A. Michelson, Thomas Edison, R.W. Wood, Percy Bridgman, Irving Langmuir, Enrico Fermi, S. Chandrasekhar, Hans Bethe, Lars Onsanger, and other highly original thinkers. The Rumford award committee recommended that the 1979 award be given to two physicists, Robert L. Mills and Chen Ning Yang, for their joint work on the theory of gauge invariance of the electromagnetic field, and to Gregorio Weber, "Acknowledged to be the person responsible for modern developments in the theory and application of fluorescent techniques to chemistry and biochemistry" (Fig. 3.6).

3.19
Proteins and Pressure

Gregorio Weber's original and life-long motivation was to use fluorescence methods to probe the nature of proteins and, in addition to his contributions to the fluorescence field, he was one of the true pioneers of protein dynamics. A study of his papers from the 1960s demonstrates that even then he regarded proteins as highly dynamic molecules. He rejected the view, common at that time after the appearance of the first X-ray structures, that proteins had a unique and rigid conformation. In an important innovation, he introduced the use of molecular oxygen to quench fluorescence in aqueous solutions [99–101], which led to the detection, for the first time and to the surprise of many, of the existence of fast fluctuations in protein structures on the nanosecond time scale. The impact of this work was shown by the increasing interest in experimental and theoretical work in protein dynamics, which followed. Weber's early description of proteins in solution as "kicking and screaming stochastic molecules" [102] has, in recent years, been fully verified both from theoretical and experimental studies. Since this chapter is meant to explore Gregorio Weber's contribution to fluorescence spectroscopy, I cannot go deeply into his far-reaching contributions to protein research – a topic requiring an additional chapter! These contributions were recognized by the American Chemical Society in 1986, which named Weber as the first recipient of Repligen Award for the Chemistry of Biological Processes. In the 1970s, initially in collaboration with H.G. Drickamer, Weber combined fluorescence and hydrostatic pressure methods and applied them to the study of molecular complexes and proteins. It is interesting to note that the initial system he thought to study was the complex formed by isoalloxazine and adenine, one of his original research interests [103]. These observations confirmed the applicability of fluorescence and high-pressure techniques to problems of structure, and particularly dynamics, at the molecular level. Weber and collaborators, in papers published from 1980 to the present, demonstrated that most proteins made up of subunits can be dissociated by the application of hydrostatic pressure, and opened, in this way, a new method to study protein-protein interactions (see [104] for a review of this area). In these studies, quite unexpected properties of protein aggregates were revealed and a new approach to problems in biology and medicine was opened by these observations. For example, Weber and his collaborators demonstrated the possibility of destroying the infectivity of viruses, without affecting their immunogenic capacity, by subjecting them to hydrostatic pressure, and thus opened the possibility of developing viral vaccines that contain, without covalent modification, all the antigens present in the original virus [105, 106].

As a result of his investigations employing fluorescence techniques in conjunction with perturbations by pressure and temperature, Weber presented, in the last few years of his life, a novel proposal, which ran contrary to the generally accepted opinion that the properties of the solvent (water) are the determinants of the folding and association of proteins. Instead, he proposed that the very large residual entropy of the protein was responsible for many of the observed properties [107, 108]. The correctness of this view, considered iconoclastic and heretical by some, remains to be determined, but Weber's approach to this topic demonstrates not only his originality of thought but also his willingness to examine critically scientific data and commonly held opinions and to draw his own conclusions. Weber's philosophy in this regard is exemplified by the dedication in his book *Protein interactions* [109]: *"Dedicated to Those Who Put Doubt Above Belief."* All scientists would do well to remember his words.

Acknowledgements. I wish to thank Bernard Valeur and Jean-Claude Brochon for the invitation to attend the 6th International MAFS conference. I am also indebted to Bernhard Nickel for reprints of his manuscripts on the history of photoluminescence, which helped to serve as an inspiration for this article. I thank Edward Voss, John Croney, and Oliver Holub for their suggestions and help with this manuscript. Finally I am grateful to all of the colleagues and friends around the world, in the extended Weber family, with whom I have interacted over the years in Gregorio Weber's laboratory and afterwards. The support of the National Science Foundation (Grant MCB 9808427) is also gratefully acknowledged.

References

1. Nickel B (1996) Pioneers in photochemistry: from the Perrin diagram to the Jablonski diagram. EPA Newsletter 58:9–38
2. Nickel B (1997) Pioneers in photophysics: from the Perrin diagram to the Jablonski diagram. Part 2. EPA Newsletter 61:27–60
3. Nickel B (1998) Pioneers in photophysics: from Wiedemann's discovery to the Jablonski diagram. EPA Newsletter 64:19–72
4. Weber G (1947) Fluorescence of riboflavin, diaphorase and related substances. Ph.D. Dissertation, Cambridge University
5. Perrin F (1926) Polarisation de la lumière de fluorescence. Vie moyenne des molécules dans l'etat excité. Jour de Phys, VIeme Série, 7:390–401
6. Weber G (1989) Final words at Bocca di Magra. In: Jameson DM, Reinhart GD (eds) Fluorescent biomolecules: methodologies and applications. Plenum Press, New York, pp 343–349
7. Weber G (1948) The quenching of fluorescence in liquids by complex formation. Determination of the mean life of the complex. Trans Faraday Soc 44:185–189

8. Weber G (1950) Fluorescence of riboflavin and flavin–adenine dinucleotide. Biochem J 47:114–121

9. Weber G (1957) Intramolecular transfer of electronic energy in dihydrodiphospho-pyridine nucleotide. Nature 180:1409

10. Weber G (1958) Transfert d'energie dans la dihydro-diphosphopyridine. J Chim Physique 55:878–886

11. Weber G (1966) Intramolecular complexes of flavins. In: Slater EC (ed) Flavins and flavoproteins. Elsevier, Amsterdam, pp 15–21

12. Scott GT, Spencer RD, Leonard N, Weber G (1970) Emission properties of NADH. Studies of fluorescence lifetimes and quantum efficiencies of NADH, AcPyADH, and simplified synthetic models. J Amer Chem Soc 92:687–695

13. Spencer RD, Weber G (1972) Thermodynamics and kinetics of the intramolecular complex in flavin adenine dinucleotide. In: Akeson A, Ehrenberg A (eds) Structure and function of oxidation reduction enzymes, Pergamon, Oxford New York, pp 393–399

14. Weigert F (1920) Über polarisiertes Fluorszenzlicht. Verh d D Phys Ges 1:100–102

15. Vavilov SI, Levschin WL (1923) Beiträge zur Frage über polarisiertes Fluoreszenzlicht von Farbstofflösungen. II. Z Physik 16:134–154

16. Shinitzky M, Dianoux AC, Gitler C, Weber G (1971) Microviscosity and order in the hydrocarbon region of micelles and membranes determined with fluorescent probes. I. Synthetic micelles. Biochemistry 10:2106–2113

17. Cogan U, Shinitzky M, Weber G, Nishida T (1973) Microviscosity and order in the hydrocarbon region of phospholipid and phospholipid–cholesterol dispersions determined with fluorescent probes. Biochemistry 12:521–527

18. Weber G (1977) Theory of differential phase fluorometry: detection of anisotropic molecular rotations. J Chem Phys 66:4081–4091

19. Weber G, Mitchell GM (1976) Detection of anisotropic rotations by differential phase fluorometry In: Birks JB (ed) Excited states of biological membranes. Wiley, London, pp 72–76

20. Mantulin WW, Weber G (1977) Rotational anisotropy and solvent fluorophore bonds: an investigation by differential polarized phase fluorometry. J Chem Phys 66:4092–4099

21. Weber G (1952) Polarization of the fluorescence of macromolecules. I. Theory and experimental method. Biochem J 51:145–155

22. Weber G (1952) Polarization of the fluorescence of macromolecules. II. Fluorescent conjugates of ovalbumin and bovine serum albumin. Biochem J 51:155–167

23. Laurence DJR (1952) A study of the adsorption of dyes on bovine serum albumin by the method of polarization of fluorescence. Biochem J 51:168–177

24. Dandliker WB, Feijen GA (1961) Quantification of the antigen-antibody reaction by the polarization of fluorescence. Biochem Biophys Res Comm 5:299–304

25. Dandliker WB, deSaussure VA (1970) Fluorescence polarization in immunochemistry. Immunochemistry 7:799–828

26. Levinson SA, Dandliker WB, Brawn RJ, Vanderlaan WP (1976) Fluorescence polarization measurement of the hormone-binding site interaction. Endocrinology 99:1129–1143

27. Jameson DM, Sawyer, WH (1995) Fluorescence anisotropy applied to biomolecular interactions. Methods Enzymol 246:283–300

28. Jablonski A (1960) On the notion of emission anisotropy. Bull. Acad Polon Sci, Série des Sci Math et Phys 8:259–264

29. Singleterry CR, Weinberger LA (1951) The size of soap micelles in benzene from osmotic pressure and from the depolarization of fluorescence. J Am Chem Soc 73:4574–4579

30. Steiner RF, McAlister AJ (1957) Use of the fluorescence technique as an absolute method for obtaining mean relaxation times of globular proteins. J Polymer Sci 24:105–123

31. Weber, G (1953) Rotational Brownian motion and polarization of the fluorescence of solutions. Adv Prot Chem 8, 415–459

32. Weber G (1954) Concentration depolarization of the fluorescence of solutions. Trans Faraday Soc 50:552–557

33. Weber G (1956) Photoelectric method for the measurement of the polarization of the fluorescence of solutions. J Opt Soc Amer 46:962–970

34. Weber G, Laurence DJR (1954) Fluorescent indicators of adsorption in aqueous solution and on the solid phase. Biochem J 56:xxxi
35. Debye P, Edwards JO (1952) A note on the phosphorescence of proteins. Science 116:143–144
36. McClure DS (1949) Triplet-singlet transitions in organic molecules: lifetime measurements of the triplet state. J Chem Phys 17:905–913
37. Shore VG, Pardee AD (1956) Fluorescence of some proteins, nucleic acids and related compounds. Arch Biochem Biophys 60:100–107
38. Teale FWJ, Weber G (1957) Ultraviolet fluorescence of the aromatic amino acids. Biochem J 53:476–482
39. Weber G, Teale FWJ (1957) Determination of the absolute quantum yield of fluorescent solutions. Trans Faraday Soc 53:646–655
40. Weber G, Teale FWJ (1958) Fluorescence excitation spectrum of organic compounds in solution. Trans Faraday Soc 54:640–648
41. Weber G, Teale FWJ (1959) Electronic energy transfer in heme proteins. Faraday Soc Discussions 27:134–141
42. Teale FWJ, Weber G (1959) Ultraviolet fluorescence of proteins. Biochem J 72:15p
43. Weber G, Teale FWJ (1959) Polarization of the ultraviolet fluorescence and electronic energy transfer in proteins. Biochem J 72:15p
44. Weber G (1960) Fluorescence polarization spectrum and electronic energy transfer in tyrosine, tryptophan and related compounds. Biochem J 75:335–345
45. Weber G (1960) Fluorescence-polarization spectrum and electronic energy transfer in proteins. Biochem J 75:345–352
46. Velick S (1958) Fluorescence spectra and polarization of glyceraldehyde-3-phosphate and lactic dehydrogenase complexes. J Biol Chem 237:1455–1467
47. Velick S, Parker CW, Eisen HN (1960) Excitation energy transfer and the quantitative study of the antibody hapten reaction. Proc Natl Acad Sci USA 46:1470–1482
48. Alpert B, Jameson DM, Weber G (1980) Tryptophan emission from human hemoglobin and its isolated subunits. Photochem Photobiol 31:1–4
49. Valeur B, Weber G (1977) Resolution of the fluorescence excitation spectrum of indole into the ^1L_a and ^1L_b excitations bands. Photochem Photobiol 25:441–444
50. Weber G (1961) Excited states of proteins. In: McElroy WD, Glass B (eds) Light and life. Johns Hopkins Press, Baltimore, pp 82–106
51. Perrin J (1926) Lumière et réactions chimiques. In, Structure et activité chimiques, rapports et discussions. Gauthier-Villars, Paris, pp 322–399
52. Weber G (1954) Dependence of the polarization of the fluorescence on the concentration. Trans Faraday Soc 50:552–555
53. Weber G, Shinitsky M (1970) Failure of energy transfer between identical aromatic molecules on excitation at the long wave edge of the absorption spectrum. Proc Natl Acad Sci USA 65:823–830
54. Hamman BD, Oleinikov AV, Jokhadze GG, Traut RR, Jameson DM (1996) Dimer/monomer equilibrium and subunit exchange of *Escherichia coli* ribosomal protein L7/L12. Biochemistry 35:16,680–16,686
55. Helms MK, Hazlett TL, Mizuguchi H, Hasemann CA, Uyeda K, Jameson DM (1998) Site-directed mutants of rat testis fructose 6-phosphate, 2-kinase:fructose 2,6-bisphosphatase: localization of conformational alterations induced by ligand binding. Biochemistry 37:14,057–14,064
56. Valeur B, Weber G (1977) Anisotropic rotations in 1-naphthylamine. Existence of a red-edge transition moment normal to the ring plane. Chem Phys Lett 45:140–144
57. Valeur B, Weber G (1978) A new red-edge effect in aromatic molecules: anomaly of apparent rotation revealed by fluorescence polarization. J Chem Phys 69:2393–2400
58. Weber G (1960) Enumeration of components in complex systems by fluorescence spectrophotometry. Nature 190:27–29

59. Christian GD, Callis JN, Davidson ER (1981) Array detectors and excitation-emission matrices in multicomponent analysis. In: Wenry EL (ed) Modern fluorescence spectroscopy. Plenum Press, Chap 4

60. Soper SA, McGown LB, Warner IM (1994) Molecular fluorescence, phosphorescence, and chemiluminescence spectrometry. Anal Chem 15:428R–444R

61. Gottlieb YY, Wahl P (1963) Étude théorique de la polarisation de fluorescence des macromolécules portant un groupe émetteur mobile autour d'un axe de rotation. J Chim Phys 60:849–856

62. Wahl P, Weber G (1967) Fluorescence depolarization of rabbit gamma globulin conjugates. J Mol Biol 30:371–382

63. Knopp JA, Weber G (1969) Fluorescence polarization of pyrenebutyric bovine serum albumin and pyrenebutyric–human macroglobulin conjugates. J Biol Chem 244:6309–6315

64. Rosen CG, Weber G (1969) Dimer formation from 1-anilino-8-naphthalene sulfonate catalyzed by bovine serum albumin – a new fluorescent molecule with exceptional binding properties. Biochemistry 8:3915–6315

65. Hudson EN, Weber G (1973) Synthesis and characterization of two fluorescent sulfhydryl reagents. Biochemistry 12:4154–4161

66. Weber G, Farris FJ (1979) Synthesis and spectral properties of a hydrophobic fluorescent probe: 2-dimethylamino-6-propionylnaphthalene. Biochemistry 18:3075–3078

67. Parasassi T, Di Stefano M, Loiero M, Ravagnan G, Gratton E (1994) Influence of cholesterol on phospholipid bilayers phase domains as detected by Laurdan fluorescence. Biophys J 66:120–132

68. Parasassi T, Gratton E, Yu W, Wilson P, Levi M (1997) Two-photon fluorescence microscopy of laurdan gp-domains in model and natural membranes. Biophys J 72:2413–2429

69. Bagatolli LA, Gratton E, Fidelio GD (1998) Water dynamics in glycosphingolipid aggregates studied by LAURDAN fluorescence. Biophys J 75:331–341

70. Bagatolli LA, Gratton E (1999) Two-photon fluorescence microscopy observation of shape changes at the phase transition in phospholipid giant unilamellar vesicles. Biophys J 77:2090–2101

71. Macgregor RB, Weber G (1981) Fluorophores in polar media: spectral effects of the Langevin distribution of electrostatic interactions. Annals NY Acad Sci 366:140–150

72. Macgregor RB, Weber G (1986) Estimation of the polarity of the protein interior by optical spectroscopy. Nature 319:70–72

73. Prendergast FG, Meyer M, Carlson GL, Iida S, Potter JD (1983) Synthesis, spectral properties, and use of 6-acryloyl-2-dimethylaminonaphthalene (Acrylodan). A thiol-selective, polarity-sensitive fluorescence probe. J Biol Chem 25:7541–7544

74. Leonard NJ (1997) The 'chemistry' of research collaboration. Tetrahedron 53:2323–2355

75. Secrist JA III, Barrio JR, Leonard NJ, Weber G (1972) Fluorescent modifications of adenosine containing coenzymes. Biological activities and spectroscopic properties. Biochemistry 11:3499–3506

76. Barrio JR, Tolman GL, Leonard NJ, Spencer, RD, Weber G (1973) Flavin 1, N^6-ethenoadenine dinucleotide: dynamic and static quenching of fluorescence. Proc Natl Acad Sci USA 70:941–943

77. Spencer RD, Weber G, Tolman GL, Barrio JR, Leonard NJ (1974) Species responsible for the fluorescence of 1, N^6-ethenoadenosine. Eur J Biochem 45:425–429

78. Gaviola E (1926) Die Abklingzeiten der Fluoeszenz von Farbstofflösungen. Z Physik 35:748–756

79. Spencer RD, Weber G (1969) Measurement of subnanosecond fluorescence lifetimes with a cross-correlation phase fluorometer. Annals New York Acad Sci 158:361–376

80. Teale FWJ (1983) Phase and modulation fluorometry. In: Cundall RB, Dale RE (eds) Time-resolved fluorescence spectroscopy in biochemistry and biology. NATO ASI Series A: Life sciences vol 69, Plenum Press, London, pp 59–80

81. French T, So PTC, Dong CY, Berland KM, Gratton E (1997) Fluorescence lifetime imaging techniques for microscopy. In, Sluder G, Wolf D (eds) Methods in cell biology. Video microscopy, vol. 56, pp 227–304

82. Franceschini MA, Wallace D, Barbieri B, Fantini S, Mantulin WW, Pratesi S, Donzelli GP, Gratton E (1997) Optical study of the skeletal muscle during exercise with a second generation frequency-domain tissue oximeter. SPIE Proc 2979:807–814
83. Franceschini MA, Moesta KT, Fantini S, Gaida G, Gratton E, Jess H, Mantulin WW, Seeber M, Schlag PM, Kaschke M (1997) Frequency-domain instrumentation techniques enhances optical mammography: initial clinical results. Proc Natl Acad Sci USA 94:6468–6473
84. Spencer RD, Weber G (1970) Influence of Brownian rotations and energy transfer upon the measurements of fluorescence lifetime. J Chem Phys 52:1654–1663
85. Weber G (1981) Resolution of the fluorescence lifetimes in a heterogeneous system by phase and modulation measurements. J Phys Chem 85:949–953
86. Jameson DM, Weber G (1981) Resolution of the pH dependent heterogeneous fluorescence decay of tryptophan by phase and modulation measurements. J Phys Chem 85:953–958
87. Jameson DM, Gratton E (1983) Analysis of heterogeneous emissions by multifrequency phase and modulation fluorometry. In: Eastwood D (ed) New directions molecular luminescence. ASTM STP 822, American Society for Testing and Materials, Philadelphia, pp 67–81
88. Jameson DM, Gratton E, Hall RD (1984) The measurement and analysis of heterogeneous emissions by multifrequency phase and modulation fluorometry. App Spectros Rev 20:55–106
89. Gadalla TWJ, Clegg RM, Jovin TM (1994) Fluorescence lifetime imaging microscopy: pixel by pixel analysis of phase-modulation data. Bioimaging 2:139–159
90. Gratton E, Limkeman M (1983) A continuously variable frequency cross-correlation phase fluorometer with picosecond resolution. Biophys J 44:315–324
91. Gratton E, Jameson DM, Rosato N, Weber G (1984) A multifrequency cross-correlation phase fluorometer using synchrotron radiation. Rev Sci Instrum 55:486–494
92. Gratton E, Mantulin WW, Weber G, Royer CA, Jameson DM, Reininger R, Hansen RWC (1996) Fluorescence dynamics of biological systems using synchrotron radiation. Rev Sci Instrum 67:1–7
93. Weber G (1971) Theory of fluorescence depolarization by anisotropic Brownian rotations: discontinuous distribution approach. J Chem Phys 55:2399–2407
94. Belford GC, Belford RL, Weber G (1972) Dynamics of fluorescence polarization macromolecules. Proc Natl Acad Sci USA 69:1392–1393
95. Ehrenberg M, Rigler R (1972) Polarized fluorescence and rotational Brownian motion. Chem Phys Lett 14:539–544
96. Weber G (1989) Perrin revisited: parametric theory of the motional depolarization of fluorescence. J Phys Chem 93:6069–6973
97. Jameson DM, Spencer RD, Weber G (1976) Construction and performance of a scanning, photon-counting spectrofluorometer. Rev Sci Instru 47:1034–1038
98. Jameson DM, Weber G, Spencer RD, Mitchell G (1978) Fluorescence polarization: measurements with a photon-counting photometer. Rev Sci Instru 49:510–514
99. Vaughan WM, Weber G (1970) Oxygen quenching pyrenebutyric acid fluorescence in water. A dynamic probe of the microenvironment. Biochemistry 9:464–473
100. Lakowicz JR, Weber G (1973) Quenching of fluorescence by oxygen. A probe for structural fluctuations in macromolecules. Biochemistry 12:4161–4170
101. Lakowicz JR, Weber G (1973) Quenching of protein fluorescence by oxygen. Detection of structural fluctuations in proteins in the nanosecond time scale. Biochemistry 12:4171–4179
102. Weber G (1975) Energetics of ligand binding to proteins. Adv Prot Chem 29:1–83
103. Weber G, Tanaka F, Okamoto BY, Drickamer HG (1974) The effect of pressure on the molecular complex of isoalloxazine and adenine. Proc Natl Acad Sci USA 71:1264–1266
104. Silva, JL, Weber G (1993) Pressure stability of proteins. Annu Rev Phys Chem 44:89–113
105. Silva JL, Luan P, Glaser M, Voss EW, Weber G (1992) Effects of hydrostatic pressure on a membrane-enveloped virus: high immunogenicity of the pressure-inactivated virus. J Virology 66:2111–2117

106. Juriekwicz E, Villas-Boas M, Silva JL, Weber G, Hunsmann G, Clegg RM (1995) Inactivation of simian immunodeficiency virus by hydrostatic pressure. Proc Natl Acad Sci USA 92:6935–6937
107. Weber G (1993) Thermodynamics of the association and the pressure dissociation of oligomeric proteins. J Phys Chem 97:7108–7115
108. Weber G (1998) Thermodynamic concepts in protein condensation. Comm Mol Cell Biophys 9:201–218
109. Weber G (1992) Protein interactions. Chapman and Hall, New York

Part 2
Fluorescence of Molecular and Supramolecular Systems

Investigation of Femtosecond Chemical Reactivity by Means of Fluorescence Up-Conversion

J.-C. MIALOCQ, T. GUSTAVSSON

Fluorescence spectroscopy is a powerful tool for the study of the reactivity of chemical and biological systems. The fluorescence spectrum, the fluorescence quantum yield, and the fluorescence lifetime of a chromophore depend on its microenvironment and the solute-solvent interactions. Fluorescence lifetimes also depend on the kinetics of possible chemical reactions. Fluorescence measurements in the time domain thus provide detailed information about intra- and inter-molecular photophysical and photochemical processes. During the past 30 years the development of laser sources and ultrafast techniques has opened new ways for the study of the fluorescent excited states of molecules in the nano-, pico-, and femtosecond time range. When analyzing the rise and the decay (often multi-exponential) of their fluorescence intensity as a function of time, the pre-exponential factors and the fluorescence time constants have to be determined accurately after a careful deconvolution using a computer-based algorithm. Several techniques are used: phase modulation spectroscopy with a modulated laser pump source on a time scale comparable to the decay times of interest, time-correlated single photon counting under picosecond laser excitation (alternatively nanosecond flashlamp excitation), ultra-fast streak cameras designed for the conversion of time dependent light emission into distance dependent electron impact on a phosphor screen. Because a sufficient resolution is not accessible with those techniques for the measurement of fast decays in the 50 fs–100 ps range, fluorescence up-conversion in nonlinear optical crystals is the technique of choice.

4.1
Nanosecond and Picosecond Time-Resolved Fluorescence Techniques

The fluorescence lifetime τ_F of a molecule is the reciprocal of the rate r_F of depopulation of the first excited singlet state following optical excitation from the ground state. In the case of a monoexponential decay τ_F is given by

$$[^1A^*] = [^1A^*]_0 \, e^{-t/\tau_F} \tag{4.1}$$

where $[^1A^*]$ and $[^1A^*]_0$ are the excited state molar concentrations at time t and $t = 0$ [1]. The data analysis is generally straightforward when the decay is monoexponential. However, when the fluorescence decay curve deviates from the simple expression at Eq. (4.1), and/or when emission lifetimes differ by a too small amount, the analysis is more difficult. More elaborate models must be used such as multi-exponential decays according to the general equation

$$I(t) = = \sum_i (1/\tau_i)\, a_i \exp(-t/\tau_i) \tag{4.2}$$

4.1.1
Phase Modulation Spectroscopy

Phase modulation spectroscopy [2] measures the phase and amplitude of the fluorescence emission relative to a periodically modulated pump beam. Using a sinusoidally modulated excitation source, the amplitude and phase of the fluorescence intensity are functions of both the input frequency ω and the fluorescence lifetime τ. As the frequency is increased, a significant delay and decrease in modulation amplitude is observed. A modulation index is defined for both the excitation and the fluorescence as the ratio of the peak-to-peak amplitude of the a.c. component to the average d.c. level. The relative modulation index $m(\omega)$ is the ratio of the modulation index for the output by that of the input. For a single exponential decay time, $m(\omega)$ and the phase shift $\phi(\omega)$ are given by

$$m(\omega) = (1 + \omega^2 \tau^2)^{-1/2} \tag{4.3}$$

$$tan\,\phi(\omega) = \omega\tau \tag{4.4}$$

In the case of multiexponential decays, it is necessary to measure $m(\omega)$ and $\phi(\omega)$ over a wide range of modulation frequencies with the center frequency close to the reciprocal of the mean fluorescence decay time. Argon-ion or dye laser sources can be modulated to 200 MHz using electro-optic modulators. The combination of mode-locked lasers with multichannel-plate photomultiplier tubes extends the frequency range up to 10 GHz.

4.1.2
Time Correlated Single Photon Counting

The time-correlated single photon counting technique has been reviewed extensively [3, 4]. An extension of the delayed-coincidence technique used in nuclear physics following gamma, neutron, or alpha excitation, this technique enables the analysis of fluorescence decays of molecular systems under nanosecond pulsed flashlamp or picosecond laser excitation. The basic principle is to excite the sample at low intensity but at a high repetition rate and to record the distribution of times of the individual emitted photons with respect to the excitation pulse. This histogram reproduces the rise and the decay of the fluorescence of the sample. The temporal analysis provides the decay parameters, the pre-exponential factors and the decay time constants of the molecular system. The experimental setup includes the excitation source, photomultipliers operating in a single photon counting mode and used in conjunction with constant fraction discriminators to define the "start" synchronization pulse and the "stop" pulse, a time-to-amplitude converter, a multichannel analyzer, and a PC computer. The stop rate has to be very low in order to detect only one photon per excitation pulse and to avoid the preferential detection of early photons, i.e., the "start" rate has to be much greater (a good ratio is 100 or 200 times greater). Otherwise the fluorescence decay would appear too fast. The measured fluorescence decay is the convolution of the instrumental response function and the real fluorescence function of the sample. Assuming a fluorescence function with one or several exponentials and convoluting with the instrumental response function, one can compare the result with the experimental decay and iterate the decay times until a good statistical agreement is obtained. Such a reconvolution least-squares analysis provides the best-fit pre-exponential factors and decay times. Fluorescence lifetimes of a few tens of picoseconds can be determined. Usually a mode-locked and cavity-dumped synchronously pumped dye laser frequency-doubled to 280–350 nm provides stable laser pulses of < 10 ps duration for excitation. More recently pulsed semiconductor diode lasers have appeared. A multichannel plate photomultiplier is used for measurements in the UV-visible or a single-photon avalanche photodiode in the infrared (down to 70 ps fwhm (full width at half-maximum) instrumental response) [5]. The use of synchrotron radiation is particularly useful below 250 nm. Standards for fluorescence decay time measurements are necessary to test the experimental technique [6].

4.1.3
Streak Cameras for Time-Domain Measurements

Streak cameras have been used for more than 30 years for the detection of ultrashort laser pulses. Fluorescence lifetime measurements in the picosecond time range have also been successful [7]. They consist of an image converter tube for the conversion of time dependent light emission into distance dependent electron impact on a phosphor screen followed by an image intensifier and a device to record the image. The time resolution is limited to about 1 ps under severe conditions. They can be used in two-dimensional spectroscopy, to record the signal as a function of time and wavelength. Three modes are nowadays possible: single pulse, synchroscan (10 kHz–100 MHz), and steady state-mode (10 Hz–10 kHz). Single shot streak cameras are restricted to strong fluorescence. The synchroscan and steady state modes allow signal averaging. Photon counting with a streak camera provides a better dynamic range [8].

4.2
Femtosecond Emission Spectroscopy by Time-Gated Up-Conversion

Although femtosecond mode-locked Ti:sapphire lasers which offer tunability and pulse duration similar to those of dye lasers but higher stability and ease of use are also available, ultrafast kinetics in the femtosecond time range cannot be resolved using the techniques mentioned in the preceding paragraph since the limiting factor of their response function is the detector. The time-gated up-conversion technique is by far the best choice in the femtosecond time range. It should not be confused with the energy up-conversion processes which occur in rare earth crystals via two photon excitation and/or energy transfer up-conversion [9].

4.2.1
Historical Background of the Time-Gated Up-Conversion Technique

The idea of measuring ultra-fast optical signals with the help of optical gates such as optical Kerr shutters based on the transient birefringence induced in a nonlinear medium (CS_2) by an intense laser pulse [10] became fruitful with the development of frequency mixing techniques. In 1968 Duguay and Hansen proposed an optical analog of the electronic sampling oscilloscope. They converted the light of a mode-locked He-Ne laser into UV light by mixing with the sampling pulses produced by a mode-locked Nd-glass laser in a KDP crystal [11]. In 1975 Mahr first demonstrated that up-conversion (optical mixing) could be used to measure very low level subnanosecond light pulses with a good spectral resolution and a time resolution only limited by the pulse duration [12]. The luminescence emitted after a laser pulse excitation is mixed with the other half of the laser pulse in a nonlinear optical crystal to generate sum frequency radiation. The crystal acts as an ultra-fast optical "light gate" that is "open" when the pump pulse is "on". This method is particularly suited for laser pulses of moderate peak powers but very high repetition rate. Very low light level emissions with picosecond lifetimes were soon measured from bacteriorhodopsin (15 ± 3 ps) [13] and the $S_2 \rightarrow S_0$ fluorescence of xanthione (14 ± 2 ps) [14]. Other early investigations using this technique include fluorescence measurements of rhodamine 6G [15], malachite green [16], cresyl violet, and HITC [17] as a function of the solvent viscosity. Beddard et al. calculated fluorescence anisotropies and rotational lifetimes by detecting the fluorescence with polarization parallel and perpendicular to the excitation polarization [17]. Luminescence up-conversion studies were also fruitful for investigating electronic properties of solids, platelets of CdSe [18], and semiconductors such as GaAs [19].

4.2.2
Principle of the Time-Gated Up-Conversion Technique

The fluorescence beam at $h\nu_F$ photon energy and a delayed laser beam at $h\nu_P$ are focused in a nonlinear optical crystal oriented at the appropriate angle with respect to the two beams. A beam at $h\nu_S$ photon energy is generated at sum or difference frequency only when both of the beams are present in the crystal. Common nonlinear optical crystals are potassium dihydrogen phosphate (KDP), β-barium borate (BBO), lithium iodate (LIO), and urea. The short wavelength absorption (at the sum frequency) is an important property. BBO is transparent till 190 nm. Phase matching conditions have to be fulfilled.

4.2.2.1
Phase Matching Conditions

The equations for phase matching are the following

$$\nu_F + \nu_P = \nu_S \tag{4.5}$$

$$\vec{k}_F + \vec{k}_P = \vec{k}_S \tag{4.6}$$

where $\vec{k}_F$, $\vec{k}_P$ and $\vec{k}_S$ are the wave vectors of the fluorescence, pump, and sum respectively [20]. In the case of collinear matching, Eq. (4.6) becomes

$$\frac{n_F}{\lambda_F} + \frac{n_P}{\lambda_P} = \frac{n_S}{\lambda_S} \tag{4.7}$$

where n is the crystal refractive index at the wavelength λ.

In uniaxial crystals for which $n_x = n_y = n_o$ (ordinary index) and $n_z = n_e$ (extraordinary index) the indexes vary with the wavelength according to the Sellmeier's equation:

$$n^2 = A + \frac{B}{C - \lambda^2} + D\lambda^2 \tag{4.8}$$

In type I crystals ($O + O \rightarrow E$), the fluorescence and the pump are polarized parallel to each other. In type II crystals ($O + E \rightarrow E$) or ($E + O \rightarrow E$), they are polarized perpendicularly. For type I crystals, the phase matching angle can be calculated using Eq. (4.9):

$$\sin^2 \theta_m = \frac{(1/n_S^2(\theta_m)) - (1/n_{o,S}^2)}{(1/n_{e,S}^2) - (1/n_{o,S}^2)} \tag{4.9}$$

where $n_S(\theta_m)$ satisfies Eq. (4.10):

$$n_S(\theta_m) = n_{O,F}\frac{\lambda_S}{\lambda_F} + n_{O,P}\frac{\lambda_S}{\lambda_P} \tag{4.10}$$

The phase mismatch is given by

$$\Delta k = k_S - k_F - k_P = (1/c)(n_S\omega_S - n_F\omega_F - n_P\omega_P) \tag{4.11}$$

where c is the speed of light and ω_i the angular frequency ($\omega_i = 2\pi c/n\lambda_i$). Since fluorescence is emitted in all directions from the excited sample, the angle of acceptance of the fluorescence by the crystal is important to consider. The acceptance angle increases inversely with the crystal length [20, 21]. The focus can also be tightened in shorter crystals [21].

Only a narrow band of the fluorescence spectrum is thus up-converted for an appropriate phase matching angle θ_m. To determine the full fluorescence spectrum the angular position of the crystal has to be varied.

4.2.2.2
Quantum Efficiency for Up-Conversion

The quantum efficiency for up-conversion varies as the ratio of the power P_P of the pump beam to its area A in the nonlinear crystal, the square of both the thickness L and the effective nonlinear coefficient of the crystal d_{eff} [20] according to

$$\eta_q = \frac{2\pi^2\, d_{eff}^2\, L^2\, (P_P/A)}{c\, \varepsilon_0^3\, \lambda_F \lambda_S\, n_{o,F}\, n_{o,P}\, n_S(\theta_m)} \tag{4.12}$$

where ε_0 is the permittivity in vacuum.

The quantum efficiency, the cut-off wavelength for the up-convertible fluorescence calculated for a 800 nm pump wavelength, and the damage threshold of a few nonlinear crystals are gathered in Table 4.1.

4.2.2.3
Group Velocity Effects

To take advantage of the availability of femtosecond laser pulses, the arrival times of the various spectral fluorescence components in the gate have to be corrected because they propagate with different velocities through the solvent, the cell window, the detection optics, and the nonlinear crystal (group velocity dispersion). The group velocities of the three waves can be calculated as a function of the refractive indices and their derivatives with respect to the central wavelength λ_i of the light pulse:

$$v_g(\lambda_i) = c\left[n(\lambda_i) - \lambda_i \left(\frac{\partial n}{\partial \lambda}\right)_{\lambda_i}\right] \tag{4.13}$$

The difference of the group velocities (more severe than that of the phase velocities) of the fluorescence and the pump deteriorates the time resolution of the experiment by introducing a time broadening in the sum frequency generation [20, 21]. The group velocity dispersion (*GVD*) through a dispersive medium is proportional to the second derivative of the index of refraction with wavelength:

$$GVD = \left(\frac{\partial^2 k}{\partial \omega^2}\right) = \frac{\lambda^3}{2\pi c^2}\frac{\partial^2 n}{\partial \lambda^2} \tag{4.14}$$

Besides obtaining kinetics at particular wavelengths, the optimization of phase matching, quantum efficiency for up-conversion, spectral bandwidth, group velocity mismatch, walk-off angle, and angle of acceptance for the optimization

Table 4.1. Relative quantum efficiencies η_q (normalized relative to KDP), damage thresholds I_{thr} (in the picosecond regime except for urea), and cut-off wavelengths for the up-convertible fluorescence assuming a pump wavelength of 800 nm

Crystal type	η_q	I_{thr} (GW/cm^2)	$\lambda_{f,min}$ (nm)
KDP (ooe)	1.0	30 (ps regime)	344
LiIO$_3$ (ooe)	44	15 (ps regime)	480
BBO (eoe)	4.1	50 (ps regime)	264
BBO (ooe)	6.5	50 (ps regime)	249
urea (eeo)	6	5 (ns regime)	313

of fluorescence up-conversion [20, 21] enabled the direct determination of time resolved fluorescence spectra [19, 22–37].

4.2.3
Experimental Setup

Time-resolved emission spectra can be obtained using the experimental setup shown in Fig. 4.1. Successive versions have been described elsewhere [27, 28, 30, 37].

The femtosecond laser source is a Ti:sapphire laser pumped by a continuous wave Ar$^+$ laser (8 W output power all lines). Typical performances of the Ti:sapphire laser are 1.2 W output power in the mode-locked regime at 788 nm and 76 MHz repetition rate. The autocorrelation trace gave a pulse duration of 125 fs (fwhm, assuming a sech^2 pulse shape). The second harmonic at 394 nm is generated in a 1 mm thick BBO crystal and separated from the residual fundamental light by a dichroic beam splitter. After passage through a delay line, the residual fundamental is focused by a 100-mm lens into a 0.2-mm Type II BBO crystal serving for the sum-frequency generation. The 394 nm pulse used for excitation is focused by a 1.5-inch off-axis parabolic mirror into the 1 mm thick flowing quartz cell containing the sample. The fluorescence is collected by a 4-inch off-axis parabolic mirror, passes through a 1-mm optical filter (Schott GG420) and is focused into the up-conversion crystal by a 4-in. off-axis parabolic mirror. The up-converted light in the UV is collected by a 150-mm lens, passes through a 1-mm optical filter (Schott UG11), and is focused onto the entrance slit of a 0.25-m monochromator. The slit width corresponds to about 10 nm bandwidth. The spectrally selected UV light is detected by a photomulti-

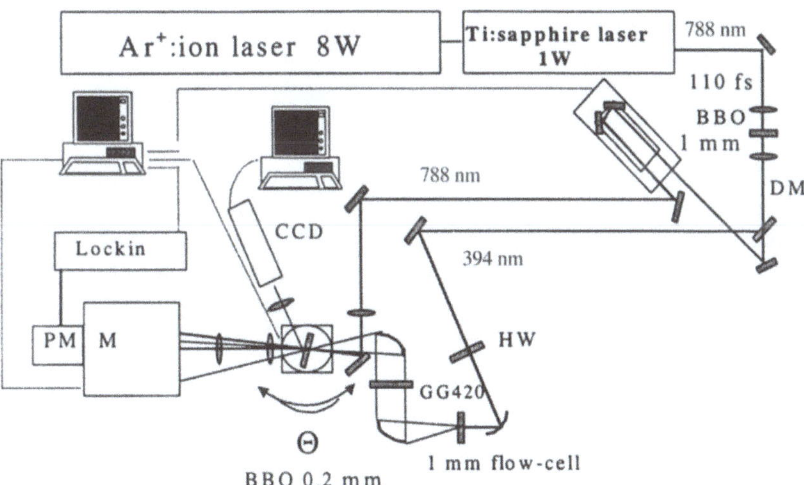

Fig. 4.1. Femtosecond fluorescence up-conversion apparatus. DM, dichroic mirror; HW, half-wave plate; CCD, video camera for the visual superposition of the beams in the BBO crystal; M, Monochromator; PM, photomultiplier

plier connected to a lock-in photon counter. All experiments are performed at a magic angle between the polarization axes of excitation and observation. The crosscorrelation trace between the laser fundamental and the second harmonic gives a fwhm value of 210 fs, confirmed by recording the Raman lines (2900 cm^{-1} for the CH and 3500 cm^{-1} for the OH vibration respectively) of pure methanol with the fluorescence up-conversion apparatus [37].

Time-resolved emission spectra are normally obtained by recording kinetics at different wavelengths [20, 23, 25] and using the spectral reconstruction method, a "numerical cut-and-paste procedure" [23]. We proposed an alternative method consisting of the direct recording of time resolved spectra which avoids the too long acquisition of the kinetic data while the long term experimental conditions may change. In principle the acquisition of one complete spectrum at a particular delay time after excitation does not suffer from long term variations of the experimental conditions. There are two points to consider. First, different spectral components of the fluorescence propagate with different group velocities in the various traversed media between the sample cell and the nonlinear crystal. The second point is the limited half bandwidth of sum frequency generation, typically 20 nm at 450 nm in a 0.2 mm thick BBO crystal. The temporal and spectral response has thus to be corrected. For the temporal correction, while scanning the monochromator, the GVD is calculated as a function of wavelength and the difference in propagation time is compensated by adjusting the delay line. For the spectral correction, the optimal phase matching angle is calculated as a function of wavelength and the crystal is rotated to the correct angular position using a step motor and a PC computer. Moreover, the spectra are corrected for the spectral response of the up-conversion apparatus. This correction is based on the comparison of the fluorescence spectrum at a long time after excitation (typically 500 ps, when the molecular system is spectrally relaxed) to the steady-state fluorescence spectrum corrected itself for the response function of the conventional spectrofluorometer [37].

4.3
Time-Resolved Spectroscopy

In the following we will make a brief overview of recent work on ultrafast relaxation processes in the condensed phase using the fluorescence up-conversion technique. This review of results obtained by this particular technique is not intended to be exhaustive. A detailed description of the various subjects is far outside the scope of the present chapter.

4.3.1
Solvation Processes

The fluorescence up-conversion technique has enabled the study of ultrafast relaxation processes of a large number of dyes used as "molecular probes" in various solvents [22–37]. The notion "molecular probes" in the context of solvation dynamics implies the absence of relaxation processes other than solvation. This is of course far from being trivial in the general case, and the rightful use of certain dyes is the subject of an ongoing discussion. It is important to point out, however, that the fluorescence up-conversion technique lends itself much better to this kind of study than the more widespread pump-probe technique. The reason for this is that, for most dyes, the $S_1 \rightarrow S_0$ transition overlaps with an $S_1 \rightarrow S_n$ transition rendering an analysis of pump-probe spectra very difficult or even impossible.

4.3.1.1
Time-Dependent Fluorescence Stokes Shift (TDFSS). Non-Specific Solvation

The "vertical" absorption transition from a solvent relaxed configuration of the dye solute to the Franck-Condon excited singlet state is followed by a solvent relaxation around the fluorescent dye molecule. This relaxation is caused by the nonequilibrium configuration of the solvent after the dipole moment change of the solute. The resulting stabilization of the fluorescent excited state can be analyzed by monitoring the Time-Dependent Fluorescence Stokes Shift (TDFSS), as defined by the shift function $S(t)$:

$$S(t) = \frac{\bar{v}(t) - \bar{v}(\infty)}{\bar{v}(0) - \bar{v}(\infty)} \tag{4.15}$$

where $\bar{v}(t)$, $\bar{v}(0)$ and $\bar{v}(\infty)$ are the characteristic frequencies at t, $t = 0$, and $t = \infty$, usually taken as the mean wave-numbers or the "first moments" (centers of gravity) of the fluorescence spectrum [38]

$$\bar{v}(t) = \int_{\text{band}} I(v, t)\, v\, dv \int_{\text{band}} (v, t)\, dv \tag{4.16}$$

$I(v, t)$ represents the time-dependent fluorescence intensity per unit wave-number at a particular wave-number v, i.e., the measured fluorescence spectra (vs wavelength) are scaled by a λ^2 factor in order to account for the change from a wavelength to a wave-number scale [38]. A difficulty in the data analysis arises from the definition of the hypothetical "time-zero" emission spectrum. A method has been introduced by Fee and Maroncelli using only steady-state spectral data [39].

The purely phenomenological shift function $S(t)$ is often compared to theoretical predictions based on the solvent longitudinal relaxation times:

$$\tau_{\text{L}} = (\varepsilon_{\infty 1}/\varepsilon_{01})\, \tau_{\text{D1}} \tag{4.17}$$

where τ_{D} is the dielectric relaxation time, $\varepsilon_{\infty 1}$ the infinite frequency dielectric constant, ε_{01} the static dielectric constant, and the subscript 1 refers to the lowest frequency dielectric dispersion regime [22]. For example, in our subpicosecond study of the LDS 751 dye molecule in ethanol, we found that $S(t)$ was well described by a single exponential decay with a 1/e time of (5 ± 1) ps for a total spectral shift of about 20 ± 3 nm [27]. Similarly, in methanol and dimethylsulfoxide, the decay times were (3.0 ± 1.0) ps and (3.9 ± 1.0) ps, respectively

[28], close to the respective values 3.3 ps and 3.1 ps obtained by Castner et al. [22] for the LDS 750 dye molecule of the same family. As expected, when using a better time resolution the appearance of the TDFSS became more complex [26, 29–32]. The evolution of the fluorescence spectrum may take the form given by a multi-exponential Eq. (4.18) [22, 25, 29–32]:

$$S(t) = \sum_{i=1}^{4} a_i \exp(-t/\tau_i) \tag{4.18}$$

By comparison with simulations of solvation dynamics, Fleming and co-workers assigned most of the TDFSS to inertial effects of the first solvation shell. Equation (4.18) has been modified by introducing a Gaussian function [26, 29, 31]:

$$S(t) = a_g \exp\left(-\frac{1}{2}\omega_g^2 t^2\right) + a_1 \exp(-t/\tau_1) + a_2 \exp(-t/\tau_2) \tag{4.19}$$

Indeed, they observed for LDS 750 in acetonitrile that a fast initial relaxation accounted for ~ 80% amplitude and was well fitted by a Gaussian of 120 fs fwhm (i.e., a decay time of 70 fs) [26]. The total spectral shift was 50 nm. In the case of the DCM laser dye, the TDFSS was found to be bi-exponential with $\tau_1 = 175$ fs ($A_1 = 0.36$) and $\tau_2 = 3.2$ ps ($A_2 = 0.64$) in methanol while the fluorescence spectrum remained practically unchanged at all times in chloroform [30].

4.3.1.2
Specific Solvation: Role of the Structure and the Charge of the Probe

The influence of both the structure and the charge of the probe has been investigated. For example, the behavior of two styrenic compounds, LDS 751, a cationic unsymmetrical polymethine cyanine dye [27, 28] and DCM, a neutral molecule [30], have been compared [40]. LDS 751 is highly polar in the ground state with a blue shift of its broad absorption spectrum in polar solvents and very slightly polar in the fluorescent excited state as shown by the feeble solvatochromism of the narrow fluorescence spectrum when changing from an apolar solvent to a very polar solvent [41]. The decrease of the dipole moment results from a large charge transfer from the dimethylaminophenyl group to the benzothiazole group upon photoexcitation. In the singlet excited state the positive charge is totally delocalized over the whole polymethinic chain as in the case of symmetrical polymethine cyanine or rhodamine dyes. The observed TDFSS results mainly from the solvation energy of the Franck-Condon (FC) ground state reached following the fluorescence transition. Indeed, immediately after photoexcitation the solvent molecules keep the same geometrical configuration around the FC excited state as in the ground state, i.e., they are still organized around the dye molecule. Later the photoinduced charge transfer which has annihilated the LDS 751 dipole moment leads to a time-dependent random reorganization of the solvent dipoles. Therefore the fluorescence transition leads initially to a well-solvated FC ground state but as time goes on the fluorescence emission leads to an FC ground state surrounded by a random solvent cage. The

TDFSS thus probes the random reorganization of the solvent (de-solvation) around a positive charge delocalized over a large molecule. Conversely, DCM undergoes a large dipole moment increase upon photoexcitation. Both the absorption spectrum and the fluorescence spectrum are broad and undergo a red shift when changing from an apolar solvent to a very polar solvent. The TDFSS thus probes the reorganization of the polar solvent molecules around the large dipole of the DCM fluorescent excited state and shows the increasing stabilization energy due to solvation.

For the coumarin 343 anion in water the initial Gaussian decay was found to be faster than 55 fs and to constitute > 50 % of the total decay [29]. For this anionic probe in a basic solution of CH_3ONa-CH_3OH obtained by embedding small pieces of sodium to pure methanol, Tominaga and Walker obtained a double-exponential function $S(t)$ with $\tau_1 = 1.0$ ps and $\tau_2 = 10.3$ ps similar to the values found for the neutral coumarin 152 with an analogous structure [32]. Coumarin 153 has been considered by Maroncelli et al. to provide a true measurement of the solvation response [31]. They concluded that intramolecular (vibrational) relaxation plays a negligible role and they fitted the TDFSS using a multiexponential function with four time constants ($\tau_1 = 30$ fs, $\tau_2 = 280$ fs, $\tau_3 = 3.2$ ps, and 15.3 ps in methanol, for example). Comparing the relaxation processes of three 7-amino-4-trifluoromethyl-coumarins differing by the alkylation degree of the amino group, NH_2 (C151), $N(C_2H_5)_2$ (C35), and N(julolidyl) (C153), Gustavsson et al. arrived at a different conclusion with respect to intramolecular relaxation. They showed that the TDFSS in methanol and dimethylsulfoxide are dominated by an ultrafast component with a characteristic time shorter than ≈ 50 fs and assigned partly to intramolecular relaxation [37]. Moreover, although the S_1 charge transfer character grows with the alkylation degree resulting in a strengthened stabilization of the fluorescent state, the dynamic spectral evolution is the same for the three molecules. After the diffusional solvent relaxation with a time constant on the order of 2.3 ± 1.2 ps (average for the three molecules), the slower component with a 10 – 20 ps time constant is assigned to the hydrogen bond formation at the carbonyl group.

Other donor-acceptor molecules have been used as molecular probes, a fluoroprobe which shows a large fluorescence solvatochromism $v_{max} = 8000$ cm^{-1} when changing from an apolar solvent to a very polar solvent [35]. Diffusional rotational motions of the solvent molecules were shown to be responsible for the major part of the fluoroprobe TDFSS in dibutylether, diisopropylether, diethylether, and ethylacetate [35].

4.3.1.3
Specific Solvation: Hydrogen Bond Dynamics

To discuss the influences of the photoinduced changes of solute-solvent hydrogen bonds on the observed dynamic spectral shifts, Gustavsson et al. first analyzed the steady-state data in 20 solvents using the empirical model of Kamlet and Taft [37]. The steady-state Stokes shift can be parametrized as

$$\Delta v = \Delta v_0 + \Delta s\,\pi^* + \Delta a\,\alpha + \Delta b\,\beta \tag{4.20}$$

where π^* is the polarity/polarizability parameter of the solvent, α the index of hydrogen bond donor (HBD) character of the solvent (acidity), β the index of hydrogen bond acceptor (HBA) character of the solvent (basicity), Δv_0 the difference of the barycenters of the absorption and emission spectra in the case of "zero" solute-solvent interaction (i.e., the purely intramolecular contribution), and Δ_s, Δa, and Δb the differences of the susceptibilities of the solute property to changing solvent polarity-polarizability, HBD character, and HBA character respectively between S_1 and S_0. Indeed, for all three coumarins the solvent HBD character at the nitrogen lone pair plays a great role ($\Delta a = (0.9 \pm 0.1) \times 10^3$ cm^{-1}). The "slow" part of the dynamic Stokes shift has an amplitude smaller by 1000 cm^{-1} than the calculated Stokes shift using the Kamlet and Taft analysis, indicating some "missed" ultrafast solvent relaxation. They thus proposed that the breaking of hydrogen bonds is very fast and mingled with the ultrafast component of the nonspecific solvent relaxation. The reformation of hydrogen bonds at the carbonyl group in the HBD methanol is believed to occur on the 10–20 ps time scale [37].

4.3.1.4
Isotope Effect

The deuterium isotope effect on the solvation dynamics of methanol has been investigated by Shirota et al. using coumarin C102 as a probe and measuring the solvation time at a single wavelength [34]. Deuterated methanols (CH$_3$OD, CD$_3$OD, and CD$_3$OD) show slower dynamics than normal methanol with an isotope effect of about 10% from the OH group and 5% from the CH$_3$ group [34]. Different solvation dynamics in water and heavy water were observed by Pant and Levinger using coumarin C343 as a probe [42]. This study aimed at interfacial solvation (see below) but the bulk isotope effect, slower dynamics in D$_2$O than in H$_2$O, was interpreted as due to stronger hydrogen bonding in D$_2$O compared to H$_2$O, slowing the reorientation of the excited-state dipoles in the bulk D$_2$O. Similar results, i.e., slower solvation dynamics in heavy water than in water, were also reported by Mialocq et al. [43], using coumarin 1 and its water-soluble derivative DATC.

4.3.1.5
Spectral Narrowing in the 10 ps Time Scale

Another important feature studied using the fluorescence up-conversion technique is the narrowing of the fluorescence spectra of dyes in solution due to molecular thermalization, i.e., the dissipation of the solute vibrational excess energy to the surrounding solvent [22, 24, 30, 31, 44, 45]. The bandwidths of the absorption and fluorescence spectra constitute a molecular thermometer. For example, in the case of DCM in methanol and chloroform the fwhm bandwidths narrow in time by a value of ≈ 250 cm^{-1} with monoexponential time constants of 10 ± 1 ps and 7 ± 1 ps respectively, which was interpreted as the loss of the excitation excess energy from the "hot" solute to the solvent [30]. Vibrational cooling has been explained by Ohta et al. [45] in terms of the thermal diffusion equation to simulate macroscopic heat conduction in the bulk solvent, rather than by energy transfer via the collisions with the solvent molecules.

4.3.2
Photoinduced Intramolecular Charge Transfer

Photoinduced charge transfer processes occur in many of the molecules which have been presented as examples in Sect. 3.1 devoted to solvation processes, LDS 750 [22, 23, 26], LDS 751 [27, 28, 40, 41], DCM [30, 33, 40], coumarins [25, 29, 31, 32, 34, 37], and fluoroprobe [35]. All these donor-acceptor molecules undergo some intramolecular charge transfer. Although coumarins without bridging alkyl groups such as C153, C343, and C102 have been said to undergo charge transfer and twisting around the C-N bond linking the amino group to the aromatic ring, Barbara and Jarzeba pointed out that "the fluorescence spectrum is not obviously due to two emitting species (the initial excited state and a charge transfer form)" [25]. Barbara et al. thus performed the time resolved spectra of C343 and used the S(t) function as input to calculations of barrierless electron transfer rates [46], the rate being roughly equal to the rate of solvation [25]. In their investigation of the ultrafast charge separation in bianthryls [46–48], they found by monitoring the fluorescence intensity decays that the intramolecular electron transfer rates are not equal to the solvent longitudinal dielectric relaxation times but are in better agreement with microscopic solvation times measured using coumarins as solvent probes [48].

Strong evidence for a fast internal charge transfer process has recently been demonstrated using the femtosecond fluorescence up-conversion technique in the case of the *trans*-isomer of 4-*N,N*-dimethylamino-4′-cyanostilbene in methanol and acetonitrile (the *cis*-isomer does not fluoresce) [36]. In acetonitrile the time-resolved spectral evolution of the fluorescence showed a clear isosbestic point related to a transition between a LE state and a CT state. The integrated fluorescence could be fitted with time constants $\tau_{LE1} = 60$ fs and $\tau_{LE2} = 250$ fs for the decay and $\tau_{CT1} = 80$ fs and $\tau_{CT2} = 510$ fs for the rise. In methanol an isosbestic point was not observed but the width of the fluorescence band and the integrated fluorescence intensity indicated that the description in terms of a single band was inadequate. The difference in spectral evolution between the two solvents was explained by the more complex dielectric response of methanol or the specific solvation effects in the protic methanol.

Nonradiative excited-state relaxation via charge transfer induced twisting has been shown by Gulbinas et al. [49] for *N,N*-dimethylaminobenzylidene-1,3-indandione (DMABI) in solution. This molecule is characterized by a large difference between the dipole moments in the ground and excited states. The characteristic times of the fluorescence dynamic Stokes shift and total intensity decay are given in Table 4.2. The roughly proportional low fluorescence quantum yield and very short excited state lifetime in all solvents and in a polymer matrix indicate that nonradiative processes are dominant in the excited-state relaxation and that the fluorescence oscillator strengths are equal. The fluorescence originates mainly from the flat configuration. From the absence in the fluorescence of a fast component decay similar to the Stokes shift dynamics, those authors

Table 4.2. Characteristic time constants for the dynamic fluorescence Stokes shift (τ_{SS}) and the total intensity decay (τ_F) of DMABI in various solvents

Solvent	Heptane	Ethanol	DMSO
τ_{SS} (ps)	0.25	0.8	2.3
$\Delta\nu$ (cm^{-1})	615	1030	1060
τ_F (ps)	1.4	4.1	9.7

concluded that molecular twisting contributes little to the observed spectral shift. They proposed that the central double bond twisting results in the very fast excited-state relaxation [49].

Glasbeek et al. have found that an adiabatic electron transfer takes place, concomitant with solvation, in organic light-emitting diodes, Al(III)- and Ga(III)-tris(8-hydroxy-quinoline) complexes (Alq$_3$ and Gaq$_3$) [50]. The band maxima of the absorption and fluorescence spectra of these complexes are almost insensitive to the solvent polarity. Their fluorescent excited states are $^1\pi\pi^*$ ligand localized states on an individual ligand where the electronic charge is transferred from the pyridyl ring to the quinoline group. Because the integrated fluorescence intensity deviates from the ν^3 dependence expected from the Einstein relation for spontaneous emission, these authors conclude that the radiative character of the emissive state and the wavefunction change as the solvation takes place [50].

4.3.3
Intermolecular Electron Transfer

The femtosecond fluorescence up-conversion technique is well-adapted to study ultrafast electron transfer (ET) which is one of the most important processes in chemistry and biology. Intermolecular photoinduced ET from electron-donating solvents (aniline, N,N-dimethylaniline, hydrazine) to excited coumarins [51–53] and oxazine-1 [54] occurred on a time scale ranging from a few nanoseconds to tens of femtoseconds depending on the structure of dyes and solvent. The ET static and dynamic parameters have been shown to depend on the alkyl substituents of aniline or hydrazine [51–54]. The group of Yoshihara recognized that in many alkylaniline-coumarin combinations the ET rate is much faster than solvation dynamics [53]. They also investigated the isotope effect on this reaction. For the fastest ET the isotope effect is small, whereas the effect is quite large ($\sim 20\%$) for slower ET. The deuterium isotope effect seems to be related with the intermolecular hydrogen-bonding [53]. Hydrazines with rather low oxidation potentials in comparison with those of anilines are much more strongly electron donating than the anilines [54].

Photoinduced electron transfer between excited dye molecules and semiconductors has applications in photography and solar energy cells. Charge injection dynamics from a photoexcited surface-bound coumarin to the conduction band of TiO$_2$ [55] and from photoexcited modified perylene chromophores anchored to a spongelike TiO$_2$ electrode [56] occurred on a time scale of ca. 200 fs.

4.3.4
Intramolecular Proton Transfer

Photoinduced intramolecular proton transfer is characteristic of enol-keto reactions and molecules capable of establishing an intramolecular hydrogen
bond. The intramolecular proton transfer in a hydroxyphenyl methyloxazole
molecule was found to be slower when caged in a β-cyclodextrin nanocavity
than in aprotic solvents (< 300 fs) due to the noticeable inhibition of the twisting
motion between the two moieties inside the nanocavity [57]. Double proton
transfer has been studied in the 7-azaindole dimer, structurally similar to hy-
drogen-bonded pairs in DNA and regarded as a model system for the study of
the photoinduced mutation of DNA [58–61]. In the case of 2,2'-bipyridyl-3,3'-
diol in cyclohexane and acetonitrile [62, 63] and in sol-gel glasses [64], two excited state intramolecular reactions occur: a concerted double proton transfer
and a two-step single proton transfer.

4.3.5
$S_2 \rightarrow S_1$ Internal Conversion

Fluorescence usually occurs from the lowest excited state S_1 of molecules due to
the rapid radiationless decay from the higher excited states (Kasha's rule). However, using the fluorescence up-conversion technique, very low level intensities
can be detected such as those of the fluorescence emitted between S_2 and S_0
singlet states. After the pioneering measurements of Hallidy and Topp on the
$S_2 \rightarrow S_0$ fluorescence of xanthione [14], other molecular systems of interest have
been investigated such as ZnTPP porphyrin and malachite green. Gurzadyan et
al. [65] monitored the depopulation of the S_2 state of ZnTPP after excitation in
the Soret band at 394 nm. They measured the decay of the $S_2 \rightarrow S_0$ fluorescence
and the rise of the $S_1 \rightarrow S_0$ fluorescence in ethanol ($\tau = 2.35$ ps) and found a high
initial anisotropy of $r \geq 0.7$. Yoshizawa et al. measured the S_2 fluorescence lifetime of the triphenylmethane dye, malachite green in water, $\tau = 0.27 \pm 0.05$ ps
[66]. Interestingly, the decay kinetics of the S_2 fluorescence was represented by a
biexponential function in high viscosity solvents. The fast component almost independent of the viscosity is assigned to the equilibrium of high frequency internal modes. The slow component is also almost independent of the solvent viscosity and mainly assigned to the $S_2 \rightarrow S_1$ internal conversion [66]. These authors
measured a decay time of $\tau = 4.2 \pm 0.1$ ps for S_1 in ethyleneglycol which is consistent with another recent observation of an initial decay time of 3 ps and a
second decay with a time constant of 6 ± 1 ps using the same technique [67]. S_2
fluorescence with a fast decay of ~250 fs has also be evidenced from a cy-
anoanthracene-dimethylaniline weak charge transfer complex [68].

4.3.6
Biological Systems

Applications of the fluorescence up-conversion technique to light-driven biological systems have been performed to elucidate the reaction mechanisms in

those systems. Only a few examples will be given which emphasize the advantage of time-resolved fluorescence spectroscopy over time-resolved absorption spectroscopy due to the well-defined ground and excited states. In relation to the biochemical importance of retinyl chromophores in photoreceptor proteins (rhodopsin and bacteriorhodopsin [13]), the fluorescence decay of all-trans retinal in hexane consists of ultrafast ($\tau = 30$ fs), fast ($\tau = 370$ fs), and slow ($\tau = 33.5$ ps) components [69]. The photoactive yellow protein from the purple bacterium *Ectothiorhodospira halophila* contains a thiol ester linked *p*-coumaric acid. The fluorescence nonexponential decays with time constants of several fs to a few ps have been interpreted as due to twisting around the vinyl bond of *p*-coumaric acid leading to a *trans/cis* isomerization similar to those of bacteriorhodopsin [70]. The blue fluorescent protein is a mutant of the green fluorescent protein from the jellyfish *Aequorea victoria*, useful for biotechnological applications, which forms spontaneously the fluorophore via internal cyclization and 1,2-dehydration of the central residue and requires molecular oxygen for the appearance of fluorescence. Boxer et al. have investigated the protein photodynamic behavior in H_2O and D_2O and concluded that an excited state proton transfer was not coupled to fluorescent emission [71]. As a final example, the ultrafast fluorescence quenching dynamics of the flavin chromophore is much faster in the riboflavin binding protein or the glucose oxidase from *Aspergillus niger* than that in solutions and has been interpreted as due to extremely fast Franck Condon relaxations, electron transfer, or electron transfer followed by proton transfer between the excited flavin and nearby tryptophan and tyrosine residues [72].

4.4
Conclusions

We have reviewed the historical background and the principle of the ultrafast fluorescence up-conversion technique. In comparison to the other time-resolved fluorescence methods, the prevalence of this technique for precise measurements in the femtosecond and subpicosecond time range is now well recognized. Following the enormous progress of femtosecond solid state laser sources, the last decade has witnessed an increasing interest for this technique. The resulting considerable improvement of the acquisition of both spectrally and time-resolved data combined with the development of modern theoretical and molecular dynamics simulation studies have led to an increasing number of applications in the field of ultrafast laser spectroscopy. Some examples of recent fluorescence up-conversion studies prove the utility of this technique in contemporary chemistry and biology.

References

1. Mialocq J-C (1997) Fundamentals of interaction between light and matter. In: Chanon M (ed) Homogeneous photocatalysis, vol 2. Wiley Series in Photoscience and Photo-Engineering, pp 15–54
2. Lakowicz JR, Gryczynski I (1991) Frequency-domain fluorescence spectroscopy. In: Lakowicz JR (ed) Topics in fluorescence spectroscopy. 1. Techniques. Plenum Press, New-York, pp 293–335
3. O'Connor DV, Phillips D (1984) Time-correlated single photon counting. Academic Press, London
4. Birch DJS, Imhof RE (1991) Time-domain fluorescence spectroscopy using time-correlated single-photon counting. In: Lakowicz JR (ed) Topics in fluorescence spectroscopy. 1. Techniques. Plenum Press, New-York, pp 1–95
5. Birch DJS, Hungerford G (1994) Instrumentation for red/near-infrared fluorescence. In: Lakowicz JR (ed) Topics in fluorescence spectroscopy. 4. Probe design and chemical sensing. Plenum Press, New-York, pp 377–416
6. Lampert RA, Chewter LA, Phillips D, O'Connor DV, Roberts AJ, Meech SR (1983): Standards for nanosecond fluorescence decay time measurements. Anal Chem 55:68–73
7. Nordlund TM (1991) Streak cameras for time-domain fluorescence. In: Lakowicz JR (ed) Topics in fluorescence spectroscopy. 1. Techniques. Plenum Press, New-York, pp 183–259
8. Davis LM, Parigger C (1992) Use of streak camera for time-resolved photon counting fluorimetry. Meas Sci Technol 3:85–90
9. Reddy BR, Venkateswarlu P (1983) Energy up-conversion in $LaF_3:Nd^{3+}$. J Chem Phys 79:5845–5850
10. Duguay MA, Hansen JW (1969) An ultrafast gate. Appl Phys Lett 15:192–194
11. Duguay MA, Hansen JW (1968) Optical sampling of subnanosecond light pulses. Appl Phys Lett 13:178–180
12. Mahr H, Hirsch MD (1975) An optical up-conversion light gate with picosecond resolution. Optics Commun 13:96–99
13. Hirsch MD, Marcus MA, Lewis A, Mahr H, Frigo N (1976) A method for measuring picosecond phenomena in photolabile species. The emission lifetime of bacteriorhodopsin. Biophysical J 16:1399–1409
14. Hallidy LA, Topp MR (1977) Picosecond luminescence detection using type-II phase-matched frequency conversion. Chem Phys Lett 46:8–14
15. Kopainsky B, Kaiser W (1978) Investigation of intra- and intermolecular transfer processes by picosecond fluorescence gating. Optics Commun 26:219–224
16. Hirsch MD, Mahr H (1979) Luminescence decay kinetics of malachite green on a picosecond time scale. Chem Phys Lett 60:299–303
17. Beddard GS, Doust T, Porter G (1981) Picosecond fluorescence depolarisation measured by frequency conversion. Chem Phys 61:17–23
18. Daly T, Mahr H (1978) Time-resolved luminescence spectra in highly photo-excited CdSe at 1.8 K. Solid State Communications 25:323–326
19. Shah J, Damen TC, Deveaud B, Block D (1987) Subpicosecond luminescence spectroscopy using sum frequency generation. Appl Phys Lett 50:1307–1309
20. Shah J (1988) Ultrafast luminescence spectroscopy using sum-frequency generation. IEEE J Quantum Electron 24:276–288
21. Kahlow MA, Jarzeba W, DuBruil TP, Barbara PF (1988) Ultrafast emission spectroscopy in the ultraviolet by time-gated upconversion. Rev Sci Instr 59:1098–1109
22. Castner EW, Maroncelli M, Fleming GR (1987) Subpicosecond resolution studies of solvation dynamics in polar aprotic and alcohol solvents. J Chem Phys 86:1090–1097
23. Castner EW, Bagchi B, Maroncelli M, Webb SP, Ruggiero AJ, Fleming GR (1988) The dynamics of polar solvation. Ber Bunsenges Phys Chem 92:363–372
24. Mokhtari A, Chesnoy J, Laubereau A (1989) Femtosecond-time and frequency resolved fluorescence spectroscopy of a dye molecule. Chem Phys Lett 155:593–598

25. Barbara PF, Jarzeba W (1990) Ultrafast photochemical intramolecular charge and excited state solvation. In: Volman DH, Hammond GS, Gollnick K (eds) Advances in photochemistry, vol 15. Wiley, New York, pp 1–68
26. Rosenthal SJ, Xie X, Du M, Fleming GR (1991) Femtosecond solvation dynamics in acetonitrile: observation of the inertial contribution to the solvent response. J Chem Phys 95:4715–4718
27. Hébert P, Baldacchino G, Gustavsson T, Mialocq JC (1993) Subpicosecond fluorescence study of the LDS 751 dye molecule in ethanol. Chem Phys Lett 213:345–350
28. Mialocq J-C, Hébert P, Baldacchino G, Gustavsson T (1994) Relaxation dynamics of a polar solvent cage around a nonpolar electronically excited solvent probe. A subpicosecond study. In: Gauduel Y, Rossky PJ (eds) Ultrafast reaction dynamics and solvent effects. AIP Conference Proceedings 298, pp 359–367
29. Jimenez R, Fleming GR, Kumar PV, Maroncelli M (1994) Femtosecond solvation dynamics of water. Nature (London) 369:471–473
30. Gustavsson T, Baldacchino G, Mialocq JC, Pommeret S (1995) A femtosecond fluorescence up-conversion study of the dynamic Stokes shift of the DCM dye molecule in polar and non-polar solvents. Chem Phys Lett 236:587–594
31. Horng ML, Gardecki JA, Papazyan A, Maroncelli M (1995) Subpicosecond measurements of polar solvation dynamics: Coumarin 153 revisited. J Phys Chem 99:17311–17337
32. Tominaga K, Walker GC (1995) Femtosecond experiments on solvation dynamics of an anionic probe molecule in methanol. J Photochem Photobiol A: Chem 87:127–133
33. van der Meulen P, Zhang H, Jonkman AM, Glasbeek M (1996) Subpicosecond solvation relaxation of 4-(dicyanomethylene)-2-methyl-6-(p-(dimethylamino)styryl)-4H-pyran in polar liquids. J Phys Chem 100:5367–5373
34. Shirota H, Pal H, Tominaga K, Yoshihara K (1996) Deuterium isotope effect on the solvation dynamics of methanol: CH_3OH, CH_3OD, CD_3OH, and CD_3OD. J Phys Chem 100:14575–14577
35. Middelhoek ER, Zhang H, Verhoeven JW, Glasbeek M (1996) Subpicosecond studies of solvation dynamics of fluoroprobe in liquid solution. Chem Phys 211:489–497
36. Eilers-König N, Kühne T, Schwarzer D, Vöhringer P, Schroeder J (1996) Femtosecond dynamics of intramolecular charge transfer in 4-dimethylamino-4'-cyanostilbene in polar solvents. Chem Phys Lett 253:69–76
37. Gustavsson T, Cassara L, Gulbinas V, Gurzadyan G, Mialocq JC, Pommeret S, Sorgius M, van der Meulen P (1998) Femtosecond spectroscopy study of relaxation processes of three amino-substituted coumarin dyes in methanol and dimethylsulfoxide. J Phys Chem A 102:4229–4245
38. Berlman IB (1971) Handbook of fluorescence spectra of aromatic molecules, 2nd edn. Academic Press, New-York, pp 21–28
39. Fee RS, Maroncelli M (1994) Estimating the time-zero spectrum in time-resolved emission measurements of solvation dynamics. Chem Phys 183:235–247
40. Gustavsson T, Hébert P, Baldacchino G, Pommeret S, Naskrecki R, Mialocq JC (1996) Rôle de la charge de la sonde moléculaire sur la solvatation. Aspects statiques et dynamiques. J Chim Phys 93:117–127
41. Hébert P, Baldacchino G, Gustavsson T, Mialocq JC (1994) Photochemistry of an unsymmetrical polymethine-cyanine dye. Solute-solvent interactions and relaxation dynamics of LDS 751. J Photochem Photobiol A: Chem 84:45–55
42. Pant D, Levinger NE (1999) Polar solvation dynamics of H_2O and D_2O at the surface of zirconia nanoparticles. J Phys Chem B 103:7846–7852
43. Mialocq JC, Gustavsson T, Pommeret S (1999) Spectroscopie laser femtoseconde dans l'eau, milieu biologique. J Phys IV 9(Pr5):101–104
44. Boczar BP, Topp MR (1982) New developments in picosecond time-resolved fluorescence spectroscopy: vibrational relaxation phenomena. In: Eisenthal KB, Hochstrasser RM, Kaiser W, Laubereau A (eds) Picosecond phenomena III. Springer, Berlin Heidelberg New York, pp 174–178

45. Ohta K, Kang TJ, Tominaga K, Yoshihara K (1999) Ultrafast relaxation processes from a higher excited electronic state of a dye molecule in solution: a femtosecond time-resolved fluorescence study. Chem Phys 242:103–114
46. Johnson AE, Tominaga K, Walker GC, Jarzeba W, Barbara PF (1993) Femtosecond electron transfer: Experiment and theory. Pure Appl Chem 65:1677–1680
47. Kahlow MA, Kang TJ, Barbara PF (1987) Electron-transfer times are not equal to longitudinal relaxation times in polar aprotic solvents. J Phys Chem 91.6452–6455
48. Kang TJ, Kahlow MA, Giser D, Swallen S, Nagarajan V, Jarzeba W, Barbara PF (1988) Dynamic solvent effects in the electron-transfer kinetics of S_1 bianthryls. J Phys Chem 92:6800–6807
49. Gulbinas V, Kodis G, Jursenas S, Valkunas L, Gruodis A, Mialocq J-C, Pommeret S, Gustavsson T (1999) Charge transfer induced excited state twisting of *N,N*-dimethylamino-benzylidene-1,3-indandione in solution. J Phys Chem 103:3969–3980
50. Humbs W, van Veldhoven E, Zhang H, Glasbeek M (1999) Subpicosecond fluorescence dynamics of organic light-emitting diode tris(8-hydroxyquinoline)metal complexes. Chem Phys Lett 304:10–18
51. Nagasawa Y, Yartsev AP, Tominaga K, Bisht PB, Johnson PB, Yoshihara K (1995) Dynamical aspects of ultrafast intermolecular electron transfer faster than solvation process: substituents effects and energy gap dependence. J Phys Chem 99:653–662
52. Shirota H, Pal H, Tominaga K, Yoshihara K (1998) Ultrafast intermolecular electron transfer in coumarin-hydrazine system. Chem Phys 236:355–364
53. Shirota H, Pal H, Tominaga K, Yoshihara K (1998) Substituent effect and deuterium isotope effect of ultrafast intermolecular electron transfer: coumarin in electron-donating solvent. J Phys Chem A 102:3089–3102
54. Rubtsov IV, Shirota H, Yoshihara K (1999) Ultrafast photoinduced solute-solvent electron transfer: configuration dependence. J Phys Chem A 103:1801–1808
55. Rehm JM, McLendon GL, Nagasawa Y, Yoshihara K, Moser J, Grätzel M (1996) Femtosecond electron-transfer dynamics at a sensitizing dye-semiconductor (TiO_2) interface. J Phys Chem 100:9577–9578
56. Burfeindt B, Hannappel T, Storck W, Willig F (1996) Measurement of temperature-independent femtosecond interfacial electron transfer from an anchored molecular electron donor to a semiconductor as acceptor. J Phys Chem 100:16463–16465
57. Douhal A, Fiebig T, Chachisvilis M, Zewail AH (1998) Femtochemistry in nano-cavities: reactions in cyclodextrins. J Phys Chem A 102:1657–1660
58. Share P, Pereira M, Sarisky M, Repinec S, Hochstrasser RM (1991) Dynamics of proton transfer in 7-azaindole. J Lumin. 48/49:204–208
59. Takeuchi S, Tahara T (1997) Observation of dimer excited-state dynamics in the double proton transfer reaction of 7-azaindole by femtosecond fluorescence up-conversion. Chem Phys Lett 277:340–346
60. Takeuchi S, Tahara T (1998) Femtosecond ultraviolet-visible fluorescence study of the excited-state proton-transfer reaction of 7-azaindole dimer. J Phys Chem A 102:7740–7753
61. Fiebig T, Chachisvilis M, Manger M, Zewail AH, Douhal A, Garcia-Ochoa I, de La Hoz Ayuso A (1999) Femtosecond dynamics of double proton transfer in a model DNA base pair: 7-azaindole dimers in the condensed phase. J Phys Chem A 103:7419–7431
62. Marks D, Zhang H, Glasbeek M, Borowicz P, Grabowska A (1997) Solvent dependence of (sub)picosecond proton transfer in photo-excited [2,2′-bipyridyl]-3,3′-diol. Chem Phys Lett 275:370–376
63. Glasbeek M, Marks D, Zhang H (1997) Femtosecond studies of double-proton transfer in [2,2′-bipyridyl]-3,3′-diol. J Lumin 72–74:832–834
64. Prosposito P, Marks D, Zhang H, Glasbeek M (1998) Femtosecond double-proton transfer dynamics in [2,2′-bipyridyl]-3,3′-diol in sol-gel glasses. J Phys Chem A 102:8894–8902
65. Gurzadyan GG, Tran-Thi T-H, Gustavsson T (1998) Time-resolved fluorescence spectroscopy of high-lying electronic states of Zn-tetraphenylporphyrin. J Chem Phys 108:385–388

66. Yoshizawa M, Susuki K, Kubo A, Saikan S (1998) Femtosecond study of S_2 fluorescence in malachite green in solutions. Chem Phys Lett 290:43–48
67. Gottfried NH, Roither B, Scherer POJ (1997) New aspects of the ultrafast electronic decay of malachite green in solution. Opt Commun 143:261–264
68. Iwai SI, Murata S, Tachiya M (1998) Ultrafast fluorescence quenching by electron transfer and fluorescence from the second excited state of a charge transfer complex as studied by femtosecond up-conversion spectroscopy. J Chem Phys 109:5963–5970
69. Takeuchi S, Tahara T (1997) Ultrafast fluorescence study on the excited singlet-state dynamics of all-trans-retinal. J Phys Chem A 101:3052–3060
70. Chosrowjan H, Mataga N, Nakashima N, Imamoto Y, Tokunoga F (1997) Femtosecond-picosecond fluorescence studies on excited state dynamics of photoactive yellow protein from *Ectothiorhodospira halophila*. Chem Phys Lett 290:267–272
71. Wachter RM, King BA, Heim R, Kallio K, Tsien RY, Boxer SG, Remington SJ (1997) Crystal structure and photodynamic behavior of the blue emission variant Y66H/Y145F of green fluorescent protein. Biochemistry 36:9759–9765
72. Mataga N, Chosrowjan H, Shibata Y, Tanaka F (1998) Ultrafast fluorescence quenching dynamics of flavin chromophores in protein nanospace. J Phys Chem B 102:7081–7084

Spectroscopic Investigations of Intermolecular Interactions in Supercritical Fluids

M. A. Kane, S. N. Daniel, E. D. Niemeyer, F. V. Bright

5.1
Introduction

Solvent choice is amongst the most important factors governing the efficiency, selectivity, and outcome of a chemical reaction, extraction, or separation. However, as a result of recent environmental legislation, the use of several common organic and especially chlorinated liquid solvents has been limited or even banned [1]. For this reason, researchers have begun intensive work to develop and exploit more "environmentally-friendly" or "green" solvents. At the forefront of this research has been the development of supercritical fluid technologies as an alternative to hazardous solvents currently in use [1–6].

When a pure substance is raised above its critical point (defined by its characteristic critical temperature (T_c) and pressure (P_c)), the liquid and gas phases coalesce into a single phase, known as a supercritical fluid (Table 5.1). Once in the supercritical state, the properties of the fluid can be tuned continuously by simply adjusting the system temperature and/or pressure. Figure 5.1 illustrates the effects of system pressure and temperature on the bulk density and dielectric constant of supercritical CO_2 ($scCO_2$) at reduced temperatures ($T_r = T_{exp}/T_c$) of 1.01 and 1.10. Examination of these isotherms shows that near the critical temperature, small changes in system pressure can lead to significant changes in the bulk fluid density. However, as one increases the system temperature ($T_r > 1.00$) and moves further from the system critical point, one sees that the changes in density are not nearly as sensitive to pressure changes (Fig. 5.1 B, C).

Table 5.1. Critical data for several common supercritical fluids

Fluid	Critical Pressure (bar)	Critical Temperature (K)	Critical Density (g/ml)
CO_2	74	304	0.472
C_2H_6	49	306	0.204
C_2H_4	50	282	0.214
CF_3H	49	300	0.506
CF_3Cl	39	302	0.585
Xe	58	290	1.096
H_2O	221	648	0.281
N_2O	72	310	0.445

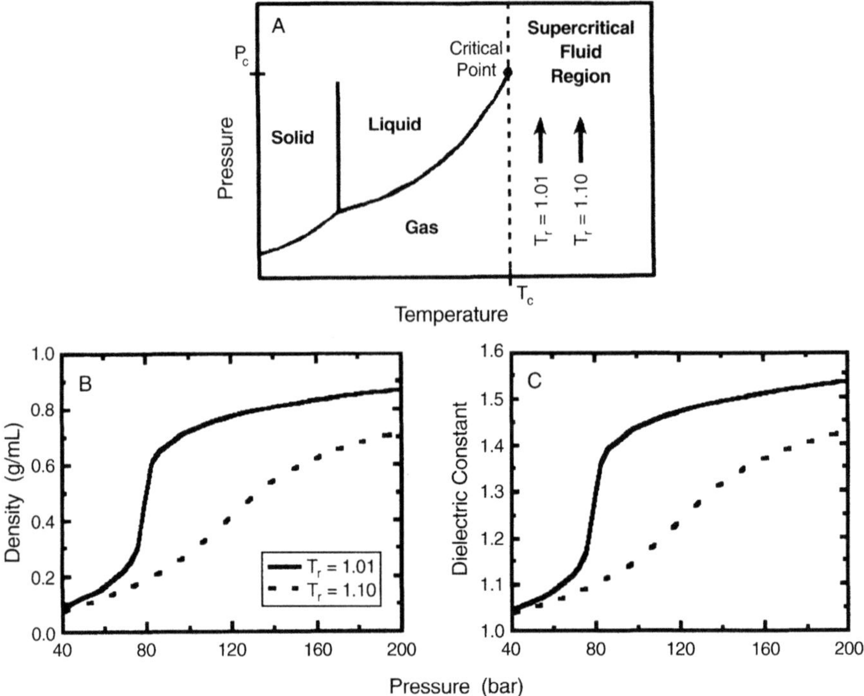

Fig. 5.1 a–c. Tunability of a supercritical fluid: **a** generic phase diagram for a pure substance; **b** density; **c** dielectric constant isotherms as a function of pressure for supercritical CO_2 at $T_r = 1.01$ and 1.10

One should also note that other physicochemical properties (e.g., refractive index, viscosity, diffusion coefficient), which scale with density, are also tunable with pressure and temperature. Thus, a supercritical fluid can be adjusted continuously between gas- and liquid-like properties without physically changing the solvent. Supercritical fluids also possess liquid-like solvation properties and have more gas-like diffusivities which can lead to higher mass transport rates in comparison to normal liquids. For these reasons, the uses of supercritical fluids continues to grow [1–6].

To date, the most commonly used supercritical fluids have been CO_2 and H_2O. Carbon dioxide is inexpensive, non-toxic, nonflammable, and it has relatively mild critical parameters (Table 5.1). Supercritical CO_2 has been used for the decaffeination of tea and coffee, the extraction of hops, spices, and flavorings, the extraction of pharmaceuticals from seeds and barks, to harden and strengthen Portland cement and building materials, and even as an alternative to perchloroethylene in the commercial dry cleaning process. Supercritical water (SCW) has attracted substantial interest because of its potential for hazardous waste disposal. Moreover, when water is raised above its critical point (Table 5.1), environmentally harmful organics can be rapidly and efficiently oxidized to benign compounds such as CO_2, H_2O, and simple inorganic acids and salts. Cur-

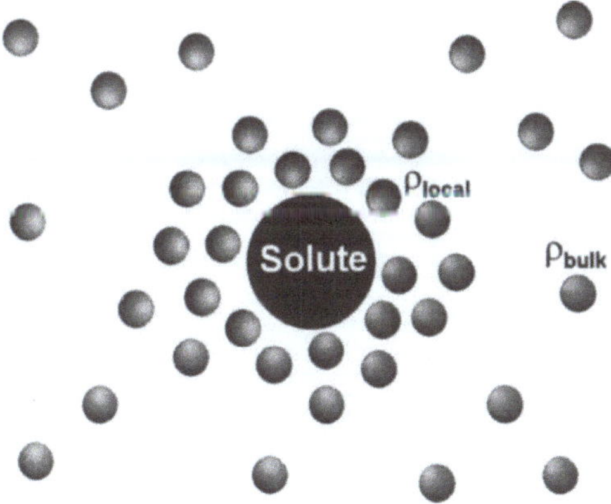

Fig. 5.2. A depiction of the solute-fluid density augmentation process with an increase in local density (ϱ_{local}) surrounding the solute compared to the bulk fluid density (ϱ_{bulk})

rently, there are several commercial waste disposal plants in operation in the US and Europe based on SCW wet oxidation technology. Supercritical water has also been used for the destruction of chemical arms and even of used automobile tires.

One of the most interesting phenomena observed in supercritical fluid systems is an apparent increase in the local fluid density, relative to the bulk density, surrounding a dissolved solute when the system is near the mixture critical point [7–10]. This increase in fluid density has been termed "solute-fluid clustering", "molecular charisma", or "local density augmentation." A simplified depiction of the solute-fluid clustering process is shown in Fig. 5.2, with an increased local fluid density (ϱ_{local}) surrounding the solute compared to the lower bulk fluid density (ϱ_{bulk}). The solute-fluid clusters observed in supercritical fluids are *dynamic* in nature, constantly exchanging fluid molecules with the bulk fluid on a sub-picosecond time scale [11]. This augmentation phenomena is thought to influence a wide variety of processes including reaction rates and outcome, solute conformational equilibria, and extractions.

The remainder of this chapter will highlight several of the more varied applications of UV-Vis absorbance spectroscopy, fluorescence spectroscopy, and laser flash photolysis as they have been applied to the study of solute-fluid interactions within supercritical fluid systems. The reader should also realize that there is an extensive literature using, for example, infrared absorbance, Raman scattering, NMR, XAFS, and thermal optical methods that are intentionally not discussed in this chapter.

5.2
Instrumentation

It is well-known that certain chromophores/dyes exhibit features that are a strong function of the physicochemical properties of their surroundings [12]. As a result, one can use these sorts of chromophores as solute probes of *their* local

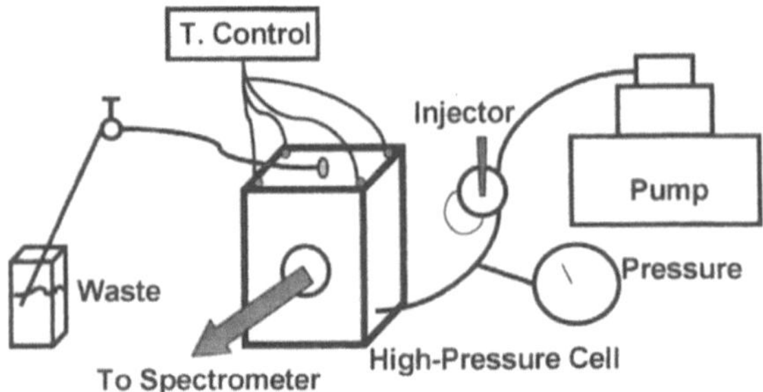

Fig. 5.3. Simplified schematic for a typical spectroscopy set-up for studying supercritical fluids

microenvironment. In this context techniques like absorbance and emission spectroscopy are almost ideal for addressing question about the local micro-environment surrounding a solute dissolved in a supercritical fluid. These techniques are also attractive because they allow one to assess the local microenvironment surrounding a solute, if the solute absorbs and/or emits, under conditions when the solute concentration is reasonably low. In this way one can minimize solute-solute interactions and focus exclusively on solute-fluid/solvent interactions.

To get a handle on the local microenvironment surrounding a solute dissolved in a supercritical fluid, one needs several pieces of equipment (Fig. 5.3). First, one needs a spectrometer. A spectrometer can range from a simple UV-Vis absorbance spectrophotometer up to a sophisticated time-resolved laser system. One needs a high-pressure optical cell to house the fluid. The high-pressure cell body is often made of stainless steel or some other suitable material (e.g., titanium). The choice of the cell material depends on the corrosive tendencies of the fluid itself and the temperature range to be studied. One also requires optical ports that will allow one to bring the electromagnetic radiation into and out of the fluid. In the UV/visible region, one might seal windows (quartz or sapphire) directly into the cell body or one might use optical fibers mounted within the cell body. Again, some care must be exercised in selecting the optical port materials depending on the sealing method that one intends to use (e.g., Teflon O-rings, Au seals), the temperature and pressure range of the experiments, and the corrosive nature of the fluid. A careful look at Fig. 5.2 shows that it is also important to control accurately and precisely *and know* the actual pressure and temperature within the high-pressure cell. A pump can serve to generate the pressure, a precision gauge is used to monitor the pressure within the high-pressure cell, and cartridge heaters are often used in concert with a thermocouple feed back system to control the cell/fluid temperature.

5.3
Sample Preparation and Precautions

There are several challenges associated with performing good spectroscopic measurements of solutes dissolved in supercritical fluids. Foremost amongst these issues is to ensure that the solute is actually dissolved completely in the fluid and not adsorbed to the interior cell walls or the optical ports. This challenge can be met by working at solute concentrations in the low micromolar range and by taking special care to look for and minimize signs of solute adsorption and/or aggregation. Unanticipated shifts in the spectral profiles with changes in fluid density, beyond continuum model predictions, are always good indicators of adsorbed species and/or aggregates, but these must always be decoupled from the intrinsic shifts induced by changes in the physicochemical properties of the fluid with changes in fluid density. A low micromolar solute concentration range also sets practical limits on the types of probes/solutes that one can effectively study. An ideal solute is one with: (1) a reasonably high *combination* of solubility in the fluid of choice over the density range in question, high molar extinction coefficient, and high fluorescence quantum yield and (2) a spectroscopically accessible signature (e. g., absorbance or emission spectrum, lifetime) that is strongly dependent on the physicochemical properties of the solvent. Examples of the more common solutes used for studying supercritical fluids include pyrene (I_1/I_3 band ratio depends on the solvent), DMABN (undergoes twisted-intramolecular charge transfer to an extent that depends on the solvent properties), or N,N'-bis-(2,5-*tert*-butylphenyl)-3,4,9,10-perylenecarboxodiimide (rotational reorientation dynamics depend on the solvent microviscosity).

In our laboratory we always make sure the solutions are thoroughly mixed and we check for adsorption or aggregation by recording the static electronic absorbance spectra in a given fluid as a function of fluid density. In general, an absorbance profile that remains more or less constant (save for the shifts induced by changes in the fluid's orientational polarizability caused by changes in fluid density with changes in pressure) in magnitude is indicative of a fully dissolved solute. In questionable situations, we run a series of control experiments over a range of solute concentrations.

5.4
Selected Applications

The electronic absorbance and emission characteristics of a solute can be used to provide information on the local environment surrounding the molecule in its ground and excited state [12]. For this reason, absorbance and fluorescence studies have been one of the most popular methods to quantify solvation phenomena in supercritical fluids. For example, solvatochromic shifts in the absorbance and fluorescence spectra of solutes dissolved in a variety of supercritical fluid systems have been used extensively to quantify solute-fluid interactions. Time-resolved measurements have also been used to provide insight into solvation in supercritical fluids.

Fluorescence-based measurements are particularly attractive because the emission process contains a wealth of information that is related to the fluorophore (i.e., the solute) *and* its surroundings. Specifically, one can conveniently divide fluorescence techniques into two forms: static and dynamic. These can in turn be divided still further into subcategories with particular features:

Static

Intensity
- Concentration of species
- Quenching of species (accessibility, conformational changes, kinetics of change)

Spectral
- Information on the local environment about the fluorophore (e.g., polarity, pH).
- Distances between sites via energy transfer techniques

Polarization/Anisotropy
- Average size of a rotationally reorienting species
- Mobility or motional restriction

Dynamic

Excited-State Decay Kinetics
- Resolve static emission into the contribution from the individual emissive centers
- Study ultrafast kinetic processes (e.g., solvation)
- Unravel excited-state decay kinetics
- Elucidate origin of quenching processes
- Study exchange processes and heterogeneity (e.g., continuous lifetime distributions)

Excited-State Decay of Anisotropy
- Detailed reorientational dynamics of non-spherical rotors (e.g., diffusion kinetics)
- Overall shape of the rotating body
- Discrimination between local and global motions in a complex system

It is common for chemists and engineers to add a small amount (1–10 mol%) of a cosolvent (e.g., an alcohol) to $scCO_2$ to improve extractions and increase solute loadings. Although this is a common practice, there is a limited amount of information on exactly how the cosolvent, fluid, and solute interact with one another. By using the $n \rightarrow \pi^*$ absorbance shift of benzophenone, Eckert and co-workers [13] recently investigated the behavior of benzophenone dissolved in binary supercritical solutions composed of C_2H_6 and a large variety of cosolvents. Because the benzophenone $n \rightarrow \pi^*$ transition is sensitive to its local solvent environment, increases in local solvent dipolarity or specific solute-fluid interactions (i.e., hydrogen bonding) result in a blue shift of the $n \rightarrow \pi^*$ absorbance. By using supercritical C_2H_6 the authors were able to eliminate any possible specific interactions with the benzophenone carbonyl and the fluid. The cosolvents that were chosen had hydrogen bonding abilities with the benzophenone. The authors found a maximum in the benzophenone $n \to \pi^*$ shift occurred when they used the cosolvent 2,2,2-trifluoroethanol (TFE) and this shift was density independent at higher TFE concentrations. Upon examination of other cosolvents, with a lesser ability to hydrogen bond to benzophenone, the authors concluded that hydrogen bonding is the primary reason for changes in the $n \to \pi^*$ absorbance shift within the supercritical C_2H_6/cosolvent/benzophenone system.

A chemical-physical model was developed to predict benzophenone $n \to \pi^*$ shifts in the TFE/C_2H_6 system. This model was based on the benzophenone $n \to \pi^*$ shifts in liquids, and it assumes that the shifts observed in liquids and supercritical fluids arises from similar changes in local solvent dipolarity. The model was used to predict the maximum benzophenone $n \to \pi^*$ shift as a function of C_2H_6 density. The model under-predicted the experimental benzophenone $n \to \pi^*$ shift at lower fluid densities and over-predicted the shift at higher fluid densities. This particular result was explained in terms of the model's inability to account for changes in local density and local cosolvent composition enhancements surrounding the benzophenone molecule in proximity to the fluid critical point. Additional integral equation analysis showed that, while this model assumed that the coordination number surrounding the benzophenone remains constant as a function of fluid density, the coordination number actually varies near the mixture critical point.

Schulte and Kauffman have exploited the fluorescence from twisted intramolecular charge transfer (TICT) molecules to quantify changes in the local solvent permittivity in a binary supercritical mixture composed of CO_2 and ethanol. A TICT compound contains both an electron donor and electron acceptor group which must twist in the excited state for charge transfer to occur [14, 15]. TICT compounds are strongly solvatochromic and can exist as two isomers, either a locally excited (LE) or charge transfer (CT) configuration. If the TICT molecule is dissolved in a dipolar solvent, after electronic excitation it may twist to form the CT complex which has a much larger dipole moment and is therefore stabilized in the polar solvent. The CT emission band is strongly red shifted with respect to the LE band and its fluorescence quantum yield is strongly dependent on the solvent dipolarity. Due to their strongly solvatochromic characteristics, TICT compounds have been very popular for use in quantifying solvation in supercritical fluids. Schulte and Kauffman used bis(4,4′-aminophenyl) sulfone (APS) [14] and bis(4,4′-dimethylaminophenyl)sulfone (DMAPS) [15] to probe changes in local permittivity in ethanol-modified CO_2 systems. Interestingly, although these compounds are structurally similar, APS is only soluble in ethanol-modified CO_2 while DMAPS is soluble in either neat or ethanol-modified CO_2.

Comparison of LE and CT spectral shifts and intensities were made between APS in a 5 mol% mixture of ethanol in CO_2 and APS dissolved in liquid n-alcohols. The authors found that in the low density fluid region, APS exists in an environment that is similar in permittivity to liquid n-butanol, but as the fluid pressure is increased (>124 bar) the permittivity becomes more similar to liquid *n*-octanol. Additionally, ratios of the CT to LE quantum yields were used to determine local solvent permittivity. Experimentally determined permittivities were compared to the empirically calculated bulk permittivities for the ethanol/CO_2 system. From this comparison, the authors showed that, in proximity to the fluid critical density, the permittivity was 8–10 times greater than the bulk permittivity. These results were discussed in terms of preferential solvation of the alcohol cosolvent surrounding the solute near the fluid mixture critical point.

When DMAPS was dissolved in neat CO_2, changes in the fluid pressure resulted in no detectable spectral shifts, indicating that the spectral shifts noted above are a result of changes in the local solvent environment/composition and are not pressure induced. The authors also noted that the DMAPS emission intensity was much higher in the ethanol-modified CO_2 than in neat CO_2 due to the increased DMAPS solubility. As observed for APS in ethanol-modified CO_2, in order for local permittivities to agree with the bulk ethanol/CO_2 permittivities, a local solvent enrichment factor, indicative of cosolvent-fluid compositional augmentation, was needed to describe the data.

The local microenvironment surrounding pyrene molecules perturbs the p-electronic orbitals via Herzberg-Teller symmetry distortions and this affects the B_u^2 and B_u^1 interstate coupling efficiency. This vibronic coupling enhances the otherwise forbidden transition from the B_u^2 excited state to the ground state, resulting in an I_1 emission band intensity (normally appearing near 373 nm) and excited-state singlet fluorescence lifetimes that are strongly dependent on the physicochemical properties of the local environment. The environmental sensitivity is manifest by changes in the pyrene I_1/I_3 band ratio and the excited-state fluorescence lifetime. Specifically, the pyrene I_1/I_3 increases with increasing solvent dipolarity. Thus, by measuring the pyrene I_1/I_3 ratio as a function of bulk fluid density and/or composition, insight can be gained into the local environment surrounding the pyrene within the supercritical fluid mixtures.

Brennecke and co-workers [16] have recently used the steady-state fluorescence emission of dilute pyrene solutions to determine the nature of solvation in sub- and supercritical fluid mixtures of CO_2 and CF_3H as a function of fluid density. Mixtures of CO_2 and CF_3H were chosen for this study because they have similar critical points (see Table 5.1) and their mixtures maintain highly compressible regions similar to neat supercritical fluids. These experiments showed that the pyrene I_1/I_3 always increased with fluid density at all temperatures and with all fluids (neat and mixtures). In addition, I_1/I_3 is linear with bulk fluid density in the higher density region (similar to pyrene in other fluids) [17]; however, there were substantial nonlinearities in I_1/I_3 in the density region near and below the critical density. This deviation in the low density region was used to estimate the local fluid composition/density surrounding the pyrene in the neat fluids as well as the fluid mixtures. The calculated ρ_{local}/ρ_{bulk} (a measure of the extent of density augmentation surrounding the pyrene molecules) for all systems was found to be smaller overall at higher temperatures and to reach a maximum well below the bulk critical density. The authors estimated that the solvent strength for CO_2 in a CF_3H rich mixture was nearly the same in the intermediate density region as pure CF_3H, indicating an exclusion of CO_2 in the pyrene solvation sphere and, hence, preferential solvation of the pyrene by the CF_3H. For the mixture in which CO_2 predominated, analysis of solvent strength also indicated preferential solvation by CF_3H in the high density region. However, in the intermediate pressure range, the solvation strength was more similar to a CO_2-like environment, indicating a depletion of CF_3H before returning to a more CF_3H rich environment at lower bulk fluid densities.

Fluorescence quantum yields and excited-state lifetimes of a model solute (9-cyanoanthracene, 9cA) dissolved in supercritical C_2H_6, CO_2, and CF_3H over a wide density range were used by Rice et al. [18] to quantify energy dissipation processes within solute-fluid "clusters." The fluorescence quantum yield (Φ) of a solute is related to its excited-state fluorescence lifetime (τ) and its radiative decay rate (k_r) by the following equation:

$$k_r = \Phi/\tau \tag{5.1}$$

where the radiative decay rate (k_r) is related to the non-radiative decay rate (k_{nr} where $k_{nr} = 1/\tau - k_r$). Thus, by determining the fluorophore quantum yield and excited-state fluorescence lifetime, one can access directly how the solute dissipates energy to the solvent bath.

The 9CA quantum yield is effectively unity in liquids. However, the experimentally determined 9CA quantum yields in supercritical C_2H_6, CO2, and CF_3H are significantly less than unity at low fluid densities before approaching unity in the high density liquid-like region. Rice et al. also found that k_r and k_{nr} were strongly density dependent, with the nonradiative rate dominating in the low density region and the radiative rate dominating in the high density region. The Strickler-Berg relationship provides a theoretical construct between the measured fluorophore radiative decay rate and the solvent physical properties:

$$k_r = 2900 \; n^2 v_0^2 \int \varepsilon \, dv \tag{5.2}$$

In this expression n denotes the solvent refractive index, v_0 is the peak frequency of the fluorophore absorption spectrum, and $\int \varepsilon \, dv$ is the integrated area under a curve of the molar extinction coefficient as a function of wavenumber. In liquids the 9CA radiative rate is well described by Eq. (5.2). Rice et al. found that significant deviations from the Strickler-Berg expression occurred in the low density regions for each of the fluids and that corrections accounting for the change in refractive index and the change in absorbance shift for 9CA could not account fully for the observed deviations. The authors proposed that changes in the 9CA molar absorptivity in the low density region below the critical point is partially responsible for the deviation from the Strickler-Berg equation. The authors also argued that the dependence of k_{nr} and k_r on bulk fluid density was ultimately a result of local solute-fluid solvation and density-dependent, fluid-induced changes in the Franck-Condon factors between the 9CA S_1 singlet state and a nearby triplet state T_2.

Anderton and Kauffman used fluorescence depolarization measurements to examine the effects of solute functionality on fluid density augmentation in supercritical CO_2 [19]. By using two fluorescent solutes that have structures differing primarily by only a hydroxyl moiety (*trans, trans*-1,4-diphenylbutadiene (DPB) and *trans*-4-(hydroxymethyl)stilbene (HMS)) the authors were able to determine how a single functional group affected the degree of local density augmentation in CO_2. The authors found that, over a CO_2 density range from 0.3 to 0.8 g/ml, DPB exhibits a slight increase in rotational reorientation time (ϕ) as a function of fluid density. In contrast, HMS (the hydroxylated analog) exhibits a much larger increase in its rotational reorientation time as the fluid density increased. Anderton and Kauffman proposed a model to interpret this rotational data based on a radial distribution function formalism in which the local solvent density, $\varrho_{1,2}^{l}(r)$, surrounding a solute is related to the bulk fluid density (ϱ) and pair distribution function, $g_{1,2}(r)$, (where r is the distance between solvent, molecule 1, and solute, molecule 2):

$$\varrho_{1,2}^{l}(r) = [1 + F(g_{1,2}(r))] \tag{5.3}$$

and $F(g_{1,2}(r))$ is an integral equation in $g_{1,2}(r)$ over the spatial coordinates that describe the local solvent density surrounding the solute molecule. By modeling the boundary conditions for the system as a function of the local fluid density, the solvent density augmentation parameter, $F(g_{1,2}(r))$, was used to predict the rotational reorientation times which were in turn used with the experimentally recovered solute rotational reorientation times to estimate the extent of solute-fluid density augmentation. In this formalism, when $F(g_{1,2}(r)) = 0$, the local density equals the bulk fluid density (i. e., no solute-fluid density augmentation is occurring). In contrast, when $F(g_{1,2}(r)) = 1$ the local fluid density is twice the bulk density. The authors found that the local fluid density surrounding HMS was approximately 40 % greater than that surrounding DPB, but there appeared to be no bulk density dependence in the degree of local density augmentation. The differences between HMS and DPB were explained in terms of an increase in degree of CO_2 local density augmentation surrounding the HMS hydroxyl moiety resulting from specific intermolecular interactions between CO_2 and HMS.

Heitz and Bright [20] used the Anderton and Kauffman model to explain the rotational reorientation of a large model solute (*N,N'*-bis(2,5-*tert*-butylphenyl)-3,4,9,10-perylenecarboxodiimide (BTBP)) in supercritical CO_2, CF_3H, and C_2H_6. The rotational reorientation time (ϕ) for BTBP in liquids was shown to follow simple hydrodynamic theory and obeyed the Debye-Stokes-Einstein (DSE) equation:

$$\phi = \eta V/RT \tag{5.4}$$

where η is the solvent viscosity, V is the volume of the reorienting species, R is the gas constant, and T is the absolute temperature. In supercritical fluids, the authors found that the BTBP measured rotational reorientation time deviated significantly from Debye-Stokes-Einstein predictions in the low density fluid region. At higher fluid densities/viscosities, the experimentally measured BTBP rotational reorientation times were well described by the DSE equation; however, at low fluid densities nearer the critical point, the BTBP rotational reorientation times were much higher than predicted. This was attributed to local density augmentation of the fluid about the BTBP which led to slower rotational reorientation times via either an increased reorienting species volume or an increased local viscosity surrounding the BTBP molecule compared to the bulk fluid. Using the Anderton and Kauffman model to determine the degree of local density augmentation surrounding the BTBP, the authors estimated that the extent of local density augmentation surrounding BTBP was approximately three times the bulk fluid density. Assuming that the deviation from DSE was a result of a larger rotating unit (i.e., BTBP + fluid molecules), the solute-fluid cluster size was estimated. The authors estimated that the BTBP-fluid cluster radius was approximately double the BTBP radius alone in CF_3H at the largest degree of density augmentation. Using the van der Waals volume for the fluid and the volume of the solute cluster, it was estimated that as many as three layers of fluid molecules were in the immediate solvation shell surrounding the BTBP in the low density region.

Although supercritical water shows tremendous promise for hazardous waste disposal (see above) there have been a limited number of fluorescence or absorbance spectroscopic studies to date on supercritical water. Most of the dearth of spectroscopic data arises from the fact that it is relatively difficult to work with and generate supercritical water. Johnston and co-workers have developed special optical cells and used absorbance spectroscopy [21] in tandem with steady-state and time-resolved fluorescence [22, 23] to characterize the acid-base behavior of a model solute in supercritical water. This work has focused on the behavior of β-naphthol, a fluorescent compound in which the lowest excited singlet state is significantly more acidic ($pK_a^* = 2.5$) than its ground state ($pK_a = 9.5$) in liquid water near ambient conditions. Initially, the equilibrium constant for the reaction between β-naphthol and the hydroxyl ion was studied using the solvatochromic shifts of the β-naphtholate anion. More recently, Johnston's team used time-resolved fluorescence measurements to quantify the excited-state deprotonation of β-naphthol in sub- and supercritical water solutions [22]. Although the proton transfer rate from β-naphthol to water follows Arrhenius behavior up to 353 K, the authors found that there were deviations from Arrhenius behavior above 383 K. On increasing the temperature to the critical point, the deviation from the Arrhenius model became much larger. This deviation was attributed to the unique properties of supercritical water that affected deprotonation at high temperatures in ways not well-described by extrapolation of the Arrhenius model. Most recently, Johnston's group extended this work to study excited-state proton transfer reactions from β-naphthol dissolved in supercritical water [23]. They determined that deprotonation of β-naphthol by acetate and borate anions in supercritical water leads to very small deviations from Arrhenius behavior. In contrast, deprotonation by ammonia or water leads to much greater deviations from the model. These results were explained in terms of the dependence of these reaction rates on changes in electric charge throughout the reaction where isocoulombic reactions differ little from Arrhenius behavior. The authors also showed that extrapolation of low temperature measurements to higher temperatures can approximate kinetic behavior well at high temperatures for the isocoulombic reactions of β-naphthol with acetate and borate anions.

5.5
Laser Flash Photolysis

Laser flash photolysis (LFP) and transient absorption measurements have been used for the quantification of energy transfer and diffusion-controlled reactions in supercritical fluids [24–31]. For example, Worrall and Wilkinson [24] have studied the rate of triplet-triplet energy transfer for the anthracene/azulene and benzophenone/naphthalene donor/acceptor systems and determined how the reaction rate is affected by cosolvent concentration in $scCO_2$. By first studying the effect of changing cosolvent (CH_3CN or n-hexane) on the anthracene/azulene system, the authors determined that the rate of energy transfer for this system was not affected by the cosolvent used. Further, by studying the rate of energy transfer for the benzophenone/naphthalene system in liquid n-hexane, the rate of energy transfer was shown to be statistically similar for both systems for a given mol fraction of n-hexane modifier. The authors also showed that above 0.5 mol fraction modifier the diffusion rates for all systems were nearly identical to those in the neat cosolvent. Rate constants for energy transfer were calculated for each of the systems, and from this, values for the probability of energy transfer (p) was calculated. As the limit of zero mol fraction was approached, the value for the probability of energy transfer exceeded the value measured in neat modifier. These deviations at low mol fractions were explained by a model wherein the local concentration of modifier (and quencher) surrounding the solute was greater than the concentration in the bulk, possibly resulting in the greater energy transfer efficiency observed.

Brennecke and coworkers have done extensive work using laser flash photolysis to study diffusion-controlled reactions in supercritical fluids [25–29]. However, the authors have had to estimate the true solute diffusion coefficients in the supercritical fluid mixtures because there is significant deviation between the diffusion coefficients predicted by hydrodynamic theories and the available true values that have been measured in supercritical fluid solutions near and below the critical points. In one series of experiments, Brennecke and coworkers used the triplet-triplet annihilation of benzophenone (^3BP) and self-termination of the benzyl radical to investigate how supercritical fluids might affect diffusion-controlled processes [25, 26]. The triplet-triplet annihilation of ^3BP and the benzyl radical recombination reactions are bimolecular processes that are diffusion controlled in liquids. The triplet-triplet annihilation of ^3BP exhibited clean second order decay kinetics in supercritical CO_2, CF_3H, and in 1 mol% CH_3CN-modified CO_2. Further, there was no indication that addition of a cosolvent in any way slowed or impeded the diffusion-controlled reaction. The benzyl radical termination reaction also exhibited clean second order kinetics at the diffusion-controlled limit in supercritical CO_2 and C_2H_6 when spin statistical factors were taken into account. In addition, the photodecomposition of dibenzylketone (DBK) and subsequent decarbonylation of the phenylacetyl radical showed that the rate constants were approximately the same between the reaction in CO_2, C_2H_6, CF_3H, and CH_3CN-modified CO_2. All of this evidence combines to show that diffusion-controlled reactions, despite increases in local solvent density surrounding a solute near the critical point, exhibit no significant change in reaction kinetics due to any local phenomena. That is, because these reactions occur at the diffusion-controlled limit in neat and/or cosolvent-modified fluids, there was no indication that solvent-solute or solute-solute interactions changed the reaction kinetics. In addition, adding small amounts of a cosolvent were shown to have no effect on the diffusion-controlled reaction rates.

5.6
Basic Picture Revealed by These Studies

Based on the results from a wide range of spectroscopic techniques, a clearer view of the molecular-level interactions occurring within these supercritical fluids has begun to emerge [32]. From initial work involving absorbance and fluorescence measurements of solvatochromic dyes in supercritical fluids, the solute-fluid density augmentation process has become better understood in simple fluid systems involving a single dissolved solute. It is known that an increase in local density with respect to the bulk occurs near the fluid critical point with a maximum in this local density occurring at around one-half the critical density. Moreover, the addition of a cosolvent has been shown to increase the local cosolvent composition surrounding a dissolved solute near the critical point which leads to preferential solvation of a solute by the cosolvent that can be controlled by adjusting the fluid pressure.

5.7
The Future

Although we have learned much about supercritical fluids over the past decade, we will continue to see new efforts to understand ever more complex, multi-component fluid systems. We can foresee three major areas where spectroscopy will impact supercritical fluid science and technology.

Several research groups have recently shown that one can form thermodynamically stable reverse micelles and microemulsions in $scCO_2$ [33–37]. The most widely reported of these new micelle systems are those based on PFPE (perfluoropolyether ammonium carboxylate) surfactants [33, 38–40]. Over the past few years several key aspects of the $PFPE/scCO_2/H_2O$ system have been investigated, including their phase equilibria, water pool and micelle shape, and internal dynamics [38], the pH within an unbuffered PFPE water pool [39] and a means to control the water pool pH within PFPE micelles [40]. Several researchers have also shown that one can use the $PFPE/CO_2/H_2O$ system to perform inorganic [41] and organic [42, 43] reactions directly within the micelle water pool. Researchers have also used these micelles to prepare tailored, uniform CdS nanoparticles [44]. Finally, researchers have performed enzymatic reactions within micelles/microemulsions formed in liquid [45] and $scCO_2$ [46]. We envisage questions on the behavior of proteins and the like within these microemulsions to be addressed by optical spectroscopic methods.

In solution, polymer behavior can be controlled by the choice of solvent, solution composition, or temperature. For example, the tail-tail segment dynamics of a polymer dissolved in a "good" solvent are significantly different to those of the same polymer dissolved in a θ or "bad" solvent. Given this, one can clearly control polymer dynamics, especially at the tail segments, by adjusting the solvent properties. DeSimone and his group have shown [47, 48] that one can perform homogeneous free radical polymerizations and the like by using environmentally friendly near- and supercritical CO_2 as the reaction solvent. The results of these efforts have been remarkable and they demonstrate the tremendous potential of supercritical CO_2 as a polymerization solvent [47, 48]. However, even though the promise and potential of polymerization reactions in supercritical CO_2 are great, there have only been a limited number of reports [49–55] on the molecular-level events that dilute polymer solutions undergo within a supercritical fluid. It seems clear that spectroscopic measurements can be used to help elucidate polymer behavior within supercritical fluid systems.

The interactions between an enzyme and a supercritical fluid are of great interest to many researchers because the inherent tunability of the fluid continuous phase (Fig. 5.2) could provide a means to tune enzyme activity. In an effort to determine how a supercritical fluid affects the conformational changes of an enzyme, Ikushima et al. [56, 57] have used FTIR spectroscopy to determine the interactions between a model enzyme (*Candida cylindracea* (CCL)) and supercritical CO_2. The enzyme amide infrared absorption bands were used to provide information about the secondary protein structure. The authors determined that there was a decrease in α-helical structure within the enzyme near the CO_2 critical point. This result was related to the opening of the "lid" within the enzyme that exposed possible active sites for the biocatalysis reaction. Thus, near the CO_2 critical point, these large conformational changes in the enzyme were likened to a "stereoselective machinery" capable of tuning the product enantiomeric excess. Again, this and other systems are amenable to spectroscopic investigation.

In summary, we envision fluorescence continuing to play an important role in the determination of fundamental processes of complex systems within supercritical fluid systems.

Acknowledgements. We thank the Division of Chemical Sciences, Office of Basic Energy Sciences, Office of Energy Research, United States Department of Energy (DEFGO290ER14143), and the American Chemical Society Analytical Division Fellowship (to EDN) sponsored by DuPont for support of our spectroscopic work in supercritical fluids.

References

1. Via J, Taylor LT (1993) Solving process problems with supercritical-fluid extraction. CHEMTECH 23:38–44
2. Johnston KP, Penninger JML (1989) Supercritical fluid science and technology, vol 406. American Chemical Society, Washington, DC
3. Bruno TJ, Ely JF (1991) Supercritical fluid technology – reviews in modern theory and applications. CRC Press, Boca Raton, FL
4. Bright FV, McNally MEP (1992) Supercritical fluid technology – theoretical and applied approaches in analytical chemistry, vol 488. American Chemical Society, Washington, DC
5. Kiran E, Brennecke JF (1993) Supercritical fluid engineering science – fundamentals and applications, vol 514. American Chemical Society, Washington, DC

6. Jessop PG, Leitner W (1999) Chemical synthesis using supercritical fluids. Wiley-VCH, Germany
7. Kauffman JF (1996) Spectroscopy of solvent clustering in supercritical fluids. Anal Chem 68:248A–253A
8. Tucker SC, Madox MW (1998) The effect of solvent density inhomogeneities on solute dynamics in supercritical fluids: a theoretical perspective. J Phys Chem B 102:2437–2453
9. Kim S, Johnston KP (1987) Clustering in supercritical fluid mixtures. AIChE J 33: 1603–1611
10. Eckert CA, Knutson BL (1993) Molecular charisma in supercritical fluids. Fluid Phase Equil 83:93–100
11. Petsche IB, Debenedetti PG (1989) Solute-solvent interactions in infinitely dilute supercritical mixtures: a molecular dynamics investigation. J Chem Phys 91:7075–7084
12. Lakowicz JR (1999) Principles of fluorescence spectroscopy, 2nd edn. Kluwer, New York, NY
13. Knutson BL, Sherman SR, Bennet KL, Liotta CL, Eckert CA (1997) Benzophenone as a probe of local cosolvent effects in supercritical ethane. Ind Eng Chem Res 36:854–868
14. Schulte RD, Kauffman JF (1994) Solvation in mixed supercritical fluids –TICT spectra of bis(4,4'-aminophenyl) sulfone in ethanol/CO_2. J Phys Chem 98:8793–8800
15. Schulte RD, Kauffman JF (1995) Fluorescence from the twisted intramolecular charge-transfer compound bis(4,4'-dimethylaminophenyl)sulfone in ethanol/CO_2 – a probe of local solvent composition. Appl Spectrosc 49:31–39
16. Zhang J, Lee LL, Brennecke JF (1995) Fluorescence spectroscopy and integral-equation studies of preferential solvation in supercritical-fluid mixtures. J Phys Chem 99: 9268–9277
17. Rice JK, Niemeyer ED, Dunbar RA, Bright, FV (1995) State-dependent solvation of pyrene in supercritical CO_2. J Am Chem Soc 117:5832–5839
18. Rice JK, Niemeyer ED, Bright FV (1996) Solute-fluid coupling and energy dissipation in supercritical fluids: 9-cyanoanthracene in C_2H_6, CO_2, and CF_3H. J Phys Chem 100: 8499–8507
19. Anderton RM, Kauffman JF (1995) Rotational relaxation in the compressible region of CO_2 – evidence for solute-induced clustering in supercritical-fluid solutions. J Phys Chem 99:13,759–13,762
20. Heitz MP, Bright FV (1996) Probing the scale of local density augmentation in supercritical fluids: a picosecond rotational reorientation study. J Phys Chem 100:6889–6897
21. Bennett GE, Johnston KP (1994) UV-visible absorbency spectroscopy of organic probes in supercritical water. J Phys Chem 98:441–447
22. Green, S, Xiang T, Johnston JP, Fox MA (1995) Excited-state deprotonation of beta-naphthol in supercritical water. J Phys Chem 99:13,787–13,795
23. Ryan ET, Xiang T, Johnston KP, Fox MA (1996) Excited-state proton transfer reactions in subcritical and supercritical water. J Phys Chem 100:9395–9402
24. Worrall DR, Wilkinson F (1996) Photochemistry in modified supercritical carbon dioxide – effect of modifier concentration on diffusion probed by triplet-triplet energy transfer. J Chem Soc Faraday Trans 92:1467–1471
25. Roberts CB, Zhang J, Chateauneuf JE, Brennecke JF (1993) Diffusion-controlled reactions in supercritical CHF_3 and CO_2/acetonitrile mixtures. J Am Chem Soc 115:9576–9582
26. Roberts CB, Zhang J, Brennecke JF, Chateauneuf, JE (1993) Laser flash-photolysis investigations of diffusion-controlled reactions in supercritical fluids. J Phys Chem 97:5618–5623
27. Roberts CB, Chateauneuf JE, Brennecke JF (1992) Unique pressure effects on the absolute kinetics of triplet benzophenone photoreduction in supercritical CO_2. J Am Chem Soc 114:8455–8463
28. Roberts CB, Brennecke JF, Chateauneuf JE (1993) Spectral shifts in the triplet-triplet absorption-spectrum of anthracene in supercritical fluids. J Chem Soc Chem Comm 868–869
29. Roberts CB, Brennecke JF, Chateauneuf JE (1995) Solvation effects on reactions of triplet benzophenone in supercritical fluids. AIChE J 41:1306–1318

30. Ji Q, Eyring EM, vanEldick R, Johnston KP, Goates SR, Lee ML (1995) Laser flash-photolysis studies of metal-carbonyls in supercritical CO_2 and ethane. J Phys Chem 99: 13,461–13,466
31. Ji Q, Lloyd CR, Eyring EM, vanEldick R (1997) Probing the solvation properties of liquid versus supercritical fluids with laser flash photolysis of $W(CO)_6$ in the presence of 2,2'-bipyridine. J Phys Chem A 101:243–247
32. Brennecke JF, Chateauneuf JE (1999) Homogeneous organic reactions as mechanistic probes in supercritical fluids. Chem Rev 99:433–452
33. Johnston KP, Harrison KL, Clarke MJ, Howdle SM, Heitz MP, Bright FV, Carlier C, Randolph TW (1996) Water in carbon dioxide microemulsions: an environment for hydrophiles including proteins. Science 271:624–626
34. McClain JB, Betts DE, Canelas DA, Samulski ET, De Simone JM, Londono JD, Cochran HD, Wignall GD, Chillura-Martino D, Triolo R (1996) Design of nonionic surfactants for supercritical carbon dioxide. Science 274:2049–2052
35. Eastoe J, Cazelles BMH, Steytler DC, Holmes JD, Pitt AR, Wear TJ, Heenan RK (1997) Water-in-CO2 microemulsions studied by small-angle neutron scattering. Langmuir 13:6980–6984
36. Ghenciu EG, Russell AJ, Beckman EJ, Steele L, Becker NT (1998) Solubilization of subtilisin in CO_2 using fluoroether-functional amphiphiles. Biotechnol Bioengr 58:572–580
37. Salaniwal S, Cui ST, Cummings PT, Cochran HD (1999) Self-assembly of reverse micelles in water/surfactant/carbon dioxide systems by molecular simulation. Langmuir 15: 5188–5192
38. Heitz MP, Carlier C, de Grazia J, Harrison KL, Johnston KP, Randolph TW, Bright FV (1997) Water core within perfluoropolyether-based microemulsions formed in supercritical carbon dioxide. J Phys Chem 101:6707–6714
39. Niemeyer ED, Bright FV (1998) The pH within PFPE reverse micelles formed in supercritical CO_2. J Phys Chem B 102:1474–1478
40. Holmes JD, Ziegler KJ, Audriani M, Lee CT Jr, Bhargava PA, Steytler DC, Johnston KP (1999) Buffering the aqueous phase pH in water-in-CO2 microemulsions. J Phys Chem B 103:5703–5711
41. Clarke MJ, Harrison KL, Johnston KP, Howdle SM (1997) Water in supercritical carbon dioxide microemulsions: spectroscopic investigation of a new environment for aqueous inorganic chemistry. J Am Chem Soc 119:6399–6406
42. Jacobson GB, Lee CT Jr, Johnston KP (1999) Organic synthesis in water carbon dioxide microemulsions. J Org Chem 64:1201–1206
43. Jacobson GB, Lee CT Jr, da Rocha RP, Johnston K P (1999) Organic synthesis in water carbon dioxide emulsions. J Org Chem 64:1207–1210
44. Holmes JD, Bhargava PA, Krogel BA, Johnston KP (1999) Synthesis of cadmium sulfide Q particles in water-in-CO_2 microemulsions. Langmuir 15:6613–6615
45. Holmes JD, Steytler DC, Rees GD, Robinson BH (1998) Bioconversions in a water-in-CO_2 microemulsion. Langmuir 14:6371–6376
46. Kane MA, Baker, GA, Pandey, S, Bright FV (2000) Performance of cholesterol oxidase sequestered within reverse micelles formed in supercritical carbon dioxide. Langmuir (in press)
47. Cooper AI, DeSimone JM (1996) Polymer synthesis and characterization in liquid/supercritical carbon dioxide. Cur Opin Solid State Mater Sci 1:761–768
48. Kendall JL, Canelas DA, Young JL, DeSimone JM (1999) Polymerizations in supercritical carbon dioxide. Chem Rev 99:543–563
49. Melnichenko YB, Kiran E, Wignall GD, Heath KD, Salaniwal S, Cochran HD, Stamm M (1999) Pressure- and temperature-induced transitions in solutions of poly(dimethylsiloxane) in supercritical carbon dioxide. Macromolecules 32:5344–5347
50. Gromov DG, de Pablo JJ (1998) Phase behavior of polymer-solvent mixtures. Fluid Phase Equil 151:657–665
51. Gromov DG, de Pablo JJ, Luna-Barcenas G, Sanchez IC, Johnston KP (1998) Simulation of phase equilibria for polymer-supercritical solvent mixtures. J Chem Phys 108:4647–4653

52. Luna-Barcenas G, Meredith JC, Sanchez IC, Johnston KP, Gromov DG, de Pablo JJ (1997) Relationship between polymer chain conformation and phase boundaries in a supercritical fluid. J Chem Phys 107:10,782–10,792
53. Luna-Barcenas G, Gromov DG, Meredith JC, Sanchez IC, dePablo JJ, Johnston KP (1997) Polymer chain collapse near the lower critical solution temperature. Chem Phys Lett 278:302–306
54. Gregg CJ (1994) PhD Thesis, Lehigh University and references cited therein
55. Poliakoff M, George MW, Howdle SM (1999) Proceeding of the 6th meeting on supercritical fluids: Chemistry and Materials, Nottingham, UK
56. Ikushima Y, Saito N, Hatakeda K, Sato O (1996) Promotion of a lipase-catalyzed esterification in supercritical carbon dioxide in near-critical region. Chem Eng Sci 51:2817–2822
57. Ikushima Y, Saito N, Arai M, Blanch HW (1995) Activation of a lipase triggered by interactions with supercritical carbon dioxide in the near-critical region. J Phys Chem 99:8941–8944

CHAPTER 6

Space and Time Resolved Spectroscopy of Two-Dimensional Molecular Assemblies

H. Laguitton Pasquier, D. Pevenage, P. Ballet, E. Vuorimaa,
H. Lemmetyinen, K. Jeuris, F.C. de Schryver, M. Van der Auweraer

The fluorescence decays of several amphiphilic dyes incorporated in Langmuir-Blodgett films were determined and analyzed globally over different dye concentrations and/or emission wavelengths. The highly non-exponential character of the fluorescence decay could be analyzed in a model based on two-dimensional energy transfer from the excited monomers to dimers of the dye. Depending upon the miscibility of the dye and the matrix, homogeneous and two-phase multilayers were distinguished. This distinction could be confirmed by the spatial distribution of the fluorescence as revealed by confocal fluorescence microscopy and near-field scanning optical microscopy. For chromophores with important internal rotation (amphiphilic crystal violet) the fluorescence decay at low concentrations is determined by a distribution of free volumes rather than by energy transfer to dimers. The homogeneous distribution assumed for an amphiphilic crystal violet in cadmium arachidate could be confirmed by the global analysis of the fluorescence decay of an amphiphilic pyronine in the presence of the amphiphilic crystal violet.

6.1
Introduction

6.1.1
Motivation

During the last few decades there has been strong interest in photophysical processes occurring in supramolecular assemblies, prepared on planar substrata by the Langmuir-Blodgett technique [1–4]. The fixed interlayer distance and restricted orientational freedom of the incorporated chromophores make those assemblies very suitable to prepare model systems for the study of excitation and electron transfer. Sophisticated devices based on multistep excitation and electron transfer were developed and investigated from both theoretical [5, 6] and experimental points of view [7–12]. Most of those experiments were performed at a relatively elevated concentration of chromophores (5–20 mol%) where molecular aggregation [1, 13, 14] and inter- and intralayer electron and energy transfer can no longer be neglected [15–22]. The latter phenomena, which are also observed in other systems, e.g., in membranes and vesicles and bilayers [23, 24], are investigated preferentially by analysis of the fluorescence decays [25–34] to elucidate the role of the chemical structure of the matrix and the incorporated chromophores. Those investigations often indicated a behavior more complex than quenching by energy transfer to dye aggregates distributed homogeneously over the films. As an inhomogeneous distribution of the dissolved chromophores was observed in Langmuir-Blodgett films of cadmium arachidate [35], we started to investigate the photophysical properties of octa-decyl substituted rhodamines in Langmuir-Blodgett films of phospholipids, where a more homogeneous dispersion can be expected [29, 36–38]. Such a homogeneous distribution which would allow for a more straightforward analysis of photophysical and photochemical processes would be the preferred situation for the development and characterization of devices based on those Langmuir-Blodgett films [1–4, 7–12].

Optical microscopy, in particular scanning confocal fluorescence microscopy [39–42], is now widely used in biological disciplines. To our knowledge, only a limited number of studies on the Langmuir-Blodgett film structure by confocal fluorescence microscopy has been reported [43]. As an alternative to confocal microscopy, near field scanning microscopy (NSOM) [44–46] can be used to investigate Langmuir-Blodgett films [47,48]. By combining the analysis of fluorescence decays with the spatial information available from scanning microscopies it will be attempted to determine the distribution of dye molecules over Langmuir-Blodgett films and to elucidate some of the molecular parameters controlling this distribution.

6.1.2
Models

In nearly all combinations of dyes and matrices a non-exponential fluorescence decay of the incorporated dye was observed. A first possibility for the non-exponential fluorescence decay is energy transfer to dimer or aggregates of the dye [24–34]. At low dye concentrations one can assume that this process will mainly occur by direct energy transfer (DET) rather than hopping of the excitation over the dye monomers [23, 49, 50] until trapping by a dimer occurs. The deviations of the fluorescence decay from the expression predicted for Förster type energy transfer in two dimensions [51, 52] were often attributed to an inhomogeneous distribution of the dye over the matrix. However often it was found that besides the stretched exponential decay the fluorescence decay contained another component that decayed significantly slower. This led to the assumption that the Langmuir-Blodgett film contained two different phases: one phase with a relatively high concentration of quenching sites and a second phase with a very low concentration of quenching sites (Eqs. 6.1 and 6.2):

$$I(t) = I(0) \left\{ \alpha \exp\left[-\left(\frac{t}{\tau_D}\right) - \gamma_2 t^{1/3} \right] + (1 - \alpha) \exp\left[-\left(\frac{t}{\tau_D}\right) \right] \right\} \tag{6.1}$$

$$\gamma_2 = \Gamma(2/3) \, \pi \, \sigma R_0^2 (\tau_D)^{-1/3} \tag{6.2}$$

τ_D (ns) represents the decay time of the donor chromophore in the absence of acceptors, while σ and R_0 correspond to the surface density (cm^{-2} or Å^{-2}) of acceptors and the critical distance R_0 (Å) for energy transfer. α corresponds to the fraction of probes in a phase with a high concentration of quenching sites, while $1 - \alpha$ corresponds to the fraction in a phase with a very low concentration of quenching sites.

An alternative explanation for the deviations from the decay expected for Förster transfer in a homogeneous two-dimensional system was based on a self-similar distribution with fractal dimension $\bar{d}$ of the acceptor around the donor. In this case one obtains for the fluorescence decays

$$I(t) = I(0) \exp\left[-\left(\frac{t}{\tau_D}\right) - \gamma_{\bar{d}} t^{\bar{d}/6}\right] \tag{6.3}$$

$$\gamma_{\bar{d}} = \left(\frac{d}{\bar{d}}\right) \Gamma\left(1 - \frac{\bar{d}}{6}\right) V_d \varrho_0 r_0^{(d-\bar{d})} R_0^{\bar{d}} (\tau_D)^{-\bar{d}/6} \tag{6.4}$$

where $\gamma_{\bar{d}}$ is a parameter, depending on the Euclidean dimensionality d, the fractal dimensionality $\bar{d}$, the Euler-gamma function Γ, $\varrho(r_0)$ the density of quenchers at a reference distance r_0 (Å) from the donor probe. V_d is the unit volume in a d-dimensional space.

As the singlet decay time of some of the probes used (e.g., the crystal violet derivative CV18 [53–56] or the monomethine cyanine THIAM18 [43]) exhibits an important dependence on viscosity or free volume one could in a heterogeneous medium take into account that the distribution of free volume can lead to a distribution of the decay rates of those flexible dyes [57]. When a Gaussian distribution is assumed for the decay rates, the following expression should be obtained for the decay in absence of energy transfer from monomers to dimers [58]:

$$I(t) = I(0) \exp\left(-k_0 t\right) \exp\left[\frac{t^2 \sigma_g^2 - 2\mu t}{2}\right] \text{erfc}\left[-\frac{\mu}{\sqrt{2}\sigma_g} + \frac{\sigma_g t}{\sqrt{2}}\right] \tag{6.5}$$

k_0 designates the lowest decay rate constant, μ the average value of the decay rate constants, and σ_g the standard deviation on the distribution of the decay rate constants. When $\mu - k_0$ is much larger than σ_g, μ can be considered as some average decay rate constant and σ_g as a standard deviation. If this condition is not met, it is not possible to give a straightforward meaning to the parameters. Combining the Gaussian distribution of decay rates with Förster type energy transfer leads to

$$I(t) = I(0) \exp\left(-k_0 t\right) \exp\left[\frac{t^2 \sigma_g^2 - 2\mu t}{2}\right] \text{erfc}\left[-\frac{\mu}{\sqrt{2}\sigma_g} + \frac{\sigma_g t}{\sqrt{2}}\right] \tag{6.6}$$

$$\{\exp\left(-\gamma_2 t^{1/3}\right) + B\}$$

where B allows one to take into account the presence of monomers not quenched by energy transfer.

By combining different dyes and matrices it was possible, as discussed below, to obtain fluorescence decays characteristic of a two-phase system, a homogeneous distribution of quenching sites, and a distribution of decay rates. While the fluorescence decays were generally determined for multilayers, the alternation of layers containing the dye by layers containing only the matrix limited the excitation transfer to intralayer transfer [29, 43, 59, 60]. For some of the combinations investigated control experiments indicated that the fluorescence decays of the multilayers and of single monolayers were – within the experimental error – identical. For some combinations of dyes and matrices the mixing behavior predicted by the analysis of the fluorescence decays was confirmed by the spatial distribution of the fluorescence in single monolayers, as determined by "near-field scanning optical microscopy" or confocal fluorescence microscopy.

6.2
Experimental

The different dyes (Fig. 6.1) were synthesized as described elsewhere [29, 59–63]. The different matrices (Fig. 6.2) were obtained from Sigma Chemical Company or Fluka and were of the best grade commercially available. The multilayer systems were prepared on a KSV 5000 ALT trough by means of the Langmuir-Blodgett deposition method, using quartz substrata. The substrata were cleaned as described earlier [64]. The deposition conditions were the same as described in detail elsewhere [29, 43, 59, 60]. Unless indicated otherwise, the monolayers containing the chromophores were separated by monolayers of the matrix so as to avoid interlayer excitation transfer in the multilayer assemblies [29, 43, 59, 60]. The probe concentration never exceeded 5 mol% to minimize excitation hopping between the probes.

The absorption spectra were recorded with a DW-2000 Aminco spectrophotometer of Perkin Elmer "Lambda 6". An uncoated quartz slide was used as a reference. The corrected steady-state fluorescence and excitation spectra were determined on a SPEX Fluorolog in a 'front-face' configuration [43, 64] at reduced pressure (1 torr), where the quartz slides were positioned normal to the excita-

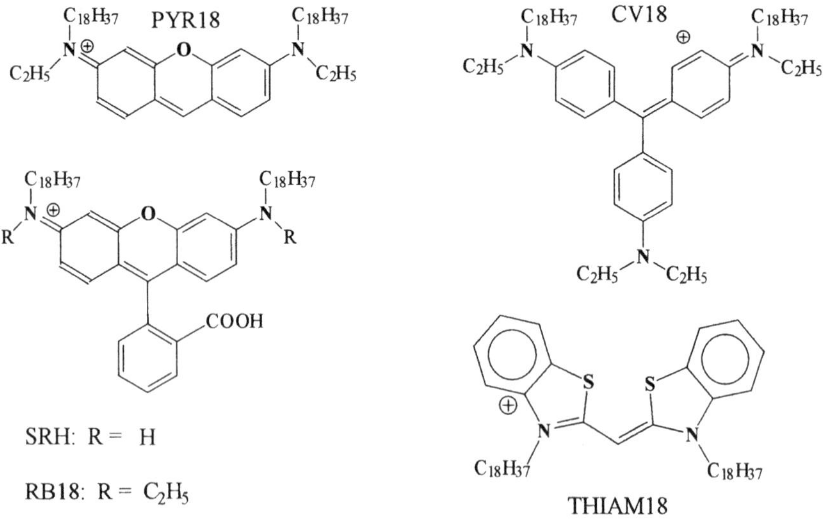

Fig. 6.1. Chemical structure of the probes and quenchers used

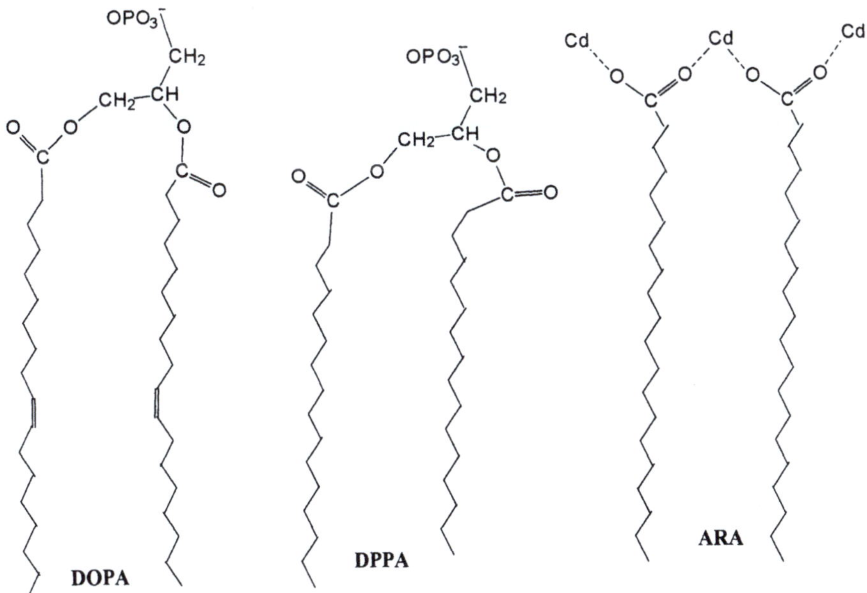

Fig. 6.2. Chemical structure of the lipids used

tion and the emission was detected at an angle of 26° with respect to the excitation.

The fluorescence decay curves were measured using time-correlated single-photon counting. For excitation at 575 nm a cavity-dumped dye laser of Rhodamine 6G (R6G Kodak), synchronously pumped by a mode-locked beam-locked Argon Ion laser (514 nm, Spectra Physics model 2080) was used as excitation source. To obtain an excitation wavelength of 545 nm, a pyrromethene-556 dye (P-556 Exciton Inc.) was used in the dye laser. The samples were also excited at 480 nm with a mode-locked frequency-doubled Titane-Saphire laser (Spectra Physics, model 3960), pumped with a continuous beam-locked Argon Ion laser (Spectra Physics, model 2080). Polarization of the excitation and emission was controlled so as to prevent a distortion of the fluorescence decays by fluorescence depolarization [29, 43, 59]. Laser pulses scattered from a ludox scatter solution in a 1-mm-cell were used for generating the instrument response functions necessary for the analysis of the fluorescence decays of the sample. The use of a reference compound, which decays very rapidly over only a few channels, is also a good alternative for the ludox suspension. The fluorescence decays were analyzed, either single curve or globally [65,66], by iterative reconvolution using a Marquardt algorithm [67].

The NSOM data were obtained with a Topometrics "Aurora" modified in the research group [46,68]. The spatial resolution amounted to 50–80 nm. The scan speed amounted to 2.2 μm/s. Fluorescence was detected through a dry 60× Nikon objective. To excite the SRH sample at 514 nm an Argon ion laser (Spectra Physics 2020) was used. A power of maximum 1 mW coupled into the tip. The fluorescence was detected through a notch filter (Kaiser Optical Systems, Inc. HNPF-514.5) and a 540 nm long-pass filter. To excite the RB18 sample at 543.5 nm a He:Ne laser was used as excitation source and a power of 0.65 mW was coupled into the tip through a cam splice with 50 % efficiency. The fluorescence was detected through a 570 nm long-pass filter.

6.3
Results and Discussion

6.3.1
Inhomogeneous Multilayers: RB18 and ARA

While the emission spectra of a mixed multilayer of RB18 and ARA shift to longer wavelengths upon increasing the dye concentration the excitation spectra shift to shorter wavelengths [59]. The red shift of the emission spectra corresponds to that observed by Tamai et al. [19] for a single monolayer and by Verschuere et al. for a multilayer assembly [31]. This blue shift of the excitation spectra does not agree with the generally observed bathochromic shift at higher concentrations, attributed to an increase of the polarizability of the multilayers [31, 62, 71]. The FWMH of the fluorescence spectrum amounts at a mixing ratio of 0.25 mol% to 1550 cm^{-1}, which corresponds to the value of 1540 cm^{-1}, found by Verschuere et al. for the same concentration [31]. The quantum yield of fluorescence decreases with increasing concentration of RB18 in the Langmuir-Blodgett film.

The fluorescence decay curves (Fig. 6.3) recorded for the RB18/ARA samples at 570 nm, 580 nm, 590 nm, 600 nm, and 620 nm) could be analyzed globally using Eqs. (6.1) and (6.2) and linking τ_D over different concentrations. At all dye concentrations the extra mono-exponential term was necessary to obtain a sa-

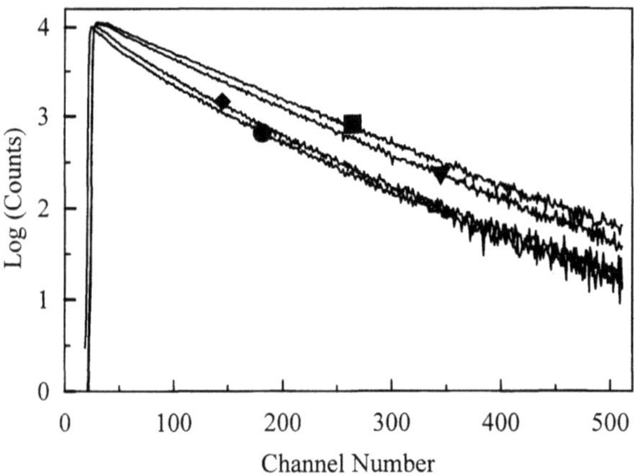

Fig. 6.3. Fluorescence decays recorded at 580 nm for different concentrations of the RB18/ARA Langmuir-Blodgett films (excitation at 540 nm). ■: 0.12 mol%, ▼: 0.26 mol%, ◆: 0.62 mol%, ●: 1.24 mol%. Time increment: 40 ps

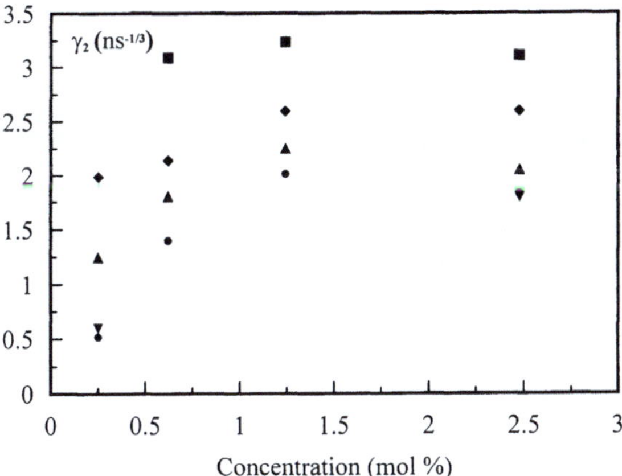

Fig. 6.4. The dependence of γ_2 upon the concentration for RB18/ARA in the Langmuir-Blodgett film (excitation at 540 nm). ■: 570 nm, ◆: 580 nm, ▲: 590 nm, l: 600 nm and ▼: 620 nm

tisfactory fit. τ_D equals the decay time of RB18 in solution (τ_{CHCl3}: 3.6 ns). No linear increase of γ_2 with the concentration can be observed from 0.1% to 2.48% (Fig. 6.4). Between 0.5 mol% and 2 mol% γ_2 is independent of concentration within experimental error. Although the concentration dependence of γ_2 at lower concentrations is less clear, there is no indication that upon decreasing the dye concentration γ_2 approaches asymptotically to zero or even to the values found for SRH in ARA. At concentrations of 0.5 mol% and higher, γ_2 decreases by about 30% when the emission wavelength is increased from 570 nm to 590 nm. Although this increase is only two or three times larger than the precision with which γ_2 is determined, it is observed at three different concentrations. Furthermore, this effect was not observed for SRH in ARA (see below).

While the pre exponential factor of the mono-exponential term in Eq. (6.1) does not change between 0.62 mol% and 2.48 mol% it increases sharply from the 0.62% over the 0.25% to the 0.12% samples [59]. The increase of $1-\alpha$ at low concentrations could suggest the presence of isolated monomers decaying mono-exponentially. It can however also be an artifact of the algorithm used to recover the parameters [72].

To study the dependence of the decay parameters recovered by global analysis on the emission wavelength, decay curves recorded at the same concentration and different emission wavelengths were analyzed globally using Eq. (6.1) and linking τ_D over the different emission wavelengths. The values of τ_D, γ_2, and $1-\alpha$ are almost identical to those obtained when the global analysis is performed with decays obtained from samples with the same emission wavelength and different concentrations The pre-exponential factor of the mono-exponential term $(1-\alpha)$ varies slightly with the emission wavelength at the lowest concentrations.

While γ_2 depends upon the critical distance for energy transfer and on the number of quenchers per unit area present in the film, $1-\alpha$ gives information on the presence of a phase containing isolated chromophores. Hence, the concentration dependence of these local parameters suggests an inhomogeneous distribution behavior of RB18 in the ARA matrix. Whether this inhomogeneous distribution relates to the presence of two phases in the true thermodynamic sense or rather of "pseudo phases" or regions is difficult to assess. In the diluted phase where monomers decay mono-exponentially, the distance between monomers must at all analytical dye concentrations exceed the critical distance for energy transfer. Extrapolating the change of $1-\alpha$ to lower concentrations indicates that a mono-exponential decay is approached asymptotically for the RB18/ARA Langmuir-Blodgett films upon decreasing the dye concentration. This is in agreement with the results of Tamai et al, but different from results of Ballet et al. where RB18 is mixed in a DPPA or DOPA matrix [29, 73]. Contrary to the fractal-like dimension invoked by Tamai et al. [19, 25], the dimension was fixed at two when fitting the fluorescence decays to Eq. (6.1) using global analysis. This means that when the heterogeneity of the layers is taken into account by the factor $1-\alpha$, the phase with a high dye concentration can be regarded as a two-dimensional mixture of monomers and dimers or higher aggregates. The wavelength and concentration dependence of γ_2 (Fig. 6.4) suggests that at concentrations of 0.5 mol% and higher the number density of quenchers, σ, does not depend upon the analytical dye concentration. This means that the "concentrated" phase is already saturated at a dye concentration of 0.5 mol%. Increasing the total dye concentration will increase the area covered by the "concentrated" phase and hence decrease $1-\alpha$.

Another possibility to explain the independence of γ_2 of the concentration, is a decrease of the critical distance of energy transfer due to a smaller overlap between the emission spectrum of the monomers or weakly coupled aggregates and the absorption spectrum of the strongly interacting dimers or aggregates, at higher concentration. This decreased overlap corresponds with the red shift of the emission spectra at higher concentration. A distribution of weakly coupled aggregates, or of monomers in different sites, implying that the fluorescence at longer emission wavelengths is due to monomers or weakly coupled aggregates, emitting at longer wavelengths can explain the decrease of R_0 and γ_2 with increasing emission wavelength [19]. The higher local concentration in the concentrated phase of the RB18/ARA-multilayer will make this effect more important than in the homogeneously distributed SRH/ARA multilayer (see below). These red-shifted monomers or random aggregates can be excited directly or by energy transfer from other more energetic monomers (the latter only at the high end of the dye concentrations used).

The inhomogeneous films postulated from the concentration and wavelength dependence of γ_2 and $1-\alpha$ can be confirmed from the spatial resolved fluorescence of an ARA monolayer containing 2 mol% RB18 (Fig. 6.5). Irregularly shaped dark spots can be observed in continuous matrix showing a light intensity that is about 2–4 times larger. It is difficult to link the fluorescence intensities of the different phases directly to the fluorescent quantum yield of the RB18 molecules, which are at a concentration of 2 mol% quenched by 80% in the "concentrated phase" compared to the diluted phase [59]. To convert those data into fluorescent intensities one should also know the light absorption of both phases. If the highly luminescent phase would be the one containing a high dye concentration the absorbance should be about 10–20 times larger than that of the

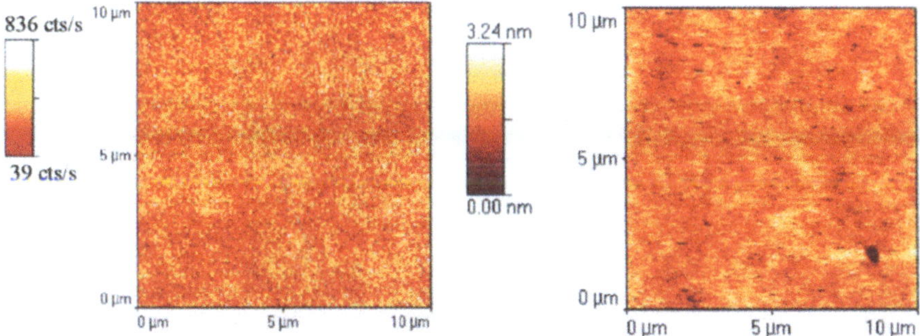

Fig. 6.5. NSOM (*left*) and topographic images (*right*) of a single monolayer of ARA containing 2 mol% RB18 deposited on a quartz slide. Excitation occurred at 543.5 nm using an He:Ne laser

"dark" spots to allow for the difference in fluorescence intensity. On the other hand, if the dark spots were the highly concentrated phase the local absorbance should be at the most 100% larger than in the highly luminescent phase. In the latter case it is however difficult to explain the pronounced quenching in the "concentrated phase." The NSOM images of a monolayer of 2 mol% RB18 in DPPA (Fig. 6.6), for which a similar concentration dependence of the decay parameters was observed [29] show a similar contrast between a luminous background and dark spots, the latter having however a sharper boundary in a DPPA matrix.

In addition, for THIAM18 in ARA and DPPA both the concentration dependence of the decay parameters recovered from the global analysis of the fluorescence decay and the spatial distribution of the fluorescence obtained using confocal microscopy suggest an inhomogeneous distribution of the THIAM18 over ARA and DPPA (Fig. 6.7) matrices [43]. For the same probe confocal microscopy suggests a homogeneous distribution in a DOPA matrix. The very small fluorescence quantum yield (< 0.01) of THIAM18 in nonviscous solvents suggests that the fluorescence decay of this dye could be influenced by the free volume and the distribution thereof. Therefore, it was also attempted to analyze the fluorescence decays of THIAM18 in ARA or DPPA, obtained at different concentrations in the

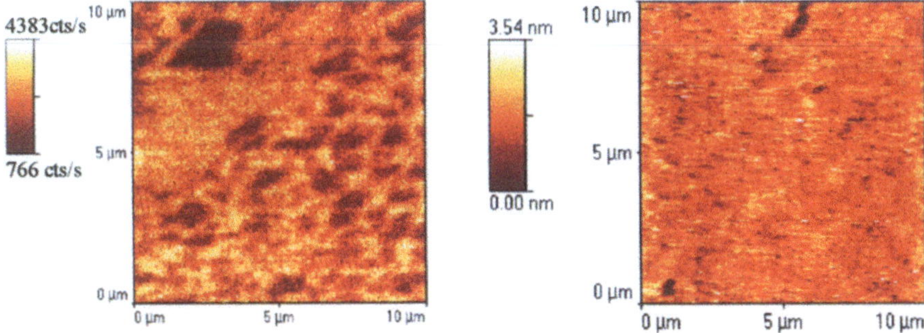

Fig. 6.6. NSOM (*left*) and topographic images (*right*) of a single monolayer of DPPA containing 2 mol% RB18 deposited on a quartz slide. Excitation occurred at 543.5 nm using an He:Ne laser

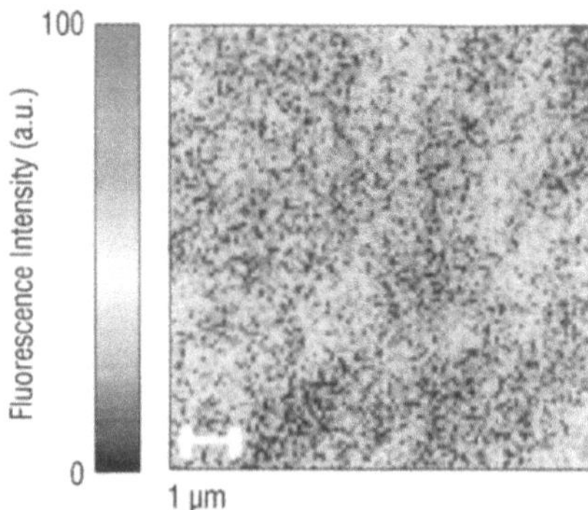

Fig. 6.7. Scanning confocal fluorescence image (dichroic mirror Chroma-435 nm and cut-off filter Chroma-435 nm, objective ×60, zoom ×10) of a mixed monolayer of THIAM18 in a DPPA matrix at a mixing ratio of 2 mol%. The fluorescence intensity is indicated by false colors. The confocal image was accumulated 20 times

framework of Eq. (6.6), linking k_0, μ, and σ_g. This analysis led however to physically unacceptable values of the decay parameters for this molecule.

6.3.2
Homogeneous Multilayers: SRH + ARA

The excitation and emission spectra of mixed multilayers of ARA and SRH resemble, with the exception of a bathochromic shift, those of a solution of SRH in ethanol [59]. For the emission spectra the shift increases with increasing dye concentration. The fluorescence decays of mixed multilayers of SRH and ARA (Fig. 6.9), excited at 537 nm, are non-exponential and become steeper upon increasing the dye concentration. Global analysis of the decays obtained at different concentrations or emission wavelengths in the framework of Eqs. (6.1) and (6.2) linking the decay time of the unquenched dye over different emission wavelengths and quencher concentrations from 0.17 mol% to 1.7 mol% yielded acceptable statistical parameters [59]. For a concentration up to 0.42 mol% $1-\alpha$ can be kept fixed at zero and even up to a concentration of 1.7 mol% $1-\alpha$ remains be-

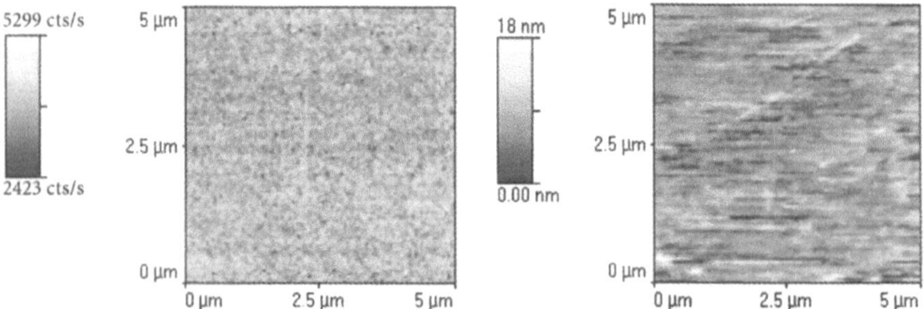

Fig. 6.8. NSOM (*left*) and topographic image (*right*) of a single monolayer of ARA containing 2 mol% SRH deposited on a quartz slide. Excitation occurred at 514 nm using an Argon laser

low 0.05. The presence of long living emission will be revealed most efficiently under conditions where its decay time differs most strongly from that of monomers susceptible to quenching by energy transfer to dimers or aggregates.

The decay times obtained from the global analyses at different emission wavelengths and linking τ over different concentrations are situated between 3.5 ns and 3.6 ns. The independence of τ_D of the emission wavelength, which is confirmed by global analysis linking decays obtained at different emission wavelengths and at two dye concentrations, suggests that the emission of the different SRH/ARA Langmuir-Blodgett films is over the complete emission band due to a single species.

γ_2 increases with increasing concentration of SRH in the Langmuir-Blodgett film (Fig. 6.10). The increase of γ_2 is almost linear up to 0.85 mol%. For the 1.7 mol% film the recovered value of γ_2 is however smaller than expected on the basis of a linear relation between γ_2 and the overall dye concentration. It should be noted that a linear or superlinear increase of γ_2 with the overall dye concentration has been observed very rarely up to now for fluorescence decays of Langmuir-Blodgett films [29, 43, 60]. Although γ_2 decreases upon decreasing the dye concentration down to 0.16%, no further decrease is observed at lower concentrations. This phenomenon has already been observed for RB18 in DOPA and is attributed to the presence of quenchers, which are inherent for the films studied, or to imperfections of the Langmuir-Blodgett film [29, 43, 69]. The independence of γ_2 of the emission wavelength indicates that the number of quenchers, which are able to accept energy from the emitting species, does not decrease for the emission observed at longer wavelengths.

This confirms, in agreement with the spectral data, that all fluorescence is due to a single species, probably the SRH monomer, over the whole emission band studied. There is furthermore no indication for a broad distribution of the excitation energies of this species. In the latter case dye molecules in a more "favorable" or

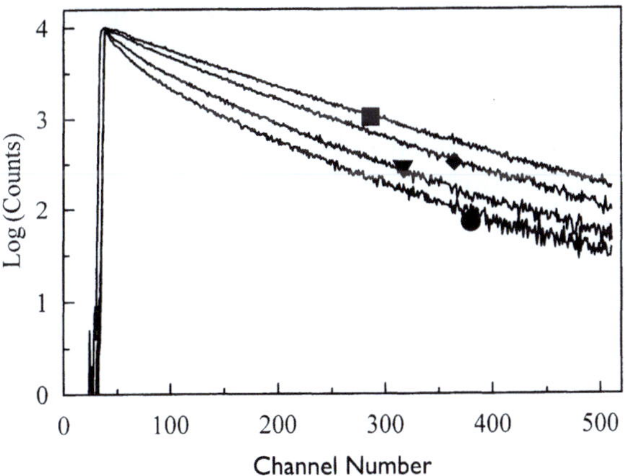

Fig. 6.9. Decay curves recorded at different concentrations of the SRH/ARA Langmuir-Blodgett films (excitation at 537 nm). ■: 0.17 mol%, ◆: 0.42 mol%, ▼: 0.85 mol%, ●: 1.7 mol%. Time increment: 38 ps/channel

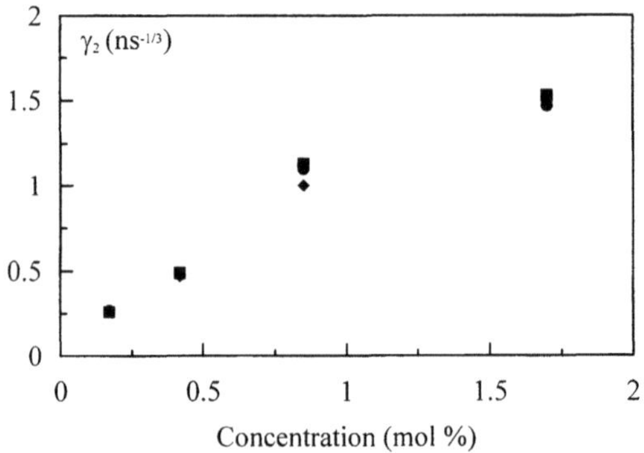

Fig. 6.10. The dependence of γ_2 upon the concentration in Langmuir-Blodgett films of SRH/ARA (excitation at 537 nm). ■: 570 nm, ◆: 580 nm, ●: 600 nm

"stabilizing" environment and emitting at longer wavelengths would be characterized by a smaller overlap between their emission spectrum and the absorption spectrum of the dimers or aggregates and hence by smaller values of R_0 and γ_2.

The concentration dependence of γ_2 observed for SRH in ARA parallels that observed for RB18 in DOPA [29] where it was considered compatible with a homogeneous distribution of the dye in the Langmuir-Blodgett film. In the DOPA samples the deviations observed at higher concentrations were attributed to a decrease of the equilibrium constant for aggregation. This is no surprise, as a concentration of 2 mol% would already correspond to 0.01 mol l^{-1}. Another possibility could be an increase of the aggregation number of the aggregates at higher dye concentration [43, 59].

Corroborating the conclusions of the analysis of the fluorescence decays the NSOM image (Fig. 6.8) gives no indication for a nonhomogeneous distribution of SRH in the Langmuir-Blodgett monolayer. No patches indicating higher or lower fluorescence intensity are obtained within the spatial resolution of the NSOM (50–80 nm). This can however not exclude that on a smaller scale, different phases or pseudophases are formed. However inhomogeneity on a scale of a few nm would make it impossible to fit the fluorescence decays to Eqs. (6.1) and (6.2). In this case fitting the decay to Eqs. (6.3) and (6.4) should yield better results with values of $\bar{d}$ between 1 and 2. There was however no evidence for this hypothesis [70].

6.3.3
Multilayers of CV18 and ARA or DPPA

6.3.3.1
CV18 in DPPA

A maximum at 540 nm and a shoulder at 585 nm for all concentrations characterize the absorption spectra of mixed multilayers of DPPA and CV18 deposited in a head-to-head fashion. The ratio of the maximum to the shoulder varies from

1.17 at 0.5 mol% to 1.28 at 5.0 mol%. Based on the molar area a molar extinction coefficient of $3.7 \times 10^4\,M^{-1}cm^{-1}$ and $3.0 \times 10^4\,M^{-1}cm^{-1}$ can be obtained at 540 nm and 585 nm. The width of the total absorption band amounts to $3600 \pm 50\,cm^{-1}$.

Those data should be compared to and absorption maximum at 590 nm and an FWMH of $1900 \pm 50\,cm^{-1}$ in $CHCl_3$. The slight spectral change and the maximum at 540 nm suggest that in the concentration range used CV18 is mainly present as dimers.

While in a dilute solution in $CHCl_3$ a fluorescence maximum is obtained at 622 nm, in Langmuir-Blodgett films of DPPA this maximum shifts, upon excitation at 580 nm, from 622 nm at 0.5 mol% to 640 nm at 5 mol% (Fig. 6.11). One should however take into account that the rising edge of the fluorescence spectrum of the 0.5 mol% sample is biased by scattered light. In solution and in DPPA films up to a concentration of 2 mol% the emission maximum is independent of the excitation wavelength (560 nm and 590 nm or 540 nm and 580 nm, respectively). However, the emission maximum of the most concentrated (5 mol%) sample shifts to 660 nm upon excitation at 540 nm. This red shift is accompanied by an increase of the FWMH from $2200 \pm 50\,cm^{-1}$ to $3000 \pm 50\,cm^{-1}$ and an increase of the Stokes shift from $900 \pm 100\,cm^{-1}$ to $1900 \pm 100\,cm^{-1}$. The excitation spectra of the emission at 630 nm of the DPPA films and of a $CHCl_3$ solution are characterized by a maximum at 585 nm. Their features are independent of the concentration of CV18 in the film (Fig. 6.12).

In analogy to the results obtained for RB18 [29] the fluorescence decays at 626 nm (Fig. 6.13) were analyzed globally in the framework of energy transfer in a 2D-system (Eq. 6.1) linking τ_D. This analysis yielded satisfactory fits while $1-\alpha$ could be kept equal to zero. Although the quality of the fits are acceptable, the independence of the parameter γ_2 of the dye concentration Table 6.1) observed for the lowest concentrations is in contradiction with Eq. (6.2), which predicts a linear increase with the concentration of acceptors. As over the whole concen-

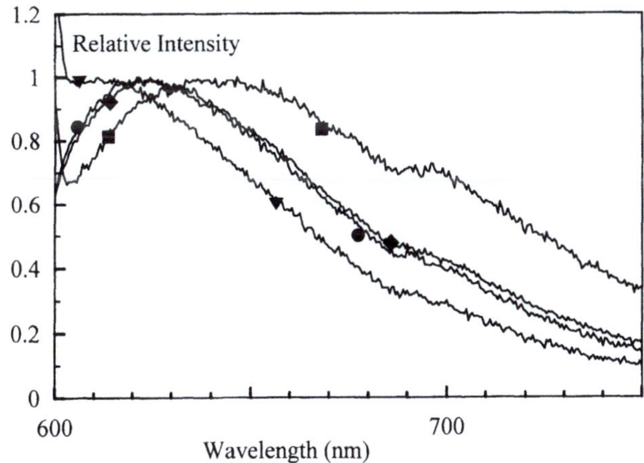

Fig. 6.11. Fluorescence emission spectra of CV18/DPPA Langmuir-Blodgett films (head to head deposition) for excitation at 590 nm. ▼: 0.5 mol%, ●: 1.0 mol%, ◆: 2.0 mol%, ■: 5.0 mol%

Table 6.1. Analysis of the fluorescence decay of mixed multilayers of CV18 and DPPA in the framework of Eq. (6.1). The time increment per channel, τ_D and χ_g^2 amounted to 9 ps, 3.27 ns, and 1.04, respectively. $1 - \alpha = 0$. $\xi\delta(t)$ is added to the convoluted Eq. (6.1) to correct for scattered light and digitization errors [72, 73]

Mol% CV18	χ^2	γ_2 (ns$^{-1/3}$)	ξ
0.5	1.06	2.77	0.027
1	0.98	2.69	0.005
2	1.07	3.03	0.001
5	1.06	3.96	−0.007

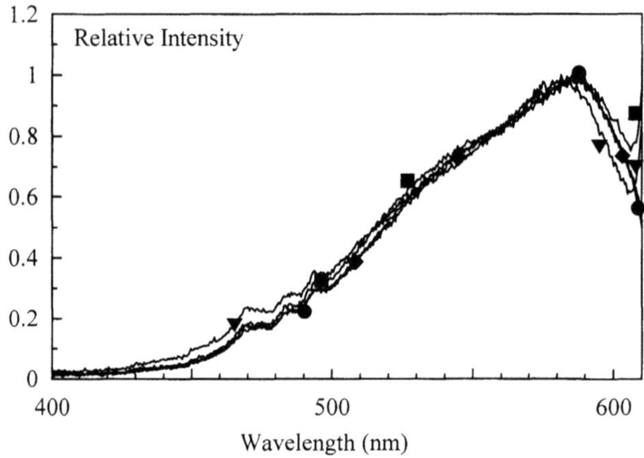

Fig. 6.12. Fluorescence excitation spectra of the emission at 630 nm of CV18/DPPA Langmuir-Blodgett films (head to head deposition). ▼: 0.5 mol%, ●: 1.0 mol%, ◆: 2.0 mol%, ■: 5.0 mol%

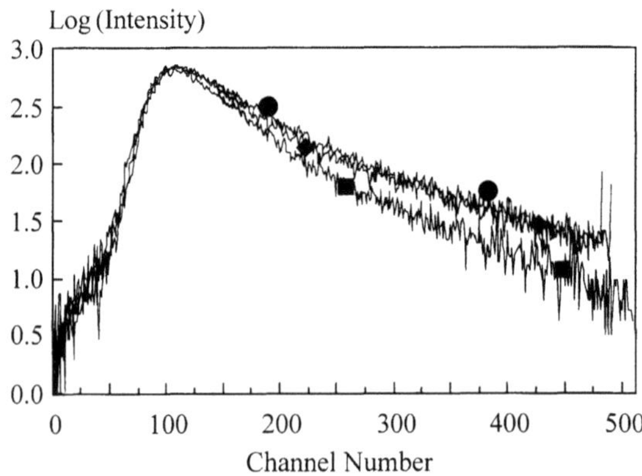

Fig. 6.13. Fluorescence decays at 630 nm of CV18/DPPA Langmuir-Blodgett films. Excitation occurred at 590 nm. ●: 0.5 mol%, ◆: 2.0 mol%, ■: 5.0 mol%

Table 6.2. The fluorescence decay of mixed multilayers of CV18 and DPPA analyzed in the framework of Eq. (6.5). The time increment per channel, k_0 and χ^2_g amounted to 9 ps, 0.433 ns^{-1} and 1.17, respectively

Mol% CV18	χ^2	μ (ns^{-1})	σ (ns^{-1})
0.5	1.15	0.125	3.20
1	1.15	0.129	3.20
2	1.10	1.124	3.20
5	1.27	4.072	4.36

tration range investigated CV18 is mainly present as dimers there one can expect a linear rather than a superlinear dependence of γ_2 on the total dye concentration. Global analysis of the fluorescence decays in the framework of energy transfer in a self-similar distribution of quenchers [74–76] (Eqs. 6.3 and 6.4), linking τ_D and $\bar{d}$ also yielded good fits ($\chi^2_g = 1.09$). However in this case it is also difficult to explain the independence of $\bar{d}\gamma_{\bar{d}}$ of the CV18 concentration, as observed for the three lowest concentrations of CV18. The independence of the fluorescence decays of the CV18 concentration can also be observed upon visual inspection of the fluorescence decays (Fig. 6.13). This suggests that the non-exponential decay of the fluorescence at low concentrations of CV18 is not due to energy transfer to dimers. As [53–55] the fluorescence decay time of CV18 in solution increases, depending upon temperature and solvent viscosity, from a few picoseconds to more than one nanosecond, one cannot exclude that the nonexponential fluorescence decay observed at low temperature is due to the presence of CV18 in a distribution of environments characterized by a distribution of free volumes. Therefore the fluorescence decays were analyzed globally in the framework of a Gaussian distribution of decay rates using Eq. (6.5) (Table 6.2), linking k_0 over all concentrations and linking σ over the three lowest concentrations. This model has been applied successfully to analyze the fluorescence decays of molecular dispersions of triphenyl benzenes in polymer films or adsorbed on silica [58, 77].

Global analysis confirms again that up to 2 mol% the fluorescence decay does not depend upon the CV18 concentration. This suggests that the non-exponential decay of CV18 is due to a distribution of free volumes rather than to energy transfer to dimers. At higher concentrations the increase of the decay rate could indicate that energy transfer to dimers also becomes an important decay process.

6.3.3.2
Cd-Arachidate Multilayers

The absorption spectra of the quartz slides on which multilayers of CV18 and cadmium arachidate were deposited in head to head fashion, were characterized by a maximum at 555 nm and a shoulder at 590 nm which became more prominent upon dilution (Fig. 6.14). Those wavelengths correspond to the absorption maximums of the monomer and the dimer, which are situated at 588 nm and 545 nm in solution. When the maximum at 555 nm and the shoulder at 590 nm in the Langmuir-Blodgett film are attributed to monomers and dimers the dye

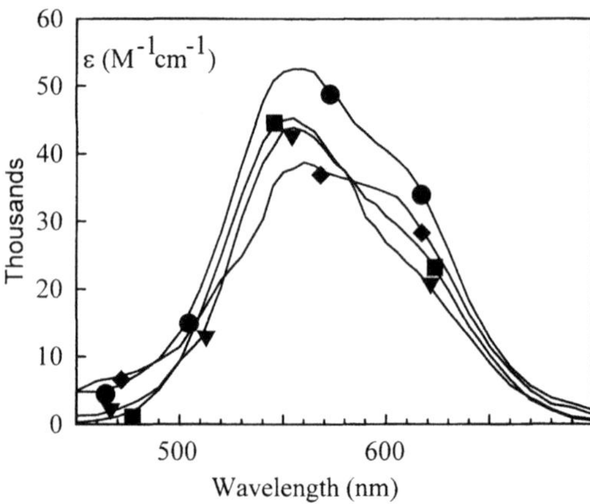

Fig. 6.14. Absorption spectra of multilayers of CV18 and cadmium arachidate. The layers are deposited in a head to head fashion. ◆: 0.38 mol% CV18, ●: 1.25 mol% CV18, ■: 3.70 mol% CV18, ▼: 19 mol% CV18

will already be mainly present as dimers at a mixing ratio of 1/364 (assuming that the oscillator strength of the dimer is twice that of the monomer [13]).

In this interpretation the shift between the monomer and the dimer would amount to 1340 ± 50 cm^{-1} and 1070 cm^{-1} in the Langmuir-Blodgett film and in solution respectively. In highly diluted samples (mixing ratio CV18/cadmium arachidate 1/476) the fluorescence maximum of the sample is situated at 640 nm. Upon increasing the concentration a second maximum at 695 nm, attributed to dimers (Fig. 6.15), is observed. While the excitation spectrum of the emission at 640 nm has a maximum at 602 nm, that obtained for an analysis wavelength of 695 nm shows a more important contribution of dimers.

For the same concentration range the absorption and emission spectra of CV18 in ARA and DPPA are similar. Although the absorption spectra of both systems show at all concentrations a major maximum at wavelengths characteristic for the dimers in solution, the dimer emission only starts to be predominant in the at mixing ratios of 5 mol% and higher. This could be due to a smaller quantum efficiency of dimer emission. The agreement between the excitation spectra of the monomer emission and the monomer absorption spectrum in solution excludes anyway that important conformational changes, modifying the absorption spectrum of the CV18 monomer [78] occur in the Langmuir-Blodgett films. The discrepancy between the concentration dependence of the absorption and emission spectra could also be related to an inhomogeneous distribution of the CV18 over the film. In this case the emission is mainly due to isolated CV18 molecules not involved in the monomer dimer equilibrium. To test this hypothesis the time-resolved fluorescence of the CV18-ARA Langmuir-Blodgett films was investigated.

The fluorescence decays of multilayers (Fig. 6.16) where the concentration of CV18 increased from 0.1 mol% to 3.3 mol% could be analyzed globally (Table 6.3) using Eqs. (6.1) and (6.2) (two-dimensional system) or Eqs. (6.3) and

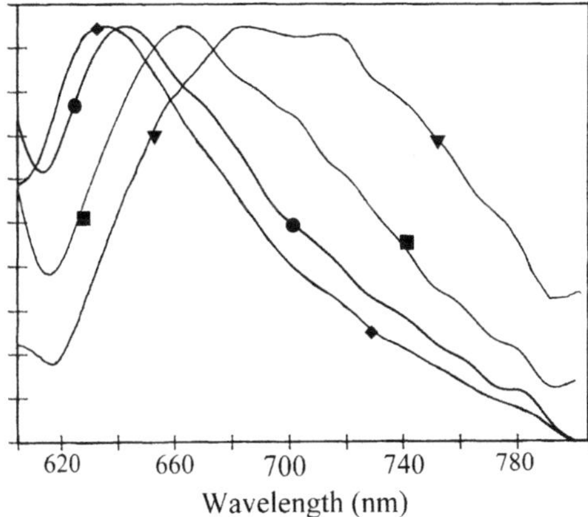

Fig. 6.15. Emission spectra of multilayers of CV18 and cadmium arachidate. The layers are deposited in a head to head fashion. ◆: 0.20 mol% CV18, ●: 0.50 mol% CV18, ●: 2.0 mol% CV18, ■: 23 mol% CV18. Excitation at 595 nm

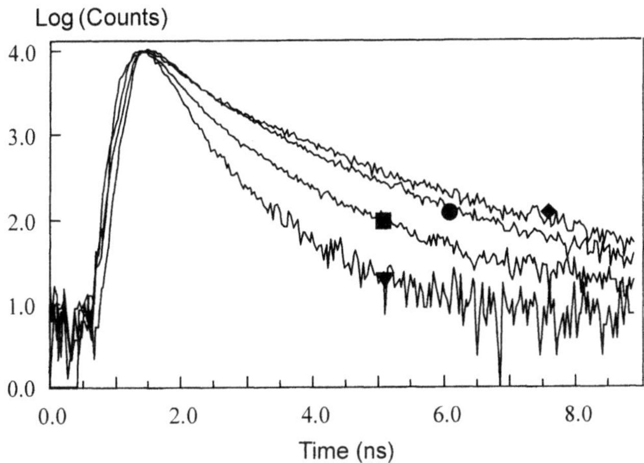

Fig. 6.16. Fluorescence decays, recorded at 630 nm, of multilayers of CV18 and cadmium arachidate. The layers are deposited in a head to head fashion. ◆: 0.10 mol% CV18, ●: 0.20 mol% CV18, ■: 0.8 mol% CV18, ▼: 3.33 mol% CV18. Excitation at 580 nm. Time increment per cannel: 50.8 ps (0.1 – 0.8 mol%) and 30.0 ps (3.33 mol%)

(6.4) (self-similar system) linking τ_D and $\bar{d}$. In agreement with the results obtained for RB18 in ARA or DPPA or for THIAM18 in ARA or DPPA the limiting value of $\gamma_2 \tau_\mathrm{D}^{1/3}$ or $\gamma_{\bar{d}}\, \tau_\mathrm{D}^{\bar{d}/6}$ at low concentrations is much larger than zero. It is about ten times larger than the value found for RB18 in DPPA [29] or for SRH in ARA [59]. On the other hand, in contrast to what is found for RB18 in ARA or for THIAM18 in ARA or DPPA $\gamma_2 \tau^{1/3}$ or $\gamma\tau_\mathrm{D}^{\bar{d}/6}$ increase significantly for CV18 concentrations above 0.4 mol%. Furthermore, in contrast to the results obtained for RB18 in ARA, $1-\alpha$ remains always very small. This makes the distribution of

Table 6.3. Global analysis of the fluorescence decays of multilayers of CV18 and ARA

Mol% CV18	2-Dimensional $\tau_D = 3.57$ ns, $\chi_g^2 = 1.11$		Self-similar $\tau_D = 4.78$ ns, $\chi_g^2 = 1.24$	
	$\gamma_2 \tau_D^{1/3}$	$1 - \alpha$	$\gamma_{\bar{d}} \tau_D^{\bar{d}/6}$	$\bar{d}/6$
0.1	2.42	0.0145	2.75	0.268
0.2	2.22	0.0044	3.07	0.268
0.4	3.39	0.0052	3.68	0.268
0.8	3.39	0.0028	4.00	0.268
1.66	3.97	0.0017	4.54	0.268
3.33	4.91	0.0005	5.70	0.268

monomers between a highly diluted phase, where no quenching occurs, and a completely or partly saturated phase, where quenching of the monomer emission by dimers is important, unlikely.

The limiting value of $\gamma_2 \tau_D^{1/3}$ or $\gamma_{\bar{d}}\, \tau_D^{\bar{d}/6}$ at low quencher concentrations is furthermore two to three times larger than the value found for THIAM18 and ARA or for RB18 and ARA at short wavelengths. If one would stick to the same framework as discussed for the inhomogeneous systems [29, 43, 59] this would mean that either the dimer concentration in the concentrated phase is much larger or that R_0 is much larger for the donor-acceptor pair CV18/(CV18)$_2$ than for RB18/(RB18)$_2$.

Another possibility is that the highly non-exponential decay, even at low concentration is due to a distribution of free volumes and hence of τ_D^{-1}. The quenching, shown by a further increase of $\gamma_2 \tau_D^{1/3}$ or $\gamma_{\bar{d}}\, \tau_D^{\bar{d}/6}$ upon increasing the concentration would than be due to a genuine monomer-dimer excitation transfer. In this case it should be possible to fit the decays obtained at different concentrations globally to Eq. (6.6), linking τ_D, σ_g, and k_0 (Table 6.4).

While the recovered value of k_0 is close to what can expected for the radiative decay rate of crystal violet, the recovered values of μ and σ_g indicate a highly skewed distribution. Contrary to the observations for CV18 in DPPA, γ_2 increases over the whole concentration range. The fluorescence analysis can be performed without a second phase with a very low CV18 concentration where no quenching

Table 6.4. Analysis of the fluorescence decays of mixed multilayers of CV18 and DPPA in the framework of Eq. (6.6). The multilayers are deposited in a head to head fashion. k_0, μ, σ_g, B, and χ_g^2 amounted to 0.25 ± 0.12 ns^{-1}, -1.1 ± 4.1 ns^{-1}, 2.5 ± 0.8 ns^{-1}, 0, and 1.11 respectively

Mol% CV18	χ^2	γ_2 (ns$^{-1/3}$)	Time increment (ps)
0.10	1.25	0.0 ± 0.2	50.8
0.20	0.92	0.45 ± 0.18	50.8
0.88	1.14	1.25 ± 0.21	50.8
0.88	1.07	0.97 ± 0.20	50.8
1.67	1.38	1.3 ± 0.2	50.8
3.33	1.16	2.4 ± 0.2	35.0

occurs. In case of a homogenous distribution of CV18, the average distance to a molecule of CV18 amounts to more than 80 Å at a mixing ratio of 1/1000 where the average surface density of CV18 amounts to 5×10^{-5} Å^{-2}. Assuming similar oscillator strength for the monomer and the dimer [43], the absorption spectra indicate that about 50% of the CV18 molecules are present as dimers. As two molecules of CV18 are necessary for a dimer the average surface density of dimers would amount to 1×10^{-5} Å^{-2}. This would correspond with an average distance to a dimer of about 160 Å. This is clearly much larger than the values possible for R_0, the characteristic distance for energy transfer. Hence if one assumes a homogeneous distribution of CV18, an efficient energy transfer to dimers is nearly impossible at a mixing ratio of 1/1000 and the non-exponential fluorescence decay observed for the diluted samples is manly due to a distribution of free volumes. The alternative would be an inhomogeneous system, as observed for THIAM18 in ARA or DPPA or for RB18 in ARA and DPPA. However analysis of the fluorescence decays in the framework of Eqs. (6.1) and (6.2) indicated that all dye molecules are in the phase where quenching by dimers occurs. However, the large values of γ_2 observed for dyes in the concentrated phase suggest that this phase occupies only a small fraction of the total monolayer area in order to obtain sufficient enrichment. For example, for R_0 equal to 40 Å [17, 60], a value of 2.4 for $\gamma_2 \tau_D^{1/3}$ requires a surface density of dimers of 3×10^{-4} Å^{-2}, which is more than six times the average concentration of CV18. Although such a model cannot be excluded, it is less likely than the combination of a homogeneous distribution and a distribution of free volumes.

6.3.4
Intralayer Quenching of PYR18 by CV18

A complete quantitative validation of the values of the decay parameter γ_2, which was attributed to energy transfer from dye monomers to dye aggregates, is quite difficult. Both the position of the monomer dimer equilibrium and the value of R_0 for monomer-aggregate transfer are not known precisely. The value of R_0^6 is proportional to the molar extinction coefficient of the dimer at the wavelengths where the monomer emits. However, in this wavelength range there is strong overlap with the tail of the absorption of the monomer. Furthermore, the low energy transition of a sandwich dimer is forbidden. A critical test of the methodology developed above is the quenching of a dye by a second dye of which the absorption spectrum overlaps strongly with the emission spectrum of the donor. The donor-acceptor couple PYR18/CV18 was preferred due to the good overlap of the absorption spectrum of CV18 (maximum of the dimer at 555 nm, maximum of monomer at 590 nm) with the emission spectrum of CV18 (maximum at 572 nm) and the small tendency of PYR18 to aggregate.

Hence, at low concentrations the fluorescence decay of PYR18 deviates only slightly from a mono-exponential decay, indicating a homogeneous distribution in the ARA-films. Furthermore the small changes over a large concentration range of the apparent CV18 extinction coefficient at the emission maximum of PYR18 suggest that the monomer-dimer equilibrium of CV18 will influence R_0 only marginally [58].

Table 6.5. Global analysis of the fluorescence decays at 590 nm of mixed multilayers of PYR18 and CV18 in ARA in the framework of Eqs. (6.1) and (6.2). The multilayers are deposited in a tail to tail fashion. τ_D and χ amount to 2.94 ± 0.12 ns and 1.11 respectively. Excitation occurred at 540 nm

Mol% CV18	γ_2	$1-\alpha$	χ^2
0.0	0.80	0.0	1.05
0.5	1.24	0.0	1.14
1.0	2.61	0.00	1.24
2.0	3.86	0.02	1.09
3.0	4.53	0.01	1.21

Tail to tail double layers of ARA containing 0.50 mol% PYR18 and 0.0–3.0 mol% CV18 were alternated with double layers of ARA. In this way interlayer dimer formation and energy transfer is minimized. When the concentration of CV18 increases gradually from 0 mol% to 3 mol%, a decrease of the fluorescence intensity of PYR18, obtained upon excitation at 540 nm, is accompanied by an increased fluorescence of CV18 [58] dimers at 680 nm.

The fluorescence decays of PYR18 could be analyzed globally over a concentration range from 0 mol% CV18 to 3 mol% CV18 in the framework of Eqs. (6.1) and (6.2) linking τ_D (Table 6.5). Global analysis in the framework of Eqs. (6.3) and (6.4) linking τ_D and $\bar{d}$ is only possible for the two samples with the lowest concentration. In that case a fractal dimension of 2.09 ± 0.24 and a value of 3.0 ± 0.2 ns for τ_D were obtained respectively. Using Eq. (6.1), the recovered value of τ_D for PYR18 is 20% longer than that obtained in solution. Even in the absence of CV18 the fluorescence decay of PYR18 was non-exponential as observed for most dye-matrix combinations studied up to now [29, 43, 59, 79, 81]. When R_0 was calculated from the spectral data for the couple PYR18/CV18 in a Langmuir-Blodgett film a value of 65 Å was obtained [58]. This allows one to calculate σ from γ_2 (after subtraction of $\gamma_{2,\mathrm{CV18=0}}$), R_0, and τ_D using Eq. (6.2) (Table 6.6). The values of σ recovered in this way are about 40% of the total CV18 dimer concentration σ_{CV18} (assuming all CV18 molecules arepresent as dimers). Those data indicate a homogeneous distribution of CV18 over the Langmuir-Blodgett film. If all CV18 molecules were concentrated in a small fraction of the film we would expect to recover a much larger value for σ using Eq. (6.2). If only a fraction of the CV18 molecules were present as dimers one would expect a smaller

Table 6.6. Calculated acceptor concentration, σ, and CV18 dimer concentration, σ_{CV18}, assuming that all CV18 molecules are present as dimers

Mol% CV18	$\sigma\,(\mathrm{nm}^{-2})$	$\sigma_{\mathrm{CV18}}\,(\mathrm{nm}^{-2})$	$\sigma/\sigma_{\mathrm{CV18}}$
0.5	0.0035	0.013	0.27
1.0	0.015	0.026	0.56
2.0	0.025	0.051	0.48
3.0	0.030	0.076	0.40

value for R_0 and hence slightly larger value for σ. However, in this case one should also use a much larger value for σ_{CV18}.

The homogeneous distribution of CV18 in ARA indicates that for the analysis of the fluorescence decays of monomers of CV18 in ARA (and probably also in DPPA) a combination of a homogenous distribution of the dye and a distribution of free volumes is a valid hypothesis.

6.4
Conclusions

For none of the systems studied was a single exponential fluorescence decay of the dye observed [29, 43, 59]. Even at the lowest concentration there is still some residual quenching. This could perhaps be due to energy transfer to metal ions in the matrix [79, 81]. For most of the samples the fluorescence decays can be analyzed in the framework of energy transfer in two dimensions (Eqs. 6.1 and 6.2) in a two-phase system. Invoking a fractal dimension [29, 43, 59] does mostly not lead to good fits or decay parameters that are physically relevant. For SRH in ARA and RB18 in DOPA the monomers and dimers are distributed homogeneously over the matrix. For most other systems THIAM18 or RB18 in ARA or DPPA two phases are present in the multilayer: a highly diluted phase where no quenching occurs and a more concentrated phase where quenching by energy transfer to dimers or aggregates occurs. The occurrence of homogeneous and nonhomogeneous monolayers is confirmed by spatial resolved fluorescence using confocal fluorescence microscopy or NSOM.

For the dye CV18, which had a very small fluorescence quantum yield and decay time [53–56, 60] in a non-viscous solvent, the values of γ_2 at low concentration and the concentration dependence of γ_2 indicate a distribution of decay rates of the unquenched CV18 monomers. At higher concentrations also energy transfer to dimers occurs and at all concentration the CV18 monomers and dimers are distributed homogeneously. This is confirmed by a global analysis of the fluorescence decays in the framework of Eq. (6.6). Although this effect of a distribution of free volumes can also occur for THIAM18 [43] its consequences for the fluorescence decay for this molecule are apparently less important than for CV18. This could be due to the smaller viscosity dependence of the fluorescence quantum yield observed for THIAM18 [43]. Another aspect could be a possible higher solubility of the less planar CV18.

Although relating the results above to molecular structure is not straightforward; the formation of a homogeneous layer seems to be favored by:

1. A liquid monolayer phase (DOPA vs DPPA and ARA).
2. Strong interactions (H-bonds) between the dye and the matrix (SRH vs RB18). The possible hydrogen bonding between the NH-group in SRH and the carboxylate or phosphate groups will enhance dye-matrix interactions vs matrix-matrix interactions. This corroborates the observed pH dependence of the aggregation of a cyanine dye in ARA [80].

3. When the decay time of the dye depends very strongly on the free volume the distribution of free volumes encountered in the Langmuir-Blodgett film by the dye will make the decay more complex.

The present study is able to rationalize the fluorescence decays of dyes in Langmuir-Blodgett films. It indicates furthermore that it will be very difficult to find an adequate model system, with an exponentially decaying probe to investigate other processes such as, e.g., electron transfer, charge injection, or interlayer electron transfer [4]. However, using global analysis [82, 83] it remained possible to investigate interlayer energy or electron transfer for a non-exponentially decaying probe.

It should also be stressed that the analysis of fluorescence decays and scanning microscopies (confocal fluorescence microscopy and NSOM) yield similar results for all combinations of a lipid matrix and a fluorescent probe to which both the methods are applied. For the study of Langmuir films, and more generally all single membranes, combination of spatial and temporal resolved fluorescence allows for a convincing interpretation of both types of experiments.

Acknowledgements. M. V. d. A. is a "Onderzoeksdirecteur" of the F.W.O.-Vlaanderen (Fonds voor Wetenschappelijk Onderzoek Vlaanderen). H. L.-P. would like to thank the DG XII for a postdoctoral grant in the framework of the TMR-network CHR-CT94–0538. P. B. and D. P. are indebted to the IWT for financial support. H.L. gratefully acknowledges the financial support of the Academy of Finland and the Technology Development Center of Finland for support of our program of photochemistry of organic films. E.V. thanks the EU for financial support in the framework of COST D4. The authors thank the "Nationale Loterij", the F. W. O.-Vlaanderen and D. W. T. C. through IUAP IV-11.

References

1. Gaines JL (1986) Insoluble monolayers at liquid-gas interfaces. Wiley, New York London
2. Roberts G (1990) Langmuir-Blodgett films. Plenum Press, New York London
3. Ito T, Yamazaki I, Nabuhiro O (1997) External electric field effect on interlayer vectorial electron transfer from photoexcited oxacarbocyanine to viologen in Langmuir-Blodgett films. Chem Phys Lett 277:125–131
4. Kuhn H, Möbius D (1993) Investigations of surfaces and interfaces, part B. In: Rossiter BW, Baetzold RC (eds) Physical methods of chemistry series, 2nd edn, vol. IXB. Wiley, New York London
5. Sienicki K (1990) Forward and reverse energy transfer in Langmuir-Blodgett multilayers. J Phys Chem 94:1944–1948
6. Sienicki K (1991) Directed energy transfer in Langmuir-Blodgett multilayers with asymmetrical forward and reverse transfer rates and energy migration. J Chem Phys 94:617–622
7. Yamazaki I, Tamai N, Yamazaki T (1990) Electronic excitation transfer in organized molecular assemblies. J Phys Chem 94:516–525
8. Nishiyama Y, Azuma T, Obata N, Kasatani K, Sato H (1991) Photoelectric conversion and fluorescence quenching in in mixed-dye monolayers; inhomogeneous distribution of dyes. J Photochem Photobiol A: Chem 59:341–355
9. Yatsue T, Miyashita T (1995) Electron-Transfer quenching accompanied by highly efficient energy migration in polymer Langmuir-Blodgett films. Langmuir 99:16,047–16,051
10. Gregory BW, Nakmin D, Gray JD, Ocko BM, Stroeve P, Cotton TM, Struve WJ (1997) Two-dimensional pigment monolayer assemblies for light-harvesting applications: structural characterization at the air/water interface with X-rays. J Phys Chem B 101:2006–2009

11. Fujihira M (1990) Photoelectric conversion by monolayer assemblies. Mol Cryst Liq Cryst 183:59–69
12. Möbius D (1978) Designed monolayer assemblies. Ber Bunsenges Phys Chem 82:848–858
13. Kasha M (1963) Energy transfer mechanisms and the molecular exciton model for molecular aggregates. Radiat Res 20:55–71
14. Akimoto S, Ohmori A, Ymazaki I (1997) Dimer formation and excitation relaxation of perylene in Langmuir-Blodgett films. J Phys Chem B 101:3753–3758
15. Bücher H, Drexhage KH, Fleck M, Kuhn H, Möbius D, Schäfer FP, Sondermann J, Sperling W, Tilman P, Wiegand J (1967) Controlled transfer of excitation through thin layers. Mol Cryst 2:199–230
16. Kemnitz K, Tamai N, Yamazaki I, Nakashima N, Yoshihara K (1986) Fluorescence decays and spectral properties of rhodamine B in submono- mono-, and multilayer systems. J Phys Chem 90:5094–5101
17. Nakashima N, Yoshihara K, Willig F (1980) Time-resolved measurements of electron and energy transfer of rhodamine B monolayers on the surface of organic crystals. J Chem Phys 73:3553–3559
18. Van der Auweraer M, Willig F (1985) Influence of dye aggregation on the photosensitized hole injection into anthracene crystals from Langmuir Blodgett monolayers containing cyanine dyes. Isr J Chem 25:274–278
19. Tamai N, Yamazaki T, Yamazaki I (1988) Excitation energy relaxation of rhodamine B in Langmuir-Blodgett monolayer films: picosecond time-resolved fluorescence studies. Chem Phys Lett 147:25–29
20. Chauvet J-P, Agrawal ML, Hug GL, Patterson LK (1985) Effects of molecular orientation on photophysical behavior: steady state and real-time behavior of chlorophyl-b fluorescence in spread monolayers of di-oleoylphosphatidylcholine. Thin Solid Films 133:227–234
21. Tweet AG, Bellamy WD, Gaines GL Jr (1964) Fluorescence quenching and energy transfer in monomolecular films containing chlorophyl. J Chem Phys 41:2068–2077
22. Taniike K, Matumoto T, Sato T, Ozika Y, Nakashima K, Iriyama K (1996) Spectroscopic studies on phase transitions in Langmuir-Blodgett films of an azobenzene-containing long chain fatty acid. Dependence of phase-transitions on the number of monolayers and transition cycles among H, J and J'-aggregates in multilayer films. J Phys Chem 100: 15,508–15,516
23. Holmes AS, Suling K, Birch DSS (1993) Fluorescence quenching by metal ions in lipid bilayers. J Phys Chem 97:189–194
24. Loura LMJ, Fedorov A, Prieto M (1996) Resonance energy transfer in a model system of membranes. Application to gel and lipid liquid crystalline phases. Biophys J 71:1823–1836
25. Tamai N, Yamazaki T, Yamazaki I (1990) Picosecond dynamics of excitation-energy relaxation of dye molecules in Langmuir-Blodgett monolayers. Can J Phys 68:1013–1022
26. Sluch M, Vitukhnovsky AG, Warren JG, Petty MC (1993) Two-dimensional energy transfer in Langmuir-Blodgett films. Mol Cryst Liq Cryst 229:37–45
27. Kemnitz K, Tamai N, Yamazaki I, Nakashima N, Yoshihara K (1987) Site-dependent fluorescence lifetimes of isolated dye molecules adsorbed on organic single crystals and other substrates. J Phys Chem 91:1423–1430
28. Gust D, Moore T, Moore A, Luttrull DK, Degraziano JM, Boldt, NJ, Van der Auweraer M, De Schryver FC (1991) Tetraarylprophyrins in mixed Langmuir-Blodgett films: steady-state and time-resolved fluorescence studies. Langmuir 7:1483–1490
29. Ballet P, Van der Auweraer M, De Schryver FC, Lemmetyinen H, Vuorimaa E (1996) Global analysis of the fluorescence decays of *N,N'*-dioctadecyl rhodamine B in Langmuir-Blodgett films of diacyl phosphatidic acids. J Phys Chem 100:13,701–13,715
30. Nakashima K, Duhamel J, Winnik M.A. (1993) Photophysical processes on a latex surface: electronic energy transfer from rhodamine dyes to malachite green. J Phys Chem 97: 10,702–10,707
31. Van der Auweraer M, Verschuere B, De Schryver FC (1988) Absorption and fluorescence properties of rhodamine B derivatives forming Langmuir-Blodgett films. Langmuir 4: 583–588

32. Vuorimaa E, Ikonen M, Lemmetyinen H (1992) Photoinduced charge transfer in rhodamine B/anthracene mixed Langmuir-Blodgett multilayers. Thin Solid Films 214:243–253
33. Vuorimaa E, Ikonen M, Lemmetyinen H (1994) Photophysics of rhodamine dimers in Langmuir-Blodgett films. Chem Phys 188:289–302
34. Kjaer K, Als-Nielsen J, Helm CA, Tippman-Krayer P, Möhwald H (1989) Synchrotron X-ray diffraction and reflection of arachidic acid monolayers at the air-water interface. J Phys Chem 93:3200–3206
35. Dick HA, Bolton JR, Picard G, Munger G, Leblanc RM (1988) Fluorescence of 5-(4-carboxyhenyl)-10,15,20-tritolylporphyrin in a mixed Langmuir-Blodgett film with dioleoylphosphatidylcholine. Langmuir 4:133–136
36. Agrawal ML, Chauvet J-P, Patterson LK (1985) Effects of molecular organization on photophysical behavior: lifetime and steady state fluorescence of chlorophyl a singlets in monolayers of dioleoylphosphatidylcholines at the air-water interface. J Phys Chem 89:2979–2982
37. Lemmetyinen H, Ikonen M, Mikkola J (1991) A kinetic study of monomer and excimer fluorescence of pyrene lecithin in Langmuir-Blodgett films. Thin Solid Films 204:417–439
38. Hall RA, Thistlethwaite PJ, Grieser F (1993) An investigation of miscibility in guest-host monolayers by surface pressure and fluorescence methods. Langmuir 9:2128–2132
39. Knoll W (1994) Dynamical lateral order in binary lipid alloys and its coupling to membrane functions. Ber Bunsenges Phys Chem 98:512–518
40. Ghiggino KP, Spizzirri PG, Smith TA (1994) Time-resolved confocal microspectroscopic imaging. In: Masuhara H, De Schryver FC, Kitamura N, Tamai N (eds) Microchemistry, spectroscopy and chemistry in small domains. North-Holland Elsevier Science B.V., Amsterdam, p 197–210
41. Sasaki K, Koshioka M (1994) Three-dimensional space- and time-resolved spectroscopy using a confocal microscope. In: Masuhara H, De Schryver FC, Kitamura N, Tamai N (eds) Microchemistry, spectroscopy and chemistry in small domains. North-Holland Elsevier Science B.V., Amsterdam p 185–196
42. Gensch T, Hofkens J, van Stam J, Faes H, Creutz S, Tsuda K, Jérôme R, Masuhara H, De Schryver FC (1998) Transmission and confocal fluorescence microscopy, and time-resolved fluorescence spectroscopy combined with a laser trap: investigation of optically trapped block copolymer micelles. J Phys Chem B 102:8440–8451
43. Laguitton-Pasquier H, Van der Auweraer M, De Schryver FC (1998) Bidimensional distribution of a cyanine dye in Langmuir-Blodgett monolayers studied by time-resolved and spatially resolved fluorescence. Langmuir 14:5172–5183
44. Betzig E, Trautman JK (1992) Near-field optics: microscopy, spectroscopy, and surface modification beyond the diffraction limit. Science 257:189–194
45. Dunn RC, Holton GR, Mets L, Sunney Xie X (1994) Near-field fluorescence imaging and fluorescence lifetime measurements of light harvesting complexes in intact photosynthetic membranes. J Phys Chem 98:3094–3098
46. Jeuris K, Vanoppen P, De Schryver FC, Hofstraat JW, van der Ven LGJ, van Velde JW (1998) Fluorescence intensity of dye containing latex particles studied by near-field scanning optical microscopy. Macromolecules 31:8579–8784
47. Hollars CW, Dunn RE (1997) Submicron fluorescence, topology, and compliance measurements of phase-separated lipid monolayers using tapping-mode near-field scanning optical microscopy. J Phys Chem B 33:6313–6317
48. Dutta AK, Vanoppen P, Jeuris K, Grim PCM, Pevenage D, Salesse C, De Schryver FC (1999) Spectroscopic, AFM and NSOM studies of 3D crystallites in mixed Langmuir-Blodgett films of N,N'-bis(2,6-dimethylphenyl)-3,4,9,10-perylenetetracarboxylic diimide and stearic acid. Langmuir 15:607–612
49. Burshtein AI (1985) Energy transfer kinetics in disordered systems. J Luminescence 34:167–188
50. Blumen A (1981) Excitation transfer from a donor to acceptors in condensed media: a unified approach. Il Nuovo Cimento 63B:50–58

51. Hauser M, Klein UKA, Gösele U (1976) Extension of Förster's theory of long-range energy transfer to donor-acceptor pairs in systems of molecular dimensions. Z Phys Chem NF 101:255–266

52. Blumen A, Manz J (1979) On the concentration and time dependence of the energy transfer to randomly distributed acceptors. J Chem Phys 4694–5004

53. Ben-Amotz D, Harris CB (1985) Ground and excited-state torsional dynamics of a triphenylmethane dye in low-viscosity solvents. Chem Phys Lett 119:305–311

54. Vogel M, Rettig W (1985) Efficient intramolecular fluorescence quenching in triphenylmethane dyes involving excited states with charge separation and twisted conformations. Ber Bunsenges Phys Chem 89:962–968

55. Martin MM, Plaza P, Meyer YH (1991) Ultrafast conformational relaxation of triphenylmethane dyes: spectral characterization. J Phys Chem 95:9310–9314

56. Vogel M, Rettig W (1987) Excited-state dynamics of triphenylmethane dyes used for investigation of microviscosity effects. Ber Bunsenges Phys Chem 91:1241–1247

57. Kemnitz K, Yoshihara K (1990) Free-volume effect of organic dye monomers in the adsorbed state. J Phys Chem 94:8805–8811

58. Verbeek G, Vaes A, Van der Auweraer M, De Schryver FC, Geelen C, Terrell D, De Meutter S (1993) Gaussian distribution of the decay times of the singlet excited state of aromatic amines dispersed in polymer films. Macromolecules 26:472–478

59. Pevenage D, Van der Auweraer M, De Schryver FC (1999) Influence of the molecular structure on the lateral distribution of xanthene dyes in Langmuir-Blodgett films. Langmuir 15:8465–8473

60. Vuorimaa E, Lemmetyinen H, Ballet P, Van der Auweraer M, De Schryver FC (1997) Intralayer and interlayer energy transfer between chromophoric pyronine and crystal violet in Langmuir-Blodgett films. Langmuir 13:3009–3015

61. Sondermann J (1971) Darstellung oberflächenaktiver polymethin cyanin-farbstoffe mit langen N-alkylketten. Liebigs Ann Chem 749:183–107

62. Verschuere B, Van der Auweraer M, De Schryver FC (1991) A spectroscopic and photoelectrochemical investigation of xanthene dyes incorporated in Langmuir Blodgett films. Chem Phys 149:385–400

63. Van der Auweraer M (1990) Study of photophysical processes in Langmuir-Blodgett films. Habilitation, Katholieke Universiteit Leuven (Belgium)

64. Verschuere B, Van der Auweraer M, De Schryver FC (1994) A fluorescence study of the miscibility of ω-arylalkanoic acids in Langmuir Blodgett films. Thin Solid Films 244:995–1000

65. Janssens LD, Boens N, Ameloot M, De Schryver FC (1990) A systematic study of the global analysis of multi-exponential fluorescence decay surfaces using reference convolution. J Phys Chem 94:3564–3576

66 Boens N, Andriessen R, Ameloot M, Van Dommelen L, De Schryver FC (1992) Kinetics and identifiability of intramolecular two-state excited-state processes. Global compartmental analysis of the fluorescence decay surface. J Phys Chem 96:6331–6342

67. Marquardt DW (1963) An algorithm for least-squares estimation of nonlinear parameters. J Soc Ind Appl Math 11:431–441

68. Hofkens J, Latterini L, Faes H, Vanoppen P, Jeuris K, De Feyter S, Kerimo J, Barbara PF, De Schryver FC (1997) Mesostructure of evaporated porphyrin thin films; porphyrin wheel formation. J Phys Chem B 101:10,588–10,598

69. Yamazaki I, Ohta N, Yoshinari S, Yamazaki T (1994) Site-selected excitation energy transport in Langmuir-Blodgett multilayer films. In: Masuhara H, De Schryver FC, Kitamura N, Tamai N (eds) Microchemistry, spectroscopy and chemistry in small domains. North-Holland Elsevier Science B.V., Amsterdam, p 431–440

70. Pevenage D (1999) Study of energy and electron transfer in Langmuir-Blodgett films using time-resolved fluorescence. PhD Dissertation, Katholieke Univeristeit Leuven, Leuven, Belgium

71. Dutta AK, Misra TN, Pal AJ (1996) A spectroscopic study of nonamphiphilic pyrene in Langmuir-Blodgett films: formation of aggregates. Langmuir 12:459–465

72. Van der Auweraer M, Ballet P, De Schryver FC, Kowalczyk A (1994) Parameter recovery and discrimination between different types of fluorescence decays obtained for dipole-dipole energy transfer in low dimensional systems. Chem Phys 187:399–416
73. Ballet P (1996) The study of molecular distributions in supramolecular structures using excitation transfer. Dissertation, Katholieke Univeristeit Leuven, Belgium
74. Klafter J, Blumen A (1985) Direct energy transfer in restricted geometries. J Lumin 34:77–82
75. Even U, Rademan K, Jortner J (1984) Electronic energy transfer on fractals. Phys Rev Lett 52:2164–2167
77. Dewey G (1991) Excitation energy transport in fractal aggregates. Chem Phys 150:45–451
77. Imans F, Viaene L, Van der Auweraer M, De Schryver FC (1996) Local properties of adsorption sites probed by the emission of fluorophores with different size and shape. Chem Phys Lett 262:583–588
78. Ishikawa M, Maruyama Y (1994) Femtosecond spectral hole burning of crystal violet in methanol. New evidence for ground state conformers. Chem Phys Lett 216:416
79. Tamai N, Matsuo H, Yamazaki T, Yamazaki I (1992) Excitation relaxation of oxacyanine in Langmuir-Blodgett monolayer films: picosecond time-resolved fluorescence studies. J Phys Chem 96:6550–6554
80. Biesmans A, Van der Auweraer M, De Schryver FC (1990) Influence of deposition circumstances on the spectroscopic properties of mixed monolayers of dioctadecyloxacarbocyanine and arachidic acid. Langmuir, 6:277–285
81. Tkachenko NV, Tauber AY, Lemmetyinen H, Hynninen PH (1996) Photoinduced energy and electron transfer of covalently linked zinc(II)-pyropheophytin-anthraquinone molecules in monolayers of Langmuir-Blodgett films. Thin Solid Films 280:244–248
82. Kristine Kilså Jensen K, Albinsson B, Van der Auweraer M, Vuorimaa E, Lemmetyinen H (1999) Interlayer energy transfer between carbazole and two 9-anthroyloxy derivatives in Langmuir-Blodgett films. J Phys Chem B 103:8514–8523
83. Pevenage D, Van der Auweraer M, De Schryver FC (1999) Photo-induced electron transfer in Langmuir-Blodgett films studied by time-resolved fluorescence. Langmuir 15:4641–4647

From Cyanines to Styryl Bases – Photophysical Properties, Photochemical Mechanisms, and Cation Sensing Abilities of Charged and Neutral Polymethinic Dyes

W. Rettig, K. Rurack, M. Sczepan

7.1
Introduction

Symmetric and unsymmetric cyanine dyes, styryl dyes, and styryl bases are among the most widely used classes of organic dyes [1, 2] and are employed in various applications including, e.g., spectral sensitization [3, 4], fluorescent labels [5, 6] and probes [7, 8] for biochemical applications, or laser dyes [9]. The general molecular structure of all three types of dyes includes an electron donating and an electron accepting moiety which are in direct electronic conjugation via either (i) an oligomethine π-system $-(CH{=}CH{-})_n$ as in the cyanines or (ii) an (elongated, for $n > 1$) styrene-type π-system $-(CH{=}CH{-})_n C_6 H_4-$ as in the ionic styryl dyes with a (formally) charged acceptor and in the neutral styryl bases. Furthermore, cyanine dyes can be subdivided according to their symmetric or unsymmetric end group substitution pattern. These interesting electronic properties along with their often largely different chemical structures such as, for instance, the presence of several bridged single and/or double bonds, have led to an intensive investigation of *trans-cis* photoisomerization processes [10–13], nonradiative deactivation pathways [14, 15], and general structure photophysics relationships [16–18] as well as many theoretical studies [19–23] over the past few years.

Upon photoexcitation, styryl dyes and styryl bases often show an excited state reaction involving an intramolecular charge transfer (ICT) process [24–26]. Thus, especially in polar solvents, many of these dyes show broad and structureless absorption and emission spectra, the latter often being strongly Stokes shifted. In contrast, symmetric and unsymmetric cyanine dyes usually show relatively narrow and structured spectra accompanied by a small Stokes shift [15, 27–29]. Accordingly, the molecular arrangement, i.e., the donor π-system acceptor design, and the resulting photophysical properties of these dyes possess an enormous potential for the construction of fluoroionophores since introduction of an ion specific receptor to the donor or acceptor position can yield fluorescent probes showing pronounced spectral shifts upon analyte binding [30–33]. To date, many such compounds are known, the great majority of them carrying a cation specific mono- or polyaza crown ether receptor as the electron donating moiety [34–45].

In the present contribution, the results and implications of various experimental and theoretical mechanistic studies of the ground and excited state pho-

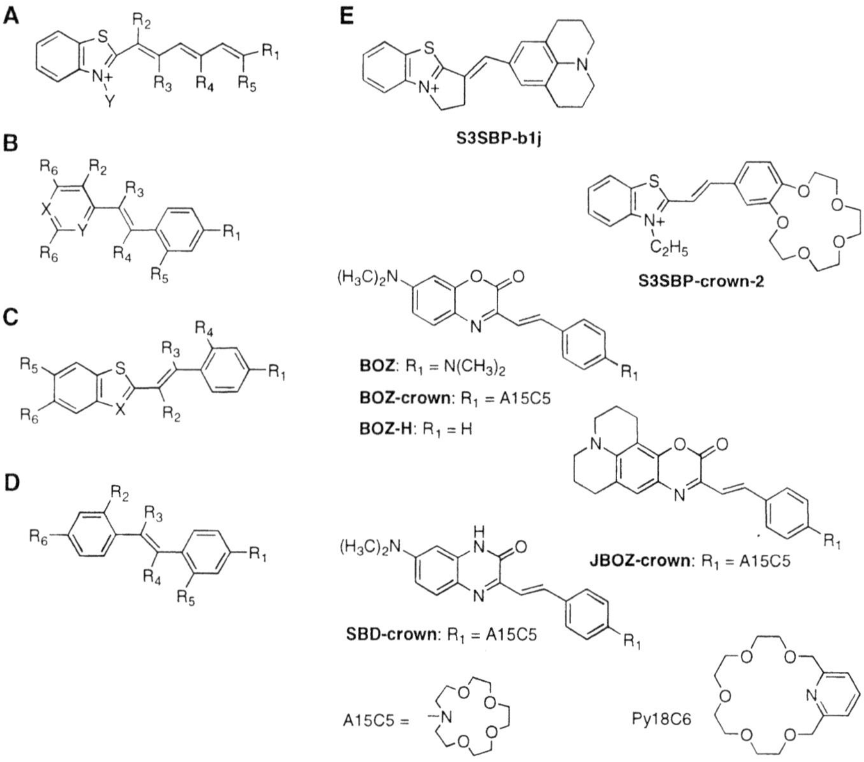

Fig. 7.1 A–E. Chemical structures of the compounds studied and discussed: **A** hemicyanine dyes, for substitution and bridging pattern, see Table 7.1; **B** styryl dyes/bases with a pyridine/ium acceptor (Table 7.2); **C** styryl dyes/bases with a benzothiazole/ium acceptor (Table 7.3); **D** styryl base related stilbene dyes (Table 7.4); **E** miscellaneous styryl dyes/bases and receptor units

Table 7.1. Substitution pattern of the hemicyanine dyes in Fig. 7.1A

	Y	R_2	R_3	R_4	R_5	R_1
S3S	CH_3	H	H	H	H	$N(CH_3)_2$
S3S-crown	CH_3	H	H	H	H	A15C5
S3S-b1	$Y+R_2=CH_2\text{-}CH_2$		H	H	H	$N(CH_3)_2$
S3S-b34	CH_3	H	$R_3+R_4=CH_2\text{-}C(CH_3)_2\text{-}CH_2$		H	$N(CH_3)_2$
S3S-b34-cr.	CH_3	H	$R_3+R_4=CH_2\text{-}C(CH_3)_2\text{-}CH_2$		H	A15C5
S3S-b3456	CH_3	H	$R_3+R_4+R_5=CH_2\text{-}CH_2\text{-}CH\text{-}CH_2\text{-}CH_2$			$N(CH_3)_2$
S3S-b3456-cr.	CH_3	H	$R_3+R_4+R_5=CH_2\text{-}CH_2\text{-}CH\text{-}CH_2\text{-}CH_2$			A15C5
S3S-b13456	$Y+R_2=CH_2\text{-}CH_2$		$R_3+R_4+R_5=CH_2\text{-}CH_2\text{-}CH\text{-}CH_2\text{-}CH_2$			$N(CH_3)_2$

tophysics of a series of dyes shown in Fig. 7.1 and Tables 7.1 – 7.4 and differing in both (donor acceptor) substitution as well as single and/or double bond bridging patterns are discussed. The resulting knowledge of the underlying photochemical mechanisms can be utilized to gain a more systematic access to the understanding of the complexation behavior of receptor substituted cyanine and styryl based fluoroionophores for future rational probe design.

Table 7.2. Substitution pattern of the styryl dyes and styryl bases in Fig. 7.1 B

	X	Y	R_1	R_2	R_3	R_4	R_5	R_6
DASPMI	N^+CH_3	CH	$N(CH_3)_2$	H	H	H	H	H
p-DPD	N^+CH_3	CH	$N(CH_3)_2$	H	H	H	H	C_6H_5
p-DPD-crown	N^+CH_3	CH	A15C5	H	H	H	H	C_6H_5
p-DPD-b1	N^+CH_3	CH	$N(CH_3)_2$	CH_2-CH_2		H	H	C_6H_5
p-DPD-b3	N^+CH_3	CH	$N(CH_3)_2$	H	H	CH_2-CH_2		C_6H_5
p-DPD-b13	N^+CH_3	CH	$N(CH_3)_2$	CH_2-CH_2		CH_2-CH_2		C_6H_5
o-DPD-crown	CH	N^+CH_3	A15C5	H	H	H	H	C_6H_5
p-DBD-n-crown	N	CH	A15C5	H	H	H	H	$C(CH_3)_3$

Table 7.3. Substitution pattern of the styryl dyes and styryl bases in Fig. 7.1 C

	X	R_2	R_3	R_4	R_5	R_6	R_1
S3SBP	$N^+C_2H_5$	H	H	H	H	H	$N(CH_3)_2$
S3SBP-crown-1	$N^+C_2H_5$	H	H	H	H	H	A15C5
S3SBP-b1	N^+CH_2-CH_2		H	H	H	H	$N(CH_3)_2$
S3SBP-b3	$N^+C_2H_5$	H	CH_2-CH_2		H	H	$N(CH_3)_2$
S3SBP-n	N	H	H	H	H	H	$N(CH_3)_2$
S3SBP-n-crown-1	N	H	H	H	Z^a		$N(CH_3)_2$
S3SBP-n-crown-2	N	H	H	H	Z		OCH_3
D-S3SBP-n	N	H	H	H	$N(CH_3)_2$	H	$N(CH_3)_2$
D-S3SBP-n-crown-1	N	H	H	H	$N(CH_3)_2$	H	A15C5

[a] $Z = O\text{-}(CH_2)_2\text{-}S\text{-}(CH_2)_2\text{-}(O\text{-}(CH_2)_2\text{-})_2S\text{-}(CH_2)_2\text{-}O.$

Table 7.4. Substitution pattern of the stilbene dyes in Fig. 7.1 D

	R_1	R_2	R_3	R_4	R_5	R_6
S	H	H	H	H	H	H
DS	$N(CH_3)_2$	H	H	H	H	H
DCS	$N(CH_3)_2$	H	H	H	H	CN
DCS-crown	A15C5	H	H	H	H	CN
DCS-b13	$N(CH_3)_2$	CH_2-CH_2		CH_2-CH_2		CN
DDS	$N(CH_3)_2$	H	H	H	H	$N(CH_3)_2$
DDS-crown	A15C5	H	H	H	H	$N(CH_3)_2$
DNS	$N(CH_3)_2$	H	H	H	H	NO_2
S-F	$N^+(CH_3)_3$	H	H	H	H	$B^-(CH_3)_3$

7.2
Cyanine Dyes

7.2.1
Photophysical Model Mechanisms

Corresponding to many other symmetric and unsymmetric cyanine dyes (see, e.g., [6, 11, 46, 47]), all the hemicyanines of the S3S series show a characteristic narrow and structured absorption band in the far visible spectral range and an only slightly Stokes shifted emission band of mirror image shape [15, 48]. Depending on the bridging pattern, the possibility of populating different ground state conformers decreases with an increasingly rigidized molecular skeleton and thus the bandwidth of the steady-state spectra of the S3S series decreases in the order of unbridged > monobridged (b34) > double-bridged (b3456) derivatives [15, 48].

In polar aprotic or protic solvents such as acetonitrile or ethanol, the fluorescence quantum yields are in the range of $0.01 < \phi_f < 0.15$ and do not correlate with the increasing rigidity of the molecule [15, 48]. Moreover, the *neo*-pentylene bridged (b34) derivatives show the weakest fluorescence for many symmetric and unsymmetric cyanines studied so far [15, 46, 48].

This unusual behavior can be more closely inspected by low-temperature fluorescence spectroscopy utilizing the viscous property of the solvent to slow down the photochemical reaction which quenches the fluorescence. At lower temperatures, the solvent viscosity increases, and therefore all the photochemical reactions are decelerated. At this point, it is instructive to consider two model cases – case *a* where the photochemical reaction proceeds over a significant activation barrier and case *b* where the reaction is barrierless. In the barrierless case *b*, only viscosity will slow down the reaction, and the rate constant of the photochemical reaction k_{nr} will vary with temperature in the same way as solvent viscosity, yielding an Arrhenius plot for k_{nr} which will show the same slope as that of solvent viscosity E_η. For case *a*, on the other hand, the barrier will additionally reduce k_{nr} for lower temperatures and therefore Arrhenius slopes will be considerably larger than E_η. Such an approach has been widely used in the discussion of the excited-state twisting reaction of dimethylaminobenzonitrile [49–51]. In this case, reaction rates reduced by low temperature could even be compared to rates slowed down by increasing viscosity at constant temperature (high-pressure experiments) enabling the quantitative determination of intrinsic activation barriers, unaffected by solvent viscosity influences [51]. Observed Arrhenius slopes equal to or smaller than E_η point to the absence of an intrinsic activation energy, i.e., to a barrierless reaction [49, 50].

Figure 7.2 shows the observed low-temperature behavior of S3S. It is obvious that the low fluorescence intensity observed at room temperature is restored at low-temperature, reaching a value close to unity [15]. From Fig. 7.2 and the corresponding fluorescence lifetime data, the temperature-dependent photochemical reaction rate constants k_{nr} can be extracted and plotted in an Arrhenius fashion (Fig. 7.3). This figure also contains the results for the selectively bridged derivatives S3S-b1, S3S-b34, S3S-b3456, and S3S-b13456. As can be seen, all

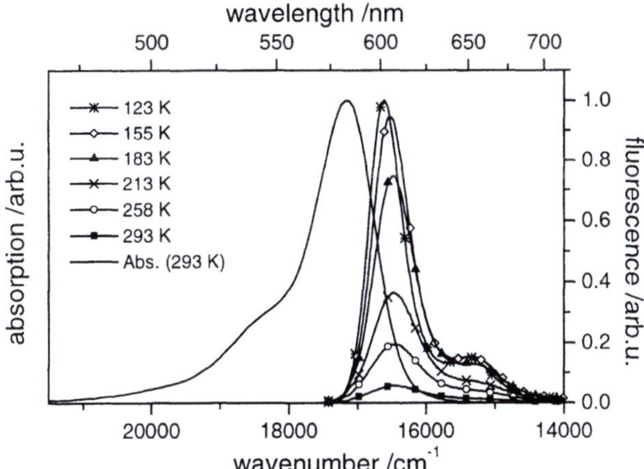

Fig. 7.2. Absorption and fluorescence spectra of S3S in ethanol at different temperatures

Arrhenius plots possess a similar slope ($E_a = 17$ kJ mol^{-1}) which is, within the margin of thermal energy ($k_B T \approx 2.5$ kJ mol^{-1}), equal to E_η ($E_\eta = 13$ kJ mol^{-1}, estimated from data from [52, 53]) and does not differ significantly for bridged and unbridged compounds. But the bridged compounds show another significant difference: The Arrhenius lines are shifted upward for these derivatives, indicating a larger preexponential Arrhenius factor. The reduced fluorescence quantum yield of the bridged compounds at room temperature as compared to the unbridged ones is thus substantiated by the low-temperature experiments and shown to be due to a photoreaction leading to a nonemissive state on an essentially barrierless excited-state reaction hypersurface.

The fact that the bridged compounds react faster than the unbridged analogues is less related to a difference in activation barrier than to differences of the preexponential factor (Fig. 7.3), i.e., to entropic effects [54–56]. The "normal" and often observed behavior of dyes is that molecular bridging entails an increase of fluorescence quantum yields by reducing the nonradiative rate constants. This occurs because flexible groups lead to an increased nonradiative coupling of the excited state with the denser vibrational manifold of the ground state (the so-called loose-bolt theory [57, 58]). In contrast to this normal behavior, the dyes of the S3S series display an opposite behavior, i.e., they behave in an "anti-loose-bolt" manner.

One possible way to understand these unusual features is the concept of conical intersections (COIs) [59–66]. These are photochemical funnels [67–71] connecting ground and excited states, with the possibility for barrierless transit, as established by femtosecond spectroscopy in a few cases [72, 73]. An example calculated for hexatriene is given in Fig. 7.4 [72]. In larger molecules like S3S, similar funnels can be present, connected with bond twisting. If the access to the COIs is provided by diffusion on a barrierless horizontal surface, the reaction rate will be position dependent and will become, according to the Einstein-Smoluchowski equation [74], inversely proportional to the square of the horizontal

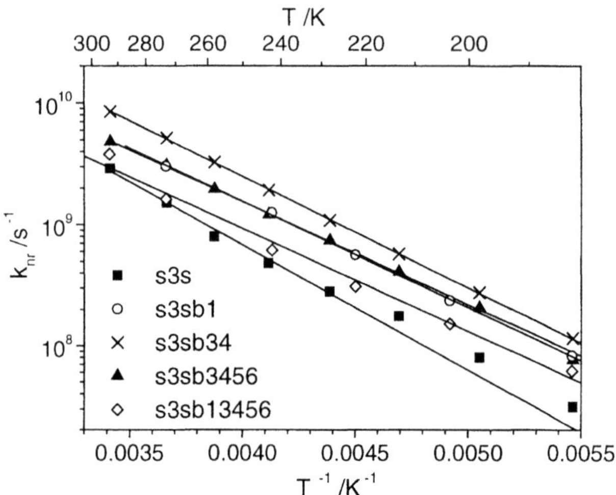

Fig. 7.3. Arrhenius plot for S3S and the bridged derivatives S3S-b1, S3S-b34, S3S-b3456, and S3S-b13456 in ethanol

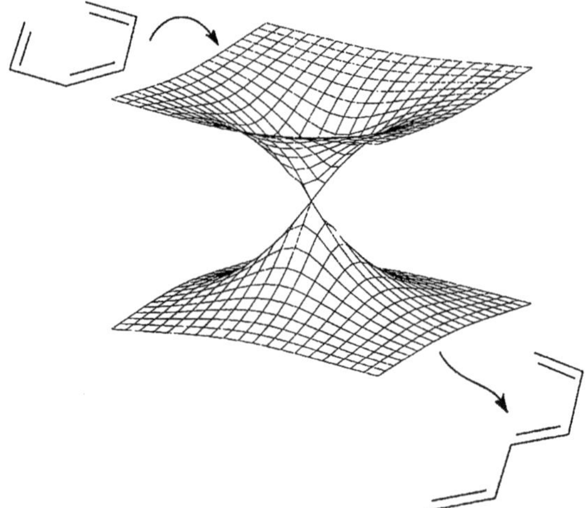

Fig. 7.4. The transition from the excited to the ground state through a conical intersection, with the example of hexatriene (adapted with permission from [72]). The central cone corresponds to a highly twisted conformation (several bonds involved)

distance between the initial position and the conical intersection. For the selectively bridged S3S derivatives, some regions of phase-space are unavailable, therefore increasing the chances of arriving at the region of the COI. This, in turn, should result in increased preexponential factors for these dyes. Similar entropic effects on essentially flat surfaces have recently been discussed in the context of femtochemistry [75].

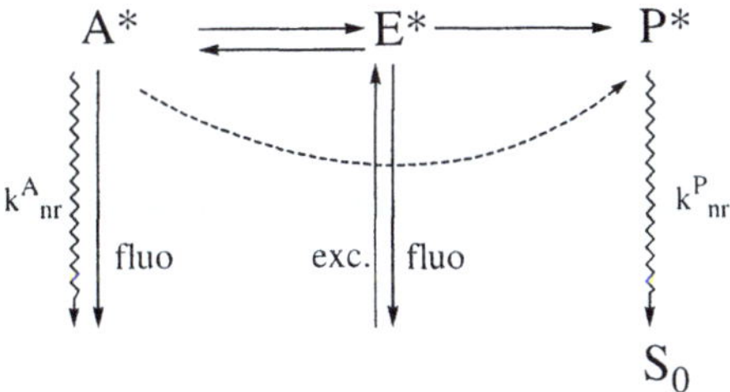

Fig. 7.5. Proposed reaction scheme involving multiple emissive states (A*, E*). E* is the planar conformer, A* corresponds to the different single-bond twisted conformers and P* to the double bond twisted conformers. Relaxation to A* can enhance the fluorescence quantum yield if k_{nr}^A is very small (case of DCS and S3S). On the other hand, if A* decays fast in a nonradiative way, P* formation can be suppressed, and the compound is then insensitive to photoisomerization (case of *p*-DPD and DASPMI). For the structures of the dyes, see Fig. 7.1 and Table 7.2

The validity of the excited state reaction scheme is also supported by recent theoretical investigations. Here, the photochemical reactivity of certain bonds in an ionic polymethine chain could be correlated with the S_0–S_1 energy gap for twisted conformations with the proximity of a possible COI leading to fast non-radiative decay [22]. These results suggest that most probably the competing population of several emissive and nonemissive twisted conformers during the lifetime of the excited state critically determines the fluorescence quantum yield. The corresponding excited-state reaction scheme (Fig. 7.5) involves the formation of nonemissive states P* (formally connected with *trans-cis* photo-isomerization) by twisting around the bonds in the hexamethine chain showing a stronger double bond character (these bonds are formally identical with the double bonds in the mesomeric structures given in Fig. 7.1) and emissive states of other twisted conformations which are accessible by rotation around a (formal) single bond [22, 76].

The behavior of the S3S dyes can be understood on the basis of this model since the most active bonds were found to be situated near the chain's end [22] and especially the second flexible bond neighboring the heteroaromatic unit was identified as the most prominent candidate for reaching a conical intersection between the ground and the excited state [22]. Moreover, since rotation of the first bond in the polymethine chain is highly improbable [22] and the second (formal) single bond is bridged in S3S-b34 and its crowned analogue, efficient P* state formation (at the first formal double bond) seems to account for the reduced fluorescence yield as compared to S3S and S3S-crown.

7.2.2
Complexation Properties

The cation-induced spectroscopic effects are also different for the series of hemicyanine crowns (Table 7.5) [48]. For S3S-crown and S3S-b34-crown, addition of $Ca(ClO_4)_2$ to an acetonitrile solution of the ionophore induces a hypso- and hypochromic shift of the absorption band at comparatively high cation concentrations indicating the formation of weak complexes (of a well-defined 1:1

Table 7.5. Spectroscopic properties of selected crowned hemicyanines, styryl dyes, and some of their cation complexes in acetonitrile at room temperature (data taken from [48], except for S3SBP-crown-1 and its Ba^{II} complex [37])

	$\tilde{\nu}_{abs}$ $(10^3\ cm^{-1})$	$\tilde{\nu}_{cm}$ $(10^3\ cm^{-1})$	$\Delta\tilde{\nu}_{Stokes}$ (cm^{-1})	$\Delta\tilde{\nu}_{cp-fp}$ [a] (cm^{-1})	ϕ_f	$\log K_S$
S3S-crown	17.09	16.34	750	–	0.100	–
+ Ca^{II}	19.60	16.31	3290	2510, 300	0.058 [b]	0.94
S3S-b34-crown	16.81	16.29	520	–	0.020	–
+ Ca^{II}	20.67	16.27	4400	3860, 200	0.009 [b]	1.41
S3S-b3456-crown	17.03	16.56	470	–	0.032	–
+ Ca^{II}	n. d. [c]	–	–	–	–	–
S3SBP-crown-1	19.23	17.18	2050	–	0.20	–
+ Ba^{II}	23.26	17.48	5780	4030, 300	0.60	1.91 [d]
p-DPD-crown	20.37	14.96	5410	–	0.021	–
+ Ca^{II}	26.61	15.28	11330	6240, 320	0.072	2.90
o-DPD-crown	20.82	14.02	6800	–	2×10^{-3}	–
+ Ca^{II}	26.67	14.90	11770	5850, 880	7×10^{-3}	2.82

[a] Shift in absorption, emission between free and complexed probe.
[b] Determined at a 4×10^5-fold excess of Ca^{II} because full complexation could not be reached.
[c] No complex detectable.
[d] Due to a better fit of cation into the cavity, $\log K_S$ of the Ca^{II} complex should be equal to or higher than $\log K_S$ of the Ba^{II} complex.

stoichiometry, see [48]). Furthermore, the cation-induced changes in emission are even smaller than those in absorption and negligible shifts accompanied by moderate fluorescence quenching are found for S3S-crown and S3S-b34-crown upon Ca^{II} binding (Fig. 7.6) [48]. Moreover, no such effects in either absorption or emission are found for the more rigidized S3S-b3456-crown even in the presence of very high Ca^{II} concentrations (~ 0.1 mol/l) [48].

The generally weak ion binding ability of the hemicyanine crowns can be understood in terms of the resonance interaction and distribution of the positive charge within the whole molecule, see Sect. 7.2.1 [20]. Thus, complex formation has to overcome the weak affinity between cation and partially positively charged receptor. The pronounced differences between the cation-induced shifts in absorption and emission for S3S-crown and S3S-b34-crown and their Ca^{II} complexes were attributed to the charge redistribution which takes place in the excited state as discussed above. The shift of the positive charge can lead to the immediate neighborhood of positively charged crown ether nitrogen atom and cation followed by a de- and recoordination reaction [77–80] with the result that the fluorescence spectra of this recoordinated species are often nearly indistinguishable from those of the uncomplexed dye [78, 79]. (Note, that for the positively charged hemicyanines, optical excitation is rather accompanied by a charge shift involving the whole chromophore as compared to an actual charge transfer from donor to acceptor in neutral merocyanines [23].)

In contrast, the cation-induced fluorescence quenching mechanism remains unclear at present. On the one hand, the increased bulkiness of the "cation-in-the-crown" receptor fragment should slow down all rotations in S_1 but should

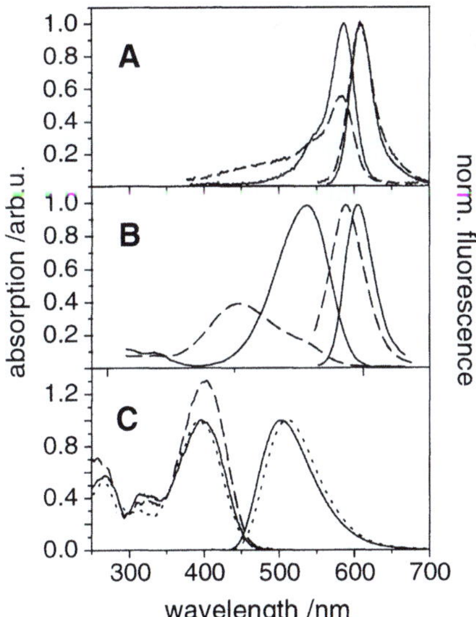

Fig. 7.6 A–C. Absorption and normalized emission spectra of: A S3S-crown (*solid line*) and its Ca^{II} complex (*dashed line*); B S3SBP-crown-1 (*solid line*) and its Ba^{II} complex (*dashed line*); C S3SBP-n (*dotted line*), S3SBP-n-crown-1 (*solid line*), and the Hg^{II} complex of S3SBP-n-crown-1 (*dashed line*, only absorption spectrum included) in acetonitrile at room temperature. Part B adapted from [37]

especially decelerate the relaxation path leading to the formation of nonemissive P* conformers [37] which, in turn, should lead to the opposite effect of enhanced emission, i.e., to the opposite effect as observed. On the other hand, the presence of a cation in the crown ether moiety, though not directly bound to the nitrogen atom, should lead to changes in the relative energetic positions of the different excited states involved in the deactivation of the photoexcited complex and should change the respective rate constants of an underlying reaction scheme such as Fig. 7.5 (for a detailed discussion, see Sect. 7.3.2). Since only comparatively weak quenching factors < 3 are observed for the crowned hemicyanines, the (slight) beneficial steric effect (slower population of P*) might be counterbalanced by the electrostatic influence of the cation bound (if A* and P* are closer lying in energy). In the case of S3S-b3456-crown, the lack of any detectable coordination was attributed to sterical crowding of bridge and cation, since the electronic features of the three crowned hemicyanine dyes are very similar [45, 48].

7.3
Styryl Dyes

7.3.1
Photophysical Model Mechanisms

Introduction of an "aromatic bridge" in the polymethine chain yields the corresponding styryl dyes of S3SBP type and, when exchanging the benzothiazolium for a 2,6-diphenyl-1-methylpyridinium moiety, the related dyes of the DPD series (Fig. 7.1, Tables 7.2 and 7.3). The latter are also closely related to the 2,6-unsubstituted DASPMI dyes (Fig. 7.1 and Table 7.2) [12, 35, 81, 82]. Within the triad principle of electronic structures of organic dyes [27, 83], the introduction of an aromatic ring into a cyanine system can be seen as a step from a more to a less ideal polymethinic structure, thus shifting absorption and emission spectra to shorter wavelengths. At the same time, the charge distribution can be more asymmetric in the ground state of the molecule leading to a pronounced intramolecular charge transfer (ICT) character upon excitation [24]. This also affects the acidity of the donor and the basicity of the acceptor fragment. Depending on solvent polarity, the more dipolar nature of the excited state of DPD as compared to the ground state can lead to pronounced Stokes shifts entailing an increase in bandwidth and a decrease in vibronic structure of both absorption and emission bands [48, 84].

This can most readily be seen if the spectra of S3S, S3SBP, and p-DPD are displayed in a comparative manner (Fig. 7.7). Several things are noteworthy:

- The absorption shifts to the red and broadens along the series.
- The fluorescence spectra broaden in the same sense.

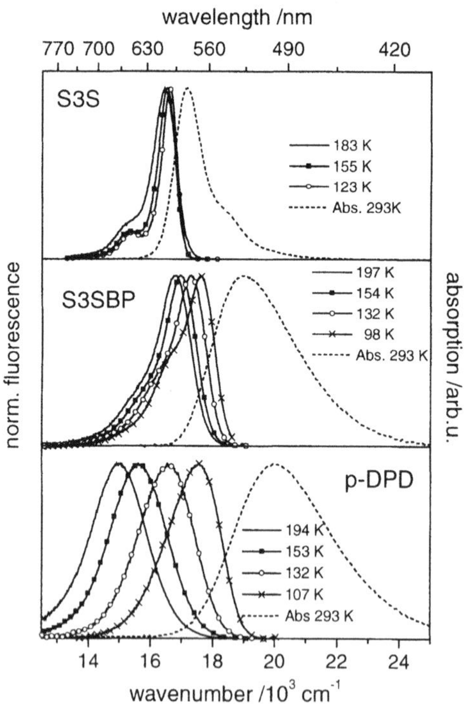

Fig. 7.7. Absorption at room temperature and fluorescence spectra at different low temperatures of S3S, S3SBP, and p-DPD in ethanol. Note the increasing Stokes shift and the broadening of the spectra along the series. The absorption shifts to the blue, although the overall length of the conjugated system remains constant

- The Stokes shift increases in the same order leading to fluorescence spectra at the highest temperature displayed (ca. 200 K) which are strongly red-shifted for *p*-DPD with respect to S3S.
- The spectra of the three compounds show a strongly different temperature dependence: whereas S3S is nearly insensitive to temperature, the spectra of *p*-DPD strongly shift to the blue upon cooling. S3SBP behaves in an intermediate way.

The temperature dependence of the fluorescence spectra can be related to the influence of solvation. In an uncharged molecule, whenever the charge distribution changes strongly upon excitation leading to a large change of the dipole moment, the solvent reorients thus stabilizing the newly created dipole. This is visible in a time-dependent fluorescence redshift which is in the time range of sub-ps to several ps at room temperature, depending on the solvent [85]. Cooling the solvent slows down its relaxation rate, and at some temperature it becomes slower than the fluorescence lifetime [86]. The fluorescence spectra then shift to the blue as observed for *p*-DPD. Although charged dyes like *p*-DPD do not possess a proper dipole moment, a dipole contribution can be defined from a multipole expansion [87], and if this dipole contribution changes upon excitation, the same effects are observed as for neutral dyes with strong dipole moment changes. In fact, the ps and sub-ps time resolved spectral changes of *p*-DPD in ethanol observed at room temperature [88] can very well be explained within the well-established theory of time-dependent solvation.

The nonradiative processes can be studied if the fluorescence quantum yields are compared as a function of temperature. A representative example is displayed in Fig. 7.8 for S3SBP and its bridged counterparts.

Using these data, the nonradiative rate constants can be derived as a function of temperature and compared for all the compounds investigated. Some compounds like S3SPB-b1j do not show measurable fluorescence at room temperature, but low temperatures produce this fluorescence (Fig. 7.8). It is therefore de-

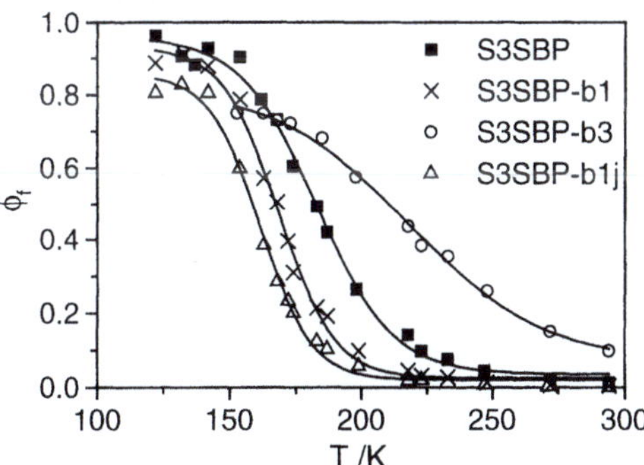

Fig. 7.8. Fluorescence quantum yields of S3SBP and bridged derivatives in ethanol as a function of temperature. The temperature dependence is sigmoid in shape, and the midpoint (inflexion point) of the sigmoid curve is an indication of how fast the photoreaction proceeds. Lower midpoint temperatures indicate that the compound has a larger nonradiative decay constant

Table 7.6. Relative changes of the nonradiative rate constant k_{nr} for selective bridging in three different compound families. The values in brackets are those for 298 K, the unbracketed values are for low temperature (200 K)

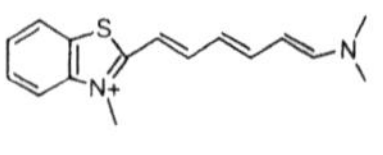 S3S: 1 (1)	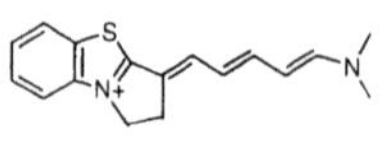 S3S-b1: 3.0 (2.1)	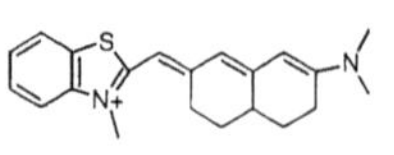 S3S-b3456: 2.6 (1.4)	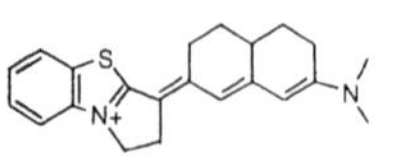 S3S-b13456: 1.9 (1.1)
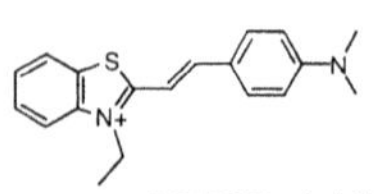 S3SBP: 1 (1)	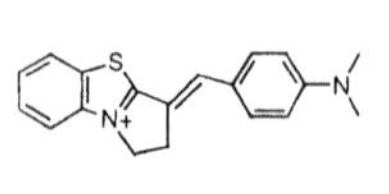 S3SBP-b1: 1.7 (3.0)	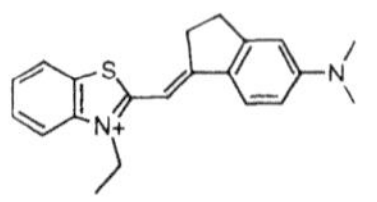 S3SBP-b3: 0.06 (0.1)	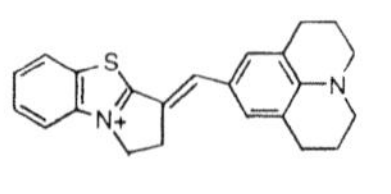 S3SBP-b1: 2.8 (-)
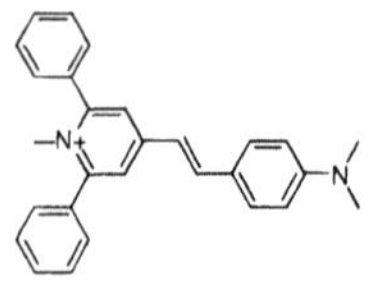 p-DPD: 1 (1)	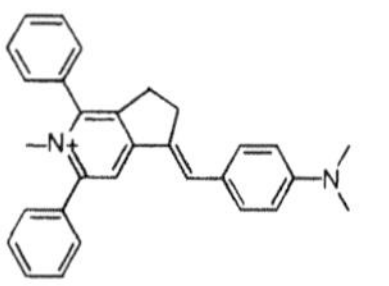 p-DPD-b1: 2.6 (1.3)	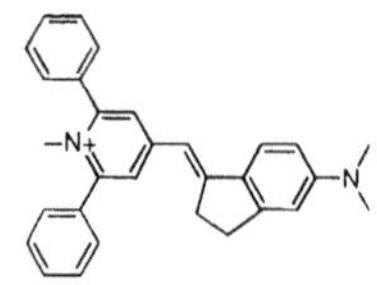 p-DPD-b3: 1.2 (0.8)	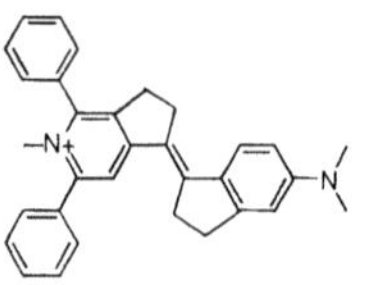 p-DPD-b13: 0.5 (0.1)

sirable to compare the rate constants for a temperature where all compounds fluoresce. Table 7.6 contains an estimation of k_{nr} normalized to the unbridged derivative of every compound family.

Some of the bridged derivatives show – contrary to the behavior expected from the loose bolt theory – higher rates of nonradiative decay ("inverse loose-bolt behavior"), while others behave in the normal way. The effect differs for the three compound families:

- In the S3S family, all bridged derivatives show higher k_{nr}.
- In the S3SBP family, the bond-1-bridged derivative shows higher k_{nr} while the bond-3-bridged analogue shows lower k_{nr}. The julolidino bridged derivative shows the highest nonradiative rate constant (lowest fluorescence quantum yield).
- In the DPD family, derivatives with only one bridged bond do not display a significant change in k_{nr} (small increase for p-DPD-b1, small decrease for p-DPD-b3), but the doubly bridged derivative p-DPD-b13 shows a strongly reduced k_{nr}.
- The stilbazolium compound p-DPD is a highly polarized stilbene derivative and can be compared to less strongly perturbed stilbenes such as DCS, see Fig. 7.1 and Table 7.4. For this compound, a family of similarly selectively bridged derivatives has been studied forming the DCS family [16, 17, 89–92]. Interestingly, the comparison of DCS with DCS-b13 (the analogue to p-DPD-b13) yields the opposite result, namely a very strong reduction of the fluorescence quantum yield for the bridged compound [89].

All the observed effects can be accounted for by the general kinetic scheme in Fig. 7.5. Most important is the feature that, in addition to states P* which are nonradiative (connected with a conical intersection), photochemical product states A* can be reached depending on the bridging pattern. A* is highly dipolar and can possess emissive properties. Bridging of (single) bonds would result in blocking of the radiative decay channel E* $\rightarrow$ A* $\rightarrow$ S$_0$ thus leading to a higher rate of nonradiative deactivation (channel E* $\rightarrow$ P* $\rightarrow$ S$_0$) if k_{nr}^A is negligible. This leads to increased accessibility of P* and to a strong (ca. 100-fold) decrease of the fluorescence quantum yield for the bridged compound in the DCS family. A* can, however, also be connected with strong nonradiative decay. This is the case for p-DPD. Here, the nonradiative decay of p-DPD via A* is much faster than that of p-DPD-b13 via P* leading to the "normal" behavior of reduced fluorescence for more flexible compounds. This explanation corresponds to that of the loose bolt theory [57, 58]. It can, however, not account for the results of S3SBP where some selective bridging leads to an increase (S3SBP-b1, S3SBP-b1j), some to a reduction of nonradiative rate constants (S3SBP-b3, Table 7.6), and the general kinetic scheme of Fig. 7.5 is much better suited than the simple loose bolt theory in this case.

Temperature dependent quantum yield changes and temperature-induced shifts of the fluorescence spectrum derive from different sources. Whereas k_{nr} depends on the compound investigated and its specific bridging pattern, the shift of the fluorescence spectrum is largely a property of the solvent used. This can be nicely seen by comparing the fluorescence shifts for the different compounds in highly polar solvents such as alcohols (Fig. 7.9). The temperature onset of the blue shift (inflexion point of the sigmoid curve) is comparable for all dye families S3S, S3SBP, and DPD, and the bridging pattern has no significant influence on the size and temperature dependence of the blue shift. However, the size depends strongly on the compound family considered, with DPD dyes showing the strongest temperature-induced changes.

The strongly different behavior in the size of the solvent-induced fluorescence red shift for S3S as compared to DPD dyes can be understood by considering the charge distribution and its changes upon excitation (charge shift between ground and excited state [93]). While for S3S HOMO and LUMO (the molecular orbitals involved in the S$_0$ $\rightarrow$ S$_1$ transition) are delocalized along the polymethinic chain, for p-DPD the frontier orbitals are localized on one of the aromatic end groups of the molecule (see Fig. 7.10). This leads to a small charge shift upon excitation for S3S but to a large degree of charge shift for p-DPD. Thus, dipolar solvent relaxation effects producing the Stokes shift are small for S3S but large for p-DPD.

It remains to elucidate the reasons why S3S and p-DPD behave so differently in terms of charge distribution and, connected with it, in the pattern of the photochemical reactivity ("photochemical spectrum" [22, 23, 94, 95]). If all the factors were known and controlled, tailor-made dyes could be constructed on the paper (or better within the computer), e.g., for analytical purposes as ion-complexing dyes (see below). One step in this direction is the observation how the ground state charge distribution depends on the molecular structure. In this respect, the electron density on amino groups of cyanine dyes is especially interesting because it determines the stability of chromophore-ion complexes of the corresponding aza crowned dye derivatives (see below). As a general tendency it can be remarked that cyanines with an increasing number of aromatic rings

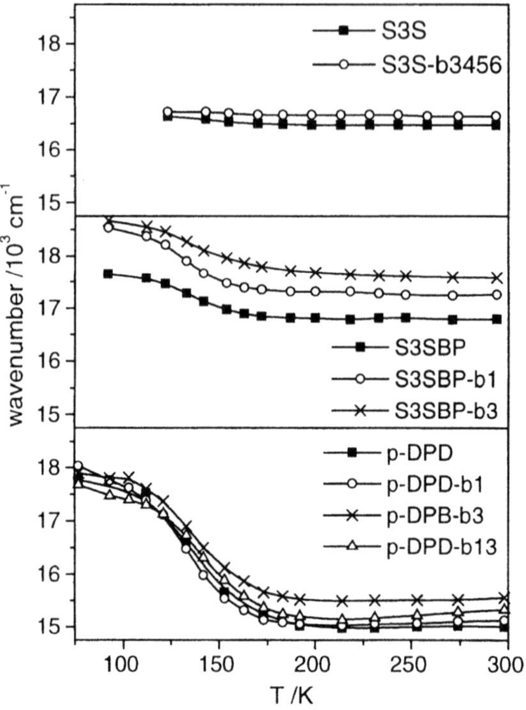

Fig. 7.9. Wavenumber of the maximum of the fluorescence bands of the investigated compounds in ethanol as a function of temperature. Note that the inflection points of the sigmoid curves neither depend on the compound family nor on the bridging pattern

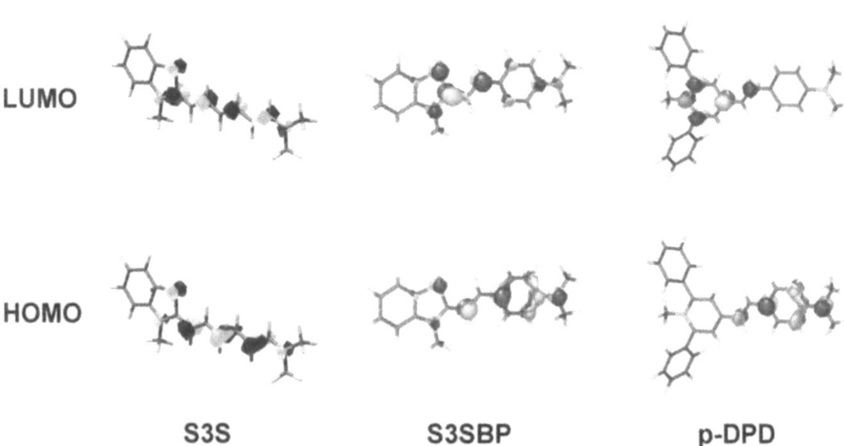

Fig. 7.10. Calculated frontier orbitals (HOMO, LUMO) for the unbridged dyes S3S, S3SBP and p-DPD (AM1 method; AMPAC 5 package, Semichem, Shawnee, 1994)

(like DPD or DASPMI dyes) show increased complexation constants because the electron density on the amino-nitrogen is increased [45]. In parallel, the charge shift upon excitation is especially large in this family, leading to the tendency of excited-state ion decoordination [77–80, 90].

Theoretical calculations can help tremendously at this point. As an example, Fig. 7.11 shows very simple ground state calculations within the stilbene family. It has been noted that the so-called bond-length alternation (BLA) is an important parameter for correlating the nonlinear optical properties of these and related dyes [96, 97]. In the stilbene family, the BLA can be defined by the difference of the bond lengths of central single and double bonds. The BLA is calculated by simple full-geometry optimization of the respective compound. Figure 7.11 shows that along the series from stilbene to the highly polarized DASPMI-structure, the BLA shows a sigmoid behavior. This is paralleled by a polarization of the orbitals [98] similar to that shown in Fig. 7.10 for *p*-DPD. It is actually the inner electric field which produces the orbital polarization as shown by the model molecule S–F, a stilbene derivative with nonconjugated positively and negatively charged attachments which has been experimentally studied in the context of nonlinear optical properties [99]. These studies can be viewed in relation to Dähne's triad theory of dyes, with the three color states of polymethinic, polyenic, and aromatic character. The bond length alternation is connected with the relative contribution of polymethinic and polyenic character or valence bond resonance structures [27, 83, 96, 97]. The aromatic rings cause the third contribution of the triad color rules, the aromatic character [27, 83], to be of importance, too. The relative weight of these triad states is connected to photochemistry and governs the "photochemical spectrum" of cyanine dyes, i.e., the availability of conical intersections leading to fluorescence quenching [22, 23, 94, 95].

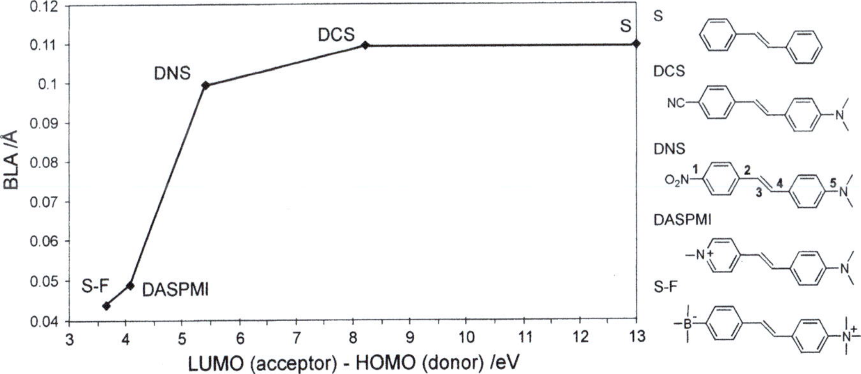

Fig. 7.11. Bond-length alternation (BLA) vs the energy gap between LUMO and HOMO in donor-acceptor stilbenes (AM1 with bond and angle optimization for planar molecules and C. I. = 8; BLA is the difference between average single and double bond lengths; the donor-acceptor difference is defined as the energy difference between LUMO and HOMO localized on the end groups by appropriate 90°-twists of acceptor and donor fragments, respectively; by courtesy of M. Dekhtyar and W. Rettig [98])

An interesting application of these color rules can be seen in the visual pigment rhodopsin or its bacterial counterpart bacteriorhodopsin, used for photosynthesis. In both cases, the active chromophore is retinal, i.e., its Schiff-base linked derivative PRSB (protonated Retinal Schiff Base). PRSB in solution shows an absorption maximum at 440 nm, which is red-shifted to 568 nm in bacteriorhodopsin [100]. This is known as the opsin shift, the origin of which is still highly controversial. We note, however, that the color rules [27, 83] predict a spectral redshift for those structures with the purest polymethine character. The opsin red shift can thus also be viewed as induction of increased polymethinic character. Various possibilities can be discussed for the cause of the opsin shift. A specific distribution of charges, induced by the opsin backbone, to adjust the BLA to a "pure polymethinic" value near zero, would be a theoretically founded possibility. Preliminary results in this direction include the reconstitution of bacteriorhodopsin with the artificial cyanine dye S3S. In this case, the absorption shifts from 576 nm (water) to 602 nm induced by the protein [101]. Similar results for other cyanines have been reported [102].

7.3.2
Complexation Properties

Most of the ionic ICT fluoroionophores known today contain their cation-sensitive receptor in the donor part of the molecule [33–38, 45, 80]. As in the crowned derivatives shown in Fig. 7.1 (Tables 7.2 and 7.3), the receptor introduced to these dyes is mostly a monoaza polyoxa crown ether. Other fluoroionophores with heteromacrocyclic receptors lacking a nitrogen atom at the terminal end of the polymethinic chain, i.e., containing a weaker donor (e.g., the benzo crown unit in S3SBP-crown-2 [103], Fig. 7.1 E), yield only considerably smaller cation-induced changes [103] with generally similar tendencies as their A15C5 substituted analogues and will be discussed in conjunction with the styryl base probes in Sect. 7.4.2.1. Some general considerations as given in Sect. 7.3.1 include an increased charge density at the crown ether nitrogen atom in the ground state for S3SBP as compared to S3S (see also Table 7.7). The opposite effect is expected for the excited state. A further important factor is the nature of charge localization on the molecular fragments (cf. S3SBP vs DPD, see Figs. 7.7 and 7.10). Thus, on the basis of the mechanistic considerations, we would expect the formation of a stronger complex in the case of o- and p-DPD-crown as compared to S3SBP-crown-1 and the complexes of all three ionic styryl probes should be more stable than the complexes of S3S-crown or S3S-b34-crown. Moreover, the stronger excited state ICT process in the DPD derivatives as compared to the S3SBP dyes, manifested in much stronger Stokes shifts of o- and p-DPD-crown (6800 cm^{-1} and 5400 cm^{-1} in acetonitrile [48]) as compared to S3SBP-crown-1 (2050 cm^{-1} [37], see also Fig. 7.7 above), should entail a larger increase in Stokes shift upon complexation for the former two dyes as well. As expected, the values of log K_S of the Ca^{II} complexes of the three crowns included in Table 7.5 support these predictions [45] and, as follows from the data in Table 7.7, correlate with the net charge of both the nitrogen donor atom and the dimethylamino (DMA) group of the corresponding DMA substituted dyes [45].

Table 7.7. Complexation-induced shifts and complex stability constants of some M^{II} complexes of S3S-crown, S3SBP-crown, and the DPD-crowns (in acetonitrile at 293 K) as well as calculated angle of pyramidalization (θ), net charges on the dimethylamino nitrogen atom (Q_N) and on the DMA group (Q_{DMA}) of corresponding S3S, S3SBP, and DPD dyes (taken from [45])

	M^{II}	$\Delta\tilde{\nu}_{cp\text{-}fp}$(abs)[a] (cm^{-1})	$\log K_S$	θ[b] (deg)	Q_N	Q_{DMA}
S3S-crown	Ca^{II}	2510	0.94	0.00	$-$ 0.2198	0.2118
S3SBP-crown	Ba^{II}	4025[c]	1.91[c]	0.00	$-$ 0.2497	0.1589
p-DPD-crown	Ca^{II}	6240	2.90	0.00	$-$ 0.2703	0.1158
o-DPD-crown	Ca^{II}	5850	2.82	0.01	$-$ 0.2707	0.1160

[a] Shift in absorption between free and complexed probe.
[b] Represented by the difference of the sum of the covalent bond angles at the N atom and 360°.
[c] Taken from [37].

Most probably, due to the essentially sp^2-hybridized nitrogen atom, pyramidalization effects do not play a role for the three positively charged compound families (Table 7.7). Furthermore, although no emission data on the Ca^{II} complex of S3SBP-crown-1 are available in the literature, comparison of the data given for the Ba^{II} complex of this compound by Lednev et al. [37] and results published by Rurack et al. [48] on the Ca^{II} complexes of o- and p-DPD-crown are in agreement with the second prediction, i.e., in the former case, a Stokes shift of 5780 cm^{-1} is found for the complex [37] and, in the latter two cases, these values amount to 11,330 cm^{-1} and 11,770 cm^{-1} [48], see Table 7.5. (Note, that, although comparing the data of a Ca^{II} and a Ba^{II} complex, the largely different shifts found for both probe families are not primarily caused by the two different cations since from numerous complexation studies of ICT fluoroionophores and alkaline-earth metal ions it is known that, for a certain probe, the spectral shifts between the different complexes are in the range of several 100 cm^{-1} [38–41, 104].)

Concerning the fluorescence quantum yields of the crowned derivatives, the trends which can be derived from the data given in Table 7.5 fit well into the model developed on the basis of the bridging studies in Sect. 7.3.1. When comparing the cation-induced fluorescence enhancement factors (FEF) of o-, p-DPD-crown, and S3SBP-crown-1 in Table 7.5, the similarity of the values is apparent. (Note that Ca^{II} normally induces slightly higher FEF than Ba^{II} [39, 104].) Accordingly, the weakening of the donor strength has a similar influence on the raising or lowering of the relative energetic positions of E*, A*, and P* states for both types of styryl dyes.

For a better understanding of these processes, it is most helpful to consider the different factors governing the influence of complexation on the deactivation of the excited dye molecule. First, depending on its size, the cation resides more or less in the cavity of the receptor and, depending on its preferred coordination geometry, is additionally coordinated to at least one solvent molecule. Thus, the donor unit of the complexed dye is much bulkier than that of the free probe. But since complexation does not specifically lead to the bridging of a certain bond, all excited state reactions connected with the rotation of an unbridged bond are possible but will be slowed down. From simplistic geometrical considerations, the space required for twisting of a double bond is larger than that required for single bond rotation and, thus, we would expect the strongest deceleration for P* state formation. As a consequence, the fluorescence of the complex should generally be higher than that of the free dye for all fluorescent probes of styryl, styryl base, or stilbene type. However, this is in contrast to many cation-induced changes of the fluorescence properties of crowned ICT-fluoro-ionophores reported in the literature so far. Here, both fluorescence quenching [36, 48, 103] and fluorescence enhancement [35, 37, 48] was observed.

The second factor coming into play upon complexation is the weakening of the donor strength of donor-acceptor-substituted probe molecules with receptor = donor. As has already been discussed in Sect. 7.3.1, the higher the charge localization in ground and excited state (cf. MOs in Fig. 7.10) the stronger the cation-induced shifts in absorption. Nevertheless, although the shifts in emission are comparatively small the influence of the cation on the excited state photochemistry is not negligible.

For a better understanding, we will relate the photochemistry of the excited complex to the excited state reaction scheme established for the dimethylamino analogues (Fig. 7.5) and will briefly recall the decoordination reaction taking place in the excited complex [77–79]. Excitation of the complex LM (L = ligand, M = metal ion) yields (LM)* which corresponds to (Franck-Condon excited) E* (see Fig. 7.5). Since for most of the ICT-probes excitation is directly connected with a considerable charge shift and E* = (LM)* is already polar, the initially excited complex undergoes ultrafast "internal dissociation" on the sub-ps time scale leading to the excited cation probe contact pair (L*M) [77]. Transient absorption measurements revealed that not only (LM)* (cf. blue-shifted absorption bands) but (L*M) as well is rather high lying in energy and thus undergoes a rapid decoordination/recoordination reaction to the main emissive species, a ternary complex (L*/S/M), in a few ps (S = solvent) [77, 79]. In (L*/S/M), the M^{n+}-crown ether nitrogen bond is broken and a solvent molecule occupies this coordination site. Nevertheless, the cation still resides in the crown and the remaining influence can be directly read from the shift in emission between the bound and the unbound probe. For aminophilic cations or cations with a high charge density, these shifts can still be in the order of ≥ 1000 cm^{-1} [104].

Returning once more to the three-state-model (Fig. 7.5), certain analogies between free and complexed probe are obvious. Whereas (LM)* corresponds to E*, the actual decoordinated species (L*/S/M) with a (formally) positively charged nitrogen atom is reminiscent of a highly polar emissive state A*. Since the relative energetic position of nonemissive P* remains unknown, the same accounts for the energetic difference between A* and P*. If these states are generally closer lying in energy in the complex than in the free dye then quenching by the slightly higher P* state should generally increase. However, most probably the delicate interplay between steric restrictions (slowing down the E* $\rightarrow$ P* path) and energetic positions of emissive (E* and A*) and nonemissive (P*) states leads to different fluorescence effects accompanying complexation of closely related styryl, styryl base, and stilbene crowns.

7.4
Styryl Bases

7.4.1
Photophysical Model Mechanisms

Styryl bases are the neutral analogues (e.g., S3SBP-*n*) of the S3SBP styryl dyes and can belong to the large group of uncharged donor acceptor substituted compounds of (R–)Ar–CH=CH–Ar–R-type (Ar=(hetero)aromatic ring system) such as, for instance, substituted benzoxazinone (BOZ) or stilbene (DS, DCS, DDS) derivatives (Fig. 7.1, Tables 7.3 and 7.4). Compared to the ionic styryl dyes, the styryl bases contain a relatively electron rich acceptor, and the reduced conjugation along the alternate single and double bonds can be directly concluded from the blue-shifted absorption bands of S3SBP-n ($25,320$ cm^{-1} in acetonitrile [105]) as compared to S3SBP ($19,230$ cm^{-1} in acetonitrile [37]) or from the spectra of the corresponding crowned derivatives shown in Fig. 7.6B, C. Photoexcitation of styryl bases is again followed by an excited state ICT reaction and, in general, the excited state reaction model derived for the styryl dyes and related to the neutral stilbene dyes, e.g., DCS in Sect. 7.3.1, is valid for their neutral analogues as well [16, 18, 25, 106]. Again, in highly polar solvents, the emission is largely Stokes shifted and, depending on the substitution pattern, the fluorescence quantum yields are moderate to high in the unbridged compounds [16, 18, 25, 48, 105, 106].

7.4.2
Complexation Properties

Generally, the effects observed for the neutral styryl bases with a receptor in the donor part of the molecule are comparable to those reported above for the corresponding ionic styryl dyes. Whereas pronounced hypsochromic shifts are found in absorption, the emission spectra of free and complexed ionophore largely overlap and both fluorescence quenching and enhancement occurs [38–42, 44, 48].

7.4.2.1
Donor Acceptor Fluoroionophores

The cation binding behavior and photophysical properties of the large number of donor (= receptor) acceptor substituted (D-A-) fluoroionophores has been the subject of many publications and reviews (e.g., [30–32, 107]) and will not be discussed in detail here. Instead, attention will be drawn to the comparatively small number of D-A-fluoroionophores capable of acceptor complexation. Keeping in mind the excited state reaction mechanism for this kind of dyes, increased charge densities are found on the receptor's heteroatoms within the acceptor part, e.g., on nitrogen atoms (particularly in S$_1$). Accordingly, complexation of a cation in the acceptor part of a probe molecule leads to bathochromic shifts in both absorption and emission [45, 48, 104, 108–112] and the complexes formed are more stable in the excited state than in the ground state, i.e., no de-coordination reaction takes place just opposite to the effects observed for the complexation in the donor part of the molecule. On the other hand, when aiming at strong complexation-induced effects careful attention has to be drawn to the exact position of the coordinating site in the acceptor part of the fluoro-ionophore. Since in the ground and the excited state the charge localization is mainly restricted to the $N_A-(CH=CH-)_m C_6 H_4-N_D$ fragment (cf. localization of MOs in Fig. 7.10), not only exchange of the nitrogen donor N_D (e.g., aza crown → benzo crown) leads to unfavorably small cation-induced effects but introduction of a benzo crown receptor to the "far end" of the acceptor (e.g., to the 5,6-positions of the benzothiazole moiety as in the S3SBP-n-crowns, see Table 7.8) as well. In the former case, the exchange of an anilino crown for a benzo crown in the donor part can be seen in accordance with the weaker ICT reaction of a methoxy donor substituted dye as compared to a dimethylamino donor substituted one (cf. S3SBP-n-crowns in Table 7.8 and [104]). Thus, especially in the case of divalent cations with a tendency for nitrogen atom coordination (e.g., most alkaline-earth metal ions), coordination with the nitrogen lone electron pair induces stronger shifts for the anilino crowned dyes. Of course, these effects are less pronounced for "hard" [113] alkali metal ions such as Li^I which show a high preference for "hard" oxygen over "soft" nitrogen donor atoms.

In the latter case, i.e., introduction of the receptor to the "far end" of the acceptor, the weak effect directly follows from a comparison of the absorption and emission band positions given for a set of dyes in Table 7.8. Regardless of 5,6-substitution pattern in the benzothiazole moiety, S3SBP-n, D-S3SBP-n, and S3SBP-n-crown-1 absorb and emit at very similar wavelengths (Table 7.8). This is supported by the charge localization in ground and excited state as derived from quantum chemical calculations (Fig. 7.10). In the case of the benzothiazole acceptor unit, the benzo moiety is not involved in the transition (Fig. 7.10, related to the S3SBP case). An experimental verification is provided by the titra-

Table 7.8. Absorption and emission band maxima ($\lambda_{abs}/\lambda_{em}$, in nm) of S3SBP and various S3SBP-n derivatives in acetonitrile at 293 K (data taken from [37, 105, 114])

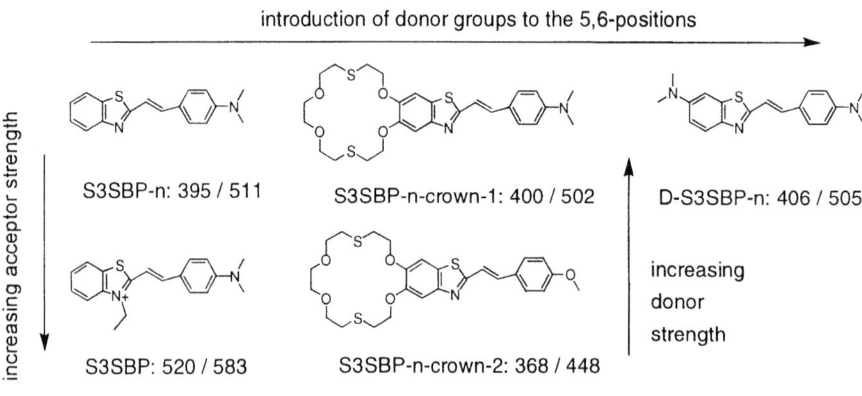

tion of S3SBP-n-crown-1 with Hg^{II}. Upon addition of this highly electrophilic ion to an acetonitrile solution of S3SBP-n-crown-1 (Fig. 7.1 and Table 7.3, spectra included in Fig. 7.6), both the absorption and emission bands are only slightly red-shifted (25,060 cm^{-1} → 24,690 cm^{-1}) indicating that the electronic effects are very weak [114].

Furthermore, when extending this concept to an appropriate combination of donor and acceptor complexation sites in a single sensor molecule (e.g., two complexation sites with largely different complex stability constants for a specific analyte such as Phenyl-A15C5 ($\log K_S = 1.70$ for K^I in methanol [115]) and Py18C6 ($\log K_S = 5.35$ [116]), see Fig. 7.1), extremely large dynamic sensing ranges or bifunctional probes (with two receptors of different selectivity and hence the possibility to detect different ions with an "AND" signaling logic) can be developed. An example of a model system combining the former feature, a large dynamic sensing range with a broad spectroscopic detection range (250–450 nm in absorption, 380–700 nm in emission) has recently been reported, namely the t-butyl styryl base analogue (p-DBD-n-crown) of p-DPD-crown [48]. At low concentrations, the cation is bound in the crown and the normal effects reminiscent of D-A-fluoroionophores are found including complexation-induced fluorescence enhancement (Fig. 7.12) [48]. But when the chelating site in the crown is saturated, the cation starts to bind to the weaker coordination site in the acceptor, the pyridino nitrogen atom. Consequently, only small changes in absorption but the appearance of a new red-shifted emission band, depending linearly on ion concentration, was found (Fig. 7.12) [48].

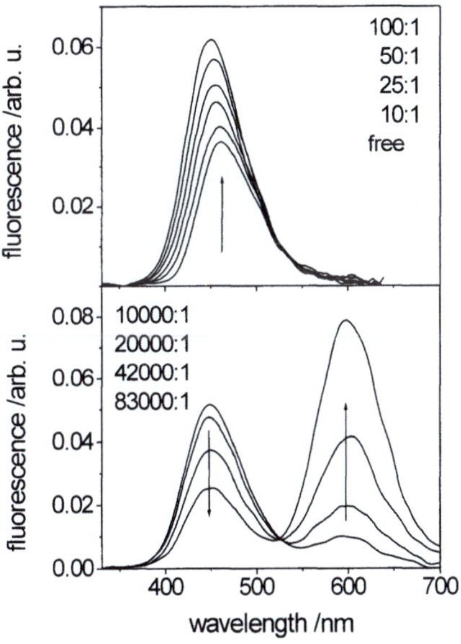

Fig. 7.12. Fluorometric titration spectra of p-DBD-n-crown and $Ca(ClO_4)_2$ in acetonitrile ($c_{dye} = 1 \times 10^{-6}$ mol/l). The titration steps and changes in the bands are indicated in the plot. The figures denote the Ca^{II}-to-crown ratios (taken from [48])

7.4.2.2
Donor Acceptor Donor Fluoroionophores

Another design concept allowing the construction of very sensitive ICT-probes follows the introduction of a second donor group to the acceptor part of a D-A-styryl base. Such systems are well-known and comprise, e.g., BOZ-crowns [39, 48] and SBD-crown [42], see Fig. 7.1. Only recently, a closely related fluoroionophore with a D-D substitution pattern (DDS-crown, Fig. 7.1 and Table 7.4 [43]) was realized. In the case of the parent compounds BOZ and DDS, Fery-Forgues et al. [25] and Létard et al. [106] have shown the validity of the photophysical reaction mechanism already introduced above (Sects. 7.2.1 and 7.3.1). Nevertheless, BOZ-crown, JBOZ-crown, and DDS-crown (Fig. 7.13) show an opposite behavior with respect to cation binding. Whereas binding of Ca^{II} leads to a fluorescence enhancement (factors of 1.4/1.2) for BOZ-crown/JBOZ-crown [39, 48], the emission of DDS-crown is quenched upon Ca^{II} binding (ninefold) [43].

For a better illustration, the photophysical implications of complex formation for the D-D- and D-A-D-crowns are schematically depicted in Fig. 7.13.

Since in the case of DDS-crown, complexation leads to a change from a D-D to a D-A arrangement (cf. cation-induced batho- and hyperchromic shift in absorption, Figs. 7.13 and 7.14A), the decrease in fluorescence quantum yield directly correlates with the acceptor strength in the series of DDS-crown > DDS-crown Ca^{II} complex > DCS [43, 89]. This suggests that most probably the change in relative energetic position of the different excited species as a function of donor and acceptor strength plays a major role for the excited stilbene derivatives and their complexes (for a discussion, see Sect. 7.3.2).

On the other hand, the influence of the second donor (the crowned aniline) is reduced through complexation for BOZ-crown, and the complexes closely resemble BOZ-H, a BOZ derivative lacking the anilino donor (Fig. 7.1) [39]. The high fluorescence yield of BOZ-H in solvents of any polarity has been attributed to very efficient formation of an intramolecular CT state of planar conformation involving partial charge transfer from the DMA group (D_1) to the benzoxazinone moiety (A) rather than the formation of an actual charge separated TICT state involving the twisting of one of the single bonds [25]. However, for BOZ and BOZ-crown (Figs 7.1 and 7.13), the presence of the anilino donor (D_2) leads to a competing excited state reaction, i.e., formation of a (probably less) emissive TICT state (cf. the quantity of k_{nr}^A in Fig. 7.5), manifested in a stronger increase in both Stokes shift and solvatochromic behavior as well as by a reduced fluorescence quantum yield in highly polar solvents as compared to BOZ-H [25, 39]. For all the BOZ derivatives, the quenching influence of P* state formation (rotation around the double bond, see Sect. 7.3.1) seems to be negligible [25, 39]. As follows from the blue-shifted emission band of the Ca^{II} complex of BOZ-crown (Fig. 7.14), complexation diminishes the donor strength of D_2, resulting in reduced competition of TICT state formation and, besides hypso- and hypochromic shifts in absorption (Fig. 7.14C), the fluorescence quantum yields of the complexes of BOZ-crown are enhanced and lie in between those of BOZ-crown (stronger D_2) and BOZ-H (weaker D_2) and, in the case of the most tightly bound ions (Ca^{II} and Ba^{II}), even exceed the fluorescence yield of BOZ-H [39]. For JBOZ-

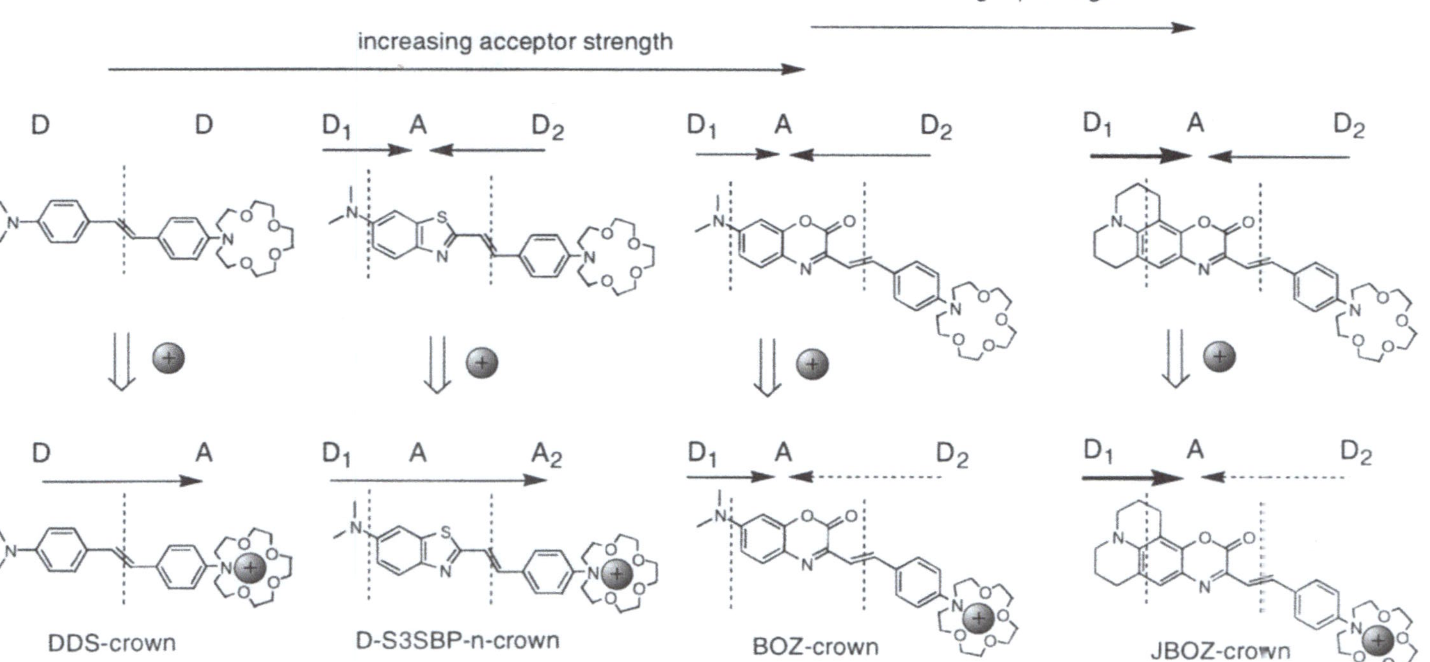

Fig. 7.13. Change of donor and acceptor strength in D-D- and D-A-D-probes upon cation complexation (as related to the complexation-induced shifts in emission shown in Fig. 7.14; *arrows* indicate CT interactions). Whereas cation binding leads to a conversion of D_2 into A_2 for excited D-S3SBP-n-crown (exemplified by the red shift in emission), D_2 remains a donor in BOZ-crown and JBOZ-crown, with the reduced donor strength in the excited complex of these two probes being manifested in a blue-shifted emission band

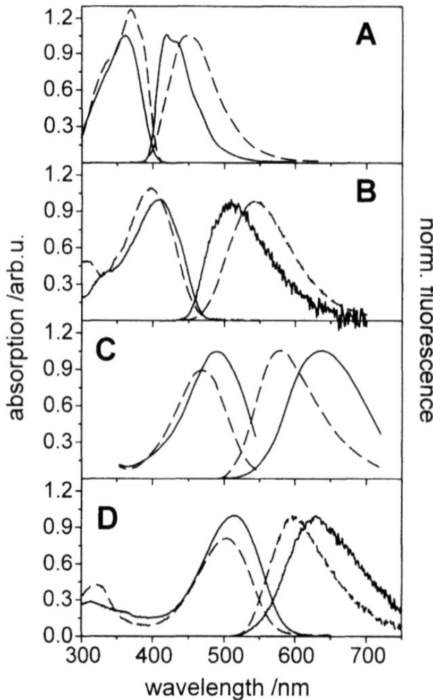

Fig. 7.14 A – D. Absorption and normalized emission spectra of free and Ca^{II}-complexed: A DDS-crown; B D-S3SBP-n-crown; C BOZ-crown; D JBOZ-crown in acetonitrile (free probes – *solid lines*, complexes – *dashed lines*; Parts A and C adapted from [43] and [39])

crown, containing the stronger julolidino donor group as D$_1$, similar but weaker complexation-induced trends are observed (cf. Fig. 7.14 C, D). Here, due to the increased donor strength of D$_1$, the competing influence of D$_2$ is less pronounced and smaller fluorescence enhancement factors are found.

Only recently, the promising results of this approach led to the design of D–S3SBP–n–crown, combining a cation-induced bathochromic shift in emission with extraordinarily high fluorescence enhancement factors, e.g., FEF = 32 for the Ca^{II} complex [105]. Here, the presence of a weaker acceptor (the benzothiazole as compared to the benzoxazinone moiety) even leads to a reversal of the D-A-D into a D-A-A arrangement, resulting in the red-shifted emission spectra (Figs. 7.13 and 7.14 B). In analogy to other D-A-A dyes [117] as well as BOZ-H, the complexes are highly fluorescent, resulting in a photoinduced electron transfer (PET)-like "switching on" process [105].

7.5
Conclusion

In conclusion, the photophysical and the complexation behavior of several dye families has been discussed and correlated. It has been shown that an understanding of the photophysics is a prerequisite for an interpretation of the cation-induced effects.

The photophysics of the dye families considered are in many cases governed by adiabatic photochemical reactions leading through twisting of single bonds to twisted excited state conformers which can live for several nanoseconds and can be highly fluorescent. In some cases, they can also predominantly be deactivated nonradiatively. In this case, the competing *trans-cis* isomerization around the double bond is quenched, and the compound becomes highly photostable, because all the absorbed photons are channeled into a nonradiative pathway leading, via the single bond twisted conformation, back to the original starting material. Selective bridging of one or more bonds can lead, depending on the type of compound, to enhanced fluorescence yields (normal loose bolt effect) or to fluorescence reduction (inverse loose bolt effect). This has been discussed in terms of competition of several excited state reaction channels and by the effect of conical intersections connecting excited and ground state.

The basic factors governing the spectroscopic complexation behavior of fluorescent probes of cyanine, styryl, and styryl base type can be rationalized as follows (R_D – nitrogen containing electron donating receptor, R_A – electron accepting receptor):

- In order to achieve strong effects for these intrinsic probes, the receptor must be part of the chromophore system and must participate in the optical transitions.
- For probes with an amino nitrogen containing R_D, the complex stability constants depend on (i) the net charge of the dye, (ii) the charge localization on the molecular fragments, and (iii) the degree of pyramidalization at the nitrogen atom and can be correlated with the net charge of the dimethylamino group of the corresponding DMA derivatives.
- The higher the degree of charge localization on donor and acceptor unit, the stronger the shifts observed in absorption.
- Due to the excited state decoordination reaction in R_D-probes, strong shifts in emission are only obtained when employing an R_A yielding stable excited complexes.
- Cation-induced fluorescence enhancement and quenching reactions critically depend on the energetic positions of the species involved in an excited state reaction and the steric effect of ion binding on bond rotation. Unfortunately, in many cases (including probes with R_D and R_A), the complexation-induced effects cannot be predicted by simple theoretical considerations.
- In particular, the combination of R_D and R_A in a single probe molecule possesses an enormous sensing potential in terms of a high dynamic range as well as 'AND' signaling.

Acknowledgements. Financial support by the Deutsche Forschungsgemeinschaft (DFG; Re 387/8 – 2 and Re 387/13 – 1) and the Bundesministerium für Bildung und Forschung (BMBF, 13N7120) is gratefully acknowledged.

References

1. Krasovitskii BM, Bolotin BM (1988) Organic luminescent materials. VCH, Weinheim
2. Okawara M, Kitao T, Hirashima T, Matsuoka M (1988) Organic colorants. Kodansha, Tokyo, Elsevier, Amsterdam
3. Fabian J, Nakazumi H, Matsuoka M (1992) Near-infrared absorbing dyes. Chem Rev 92:1197–1226
4. Dähne S (1994) The evolution of thinking on the mechanism of spectral sensitization. J Imaging Sci 38:101–117
5. Patonay G, Antoine MD (1991) Near-infrared fluorogenic labels: new approach to an old problem. Anal Chem 63:321A–327A
6. Mujumdar SR, Mujumdar RB, Grant CM, Waggoner AS (1996) Cyanine-labeling reagents: sulfobenzindocyanine succinimidyl esters. Bioconjugate Chem 7:356–362
7. Fromherz P, Dambacher KH, Ephardt H, Lambacher A, Müller CO, Neigl R, Schaden H, Schenk O, Vetter T (1991) Fluorescent dyes as probes of voltage transients in neuron membranes. Ber Bunsenges Phys Chem 95:1333–1345
8. Leung W-Y, Mao F, Haugland RP, Klaubert DH (1996) Lipophilic sulfophenylcarbocyanine dyes: synthesis of a new class of fluorescent cell membrane probes. Bioorg Med Chem Lett 6:1479–1482
9. Brackmann U (1994) Lambdachrome laser dyes, 2nd edn. Lambda Physik GmbH, Göttingen
10. Schöffel K, Dietz F, Krossner T (1990) Model mechanisms for the thermal *cis-trans* isomerization of cyanines. Chem Phys Lett 172:187–192
11. Chibisov AK, Zakharova GV, Görner H, Sogulyaev YA, Mushkalo IL, Tolmachev AI (1995) Photorelaxation processes in covalently linked indocarbocyanine and thiacarbocyanine dyes. J Phys Chem 99:886–893
12. Ephardt H, Fromherz P (1989) Fluorescence and photoisomerization of an amphiphilic aminostilbazolium dye as controlled by the sensitivity of radiationless deactivation to polarity and viscosity. J Phys Chem 93:7717–7725
13. Alfimov MV, Kamalov VF, Struganova IA, Yoshihara K (1992) Excited state relaxation of crown ether styryl dyes: photoisomerization. Chem Phys Lett 195:262–266
14. Soper SA, Mattingly QL (1994) Steady-state and picosecond laser fluorescence studies of nonradiative pathways in tricarbocyanine dyes: implications to the design of near-IR fluorochromes with high fluorescence efficiencies. J Am Chem Soc 116:3744–3752
15. Sczepan M, Rettig W, Bricks YL, Slominski YL, Tolmachev AI (1999) Unsymmetric cyanines: chemical rigidization and photophysical properties. J Photochem Photobiol A Chem 124:75–84
16. Rettig W, Majenz W (1989) Competing adiabatic photoreaction channels in stilbene derivatives. Chem Phys Lett 154:335–341
17. Létard J-F, Lapouyade R, Rettig W (1994) Multidimensional photochemistry in 4-(*N,N*-dimethylamino)stilbene. Chem Phys 186:119–131
18. Létard J-F, Lapouyade R, Rettig W (1993) Structure-photophysics correlations in a series of 4-(dialkylamino)stilbenes: intramolecular charge transfer in the excited state as related to the twist around the single bonds. J Am Chem Soc 115:2441–2447
19. Baraldi I, Carnevali A, Momicchioli F, Ponterini G (1993) Electronic spectra and *trans-cis* photoisomerism of carbocyanines. A theoretical (CS INDO CI) and experimental study. Spectrochim Acta 49A:471–495
20. Kachkovskii AD (1997) The nature of electronic transitions in linear conjugated systems. Russ Chem Rev 66:647–664
21. Cao X, Tolbert RW, McHale JL, Edwards WD (1998) Theoretical study of solvent effects on the intramolecular charge transfer of a hemicyanine dye. J Phys Chem A 102:2739–2748
22. Dekhtyar ML, Rettig W, Rozenbaum V (1999) Origin of states connected with twisted intramolecular charge shift in polymethine cations: a simple analytical treatment. J Photochem Photobiol A Chem 120:75–83

23. Dekhtyar ML, Rettig W (1999) Photochemical switching through protonation in merocyanines. J Photochem Photobiol A Chem 125:57–62
24. Liptay W (1965) Die Lösungsmittelabhängigkeit der Wellenzahl von Elektronenbanden und die chemisch-physikalischen Grundlagen. Z Naturforsch 20a:1441–1471
25. Fery-Forgues S, Le Bris M-T, Mialocq J-C, Pouget J, Rettig W, Valeur B (1992) Photophysical properties of styryl derivatives of aminobenzoxazinones. J Phys Chem 96:701–710
26. Narang U, Zhao CF, Bhawalkar JD, Bright FV, Prasad PN (1996) Characterization of a new solvent-sensitive two-photon-induced fluorescent (aminostyryl)pyridinium salt dye. J Phys Chem 100:4521–4525
27. Dähne S, Leupold D (1966) Kopplungsprinzipien organischer Farbstoffe. Angew Chem 78:1029–1039
28. Dähne S, Moldenhauer F (1985) Structural principles of unsaturated organic compounds: evidence by quantum chemical calculations. Prog Phys Org Chem 15:1–130
29. Ishchenko AA (1994) The length of the polymethine chain and the spectral-luminescent properties of symmetrical cyanine dyes. Russ Chem Bull 43:1161–1174
30. Löhr H-G, Vögtle F (1985) Chromo- and fluoroinophores. A new class of dye reagents. Acc Chem Res 18:65–72
31. Valeur B (1994) Principles of fluorescent probe design for ion recognition. In: Lakowicz JR (ed) Probe design and chemical sensing (Topics in fluorescence spectroscopy, vol 4). Plenum, New York, pp 21–48
32. Rettig W, Lapouyade R (1994) Fluorescence probes based upon twisted intramolecular charge transfer TICT states and other adiabatic photoreactions. In: Lakowicz JR (ed) Probe design and chemical sensing (Topics in fluorescence spectroscopy, vol 4). Plenum, New York, pp 109–149
33. Alfimov MV, Gromov SP (1999) Fluorescence properties of crown-containing molecules. In: Rettig W, Strehmel B, Schrader S, Seifert H (eds) Applied fluorescence in chemistry, biology, and medicine. Springer, Berlin Heidelberg New York, pp 161–178
34. Barzykin AV, Fox MA, Ushakov EN, Stanislavsky OB, Gromov SP, Fedorova OA, Alfimov MV (1992) Dependence of metal ion complexation and intermolecular aggregation on photo-induced geometric isomerism in a crown ether styryl dye. J Am Chem Soc 114:6381–6385
35. Thomas KJ, Thomas KG, Manojkumar TK, Das S, George MV (1994) Cation binding and photophysical properties of a monoaza-15-crown-5-ether linked cyanine dye. Proc Indian Acad Sci (Chem Sci) 106:1375–1382
36. Alfimov MV, Churakov AV, Fedorov YV, Fedorova OA, Gromov SP, Hester RE, Howard JAK, Kuz'mina LG, Lednev IK, Moore JN (1997) Structure and ion-complexing properties of an aza-15-crown-5 ether dye: synthesis, crystallography, NMR spectroscopy, spectrophotometry and potentiometry. J Chem Soc Perkin Trans 2:249–2256
37. Lednev IK, Ye T-Q, Hester RE, Moore JN (1997) Photocontrol of cation complexation with a benzothiazolium styryl azacrown ether dye: spectroscopic studies on picosecond to kilosecond timescales. J Phys Chem A 101:4966–4972
38. Rurack K, Bricks JL, Slominskii JL, Resch-Genger U (1998) Long wavelength emitting fluorescence probes for metal ions. In: Dähne S, Resch-Genger U, Wolfbeis OS (eds) Near-infrared dyes for high technology applications (NATO ASI Series, ser 3, vol 52). Kluwer Academic, Dordrecht, pp 191–200
39. Fery-Forgues S, Le Bris M-T, Guetté J-P, Valeur B (1988) Ion-responsive fluorescent compounds. 1. Effect of cation binding on photophysical properties of a benzoxazinone derivative linked to monoaza-15-crown-5. J Phys Chem 92:6233–6237
40. Bourson J, Valeur B (1989) Ion-responsive fluorescent compounds. 2. Cation-steered intramolecular charge transfer in a crowned merocyanine. J Phys Chem 93:3871–3876
41. Létard J-F, Lapouyade R, Rettig W (1993) Synthesis and photophysical study of 4-(N-monoaza-15-crown-5) stilbenes forming TICT states and their complexation with cations. Pure Appl Chem 65:1705–1712
42. Cazaux L, Faher M, Lopez A, Picard C, Tisnes P (1994) Styrylbenzodiazinones 3. Chromo- and fluoroionophores derived from monoaza-15-crown-5. Photophysical and complexing properties. J Photochem Photobiol A Chem 77:217–225

43. Delmond S, Létard J-F, Lapouyade R, Mathevet R, Jonusauskas G, Rullière C (1996) Cation-triggered photoinduced intramolecular charge transfer and fluorescence red-shift in fluorescence probes. New J Chem 20:861–869

44. Rurack K, Bricks JL, Kachkovskii AD, Resch U (1997) Complexing fluorescence probes consisting of various fluorophores linked to 1-aza-15-crown-5. J Fluoresc 7:63S–66S

45. Rurack K, Sczepan M, Spieles M, Resch-Genger U, Rettig W, (1999) Correlations between complex stability and charge distribution in the ground state for Ca^{II} and Na^{I} complexes of charge transfer (CT) chromo- and fluoroionophores. Chem Phys Lett 320:87–94

46. Benson RC, Kues HA (1977) Absorption and fluorescence properties of cyanine dyes. J Chem Eng Data 22:379–383

47. Noukakis D, Van der Auweraer M, Toppet S, De Schryver FC (1995) Photophysics of a thiacarbocyanine dye in organic solvents. J Phys Chem 99:11,860–11,866

48. Bricks JL, Slominskii JL, Kudinova MA, Tolmachev AI, Rurack K, Resch-Genger U, Rettig W (2000) Syntheses and photophysical properties of a series of cation-sensitive polymethine and styryl dyes. J Photochem Photobiol A Chem 132:193–208

49. Lippert E, Rettig W, Bonacic-Koutecky V, Heisel F, Miehé JA (1987) Photophysics of internal twisting. Adv Chem Phys 68:1–173

50. Heisel F, Miehé JA (1985) p-Dimethylaminobenzonitrile in polar solution. I. Time-dependent rate in intramolecular electron transfer reaction. Chem Phys 98:233–241

51. Rettig W, Fritz R, Braun D (1997) Combination of pressure and temperature dependent measurements: a simple access to intrinsic thermal activation energies. J Phys Chem A 101:6830–6835

52. Lide DR (1993) Handbook of chemistry and physics, 73rd edn. CRC Press, Boca Raton, pp 6–167

53. Yaws CL (1995) Handbook of viscosity. Gulf Publishing, Houston

54. Hurst JR, Schuster GB (1982) Ene reaction of singlet oxygen: an entropy-controlled process determines the reaction rate. J Am Chem Soc 104:6854–6856

55. Moss RA, Lawrynowicz W, Turro NJ, Gould IR, Cha Y (1986) Activation parameters for the additions of arylhalocarbenes to alkenes. J Am Chem Soc 108:7028–7032

56. Jagannadham V, Steenken S (1988) Reactivity of alpha-heteroatom-substituted alkyl radicals with nitrobenzenes in aqueous solution: an entropy-controlled electron-transfer/addition mechanism. J Am Chem Soc 110:2188–2192

57. Lewis GN, Calvin M (1939) The color of organic substances. Chem Rev 25:273–328

58. Hofer LJE, Grabenstetter RJ, Wiig EO (1950) The fluorescence of cyanine and related dyes in the monomeric state. J Am Chem Soc 72:203–209

59. Teller E (1937) The crossing of potential surfaces. J Phys Chem 41:109–116

60. Herzberg G, Longuet-Higgins HC (1963) Intersection of potential energy surfaces in polyatomic molecules. Discussions Faraday Soc 35:77–82

61. Longuet-Higgins HC (1975) Intersection of potential energy surfaces in polyatomic molecules. Proc R Soc London Ser A 344:147–156

62. Celani P, Garavelli M, Ottani S, Bernardi F, Robb MA, Olivucci M (1995) Molecular "trigger" for radiationless deactivation of photoexcited conjugated hydrocarbons. J Am Chem Soc 117:11,584–11,585

63. Bernardi F, Olivucci M, Robb MA (1996) Potential energy surface crossings in organic photochemistry. Chem Soc Rev 25:321–328

64. Bernardi F, Olivucci M, Robb MA (1997) The role of conical intersections and excited state reaction paths in photochemical pericyclic reactions. J Photochem Photobiol A Chem 105:365–371

65. Garavelli M, Bernardi F, Olivucci M, Vreven T, Klein S, Celani P, Robb MA (1998) Potential-energy surfaces for ultrafast photochemistry. Static and dynamic aspects. Faraday Discuss Chem Soc 110:51–70

66. Zilberg S, Haas Y (1999) Molecular photochemistry: a general method for localizing conical intersections using the phase-change rule. Chem Eur J 5:1755–1765

67. Michl J (1974) Physical basis of qualitative MO arguments in organic photochemistry. Top Curr Chem 46:1–59

68. Michl J (1975) Model calculations of photochemical reactivity. Pure Appl Chem 41: 507–534
69. Bonacic-Koutecky V, Michl J (1985) Charge-transfer-biradicaloid excited states: relation to anomalous fluorescence. "Negative" S_1–T_1 splitting in twisted aminoborane. J Am Chem Soc 107:1765–1766
70. Michl J, Bonacic-Koutecky V (1990) Electronic aspects of organic photochemistry. Wiley, New York
71. Klessinger M, Michl J (1995) Excited states and photochemistry of organic molecules. VCH, New York
72. Müller AM, Lochbrunner S, Schmid WE, Fuß W (1998) Low-temperature photochemistry of previtamin D: a hula-twist isomerization of a triene. Angew Chem Int Ed Engl 37:505–507
73. Fuß W, Lochbrunner S, Müller AM, Schikarski T, Schmid WE, Trushin SA (1998) Pathway approach to ultrafast photochemistry: potential surfaces, conical intersections and isomerizations of small polyenes. Chem Phys 232:161–174
74. Atkins PW (1978) Physical chemistry. Oxford University Press, Oxford, p 912
75. De Feyter S, Diau EW-G, Scala AA, Zewail AH (1999) Femtosecond dynamics of diradicals: transition states, entropic configurations and stereochemistry. Chem Phys Lett 303:249–260
76. Rettig W (1994) Photoinduced charge separation via twisted intramolecular charge transfer states. Top Curr Chem 169:253–299
77. Martin MM, Plaza P, Meyer YH, Badaoui F, Bourson J, Lefevre J-P, Valeur B (1996) Steady-state and picosecond spectroscopy of Li^+ or Ca^{2+} complexes with a crowned merocyanine. Reversible photorelease of cations. J Phys Chem 100:6879–6888
78. Martin MM, Plaza P, Dai Hung N, Meyer YH, Bourson J, Valeur B (1993) Photoejection of cations from complexes with a crown-ether-linked merocyanine evidenced by ultrafast spectroscopy. Chem Phys Lett 202:425–430
79. Mathevet R, Jonusauskas G, Rullière C, Létard J-F, Lapouyade R (1995) Picosecond transient absorption as monitor of the stepwise cation-macrocycle decoordination in the excited singlet state of 4-(N-monoaza-15-crown-5)-4'-cyanostilbene. J Phys Chem 99: 15,709–15,713
80. Druzhinin SI, Rusalov MV, Uzhinov BM, Gromov SP, Sergeev SA, Alfimov MV (1999) Fluorescence of crowned butadienyl dye and its metal complexes. J Fluoresc 9:33–36
81. Strehmel B, Rettig W (1996) Photophysical properties of fluorescence probes. I. Dialkylamino stilbazolium dyes. J Biomedical Opt 1:98–109
82. Strehmel B, Seifert H, Rettig W (1997) Photophysical properties of fluorescence probes. II. A model of multiple fluorescence for stilbazolium dyes studied by global analysis and quantum chemical calculations. J Phys Chem B 101:2232–2243
83. Dähne S (1978) Color and constitution: one hundred years of research. Science 199: 1163–1167
84. Sczepan M, Rettig W, Tolmachev AI, Slominski JL (1999) From asymmetric cyanines to stilbazolium dyes – fluorescence behaviour of different ionic dyes and their bridged derivatives. In: Book of Abstracts, VI International Conference on Methods and Applications of Fluorescence Spectroscopy, Paris, CNAM/CNRS, Paris, Abstract P63
85. Horng ML, Gardecki JA, Papazyan A, Maroncelli M (1995) Subpicosecond measurements of polar solvation dynamics: coumarin 153 revisited. J Phys Chem 99:17,311–17,337
86. Lippert E (1975) Laser-spectroscopic studies of reorientation and other relaxation processes in solution. In: Birks JB (ed) Organic molecular photophysics, vol 2. Wiley, London, pp 1–31
87. Böttcher CJF (1973) Theory of electric polarization, 2nd edn, vol 1. Elsevier, Amsterdam
88. van der Meer MJ, Zhang H, Rettig W, Glasbeek M (2000) Femto- and picosecond fluorescence studies of solvation and non-radiative deactivation of ionic styryl dyes in liquid solution. Chem Phys Lett 320:673–680
89. Lapouyade R, Czeschka K, Majenz W, Rettig W, Gilabert E, Rullière C (1992) Photophysics of donor-acceptor substituted stilbenes. A time-resolved fluorescence study using

selectively bridged dimethylamino cyano model compounds. J Phys Chem 96:9643–9650

90. Jonusauskas G, Lapouyade R, Delmond S, Létard J-F, Rullière C (1996) Picosecond observation of cation-stepwise delayed and cation-triggered photoinduced intramolecular charge transfer in fluorescent cation probes. J Chim Phys Phys-Chim Biol 93:1670–1696

91. Abraham E, Oberle J, Jonusauskas G, Lapouyade R, Rullière C (1997) Photophysics of 4-dimethylamino 4'-cyanostilbene and model compounds: dual excited states revealed by sub-picosecond transient absorption and Kerr ellipsometry. Chem Phys 214:409–423

92. Abraham E, Oberle J, Jonusauskas G, Lapouyade R, Minoshima K, Rullière C (1997) Picosecond time-resolved dual fluorescence, transient absorption and reorientation time measurements of push-pull diphenyl-polyenes: evidence for "loose" complex and "bicimer" species. Chem Phys 219:73–89

93. Fromherz P (1995) Monopole-dipole model for symmetrical solvatochromism of hemicyanine dyes. J Phys Chem 99:7188–7192

94. Rurack K, Dekhtyar ML, Bricks JL, Resch-Genger U, Rettig W (1999) Quantum yield switching of fluorescence by selectively bridging single and double bonds in chalcones: involvement of two different types of conical intersections. J Phys Chem A 103:9626–9635

95. Dekhtyar M, Rettig W, Sczepan M (2000) Small S_0–S_1 energy gaps for certain twisted conformations of unsymmetric polymethine dyes: quantum chemical treatment and spectroscopic manifestations. Phys Chem Chem Phys 2:1129–1136

96. Marder SR, Gorman CB, Meyers F, Perry JW, Bourhill G, Brédas J-L, Pierce BM (1994) A unified description of linear and nonlinear polarization in organic polymethine dyes. Science 265:632–635

97. Gorman CB, Marder SR (1993) An investigation of the interrelationships between linear and nonlinear polarizabilities and bond-length alternation in conjugated organic molecules. Proc Natl Acad Sci 90:11,297–11,301

98. Dekhtyar M, Rettig W (unpublished results)

99. Lambert C, Stadler S, Bourhill G, Bräuchle C (1996) Polarized π-electron systems in a chemically generated electric field: second-order nonlinear optical properties of ammonium/borate zwitterions. Angew Chem Int Ed Engl 35:644–646

100. Albeck A, Livnah N, Gottlieb H, Sheves M (1992) 13-C NMR studies of model compounds for bacteriorhodopsin: factors affecting the retinal chromophore chemical shifts and absorption maximum. J Am Chem Soc 114:2400–2411

101. Rettig W, Sheves M, Ottolenghi M, Tolmachev AI (unpublished results)

102. Friedman N, Sheves M, Ottolenghi M (1989) Model systems for rhodopsins: the photolysis of protonated retinal Schiff bases, cyanine dye, and artificial cyanine-bacteriorhodopsin. J Am Chem Soc 111:3203–3211

103. Druzhinin SI, Rusalov MV, Uzhinov BM, Alfimov MV, Gromov SP, Fedorova OA (1995) Excited state relaxation processes of crowned styryl dyes and their metal complexes. Proc Indian Acad Sci (Chem Sci) 107:721–727

104. Rurack K, Bricks JL, Reck G, Radeglia R, Resch-Genger U (2000) Chalcone-analogue dyes emitting in the near-infrared (NIR): influence of donor-acceptor substitution and cation complexation on their spectroscopic properties and X-ray structure. J Phys Chem A 104:3087–3109

105. Rurack K, Rettig W, Resch-Genger U (2000) Unusually high cation-induced fluorescence enhancement of a structurally simple intrinsic fluoroionophore with a donor acceptor donor constitution. Chem Commun 407–408

106. Létard J-F, Lapouyade R, Rettig W (1994) Relaxation pathways in photoexcited electron-rich stilbenes (D-D stilbenes) as compared to D-A stilbenes. Chem Phys Lett 222:209–216

107. de Silva AP, Gunaratne HQN, Gunnlaugsson T, Huxley AJM, McCoy CP, Rademacher JT, Rice TE (1997) Signaling recognition events with fluorescent sensors and switches. Chem Rev 97:1515–1566

108. Bourson J, Pouget J, Valeur B (1993) Ion-responsive fluorescent compounds. 4. Effect of cation binding on the photochemical properties of a coumarin linked to monoaza- and diaza-crown ethers. J Phys Chem 97:4552–4557
109. Bourson J, Badaoui F, Valeur B (1994) Coumarinic fluorescent chemosensors for the detection of transition metal ions. J Fluoresc 4:275–277
110. Bourson J, Borrel M-N, Valeur B (1992) Ion-responsive fluorescent compounds. Anal Chim Acta 257:189–193
111. Roshal AD, Grigorovich AV, Doroshenko AO, Pivovarenko VG, Demchenko AP (1998) Flavonols and crown-flavonols as metal cation chelators. The different nature of Ba^{2+} and Mg^{2+} complexes. J Phys Chem A 102:5907–5914
112. Marcotte N, Fery-Forgues S, Lavabre D, Marguet S, Pivovarenko VG (1999) Spectroscopic study of a symmetrical bis-crown fluoroionophore of the diphenylpentadienone series. J Phys Chem A 103:3163–3170
113. Pearson RG (1963) Hard and soft acids and bases. J Am Chem Soc 85:3533–3539
114. Rurack K (unpublished results)
115. Buschmann H-J (1985) Stabilitätskonstanten und thermodynamische Werte für die Bildung von 1:1- und 2:1-Komplexen von Kronenethern mit Alkali- und Erdalkali-Ionen in Methanol. Chem Ber 118:2746–2756
116. Bradshaw JS, Maas GE, Lamb JD, Izatt RM, Christensen JJ (1980) Cation complexing properties of synthetic macrocyclic polyether-diester ligands containing the pyridine subcyclic unit. J Am Chem Soc 102:467–474
117. Le Bris M-T, Mugnier J, Bourson J, Valeur B (1984) Spectral properties of a new fluorescent dye emitting in the red: a benzoxazinone derivative. Chem Phys Lett 106:124–127

Phototunable Metal Cation Binding Ability of Some Fluorescent Macrocyclic Ditopic Receptors

J.-P. Desvergne, E. Perez-Inestrosa, H. Bouas-Laurent,
G. Jonusauskas, J. Oberlé, C. Rullière

The fluorescence technique has been applied to the study of the metal cation binding of three photoresponsive complexing systems able to form 1:1 and 1:2 (ligand:cation) complexes. These are constituted of bisarylcyclophanes denoted: AAO5O5 (bis 9,10-dioxyanthrylcoronand), BBO5O5 (bis 1,4-dioxyphenylcoronand) and TTO5O5 (bis 1-methylene-4-oxyphenylcoronand). The metal cation binding ability in the singlet excited state is shown to *increase* for AAO5O5 owing to the formation of a long lived excimer and *decrease* for BBO5O5 and TTO5O5; this diminution is interpreted as a transitory photodecomplexation between the metal cations and the phenolic oxygen atoms, but the metal cations are believed to remain in close proximity to the macrocycle as shown by fluorescence anisotropy measurements. The determination of the association constants K_{11} and K_{12} allow a discussion of the cooperative effects (positive or negative) found in the ground state and in the excited state.

8.1
Introduction

There is a continuing interest in fluorescent molecules which respond to metal cations especially for their application to trace metal detection [1] and molecular recognition [2].

The design of fluorescent probes rests on the combination of a metal cation or molecule complexing center (recognition subunit) with a light emitting chromophore (signaling subunit) (Scheme 8.1, [3, 4]). The recognition of the analyte is usually accompanied by fluorescence modifications (wavelengths, intensities, lifetimes) detectable at very low concentration.

Among the light emitting chromophores, aromatic hydrocarbons are particularly useful since their strong, and sometimes dual, fluorescence occurs in wavelength ranges convenient for the usual commercial instruments [5].

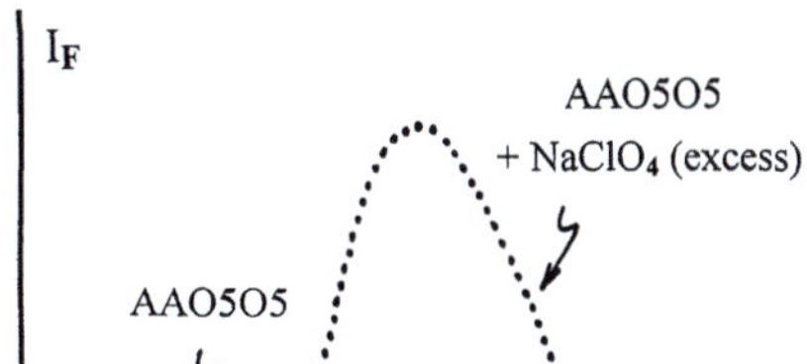

Scheme 8.1. Sketch of a fluorescent probe constituted of a complexing center (a) (metal cation recognition subunit) and a fluorescent center (b) (signaling subunit) which emits light on complexation

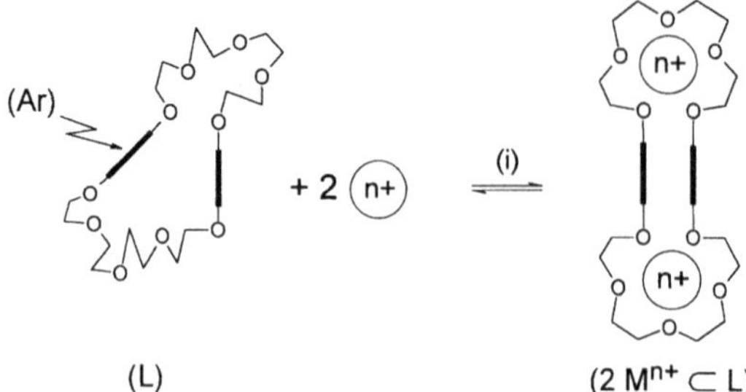

Scheme 8.2. Macromonocyclic bisaryl cyclophane (L) incorporating both the binding and signaling subunits in one single body. L are denoted AAO5O5, BBO5O5, TTO5O5 where A = 9,10-anthracenediyl, B = 1,4-benzenediyl, T = 4,7-toluenediyl; (i) the intermediate 1:1 complex M^{n+}⊂L is not shown

The specificity to analytes is often conferred by rigid receptors whose shapes are adequate to that of the guest as sketched in Scheme 8.1. However, some fluorescent probes are flexible, with the potential to shape up around the guest [6]. Of special interest are some *macrocyclic bisaryl cyclophanes* which combine rigidity and flexibility. As shown in Scheme 8.2, the polyoxyethylene fragments, by coiling up around the metal cations of the appropriate size, transform the floppy macro*monoc*ycle into a *di*topic receptor.

We report on the spectroscopic study (UV absorption, fluorescence) and the binding constant determination of three macromonocyclic bisaryl cyclophanes: AAO5O5, BBO5O5, and TTO5O5; the 1:2 (ligand: metal cation) complex fluorescence spectra of the two latter receptors were found to undergo very little shifts in comparison with those of the corresponding absorption spectra; these results are interpreted by a *transitory metal cation photodisplacement*, previously reported [7, 8] for dissymmetrical systems but *unexpected* for *symmetrical derivatives*. In the case of BBO5O5, fluorescence anisotropy relaxation measurements allow a deeper insight into the interpretation of the stationary spectra. In addition, the positive or negative *cooperativities* (revealed from the ratio K_{11}/K_{12} of the 1:1 and 1:2 (ligand:metal cation) binding constants) are commented upon.

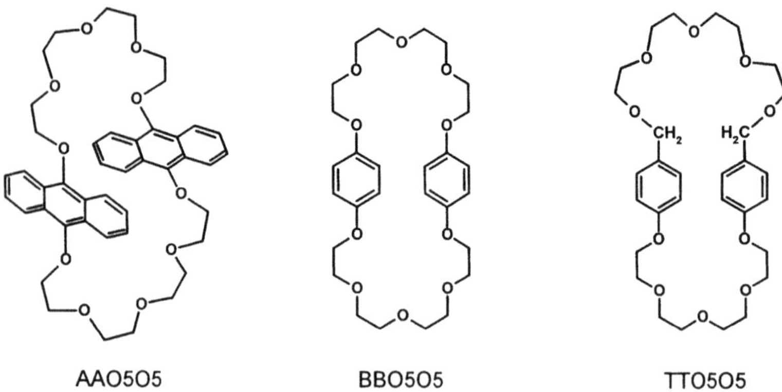

8.2
Anthraceno Coronands AAO5O5

8.2.1
Free Ligand

The ditopic receptor AAO5O5 was designed in order to combine the well-known complexing ability of the coronands [9] and the remarkable photophysical and photochemical properties of the anthracene ring [10]. Indeed, the anthracene ring is an efficient fluorophore and the dual fluorescence emission (monomer-like and excimer) is particularly suited to reveal intramolecular interactions in the excited state which reflect conformational mobility [11].

Thus, in degassed methanol ($\cong 10^{-5}$ mol/l), both structured monomer-like and nonstructured excimer fluorescence emission ($\lambda_{max} \cong 520$ nm) were observed in contrast to the reference monochromophoric compound (R) which only displays a pure monomer emission [12] (see Fig. 8.1). The presence of an excimer band with (λ_{max}: 520 nm) indicates a partial overlap of the two anthracenes [13].

A study of the time-dependent fluorescence intensities vs temperature recorded at 420 nm (monomer like emission) and 560 nm (excimer emission) demonstrated the great flexibility of the coronand AAO5O5 and showed the complexity of the system [12]. Indeed, the fluorescence emission decays (fitted with a linear combination of 3 exponentials) were clearly not compatible with the Birks classical kinetic scheme [5] for excimer formation in which the "monomer" decay is described with a sum of 2 exponentials and the "excimer" decay by a difference of 2 exponentials with the same kinetic parameters. The decays observed for AAO5O5 were found to be consistent with the occurrence of two distinguishable sets of monomers M_1 and M_2 and one excimer species E. The latter, must be formed from a geometrically close ground state conformer M_2 (see Scheme 8.3).

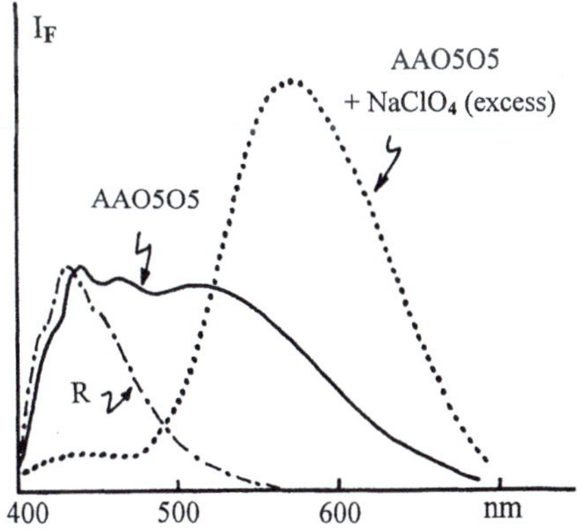

Fig. 8.1. Fluorescence emission spectra of reference compound **R** [9,10-bis(1-methoxy-3,6,9-trioxaundecyloxy)anthracene] and AAO5O5 in the absence and in the presence of NaClO₄

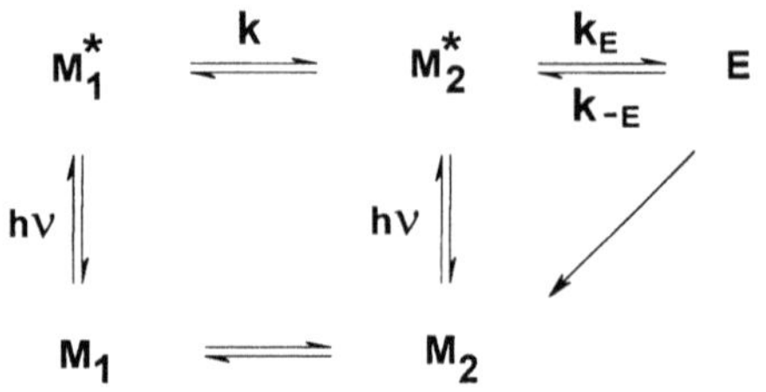

$$M_1^* \; \xrightleftharpoons{\;\;k\;\;} \; M_2^* \; \xrightleftharpoons[k_{-E}]{\;\;k_E\;\;} \; E$$

Scheme 8.3. Formation of excimer E from two kinetically distinguishable conformers M_1 and M_2 in AAO5O5

The determination of the rate constant k and the related activation energy E_a involved in the molecular folding $M_1^* \rightarrow M_2^*$ underlined the great geometrical mobility of the receptor ($E_a < 2$ kcal/mol, $k > 10^8$ s^{-1}). Thus, due to its high flexibility, AAO5O5 should readily shape up around appropriate guests.

8.2.2
In the Presence of Metal Cation

The spectroscopic properties (electronic absorption and fluorescence) of AAO5O5 are strongly modified in the presence of Na$^+$ cations as compared with the reference compound **R** (see Fig. 8.1) [12].

The peculiar redistribution of the two vibronic bands in the UV absorption spectrum (^1B_b transition) not shown here and the emergence of a new broad emission culminating at 570 nm (characteristic of a sandwich excimer), in the presence of Na$^+$ in excess, was found to be indicative of a strong conformational reorganization of the receptor into a double crown where the two anthracenes experience a large degree of overlap [13]. X-ray analysis [14] supported this picture (Fig. 8.2): in the crystal the two aromatic rings are aligned forming a quasi sandwich with an interplanar distance of 3.4 Å. The complex presents two *five*

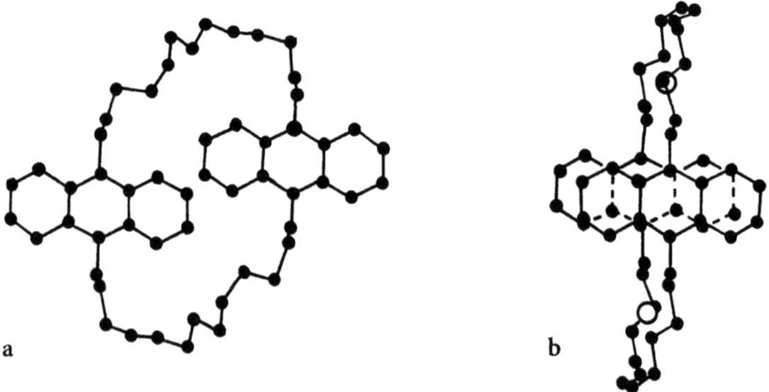

Fig. 8.2 a, b. X-ray molecular structures: **a** the free ligand AAO5O5 (the aromatic nuclei do not overlap in the crystal); **b** the 2 Na$^+ \subset$ AAO5O5 complex (the aromatic nuclei superimposed upon each other are slightly staggered in the crystal)

Table 8.1. Binding constants of AAO5O5 for Na^+ (perchlorate) at RT determined in acetonitrile from UV and fluorescence data; $\beta = K_{11} \times K_{12}$

	K_{11} (M^{-1})	K_{12} (M^{-1})	β (M^{-2})
UV	155	234	36,300
Fluorescence	200	500	100,000

oxygen cavities well suited for binding Na^+. The crystal exhibits a pure excimer fluorescence similar to that of the Na^+ saturated methanolic solution, proving the structural similarities of the complexes in the solid and in the solution. Besides, the fluorescence of a crystal of the free ligand displays only monomer-like emission, as no significant interaction occurs between the anthracenes.

In fluid medium the formation of a 1:2 complex (2 sodium for 1 ligand) was definitively demonstrated by UV and fluorescence data titrations (following equations).

The analyses revealed, along the stepwise process, a positive cooperative effect [15] in methanol as well as in acetonitrile ($K_{12} > K_{11}$) (see Table 8.1).

$$Na^+ + L \; \underset{}{\overset{K_{11}}{\rightleftharpoons}} \; Na^+, L$$

$$\beta = K_{11} \cdot K_{12}$$

$$Na^+, L + L \; \underset{}{\overset{K_{12}}{\rightleftharpoons}} \; 2Na^+, L$$

Amazingly, in acetonitrile, the fluorescence data show, compared with the UV measurements, a larger value for the overall binding constant β. This can be ascribed to the formation in the excited state of long lived conformers (excimers $\tau \approx 200$ ns) having stronger binding properties. The fluorescence lifetimes experiments showed that the excited free ligand L^* produced the excited complex C^* (whatever the stoichiometry considered) but the latter did not give back L^*, in contrast to the ground state equilibrium (Scheme 8.4).

Consequently, the binding properties of AAO5O5 increase upon light irradiation, creating a transitory reduction of Na^+ concentration in the solution.

Surprisingly, in contrast to other elements such alkaline-earth cations which did not bring any strong optical response, addition of K^+ to a solution (methanol or acetonitrile) of AAO5O5 showed spectral perturbations similar to those

$$L^* \; \overset{}{\underset{}{\rightleftharpoons\!\!\!\!\times}} \; C^* \qquad \textit{excited state}$$

$$h\nu \qquad\qquad\qquad h\nu$$

$$L \; \rightleftharpoons \; C \qquad \textit{ground state}$$

Scheme 8.4. Simplified diagram for the formation of the excited state of the free ligand (L) of AAO5O5 and the Na^+ complexes (C)

recorded with Na^+, in spite of the bigger size of K^+ (ionic radii: $K^+ = 1.36$ Å; $Na^+ = 0.99$ Å) [16, 17]. A fine analysis of the spectra combined with titration experiments demonstrated the sole formation of a 1:1 inclusion complex with K^+; for steric reasons, this complex could not imprison a second cation (of note, the association constant for K^+ was found significantly smaller than with Na^+, presumably in connection with the conformational constraints).

8.3
Benzeno Coronands

8.3.1
BBO5O5

It was of interest to examine the binding behavior of benzeno coronand BBO5O5, the simplest photoactive host in the series, for revealing the part played by the fluorescent aromatic moieties (electronic and steric effects) on the coordinating properties of the receptor [18]. As registered for AAO5O5, addition of metal cations to a fluid solution of BBO5O5 produced spectral perturbations, their amplitudes being dependent on the cation (alkali and alkali-earth metals). Thus, as listed in Table 8.2, significant changes were recorded with cations of radii ranging from 0.95 Å to 1.35 Å ($Na^+ \rightarrow Ba^{2+}$), whereas very weak or no effects occurred for those of radii equal to 0.65–0.70 Å (Li^+ and Mg^{2+}). The metal coordination, as shown by X-ray structure examination, involved the phenolic oxygens of the aromatics producing a *blue-shift of the UV spectra*, since the phenolic oxygen lone pairs were less conjugated with the para-phenylene subunits).

Titration experiments (from UV data) demonstrated the presence of both 1:1 and 1:2 inclusion complexes (ligand:cation), the latter being predominant at high cation concentrations. In contrast to AAO5O5, a clear negative cooperative effect ($4 K_{12} \ll K_{11}$ [15]) was revealed for Na^+, Ca^{2+}, Ba^{2+}, and Sr^{2+}, presumably because of electrostatic repulsions between the two bound cations combined with adverse conformational effects.

The fluorescence of single crystals of pure BBO5O5 and of its Sr^{2+} bicomplex (Fig. 8.3) does not display (as expected from the molecular structure in the so-

Table 8.2. Hypsochromic shifts (Δv/cm^{-1}) of the UV spectrum (first electronic transition) of BBO5O5 upon addition of various metal cations and binding constants (1:1 and 1:2 stoichiometry) in acetonitrile at RT. $\beta = K_{11} \times K_{12}$

Cation	Ionic radius Å	λ nm	Δv (cm^{-1})	β (10^{-2})	K_{11}	K_{12}
–	–	291	–	–	–	–
Li^+	0.78	290	142	a	a	a
Na^+	0.98	286	601	160	700	23
Ca^{2+}	1.06	284	847	46	387	12
Sr^{2+}	1.27	282	1097	455	1820	25
Ba^{2+}	1.43	282	1097	6000	5900	102

[a] Not determined.

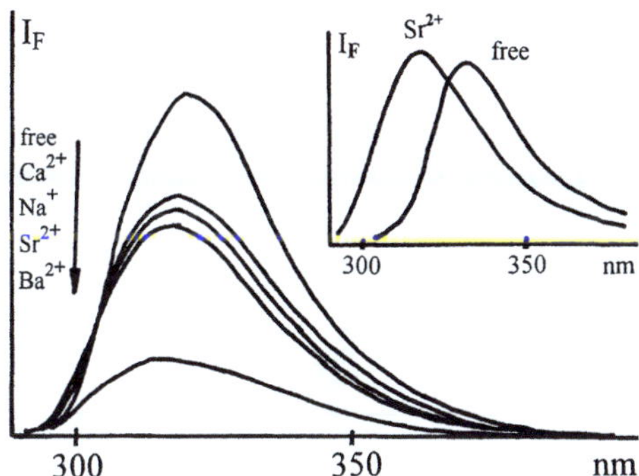

Fig. 8.3. Fluorescence spectra of BBO5O5 (acetonitrile, 20 °C) in the absence and in the presence of various cations. (*insert*: fluorescence spectra of single crystals of BBO5O5 and $2\,Sr^{2+} \subset BBO5O5$ complex)

lid, where the phenyl groups do not overlap [18, 19]) any excimer emission: the fluorescence is characteristic of 1,4-dialkoxybenzenes and blue-shifted for the Sr^{2+} complex in agreement with the UV spectrum displacement observed in solution.

Unexpectedly, the fluorescence spectra of the fluid solutions (CH_3CN, MeOH) were significantly not modified in shape and position upon addition of cations; they resembled that of the pure ligand (Fig. 8.3) whereas the *excitation spectra matched the UV absorption of the complexes*. Only a quenching was registered in the presence of salts; this effect was found to be stronger with larger elements such as Sr^{2+} and Ba^{2+} (ϕ_F: 0.18/free ligand, 0.16/Ca^{2+}, 0.15/Na^+, 0.09/Sr^{2+}, 0.05/Ba^{2+}).

These results can be interpreted by a decomplexation of the cation during the excited state of the host, which might explain the free ligand-like fluorescence of the salt saturated solutions [7, 8, 19].

This explanation is consistent with the values of association constants calculated from fluorescence titrations which were systematically found to be lower than those obtained from UV data (Table 8.3). This is in contrast with AAO5O5 and Na^+, presumably because in the meso anthracenic positions the conjugation between the ring and the oxygen lone pairs is negligible, owing to the presence of peri hydrogen atoms.

Table 8.3. Binding constants of BBO5O5 and TTO5O5 for Sr^{2+} and Ba^{2+} cations (perchlorates) at 20 °C in acetonitrile at RT. $\beta = K_{11} \times K_{12}$

		$\log K_{11}$		$\log K_{12}$		$\log \beta$	
		UV	Fluor.	UV	Fluor.	UV	Fluor.
BBO5O5	Sr^{2+}	3.26	1.89	1.40	1.89	4.66	3.78
BBO5O5	Ba^{2+}	3.77	3.83	2.01	–	5.78	3.83
TTO5O5	Ba^{2+}	4.75	4.45	5.51	4.66	10.26	9.11

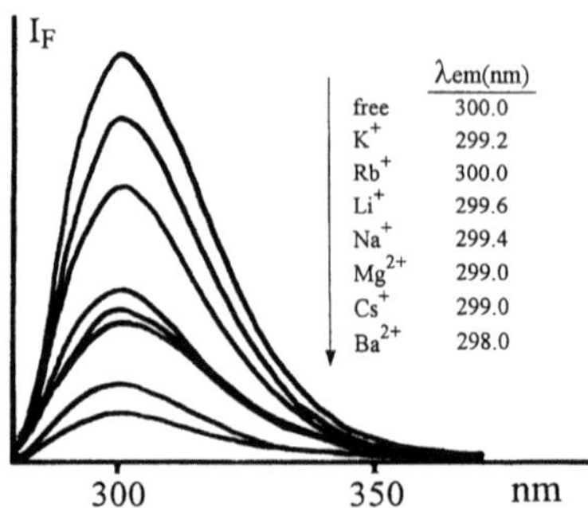

Fig. 8.4. Fluorescence spectra of TTO5O5 in the absence and in the presence of various metal cations (perchlorate)

8.3.2
TTO5O5

The related compound TTO5O5 was designed and synthesized in order to enhance the binding ability of the coronands by inserting a methylene group between the phenyl rings and the phenolic oxygen atom and to verify whether the "photodecomplexation" speculated upon for BBO5O5 was a general process in this series [21].

TTO5O5 was found to behave similarly to BBO5O5: addition of cations do not induce any significant blue-shift of the fluorescence spectrum with respect to the fluid solution spectrum as depicted in Fig. 8.4; only a quenching was observed upon complexation. The fluorescence quenching has been demonstrated to be due only to metal complexation and not to ion pair effect [22], because $Bu_4N^+ClO_4^-$ which could not be accommodated by the host had no influence on the emission spectrum. Moreover, no anion effect was noted as different barium salts (ClO_4^-, SCN^- and I^-) gave identical spectral changes and the same association constants. Titration experiments showed (as observed with BBO5O5) lower association constants from fluorescence data, suggesting here again a decrease of the complexing ability of TTO5O5 upon irradiation (Table 8.3). Of note is the enhancement of the overall binding constant β for TTO5O5 and the *positive cooperative effect* in comparison with BBO5O5 (in the ground state as well as in the excited state). The behavior of TTO5O5 probably originates in the presence of only two phenolic oxygens and some favorable conformational effect for the binding of the second cation by the less coordinating loop 2 as suggested by molecular modeling [23] (Scheme 8.5).

8.3.3
Fluorescence Anisotropy Experiments with BBO5O5

As mentioned above, the fluorescence data suggested that BBO5O5 and TTO5O5 might release cations under light irradiation as reported for other, but dissymmetrical, receptors [7, 8, 20].

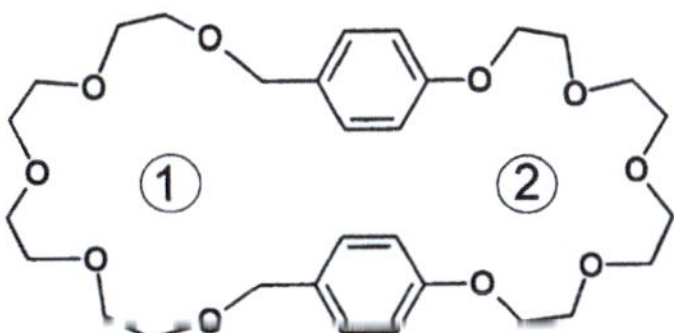

Scheme 8.5. Schematic representation of the sequential complexation of Ba^{2+} by TTO5O5 as anticipated from PC MODEL molecular modeling [23]

However, no compelling evidence supporting a possible photoejection of cation was available using nanosecond time scale fluorescence techniques: if the decays of BBO5O5 in Ba^{2+} saturated CH_3CN were found to be biexponential $[I_F(E)\alpha\, A_1\, exp\,(-\lambda_1 t) + A_2\, exp\,(-\lambda_1 t)]$ as observed for pure BBO5O5 solutions, the kinetic parameters were not identical discarding the similarity of the fluorophores (pure BBO5O5: A_1 $(1/\lambda_1)$: 0.8 (3.75 ns); A_2 $(1/\lambda_2)$: 0.7 (0.8 ns) Ba^{2+}/ BBO5O5: $A_1(1/\lambda_1)$: 0.05 (5.5 ns); A_2 $(1/\lambda_2)$: 1.4 (0.8 ns)) (the reference compound bis 1,4-pentaoxa-1,4,7,10,13-tetradecane displayed a single exponential decay $1/\lambda$: 2.7 ns). Moreover, time-resolved fluorescence spectra showed, on the picosecond time scale, no changes of the profiles whatever the delays applied for the measurements indicating that the emitting species remains unchanged along the deactivation process.

Fluorescence anisotropy experiments carried out on solutions of pure BBO5O5 and in the presence of Na^+ and Ba^{2+} (in large excess in order to form mainly the 1:2 complex) gave single exponential relaxation profiles; the corresponding time constant τ increased with the atomic mass of the added cation (free ligand: 50 ps, Na^+: 70 ps, Ba^{2+}: 140 ps). The geometrical preservation of the fluorescent species, within the picosecond time scale, could account for these single exponential fluorescence anisotropy relaxations.

Besides, the function of reorientation $r_0(t) = (I_\| - I_\perp)/(I_\| + 2\,I_\perp)$ correlated [22] with the moment of inertia J of the fluorescent molecular system was found to be

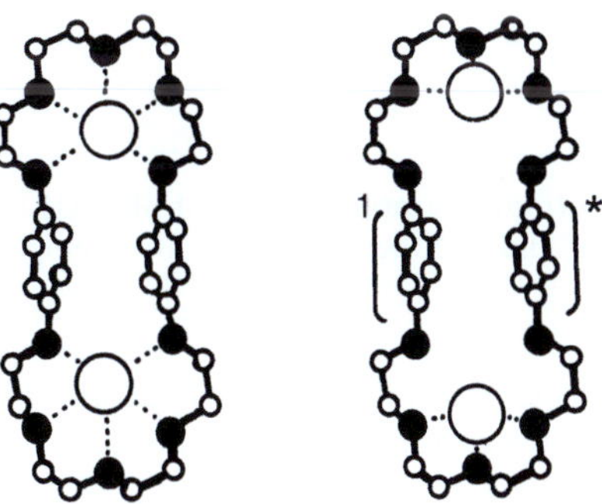

Scheme 8.6. Schematic representation of BBO5O5 in the presence of Na^+ or Ba^{2+} (perchlorate) in excess, at ambient temperature, in the ground state, and the relaxed excited singlet state

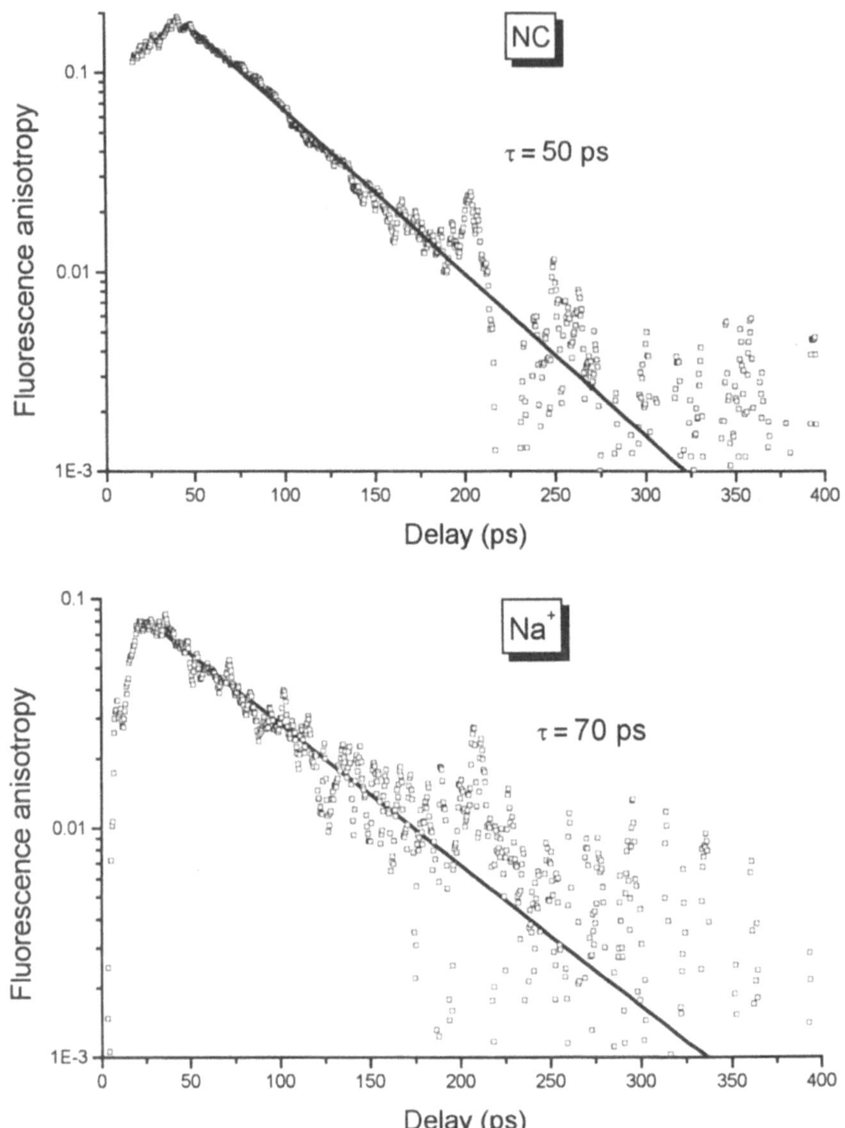

Fig. 8.5. Fluorescence anisotropy relaxation profiles of BBO5O5 in the absence [NC] and in the presence (excess) of Na^+ and Ba^{2+} (perchlorate) at room temperature

proportional to J values corresponding to the inclusion of two metals inside the host ($J = \sum_i r_i^2 m_i$; r_i being the distance of each atom to the center of gravity and m_i the corresponding atomic mass); thus $\tau_{\text{free}}/J_{\text{free}} \approx \tau_{Na+}/J_{2\,Na+} \approx \tau_{Ba++}/J_{2Ba++}$.

This suggests [24, 25] that during the fluorescence anisotropy relaxation time τ, the molecular system evolves as a complex; the binding between the host and the metal cations is not totally broken.

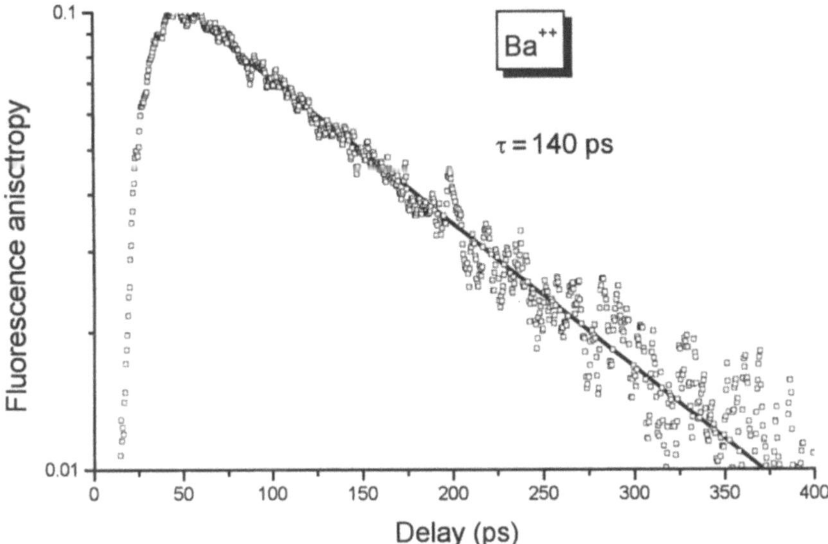

Fig. 8.5 (continued)

Thus on the ps time scale (as well as the ns scale), the cations seem to remain inside the host and are probably not released as free elements in the fluid solution. The scheme 8.6, in which the cation is partially photodisplaced within the host is proposed to account for the fluorescence properties of BBO5O5 (and TTO5O5).

The apparent decrease of the association constants K_a calculated from the fluorescence data would originate in the partial decomplexation of the metal with the phenolic oxygens which are conjugated with the aromatics. The cation maintained during the excited state lifetime in the proximity of the host should not be in direct interaction with the fluorescing subunit (Scheme 8.6).

8.4
Conclusion

In addition to the UV absorption spectroscopy which reveals the molecular behavior in the ground state, the use of the fluorescence technique allows one to show differences in the singlet excited state binding ability and cooperativity of some photoresponsive systems.

Thus the association constants K_{11} and K_{12} of AAO5O5 with Na^+ increase in the excited state, presumably because of the formation of a better host owing to the sandwich excimer configuration; moreover the positive cooperative effect, experienced in the ground state, is maintained.

One observed, however, a decrease of K_{11} and K_{12} for BBO5O5 with Sr^{2+} and Ba^{2+} together with stationary fluorescence spectra close to those of the free ligand; it is interpreted by a partial photodecomplexation affecting the phenolic oxygen atoms which become more positive in the S_1 state; nevertheless the metal cations are still in close proximity to the host as shown by fluorescence anisotropy measurements. So is the case for TTO5O5 with Ba^{2+} where the cooperative effect is observed both in the ground and in the excited state.

These observations were unexpected for symmetrical photoresponsive systems.

Acknowledgement. We acknowledge the partial contribution of Dr D. Marquis to this research work. "La Région Aquitaine" is warmly thanked for financial support. E. P-I. is indebted to the University of Malaga and Junta de Andalucia for a postdoctoral fellowship.

References

1. Wolfbeis OS (1991) Fiber optic chemical sensors and biosensors. CRC Press, vols 1 and 2
2. Desvergne JP, Czarnik AW (1997) Chemosensors of ion of molecule recognition. NATO ASI Series C, 492. Kluwer Academic Publishers, Dordrecht
3. De Silva AP, Ghunaratne HQN, Gunnlaugsson T, Huxley AJM, Rademacher CP, Rice TE (1997) Signaling recognition events with fluorescent sensors and switches. Chem Rev 97:1515–1566 and references cited therein
4. Fabbrizzi L, Poggi A (1995) Sensors and switches from supramolecular chemistry. Chem Soc Rev 197–202
5. Birks JB (1970) Photophysics of aromatic molecules. Wiley-Interscience, London
6. Lehn JM (1995) Supramolecular chemistry – concepts and perspectives. VHC, Weinheim
7. Létard JF, Delmond S, Lapouyade R, Braun D, Rettig W, Kreissler M (1995) New intrinsic fluoroionophores with dual fluorescence: DMABN-crown4 and DMABN-crown5. Rec Trav Chim Pays-Bas 114:517–527
8. Martin MM, Plaza P, Meyer YH, Badaoui F, Bourson J, Lefèvre JP, Valeur B (1996) Steady-state and picosecond spectroscopy of Li^+ or Ca^{2+} complexes with a crowned merocyanine. Reversible photorelease of cations. J Phys Chem 100:6879–6888
9. Izatt RM, Bradshaw JS, Nielsen SA, Lamb JD, Christensen JJ, Sen D (1985) Thermodynamic and kinetic data for cation-macrocycle interaction. Chem Rev 85:271–339
10. Bouas-Laurent H, Castellan A, Desvergne JP (1980) From anthracene photodimerization to jaw photochromic materials and photocrowns. Pure Appl Chem 52:2633–2648 and references cited therein
11. Castellan A, Desvergne JP, Bouas-Laurent H (1980) Kinetic study of photophysical and photochemical processes in α,ω-(bis-9-anthryl) n alkanes (ethane to decane) at room temperature. Chem Phys Lett 76:390–397
12. Marquis D, Desvergne JP, Bouas-Laurent H (1995) Photoresponsive supramolecular systems: synthesis, photophysical and photochemical study of bis-(9,10-anthracenediyl) coronands AAOnOn. J Org Chem 60:7984–7996
13. Ferguson J (1986) Absorption spectroscopy of sandwich dimers and cyclophanes. Chem Rev 86:957–982
14. Bouas-Laurent H, Desvergne JP, Fages F, Marsau P (1991) Tunable fluorescence of some macrocyclic anthracenophanes. In: Schneider HJ, Dürr H (eds) Frontiers in supramolecular organic chemistry and photochemistry. VCH, Weinheim, pp 265–286
15. Connors KA (1987) Binding constants. The measurement of molecular complex stability. Wiley, New-York
16. Lehn JM (1973) Design of organic complexing agents. Strategies towards properties. Structure and Bonding 16:2–112
17. Pauling L (1945) The nature of the chemical bond, 2nd edn. Cornell University Press, Ithaca, New York

18. Marquis D, Greiving H, Desvergne JP, Lahrahar N, Marsau P, Hopf H, Bouas-Laurent H (1997) From p-dimethoxybenzene to crown benzenophanes (4). Cation-complexing properties of bis-(p-phenylene-34-crown-10); a structural and spectrophotometric study. Liebigs Ann Recueil 97–106
19. Slawin AMZ, Spencer N, Stoddart JF, Williams DJ (1987) Complexation of diquat by a bisparaphenylene-34-crown-10 derivative. JCS Chem Commun 1061–1064
20. Kimura K, Kaneshige M, Yokoyama M (1995) Cation complexation, photochromism, and photoresponsive ion-conducting behavior of crowned malachite green leuconitrile. Chem Mater 7:945–950
21. Desvergne JP, Bouas-Laurent H, Perez-Inestrosa E, Marsau P, Cotrait M (1999) Photoinduced control of cation binding ability of non-conjugated bichromophoric receptors. Coord Chem Rev 185/186:357–371
22. McCullough JJ, Yeroushalmi S (1983) Quenching of fluorescence by substituted ethylenes. Substituent and salt effects as criteria of quenching mechanism. JCS Chem Commun 254–256
23. PC MODEL for IBM 386 or 486, Serena Software, Bloomington, USA
24. Chang YJ, Castner EW Jr (1994) Deuterium isotope effects on the ultrafast relaxation of formamide and N,N-dimethylformamide. J Phys Chem 98:9712–9722
25. Buchhauser J, Groß T, Karger N, Lüdemann HD (1999) Self diffusion in CD_4 and ND_3: with notes on the dynamic isotope effect in liquids. J Chem Phys 110:3037–3042

Part 3
Fluorescence in Sensing Applications

The Design of Molecular Artificial Sugar Sensing Systems

S. SHINKAI, A. ROBERTSON

This article is concerned with the development of new receptor molecules that can precisely recognize sugar molecules by making use of the reversible formation of boronate esters from suitable diols and boronic acids. Since one boronic acid can react with *cis*-1,2-diols or *cis*-1,3-diols to form a boronate ester, one *di*boronic acid can immobilize two suitably positioned diol units to form a sugar-containing macrocycle that can discriminate between the relative positions of *cis*-diol moieties on the guest saccharide. When a boronic acid-based receptor contains an aminomethylfluorophore, the complexation event can be conveniently read out by fluorescence spectroscopy. This is a novel application of PET (photoinduced electron transfer) sensors: sugar binding changes the strength of the B···N interaction which consequently changes the fluorescence quenching efficiency of the amine. We have demonstrated, using a chiral 1,1′-binaphthyl group as a fluorophore, that even discrimination between enantiomeric saccharides is possible. These abundant examples support the superiority of boronic acid-based covalent-bond recognition over hydrogen bond-based noncovalent-bond recognition for sugars in water.

9.1
Introduction

The molecular design of artificial receptors, which show high affinity and high selectivity comparable with natural systems, has recently become a very active area of endeavor and we are currently interested in sugar recognition with consequent detection of the recognition process [1–5]. An overview of past literature teaches us that hydrogen-bonding interactions are widely used for recognition of guest molecules but the effect is exerted only in aprotic organic solvents. Hence, although hydrogen-bonding interactions are useful for sugar recognition

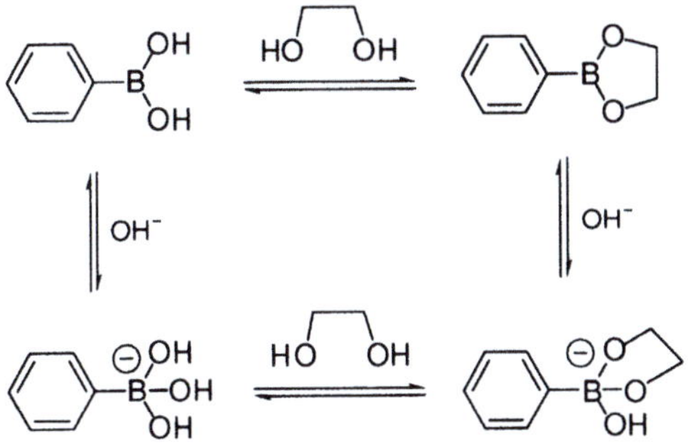

Scheme 9.1. Boronate ester formation between a boronic acid and a diol in water

in several systems [6–8], they are generally useless for sugar recognition in water. Then, how can we "touch" sugars and "recognize" them in this vital solvent? As an attempt to solve this dilemma we have proposed to use a boronic acid which reversibly forms covalent-bonds with a variety of sugar molecules in water (Scheme 9.1) [2–5, 9–11] and, since this process is much faster than the human time-scale, one can treat the process in a similar manner to the noncovalent interactions frequently used for molecular recognition. Although this strategy is quite different from that employed by nature (using multiple hydrogen-bonding interactions) [1, 6–8], this is undoubtedly a practical (and probably the sole) way to "touch" sugars in water.

9.2
Fluorescent Monoboronic Acids

It is known that the acidity of monoboronic acids is increased when they form covalent complexes with diols; hence, at constant pH, saccharide addition can change neutral boronic acids to anionic boronate esters. In fluorescent monoboronic acids this change is reflected as a decrease in the fluorescence intensity (I) of the neighboring fluorophore [10, 12, 13]; typical examples are compounds 1–5 [10, 12]. Among these compounds, 2b and 4b have three important features for saccharide sensing: that is, strong fluorescence intensity, large pH-dependent change in I, and shift of the pH-I profile to a lower pH region in the presence of saccharides [12]. However, the affinity order of monoboronic acids for saccharides is always the same: e. g., D-fructose>D-arabinose>D-mannose>D-glucose for monosaccharides and so we decided to devote our research effort toward recognition of saccharides with alternative selectivities.

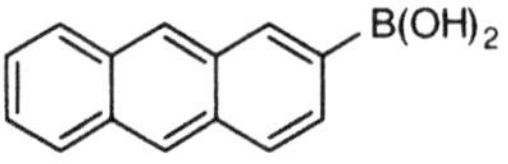

1

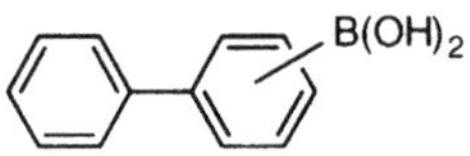

2a : 2-B(OH)$_2$
2b : 3-B(OH)$_2$
2c : 4-B(OH)$_2$

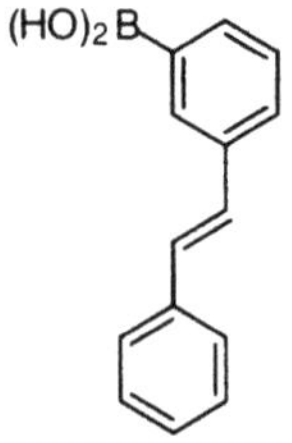

3

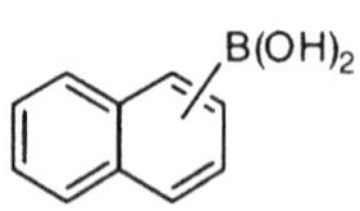

4a : 1-B(OH)$_2$
4b : 2-B(OH)$_2$

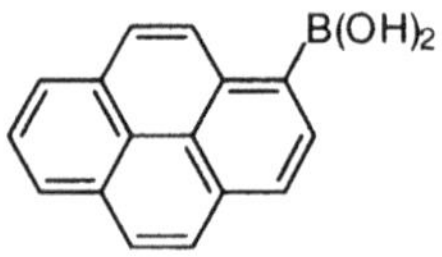

5

9.3
Selective Recognition of Saccharides by Diboronic Acids

Regular monosaccharides have five OH groups and, since a boronic acid reacts with a 1,2-diol or 1,3-diol, diboronic acids can potentially immobilize four of these five OH groups. We thus expected diboronic acids to show selectivity toward saccharides which would depend on the relative spatial position of the two boronic acids in relation to diol functionalities on the coordinating sugars.

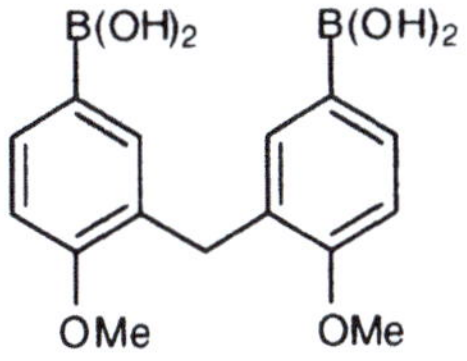

6 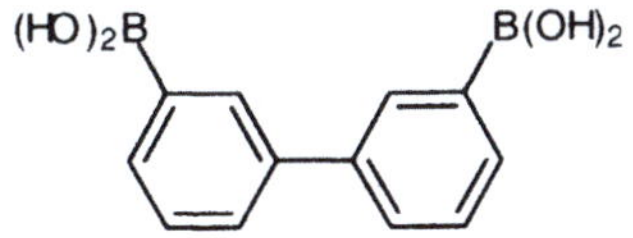   **7**

Compound **6** is a flexible diboronic acid but when it adopts a folded *syn* conformation the distance between the two boronic acids is comparable with that between the 1,2-diol and the 4,6-diol in monosaccharides (ca. 6 Å). It was shown that, at 25 °C and pH 11.3, **6** can complex several monosaccharides such as glucose, mannose, galactose, and talose to form intramolecular 2:1 boron/saccharide complexes (Fig. 9.1) [2]. The highest affinity (K_{ass} = 19,000 M^{-1}) was observed for glucose and so this was the first example of a boronic acid derivative showing the highest affinity for a saccharide other than fructose. With the exception of D-galactose, the D-mono- and D-disaccharides tested gave CD spectra with positive exciton coupling while L-glucose demonstrated a negative exciton coupling. The results indicate that the absolute configuration of saccharides can be conveniently deduced from the sign and the strength of the CD spectra of **6** and this means that the CD spectroscopic method, using **6** as a receptor probe, can serve as a new sensory system for sugar molecules.

Compound **7** was designed for disaccharides, since the spacing between the two boronic acid units is similar to the spacing between the 1,2 -diol and 4′-OH and 5′-OH of disaccharides (ca. 7.4 Å) [4]. In the presence of D-maltose a distinct CD band which crosses the $[\theta] = 0$ line at 210 nm (λ_{max} = 207 nm in the absorp-

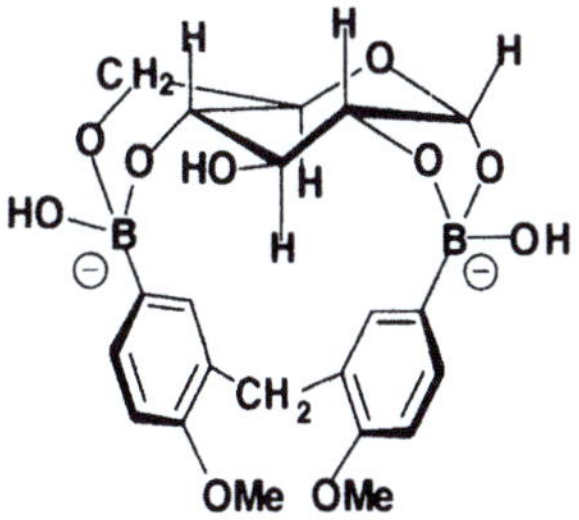

Fig. 9.1. Proposed structure of 6-D-glucose complex. Here the pyranose form of D-glucose is given, but the furanose complex cannot be ruled out

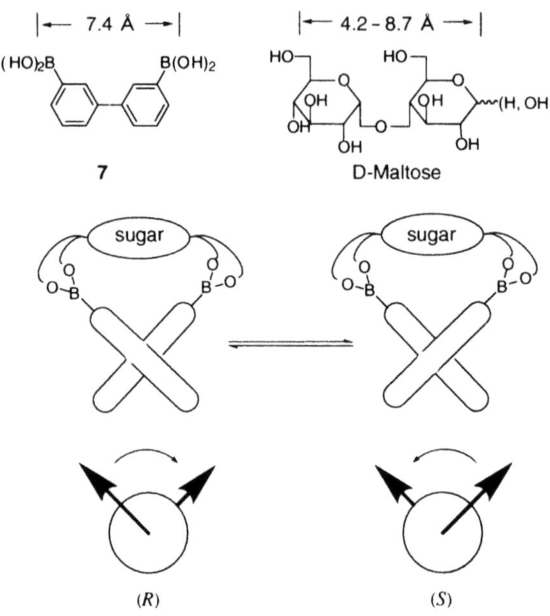

Fig. 9.2. Illustration of the macrocyclization, upon complexation of 7 and D-maltose, to produce a pair of atropisomers

tion spectrum) appeared, which is ascribed to exciton coupling. The negative sign for the first Cotton effect (223 nm) and the positive sign for the second Cotton effect (201 nm) indicate that the two dipoles along the phenylboronic acid molecular axis are oriented in a chiral, anti-clockwise direction when they interact in the excited state (Fig. 9.2). These findings reveal that when 7 forms a complex with D-maltose, the two dipoles favorably adopt (S)-chirality. Interestingly, D-cellobiose induced a positive sign for the first Cotton effect whereas D-lactose induced a negative sign for the first Cotton effect although the wavelength for the second Cotton effect could not be determined precisely because of strong background noise. These results imply that the complexes with D-cellobiose and D-lactose employ (R)- and (S)-chirality, respectively, whereas D-saccharose was totally CD-silent.

When Fe^{2+}, in the form of $FeCl_2$, is added to a solution containing a disaccharide·8 complex, a CD spectrum with an exciton coupling band appears at the metal-ligand charge-transfer (MLCT) band region (ca. 500 nm) [15]. The same CD spectrum could be obtained by adding disaccharides to a solution containing the Fe^{2+}· 8_3 complex. Separately, we confirmed that the Fe^{2+}·(2,2′-bipyridine)₃ complex is CD silent even in the presence of disaccharides and so the CD-activity arising after the disaccharide-diboronic acid interaction generates a chiral Fe^{2+} complex. It was shown that Fe^{2+}·(8·D-maltose)₃ adopts L chirality while Fe^{2+}·(8·D-cellobiose)₃ adopts D chirality.

The diboronic acids 8 and 9 form rigid, cyclic, chiral complexes with mono- and disaccharides which can be monitored by CD spectroscopy and this rigidification process can be utilized in the design of spectroscopic sensors. The main path of nonradiative deactivation of the lowest excited singlet state of stilbene, for instance, is known to be via rotation of the ethylenic double bond and enhanced fluorescence emission, following inhibited bond rotation, is known for stilbenes in solid matrices, viscous solvents, and cyclodextrin inclusion complexes. The fluorescence of stilbene-3,3′-diboronic acid 9 increases upon binding to disaccharides in basic aqueous media and large fluorescence increases were observed specifically for the disaccharide D(+)-melibiose in basic aqueous media compared to small increases observed for monosaccharides such as D(+)-glucose, D(+)-mannose and D(−)-arabinose [16]. This fluorescence increase was attributed to the formation of a cyclic complex of diboronic acid with disaccharide and subsequent freezing of ethylenic bond rotation in the excited state. Among the disaccharides which have a 1,6′-ether link between sugar monomers, D(+)-melibiose has the best fit for the diboronic acid receptor. Although maltose, which has a 1,4′-ether link between saccharide monomers, did not give any fluorescence enhancement, and maltotriose, which has an extended monomer unit, gave some fluorescence enhancement. Other disaccharides studied which have shorter distances between suitable diol groups gave no fluorescence enhancement and this suggests that the length of the disaccharide is very important.

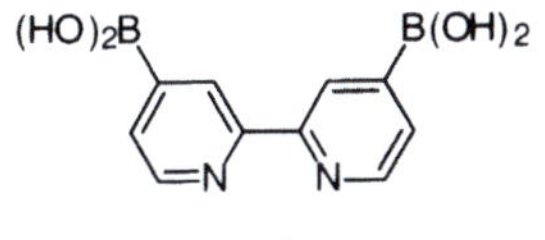

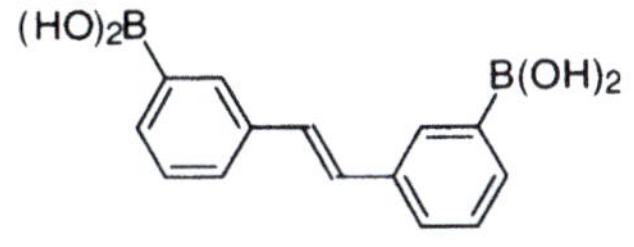

These concepts can be developed to allow the receptors to discriminate between enantiomers of the same sugar [17]. An enantio-specific receptor was produced by linking two boronic acid binding sites to a binaphthyl moiety, thus rendering the receptor chiral overall. While both enantiomers of suitable monosaccharides can bind to **10**, the chirality of the receptor naturally renders complexes with these enantiomers diastereomeric and CPK models suggested that one diastereomer is usually more flexible than its counterpart. The differences in flexibility of these diastereomeric complexes can be followed by use of spectroscopy, notably through fluorescence and CD spectroscopies.

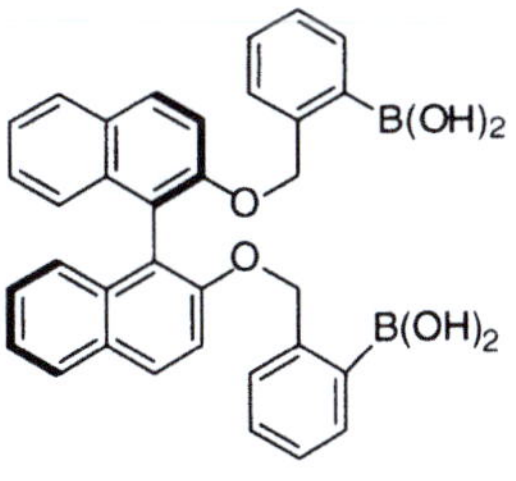

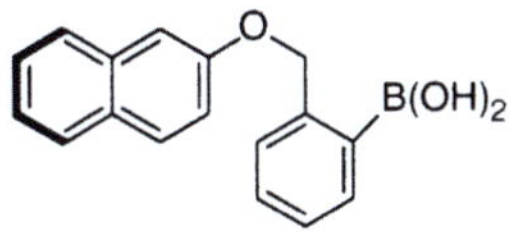

UV spectroscopy did not show any change in the absorption spectrum of **10** upon addition of saccharides which is considered to be the result of a relatively large separation between the absorbing binaphthyl group and the boronic acid moieties. Fluorescence spectroscopy, however, did show a distinct change in intensity and peak position upon complexation with peak position being enantiomer-specific. Since the increase in the intensity of the fluorescence is not observable when using methyl-α-D-glucopyranoside, which lacks the 1-OH binding site, or when using the monodentate compound **11**, it can be concluded that the fluorescence change is brought about by the formation of a cyclic, bidentate interaction between the receptor and the saccharide. This increase in fluorescence is considered to be a result of reduced rotation of the naphthalene-naphthalene bond upon complexation, i.e., greater rigidity. Assuming a 1:1 binding stoichiometry, saturation curves indicated an association constant of 2.1×10^{-3} M^{-1} for D-glucose and 1.9×10^{-3} M^{-1} for L-glucose. Similar behavior was observed for fructose, talose, and galactose whereas changes for allose and mannose were relatively small. The saturation curves for xylose demonstrated a biphasic behavior, however, indicative of a change in sugar:receptor stoichiometry from 1:1 to 2:1 at high relative concentrations of saccharide.

For glucose, fructose, galactose, and allose, the enantiomer that induced the largest fluorescence increase also had the lower binding constant suggesting that there is a stability penalty for increasing the rigidity of **10**. The differences in binding constant between enantiomers for this series was rather small, however (D/L = 1.1/1.0 ~ 1.0/1.7) whereas the largest difference was found for xylose (D/L = 1.0/8.7) and talose (D/L = 1.0/2.0) where the relationship between fluorescence increase and stability was reversed. The binding constants for these complexes were rather small for the unfavored enantiomer (0.3×10^{-3} M^{-1} for D-xylose) and so the change in trend appears to be a result of decreased stability of the flexible complex rather than extra stability of the rigid complex. The special behavior of xylose is thought to arise because xylose can only form the bidentate complex in its furanose form instead of the pyranose form, contrary to the other sugars mentioned above.

The binding behavior of **10** could be further investigated by CD spectroscopy. Although **10** is a priori CD active, saccharide binding induces a change in the CD spectrum which acts as a second useful tool for probing these events. Upon complexation with D-glucose, the θ minimum shifts to a shorter wavelength and decreases (i.e., an increase in magnitude) whereas L-glucose causes a shift of the θ minimum to longer wavelengths with an increase in value. A possible explanation for the spectral change of **10** upon complexation with L-glucose is that the binaphthyl moiety flattens out with a concomitant gain in conjugation that changes the position and intensity of the θ minimum. The addition of methyl-α-D-glucopyranoside did not produce any change in the CD spectrum of **10**, once again indicating that formation of a bidentate complex is essential in generating these spectral changes. A Job plot constructed from the CD spectral studies confirmed the 1:1 binding stoichiometry of D-glucose with **10**, which proved to be common for all binding saccharides.

9.4
Introduction of the Concept of PET (Photoinduced Electron Transfer) Sensors

Photoinduced electron transfer (PET) has been wielded as a tool of choice in fluorescent sensor design for protons and metal ions but the design of fluorescent sensors for neutral organic species presents a harsher challenge due to the lack of electronic changes upon inclusion [18]. The design of a fluorescent sensor based on the boronic acid-saccharide interaction has been difficult due to the lack of sufficient electronic changes found in either the boronic acid moiety or in the saccharide moiety and, furthermore, facile boronic acid saccharide complexation occurs only at the high pH conditions required to create a boronate anion. It is known, however, that saccharide complexation changes the pK_a of the boronic acid moiety, as is the case for 2- and 9-anthrylboronic acids which display enhanced acidity upon binding to saccharides and consequent fluorescent suppression via a PET mechanism [10]. However, the photoinduced electron transfer from the boronate anion was not efficient despite the fact that it is directly bound to the chromophore (I (in the presence of saccharide)/I_0 (in the absence of saccharide) = ca. 0.7)

In order to overcome the above mentioned disadvantages of boronic acid-saccharide interactions we have modified the boronic acid binding site to create a better electron center around the boronic acid moiety [19]. In compound **13** the basic skeleton of a known PET sensor, **12**, has been preserved with the addition of the improved binding site with a tertiary amine as in **14** [18]. The amine

12   13   14

can interact intramolecularly with the boronic acid, creating a boronate anion-containing five-membered ring. The fluorescence pH profile of **13**, in unbuffered aqueous media, gave one large step at low pH (pK_a = 2.9) and a possible small step at high pH which is interesting given that the pK_a of **12** is known to be 9.3 (fluorescence measurements in ethanolic aqueous media) [18]. The large shift of the pK_a is due to the interaction found between the boronic acid moiety and the amine group but this does not inhibit the photoinduced electron transfer quenching process in the complex (Scheme 9.2) [19]. Complete separation of the amine and the boronic acid moiety at very high pH further quenched the anthracene fluorescence but the fluorescence decrease is insufficient for the calculation of the pK_a of this process. The introduction of saccharides (D-glucose and D-fructose) strongly changes the fluorescence of **13** over a large pH range; Scheme 9.2 is suggestive of the most important species involved in the fluores-

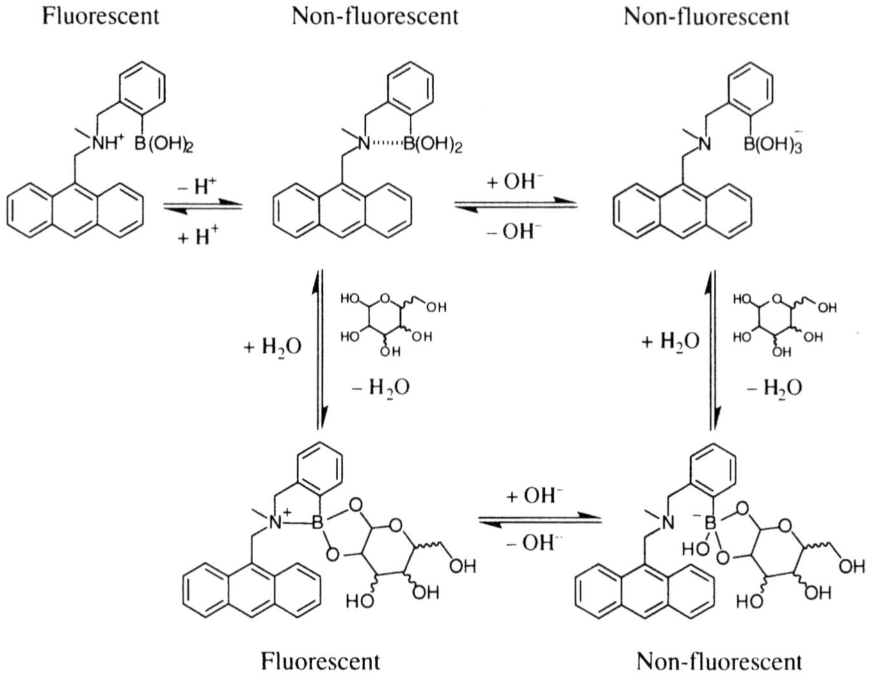

Scheme 9.2. Species and equilibria involved in the action of a PET sensor for saccharides

cence changes. Saccharide binding increases the acidity of the boronic acid group, thus intensifying the B···N interaction and this consequently inhibits the electron transfer process giving greater fluorescence. From the pH-fluorescence intensity plots it is now obvious that **13** can bind saccharides even over the neutral pH region, with the aid of the amine, in the same way that simple boronic acids do at the alkaline pH region. The more flexible and less bulky ethylene glycol gave very low fluorescence enhancements, however, suggesting the importance of some steric factors. The pK_a for the saccharide complex, as calculated from fluorescence measurements at high pH ($pK_a = 11.1$), is in line with the second pK_a of **14** in the absence of sugar ($pK_a = 11.8$) which is the parent binding site of **13**.

9.5
A Glucose Sensor and an Enantioselective Sensor

By combining the concepts of selective saccharide binding with those of PET sensors, as outlined above, it should be possible to design a monosaccharide selective sensor, a suitable design strategy for which is exemplified by **15** [20, 21]. The formation of a large macrocyclic structure upon 1:1 binding of glucose to **15** holds glucose close to the anthracene aromatic face (Fig. 9.3) with the H_3 proton of D-glucose, in particular, pointing towards the π-electrons of the anthracene moiety, giving a very large paramagnetic shift in the ^1H NMR spectrum

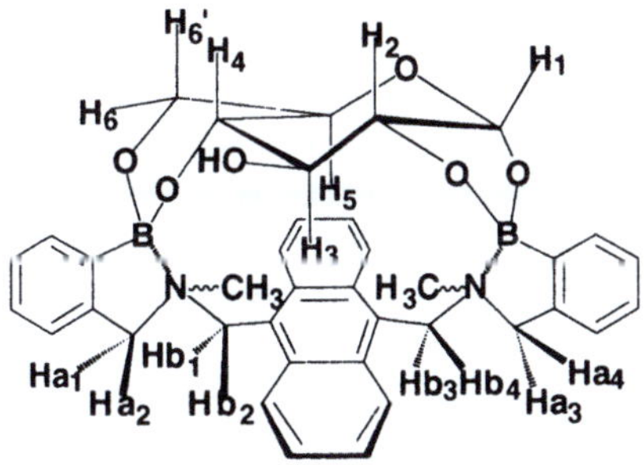

Fig. 9.3. 15-D-glucose complex. ¹H NMR studies have established that **15** immobilizes the pyranose form of D-glucose, which may isomerize to the furanase form in protic media

($H_3 = -0.3$ ppm). The coupling constant $J_{2,3} = 7.5$ Hz implies that the pyranose form of glucose is complexed in the cleft of **13** while the existence of a 1:1 complex of **15** and D-glucose was further confirmed by mass spectral data of the complex [20, 21].

Because both binding sites must be occupied in order to prevent fluorescence quenching, non-cyclic 1:1 bound species could not be detected by fluorescence spectroscopy. Only the 1:1 cyclic and 1:2 complexes gave signals with the "switch-on" factor (ratio of maximum to minimum fluorescence intensity) for **15** being greater than that for **13**.

The most important species involved in the equilibrium process are shown in Scheme 9.3 and such cooperative binding of saccharides, specifically glucose, occurs at very low saccharide concentrations. In human blood three main monosaccharides are present: D-glucose (0.3–1.0 mmol/l), D-fructose (≤ 0.1 mmol/l), and D-galactose (≤ 0.1 mmol/l), and competitive binding studies show that **15** is suitable for the selective detection of glucose at physiological levels.

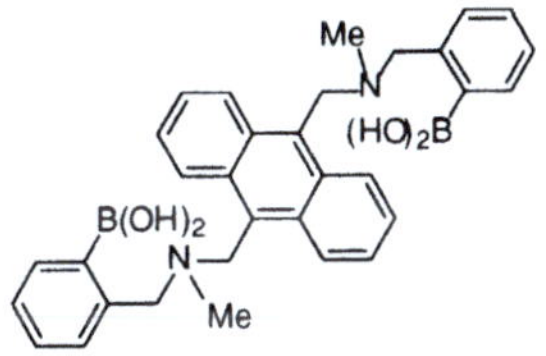

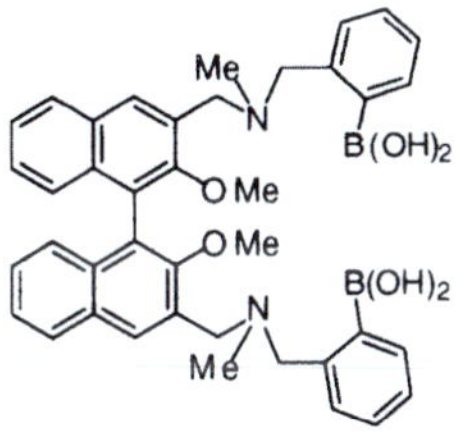

Enantioselective recognition of saccharides by **16** utilizes both steric and electronic factors [22]. The asymmetric immobilization of the amine groups, relative to the binaphthyl moiety, upon 1:1 complexation of saccharides by D- or L-isomers, creates a difference in PET and this difference is manifested in the maximum fluorescence intensity of the complex. Steric factors arising from the chiral binaphthyl building block are chiefly represented by the stability constant of the complex; however, the interdependency of electronic and steric factors upon each other is not excluded. This molecular cleft, with a longer spacer unit compared to the anthracene-based diboronic acid **15**, gave the best recognition

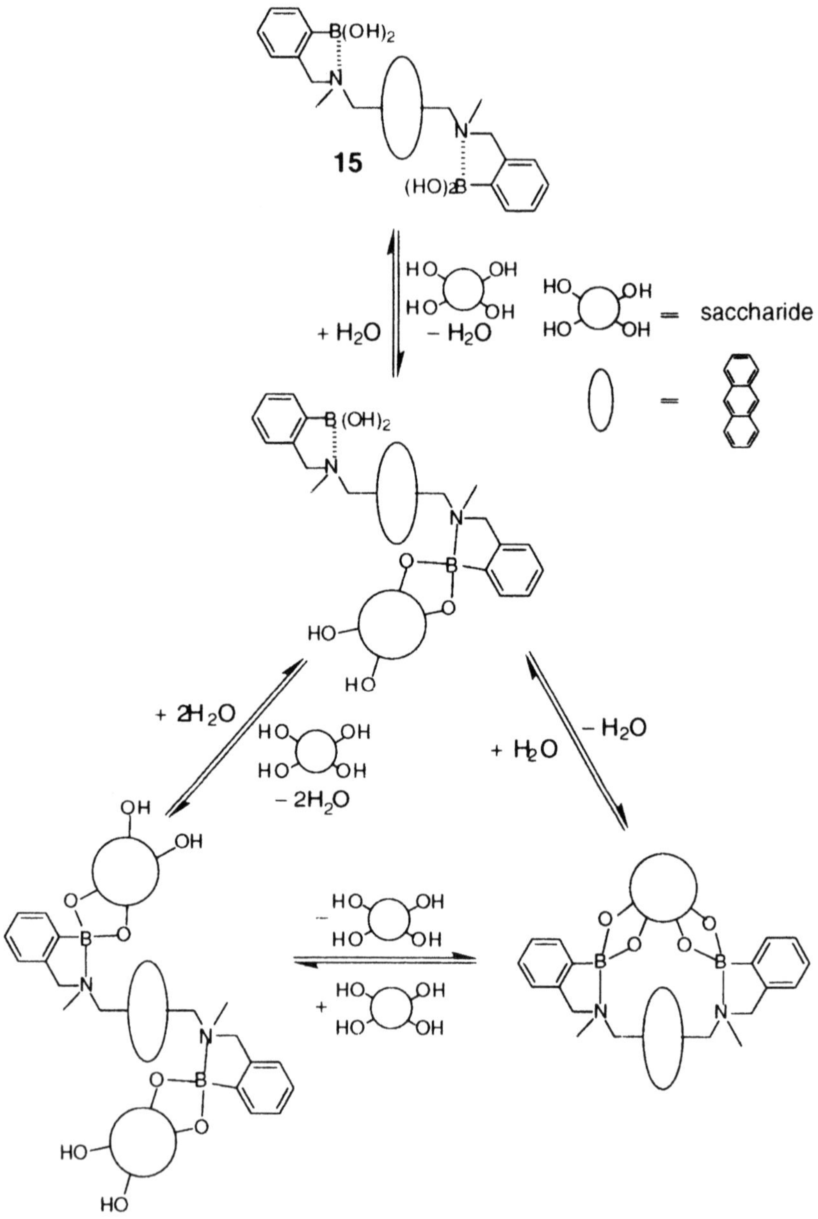

Scheme 9.3. Equilibria between different binding stoichiometries of glucose and PET sensor **15**

for fructose which was best bound by (R)-**16** and gave a large fluorescence increase. In this system steric factors and electronic factors bimodally discriminate between enantiomers, and competitive studies with D- and L-monosaccharides show the possibility of enantioselective detection of saccharides. Furthermore, the availability of both (R)- and (S)-isomers of this molecular sensor is an important advantage since complementary detection using either form is possible.

An alternative binding site was employed in the design of **17** to allow complexation and detection of uronic and sialic carboxylates (Fig. 9.4) even in the presence of neutral saccharides [23]. In this case, the phenanthroline moiety acts as a source of fluorescence and as a ligand for zinc ions which themselves are capable of further coordinating to carboxylate anions. Uronic and sialic acids, therefore, can coordinate to the boronic acid through their hydroxyl groups at one end of the sensor and to the phenanthroline-coordinated zinc ion at the other end through their carboxylate function.

Fluorescence binding studies demonstrated that **17** has the strongest affinity for D-galacturonic acid and a much weaker affinity for neutral saccharides. The most important feature of this binding behavior, however, was that cooperative binding was required to achieve strong binding for the acids – although the binding constants for the acids were equal to, or higher than, the neutral sugars in the presence of Zn (II), the constants were too weak to be measured in the absence of Zn (II). The binding constants of the simple saccharides were unchanged by the addition of Zn (II), however. Furthermore, titration of **17** with Zn (II) only produced a decrease in fluorescence whilst titration with Zn (II) in the presence of D-galacturonic acid produced a fluorescence increase. This unusual behavior has been attributed to the greater fluorescence quenching strength of

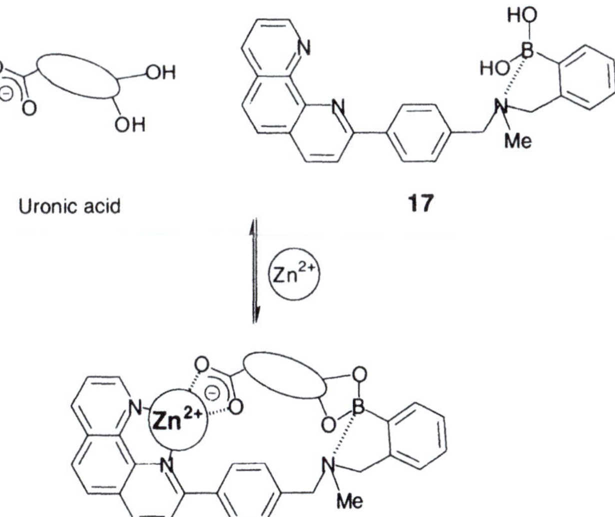

Fig. 9.4. Allosteric interaction between **17** and Zn²⁺ for uronic acid complexation with detection by fluorescence spectroscopy

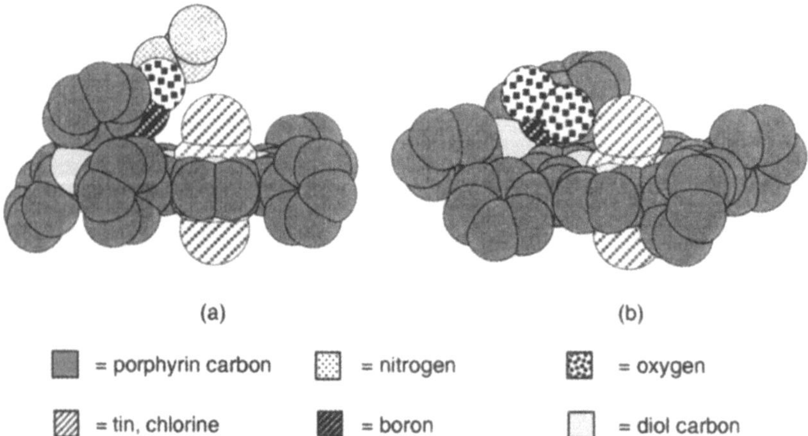

18

the amine nitrogen compared to that of the coordinating Zn (II) ion so that, although Zn (II) itself contributes to the quenching, it simultaneously increases the acid coordinating power of **17** which reduces the quenching effect of the amine. This separation of the boron and amine leads to an overall reduction in quenching, i. e., a gain in fluorescence.

The latest development in this series of boronic acid-based PET sensors has been a sensor, **18**, that actually reverses the action of this sensing motif so that the fluorescence decreases upon coordination to diols (Fig. 9.5) [24]. In the uncomplexed state, **18** is able to fluoresce weakly as a result of residual boron-nitrogen interactions that prevent complete quenching from the amine. In this case, however, the boronic acid is positioned within a sterically crowded cleft of a tin-metallated porphyrin which renders it slightly inaccessible to diols. When simple diols are added to the solution the boronic acid moiety is pulled away from the cleft which extends the linkage that joins it to the porphyrin. This extension separates the boron and amine-nitrogen atoms so that the boron-nitrogen interaction is actually weakened rather than strengthened by diol com-

(a)   (b)

Fig. 9.5a, b. **a** Complexation of **18** by diols pulls the boron atom and amine nitrogen apart, increasing the fluorescence compared to; **b** that of the uncomplexed sensor

plexation. The general trend is for the greatest fluorescence reduction to be caused by the most bulky diols.

9.6
Conclusion

The recognition of saccharides by boronic acid-based molecular receptors has shown tremendous growth during the last few years – from inherent saccharide selectivity with monoboronic acids and controlled selectivity with simple diboronic acids through to the enantioselective recognition of saccharides. The biggest breakthrough in this study was a combination of the PET sensor concept with the boronic-acid sugar-binding, which enabled us to solve two difficult problems at one time, sugar-binding at the neutral pH region and detection of the sugar-binding process. We believe that such sensors will find many applications in biological systems for both the monitoring and mapping of biologically important saccharides. This relatively new field will attract many scientists' attention in the years to come.

References

1. For comprehensive reviews see: Rebek J Jr (1990) Angew Chem Int Ed Engl 29:245; Hamilton AD (1991) Bioorg Chem Front 2:115
2. Tsukagoshi K, Shinkai S (1991) J Org Chem 56:4089; Shiomi Y, Saisho M, Tsukagoshi K, Shinkai S (1993) J Chem Soc, Perkin Trans 1:2111
3. Shinkai S, Tsukagoshi K, Ishikawa Y, Kunitake T (1991) J Chem Soc, Chem Commun 1039
4. Kondo K, Shiomi Y, Saisho M, Harada T, Shinkai S (1992) Tetrahedron 48:8239
5. James TD, Harada T, Shinkai S (1993) J Chem Soc, Chem Commun 857
6. Kano K, Yoshiyasu K, Hashimoto S (1988) J Chem Soc, Chem Commun 801
7. Aoyama Y, Tanaka Y, Toi H, Ogoshi H (1988) J Am Chem Soc 110:634
8. Kikuchi Y, Kobayashi K, Aoyama Y (1992) J Am Chem Soc 114:1351
9. Lorand JP, Edwards JO (1959) J Org Chem 24:769
10. Yoon J, Czarnik AW (1992) J Am Chem Soc 114:5874
11. Mohler LK, Czarnik AW (1993) J Am Chem Soc 115:2998
12. Suenaga H, Mikami M, Sandanayake KRAS, Shinkai S (1995) Tetrahedron Lett 36:4825
13. Nagai Y, Kobayashi K, Toi H, Aoyama Y (1993) Bull Chem Soc Jpn 66:2965
14. Norrild JC, Eggert H (1995) J Am Chem Soc 117:1479
15. Nakashima K, Shinkai S (1994) Chem Lett:1267
16. Sandanayake KRAS, Nakashima K, Shinkai S (1994) J Chem Soc, Chem Commun 1621
17. Takeuchi M, Yoda S, Imada T, Shinkai S (1997) Tetrahedron 53:8335
18. Bryan AJ, de Silva AP, Rupasingha RAD, Sandanayake KRAS (1989) Biosensors 4:169; de Silva AP, Rupasingha RAD (1985) J Chem Soc, Chem Commun 1669
19. James TD, Sandanayake KRAS, Shinkai S (1994) J Chem Soc, Chem Commun 477
20. James TD, Sandanayake KRAS, Shinkai S (1994) Angew Chem Int Ed Engl 33:2207
21. James TD, Sandanayake KRAS, Iguchi R, Shinkai S (1995) J Am Chem Soc 117:8982; more recently, it has been shown that the bound species may be changed from α-D-glucopyranose to α-D-glucofuranose: Bielecki M, Eggert H, Norrild JC (1999), J Chem Soc, Perkin Trans 2:449
22. James TD, Sandanayake KRAS, Shinkai S (1995) Nature 374:345
23. Takeuchi M, Yamamoto M, Shinkai S (1997) J Chem Soc, Chem Commun:1731; Takeuchi M, Yamamoto M, Shinkai S (1998) Tetrahedron 54:3125
24. Kijima H, Takeuchi M, Robertson A, Shinkai S, Cooper C, James TD (1999) J Chem Soc, Chem Commun 2011

PCT (Photoinduced Charge Transfer) Fluorescent Molecular Sensors for Cation Recognition

B. Valeur, I. Leray

10.1
Introduction

Considerable efforts are being made to develop selective fluorescent sensors for cation detection in analytical chemistry, biology, medicine (clinical diagnosis), environment, chemical oceanography, etc. [1–6].

A fluorescent molecular sensor for cation recognition consists of a *fluorophore* (the *signaling moiety*) linked to an *ionophore* (the *recognition moiety*) and is thus often called *fluoroionophore*. The design of such sensors requires special care [7, 8]. The *signaling moiety* acts as a signal transducer, i.e., it converts the information (recognition event) into an optical signal expressed as the changes in the photophysical characteristics of the fluorophore. These changes are due to the perturbation (by the bound cation) of photoinduced processes such as electron transfer, charge transfer, energy transfer, excimer or exciplex formation or disappearance, etc. The *recognition moiety* is responsible for selectivity and efficiency of binding which depend on the ligand topology, on the characteristics of the cation (ionic radius, charge, coordination number, hardness, etc.), and on the nature of the solvent (and pH, ionic strength in the case of aqueous solutions).

There are three main classes of fluorescent molecular sensors for cation recognition that differ by the nature of the cation-controlled photoinduced processes:

1. Sensors based on cation control of photoinduced electron transfer (PET sensors).
2. Sensors based on cation control of photoinduced charge transfer (PCT sensors).
3. Sensors based on cation control of excimer formation or disappearance.

In each class, distinction is to be made according to the structure of the complexing moiety: chelators, podands, coronands (crown ethers), cryptands, calixarenes [8].

A distinct advantage of PET sensors is the very large change in fluorescence intensity usually observed upon cation binding, so that the expression „off-on" and „on-off" fluorescent sensors is often employed. Another characteristic is the absence of shift of the fluorescence or excitation spectra which precludes the possibility of intensity-ratio measurements at two wavelengths. Furthermore, PET often arises from a tertiary amine whose pH sensitivity may affect the response to cations.

In PCT sensors, the changes in fluorescence quantum yield on cation complexation are generally not very large as compared to those observed with PET sensors (with noticeable exceptions included in the present review). However, the absorption and fluorescence spectra are shifted upon cation binding so that an appropriate choice of the excitation and observation wavelengths often allows one to observe quite large changes in fluorescence intensity. Moreover, ratiometric measurements are possible: the ratio of the fluorescence intensities at two appropriate emission or excitation wavelengths provides a measure of the cation concentration which is independent of the probe concentration (provided that the ion is in excess) and is insensitive to intensity of incident light, scattering, inner-filter effects, and photobleaching. Ratiometric measurements are also possible with excimer-based sensors.

Several reviews have already been partially devoted to PCT fluorescent molecular sensors [7–12]. The particular case where the fluorophore is a cyanine or a styryl base is the subject of Chap. 7 of the present volume. This review will focus on the various topologies of PCT sensors with special attention to selectivity which must be viewed in terms of both selectivity of bonding and selectivity of photophysical effects.

10.2
Principles

When a fluorophore contains an electron-donating group (often an amino group) conjugated to an electron-withdrawing group, it undergoes intramolecular charge transfer from the donor to the acceptor upon excitation by light. The consequent change in dipole moment results in a Stokes shift that depends on the microenvironment of the fluorophore; polarity probes have been designed on this basis [13]. It can thus be anticipated that cations in close interaction with the donor or the acceptor moiety will change the photophysical properties of the fluorophore because the complexed cation affects the efficiency of intramolecular charge transfer.

When a group (like an amino group) playing the role of an electron donor within the fluorophore interacts with a cation (Fig. 10.1), the latter reduces the electron-donating character of this group; owing to the resulting reduction of conjugation, a blue shift of the absorption spectrum is expected together with a decrease of the molar absorption coefficient. Conversely, a cation interacting with the acceptor group (Fig. 10.1) enhances the electron-withdrawing character of this group; the absorption spectrum is thus red-shifted and the molar absorption coefficient is increased. The fluorescence spectra are in principle shifted in the same direction as those of the absorption spectra. In addition to these shifts, changes in quantum yields and lifetimes are often observed. All these photophysical effects are obviously dependent on the above described characteristics of the cation, and selectivity of these effects are expected.

The photophysical changes upon cation binding can also be described in terms of charge dipole interaction [14]. Let us consider only the case where the dipole moment in the excited state is larger than that in the ground state. Then, when the cation interacts with the donor group, the excited state is more strongly destabilized by the cation than the ground state (Fig. 10.2), and a blue shift of

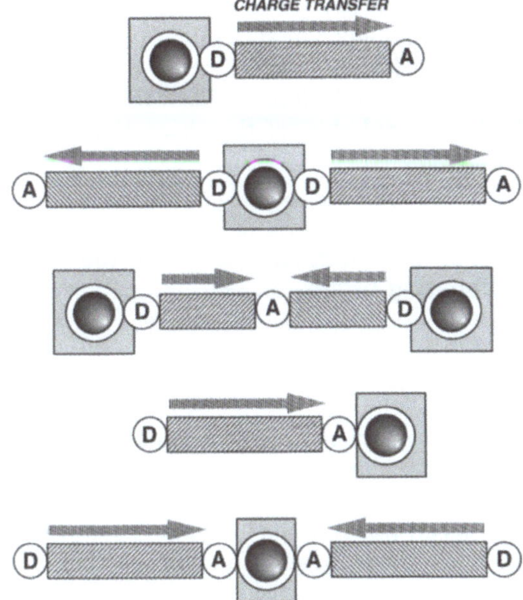

Fig. 10.1. Topology of PCT fluorescent molecular sensors

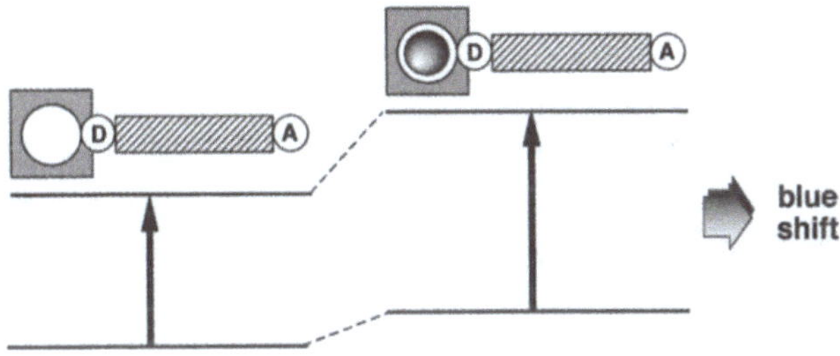

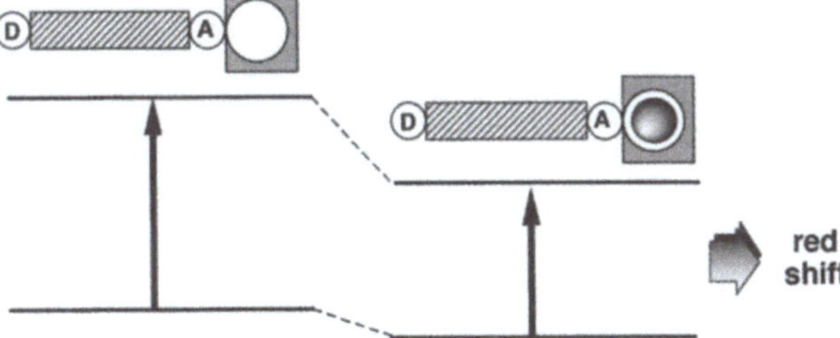

Fig. 10.2. Spectral displacements of PCT sensors resulting from interaction of a bound cation with an electron-donating or electron-withdrawing group

the absorption and emission spectra is expected (however, the fluorescence spectrum undergoes only a slight blue shift in most cases; this important observation will be discussed below). Conversely, when the cation interacts with the acceptor group, the excited state is more stabilized by the cation than the ground state, and this leads to a red shift of the absorption and emission spectra (Fig. 10.2).

10.3
PCT Sensors Based on the Interaction Between the Bound Cation and an Electron-Donating Group

10.3.1
Crown-Containing PCT Sensors

Many fluoroionophores contain an azacrown whose nitrogen atom is conjugated to an electron-withdrawing group (Fig. 10.3). Compounds 1–7 exhibit a common feature: the blue shift of the absorption spectrum is much larger than that of the emission spectrum on cation binding. The wavelengths corresponding to the maximum of absorption and emission spectra in the absence and in the presence of calcium ion are reported in Table 10.1 for 1 [15], 2 and 3[16], 4

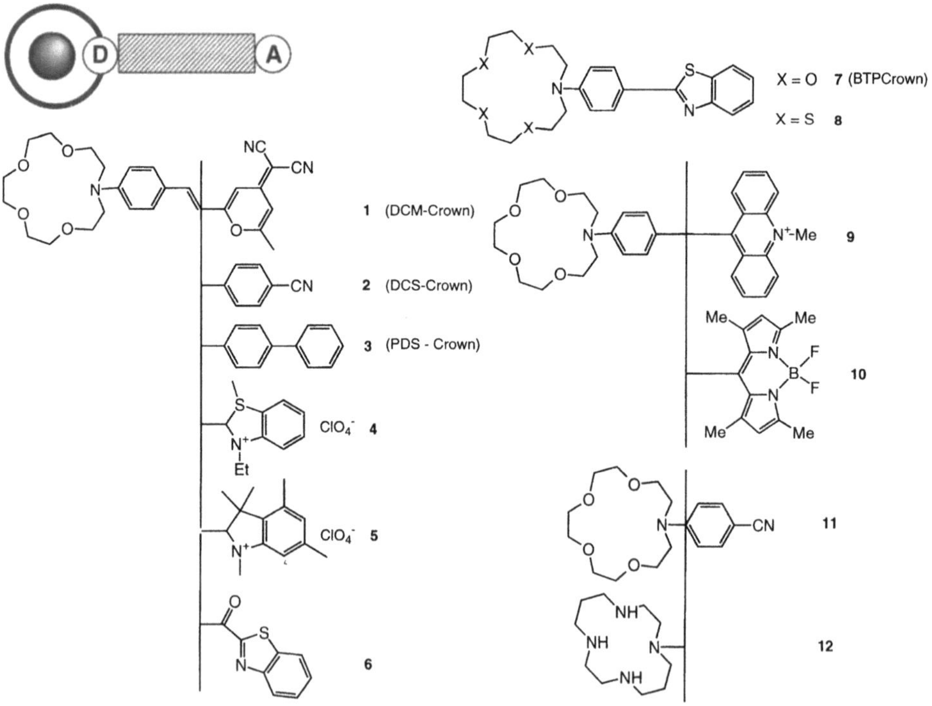

Fig. 10.3. Crown-containing PCT sensors in which the bound cation interacts with the donor group

Table 10.1. Wavelength of the absorption and emission maximum in acetonitrile for some crown-ether-linked cation sensors (the formulae are given in Fig. 10.12). $\Delta\lambda$ represents the wavelength difference between the maximum for the free ligand and the maximum in the presence of an excess of calcium (full complexation)

Compound	Absorption max (nm)		$\Delta\lambda$ (nm)	Emission max. (nm)		$\Delta\lambda$ (nm)
	Free	Comp. Ca^{2+}		Free	Comp. Ca^{2+}	
1[a]	464	398	66	621	608	13
2[b]	392	330	62	525	503	22
3[b]	372	332	40	483	469	14
4[c]	521	411	110	600	591	9
5[d]	585	450	135	702	691	11
6[e]	463	360	103	634	606	28
7[f]	360	315	45	420	~418	~2

[a] [15]; [b] [16]; [c] [17]; [d] [18]; [e] [19]; [f] [20].

[17], **5** [18], **6** [19], **7** [20]. Similar photophysical effects were observed with compound **8** consisting of benzothiazyl group linked to polythiaazaalkane as complexing unit in order to promote complexation of Ag^+. Such a small shift of the fluorescence spectrum – which is surprising at first sight – can be interpreted as follows. The photoinduced charge transfer reduces the electron density on the nitrogen atom of the crown, and this nitrogen atom becomes a noncoordinating atom because it is positively polarized. Therefore, excitation induces a photodisruption of the interaction between the cation and the nitrogen atom of the crown. The fluorescence spectrum is thus only slightly affected because most of the fluorescence is emitted from species in which the interaction between the cation and the fluorophore does not exist any more or is much weaker.

This interpretation is supported by a thorough study of the photophysics of **1** and its complexes with Li^+ and Ca^{2+} [21–23]. In particular, subpicosecond pump-probe spectroscopy provided compelling evidence for the disruption of the link between the crown nitrogen atom and the cation. A photodisruption was also demonstrated in complexes of **2** and **3** [24, 25]. The cation-induced spectral changes in **4** [17], **5** [18], and in another crowned styryl dye [26] were interpreted in the same way.

Such a photodisruption results in a lower stability of the complexes in the excited-state. Therefore, excitation of these complexes by an intense pulse of light is expected to cause some cations to leave the crown and diffuse away provided that the time constant for total release of the cation from the crown is shorter than the lifetime τ of the excited state (for a discussion on this point, see [11]).

Intramolecular charge transfer in conjugated donor-acceptor molecules may be accompanied with internal rotation leading to TICT (twisted intramolecular charge transfer) states [27]. A dual fluorescence may be observed as in **11** [28] (which resembles the well-known DMABN containing a dimethylamino group instead of the monoaza-15-crown-5): the short-wavelength band corresponds to the fluorescence from the locally excited-state and the long-wavelength band arises from a TICT state. The fluorescence intensity of the latter decreases upon cation binding because interaction between a bound cation and the crown nitrogen disfavors the formation of a TICT state, which leads to a concomitant increase of the short-wavelength band.

Compound **12** [29] is an analog of **11** in which the monoazacrown has been replaced by a tetraazacrown (cyclam) in order to promote complexation of transition metal ions. In this case, a triple fluorescence is observed: in addition to the emission from the locally excited and the TICT state, fluorescence is also emitted from an intramolecular exciplex (sandwich complex formed in the excited state thanks to the flexibility of cyclam). Such a triple fluorescence is solvent and pH dependent and is perturbed by cation binding. The relative changes of the bands depend on the nature of the cation.

The formation of a TICT state is often invoked even if no dual fluorescence is observed. For donor-acceptor stilbenes (**2** and **3**), the proposed kinetic scheme [16] contains three states: the planar state E* reached upon excitation can lead to state P* (nonfluorescent) by double-bond twist, and to TICT state A* by single-bond twist, the latter being responsible for the main part of the emission. The existence of a fluorescent TICT state was also stated to be responsible for the photophysical properties of **6** [19].

The absence of fluorescence of **9** [30] may be due to the formation of a nonfluorescent TICT state, and an acridinium type fluorescence is recovered upon binding of H^+ and Ag^+. The same explanation may hold for the low fluorescence of **10** [31] whose electron-withdrawing group, boron-dipyrromethen, must be twisted due to steric interactions with the phenyl ring: the fluorescence enhancement factor varies from 90 for Li^+ to 2250 for Mg^{2+}. Both **9** and **10** compounds undergo much larger fluorescence enhancements than most of PCT molecular sensors (factors of ca. 2–5 are generally observed). Since they undergo almost complete charge transfer (i.e., electron transfer) in the absence of cation, they can be considered as limiting cases closely resembling PET sensors but with a virtual spacer.

In contrast to the PCT fluoroionophores described above, those shown in Fig. 10.4 undergo a large blue shift of the fluorescence spectrum upon cation binding. Compound **13** [32–34] is one of the first crown-containing fluorescent PCT sensors that has been designed. The fluorescence maximum shifts from 642 nm for the free ligand to 574 nm for the calcium complex in acetonitrile. Such a blue shift means that there is no photodisruption of the interaction between the cation and the nitrogen atom of the crown in this case in contrast to **1–8**, because the nitrogen atom of the crown plays the role of a second electron-donating atom with respect to the benzoxazinone moiety which is itself a donor-acceptor system. The existence of a fluorescent TICT state was shown to be likely in this compound [35]. Compound **13** is not only responsive to alkaline-earth cations but also to divalent heavy metal ions Hg^{2+} and Pb^{2+} [36]. Environmental analytical applications are thus possible.

The structure of **14** [37] is very similar to that of **13**: the heterocyclic oxygen atom has been replaced by a nitrogen atom so that the dye is a benzodiazinone instead of a benzoxazinone. The cation-induced changes in photophysical properties and the complexing ability are comparable.

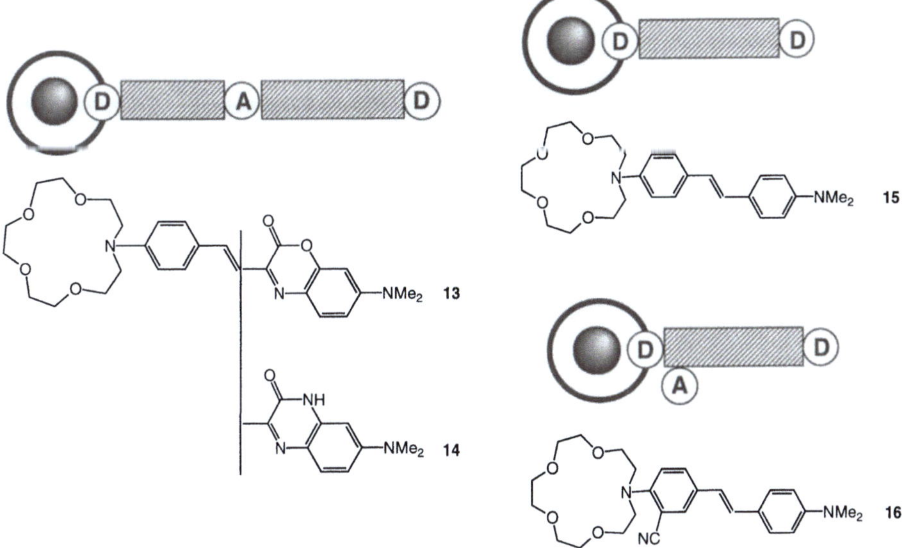

Fig. 10.4. Crown-containing PCT sensors with various arrangements of donor and acceptor groups

It is interesting to note that in the symmetrical fluorophore 15 [38] with two electron-donating groups at the two ends, one of which belonging to an aza-crown, cation binding induces photoinduced charge transfer. This results in a red shift of the fluorescence spectrum in contrast to the fluoroionophores described above. The presence of an additional electron-withdrawing group -CN in 16 [38] enhances the charge transfer with respect to 15 and an even larger red-shift is observed upon cation binding.

The crown-containing fluoroionophores described above are of great interest for the understanding of cation-dipole interactions. They offer a large variety of photophysical changes upon cation binding that can be used for cation recognition. Their poor solubility in water precludes applications in aqueous solutions but they can be used in extraction processes or for doping the sensitive part of optical sensor devices. However, the selectivity of azacrowns towards metal ions is not good enough when stringent discrimination between cations of the same chemical family is required.

Improvement of selectivity can be achieved by the participation of external groups (lariat crown ether), as shown in Fig. 10.5. In 17 (PBFI) [39] and 18 (SBFI) [40], the oxygen atom of the methoxy substituent of the fluorophore can interact with a cation; binding efficiency and selectivity are thus better than those of the crown alone. SBFI has been designed for probing intracellular sodium ions and PBFI for potassium ions. In both compounds, the photophysical changes are likely to be due to the reduction of the electron-donating character of the nitrogen atoms of the diazacrown by the complexed cation. Further improvement of the

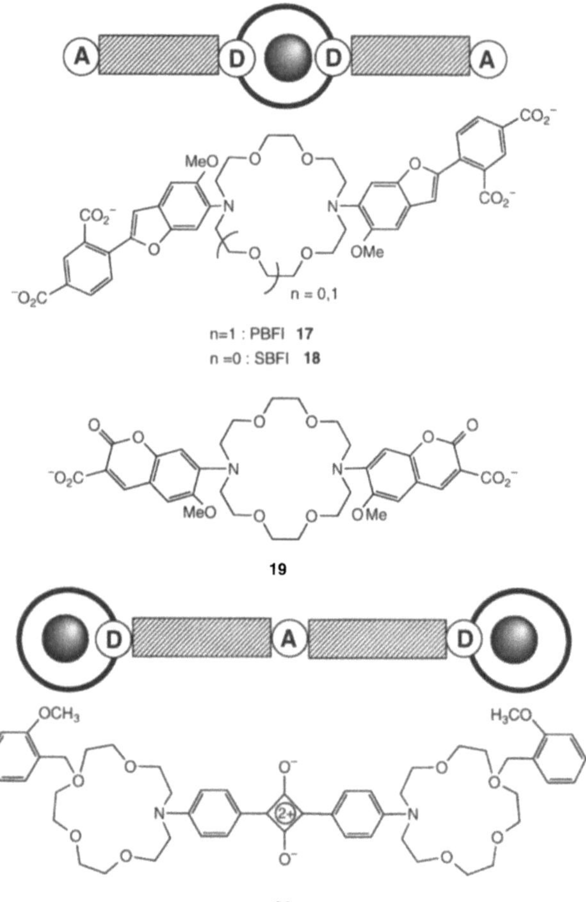

Fig. 10.5. PCT sensors with lariat crown ethers

selectivity towards K$^+$ with respect to Na$^+$ is desirable. Compound **19** [41] resembles PBFI but shows greater selectivity for potassium over sodium than PBFI.

The squaraine-based fluoroionophore **20** [42] has been designed for the selective detection of sodium ion thanks to appropriate lariat crown ethers

10.3.2
Chelating PCT Sensors

In light of the preceding considerations on crowned charge-transfer compounds, it is worth examining the photophysical properties of well-known chelators in the BAPTA series used for the recognition of cytosolic calcium [43]. Examples are given in Fig. 10.6 (top). For instance **21** (Indo-1) and **25** (Fura-2) are widely used as calcium indicators. In this series, the fluorophore is a donor-ac-

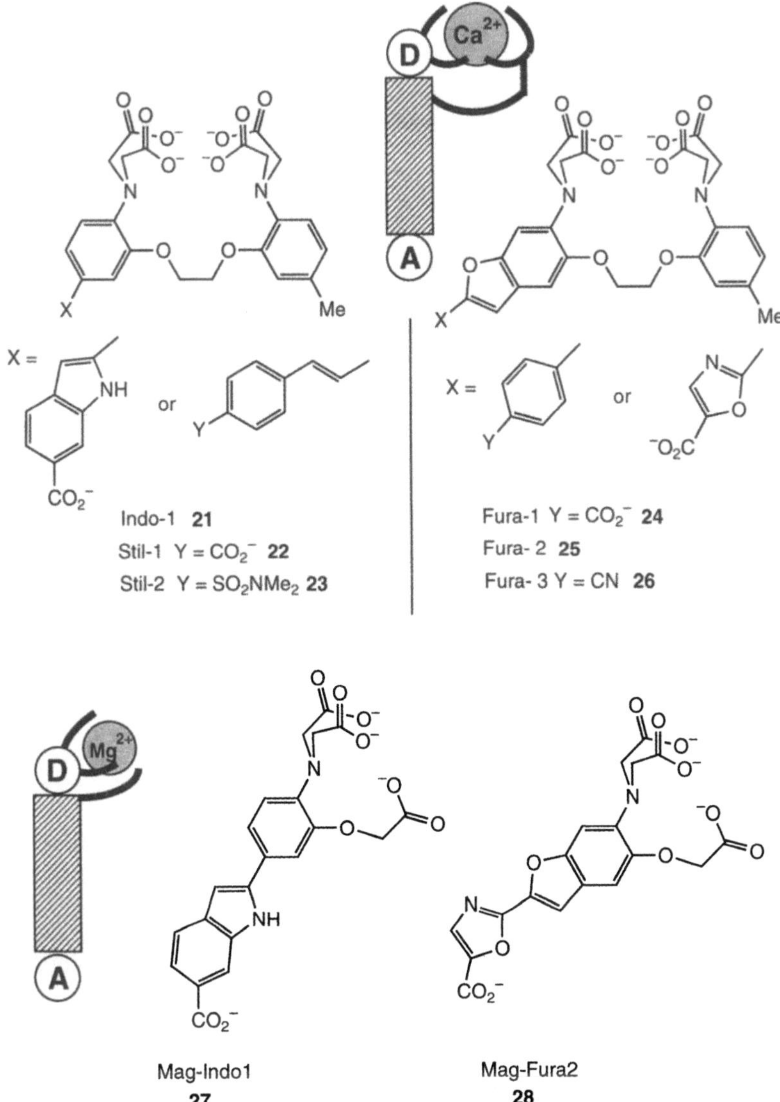

Fig. 10.6. Chelating PCT sensors in which the bound cation interacts with the donor group

ceptor molecule with an amino group as the electron-donating group which participates in the complexation, the ionophore being a chelating group of the BAPTA type. Upon complexation by Ca^{2+} in water, the absorption spectrum is blue shifted, whereas there is almost no shift of the fluorescence spectrum, except for Indo-1.

The same interpretation as for the crown-ether-linked compounds described above can be proposed: the electron density of the nitrogen atom conjugated with the electron-withdrawing group of the fluorophore is reduced upon excitation and might even become positively polarized; this causes disruption of the interaction between this nitrogen atom and a bound cation. Consequently, fluorescence emission closely resembles that of the free ligand. The possibility of photoejection has not been examined yet. Along this line, information on the stability of the complexes in the excited state, as compared with the ground state, is again of the utmost importance. A study of the fluorescence decay of Fura-2 with global compartmental analysis [44] showed that the stability constant of the calcium complex in the excited state is more than three orders of magnitude smaller than in the ground state.

Regarding the exception of Indo-1, the photoinduced charge transfer may not be sufficient to cause nitrogen-Ca^{2+} bond breaking. This interpretation is consistent with the fact that the fluorescence maximum of free Indo-1 is located at a shorter wavelength than all the other ligands by $\sim 30 – 60$ nm, thus indicating a less polar charge-transfer state.

By keeping the same fluorophores, but reducing the „cavity" size of the ionophore, one obtains **27** (Mag-Indo1) and **28** (Mag-Fura2) (Fig. 10.6) which are selective for magnesium [45]. Most of the chelators of Fig. 10.6 are commercially available in the nonfluorescent acetoxymethylester form so that they are cell permeant and they recover their fluorescence upon hydrolysis by enzymes [45].

10.3.3
Cryptand-Based PCT Sensors

The chelators described above are well suited for the detection of alkaline-earth cations but not alkali cations. In contrast, cryptands are very selective towards the latter.

Compound **29** (FCryp-2) [46] (Fig. 10.7) is a nice example of a fluorescent signaling receptor in which the ionophore moiety has been specially designed for determination of intracellular free sodium concentration. An indole derivative acts as the fluorophore: upon sodium binding, the emission maximum shifts from 460 nm to 395 nm and the fluorescence intensity increases 25-fold. The origin of these photophysical changes has not been studied so far. The large Stokes shift of the free ligand may be accounted for by photoinduced charge transfer with concomitant internal rotation in the excited state leading to a TICT state. The blue shift of the emission spectrum upon sodium binding is likely to be due to the reduction of the electron-donating character of the dye-bound nitrogen atom of the cryptand.

Another example of cryptand is **30** [47] (Fig. 10.7) which has potential applications as an extracellular probe of potassium.

10.3.4
Calixarene-Based PCT Sensors

Examples of PCT sensors based on calixarene are given in Fig. 10.15. Compound **31** [48] (Fig. 10.8) containing a benzothiazole group linked to calix[4]arene was found to be very selective towards Li^+. In a medium in which the phenolic group

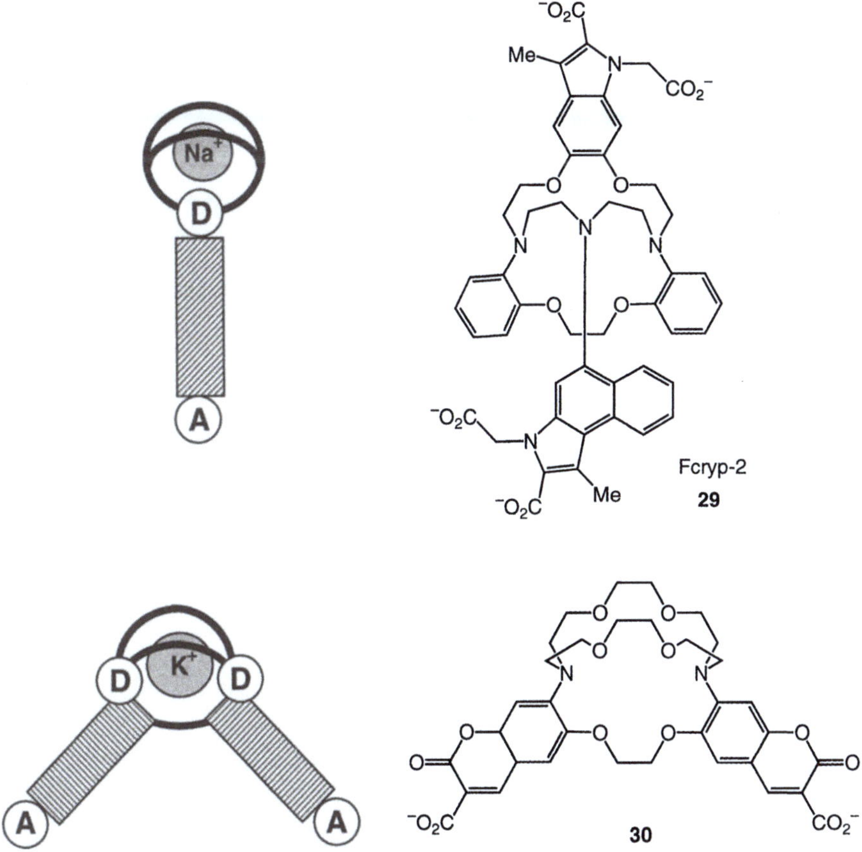

Fig. 10.7. Cryptand-based PCT sensors in which the bound cation interacts with the donor group

is not deprotonated, complexation with Li$^+$ induces a red shift of the emission spectra as a result of cation-induced proton ejection leading to a phenolate-cation pair. In the complex, Li$^+$ is also coordinated to two methoxy groups. No change of the emission spectra was observed with Na$^+$ and K$^+$.

The fluorescence of thiacalix[4]arene **32** [49] with two appended dansyl moieties was shown to increase upon complexation of Cd^{2+}. Significant effects are also observed with Al^{3+}, Cr^{3+}, Zn^{2+}, and Cu^{2+}. The dansyl fluorophore is well known for its sensitivity to polarity of the medium owing to intramolecular PCT. Consequently, the authors explained the cation-induced enhancement of fluorescence intensity by a change in polarity of the microenvironment of the dansyl moieties which are supposed to move from the outside bulk water towards the hydrophobic interior of the thiacalix[4]arene cavity. However, this explanation is not compatible with the small size of the cavity. The observed effects are more likely to be due to cation-dipole interaction but further investigation is necessary to understand the excited-state processes.

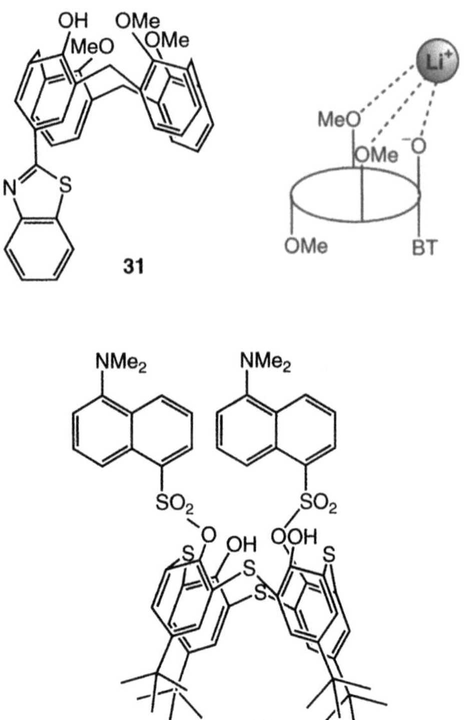

Fig. 10.8. Calixarene-based PCT sensors in which the bound cation interacts with the donor group

10.4
PCT Sensors Based on the Interaction Between the Bound Cation and an Electron-Withdrawing Group

10.4.1
Crown-Containing PCT Sensors

In contrast to the systems described above, there are few systems in which the bound cation can interact with the acceptor part of charge-transfer fluorophores. The case of coumarins linked to crowns is of special interest because the cation interacts directly with the electron-withdrawing group, i.e., the carbonyl group [50–54], in spite of the spacer between the fluorophore and the crown (Fig. 10.9). Regarding the cation-induced photophysical changes, it should be kept in mind that the dipole moment of aminocoumarins in the excited state is larger than in the ground state because of the photoinduced charge transfer occurring from the nitrogen atom of the julolidyl ring to the carbonyl group. Therefore, when a cation is coordinated with the carbonyl group, the excited state is more stabilized than the ground state so that both the absorption and emission spectra are red shifted.

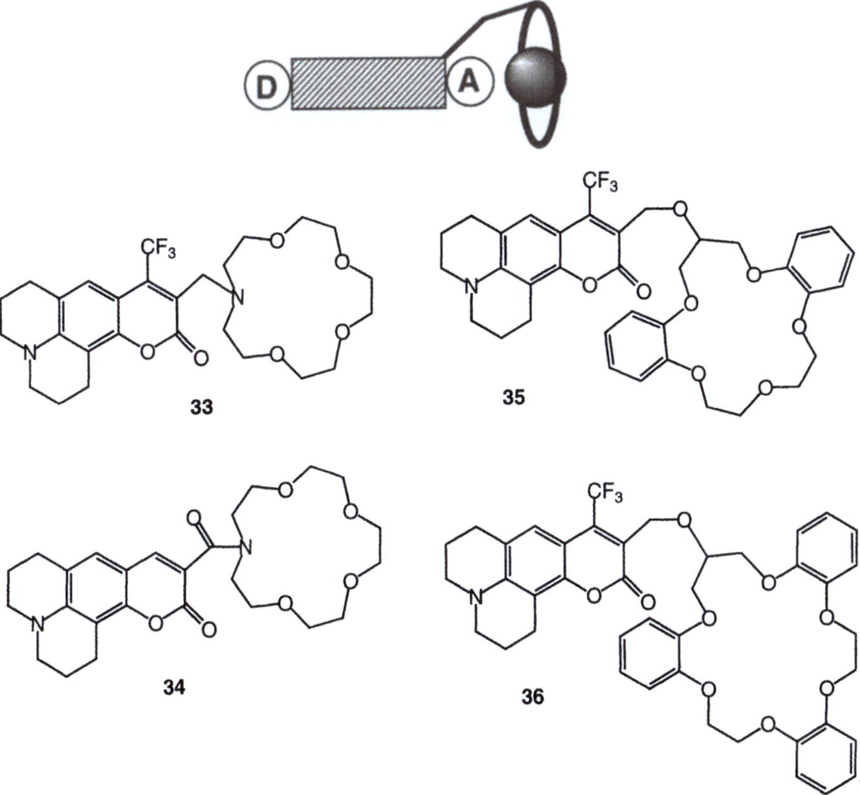

Fig. 10.9. Coumarin PCT sensors in which the bound cation interacts with the acceptor group

The spectral shifts upon cation binding are reported in Fig. 10.10. It should be noted that the red shift of the fluorescence spectrum are larger for **35** and **36** than for **33** and **34**, which can be explained by the length of the bridge. One may indeed expect that compounds **35** and **36** can adopt a conformation allowing the cation to move closer to the carbonyl group upon excitation as a result of the increasing electron density in the excited state. In contrast, the shifts of the absorption spectra appear to be less dependent on the fluoroionophore structure.

The increase of the molar absorption coefficient is much more pronounced for **34** than for **33** [53]. Those two compounds differ only by the bridging group separating the coumarin from the azacrown. In the case of **33**, when the azacrown is occupied by a cation, the linking amide behaves as an electron-withdrawing group and generates hyperchromicity as well as additional spectra shifts.

For all crowned coumarins, the cation-induced changes in fluorescence quantum yield upon cation complexation are not very large but, thanks to an appropriate choice of excitation and emission wavelengths, quite large variations in fluorescence intensity can be observed. Moreover, ratiometric measurements are possible.

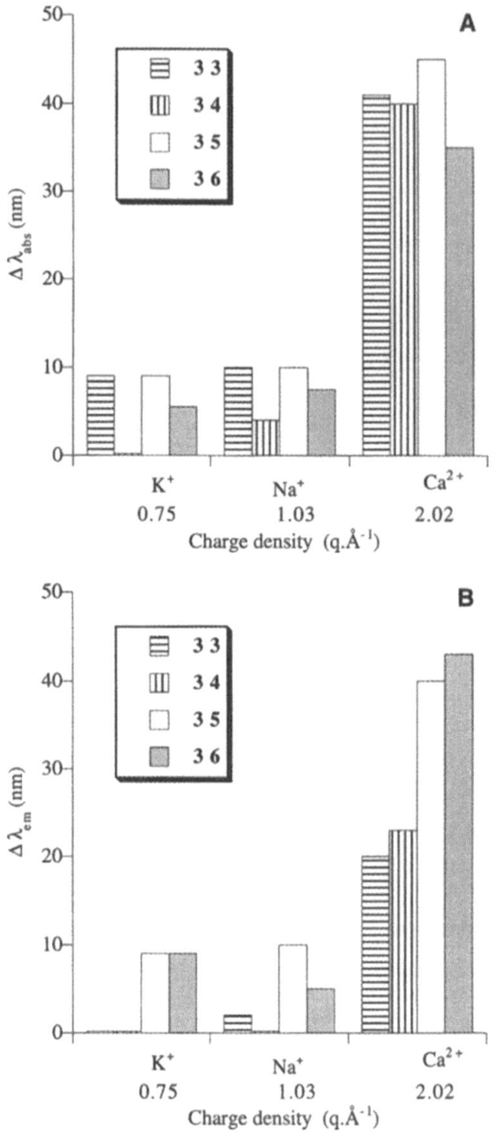

Fig. 10.10 A, B. Shifts of: **A** the absorption spectra; **B** the fluorescence spectra for crowned coumarins in acetonitrile upon cation binding

The stoichiometry and the stability constants of the complexes were determined by fitting the variations of the absorbance or the fluorescence intensity as a function of perchlorate salt concentration at constant ligand concentration as previously described [51] or by using the SPECFIT software. For most of the complexes, the stoichiometry in acetonitrile is 1:1 but there are a few exceptions. The stoichiometry can be 2:1 (ligand:metal) for **36** with Ba^{2+} and Ca^{2+}. Similar complexes have already been reported in the literature for the binding of benzocrown linked to pyrene [55] or to bistyrylbenzene [56]. For calcium complexation of **34**, analysis of the variations in absorbance vs salt concentration at a single wavelength led us to conclude that two types of complexes are formed, but the quality of the fit of the titration curve was equally good for $(1:1)+(1:2)$ (ligand:metal) and $(1:1)+(2:3)$ [53]. Later on, further analysis of the evolution of the whole absorption spectra (350 wavelengths) by means of the SPECFIT software allowed us to demonstrate that the stoichiometry of the two complexes are 1:1 and 1:2 [57].

The stability constants of the complexes in acetonitrile are given in Fig. 10.11. Thanks to the participation of the carbonyl group in the complexation, the stability constants are higher than those containing the same crown but without the participation of external coordinating atom. This is a general feature of crown-ether lariats compared with the analogous crown ether. The selectivity of **33** was found to be poor owing to the flexibility of the crown and the bridge. In the case of **34**, the replacement of the methylene bridge by an amide bridge precludes pH sensitivity and leads to an improvement of the selectivity. The selectivity towards alkaline earth-metal ions with respect to alkali cations is much better, but the selectivity between calcium and magnesium is to be improved. As expected, the selectivity is higher by using a more rigid crown incorporating phenyl groups (**35** and **36**). The selectivity for Ca^{2+}/Mg^{2+}, expressed as the ratio of the stability constants, was found to be 12,500 for **35** and 58,000 for **36**. Mg^{2+} is in fact too small to fit the crown cavity. In comparison, the selectivity Ca^{2+}/Mg^{2+} was only 3.5 for **33**. Regarding the distinction between Na^+ and K^+, the selectivity of **35** for Na^+ vs K^+ is $Na^+/K^+ = 16$.

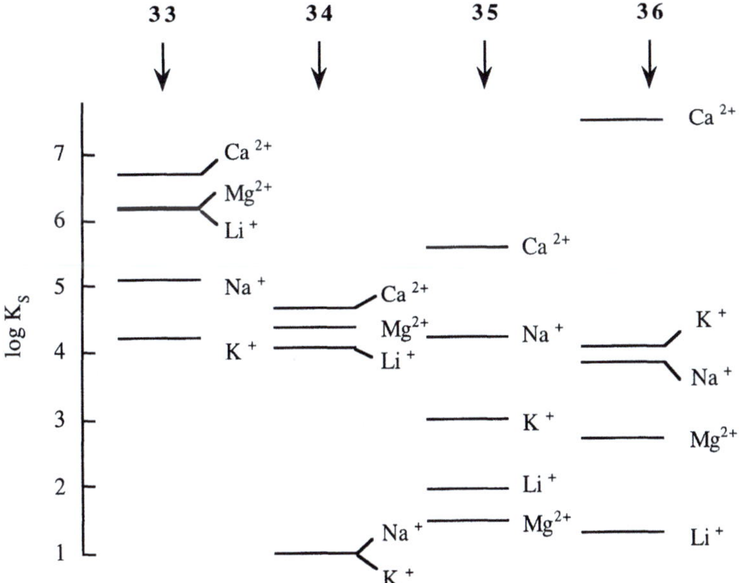

Fig. 10.11. Stability constants of complexes with crowned coumarins in acetonitrile

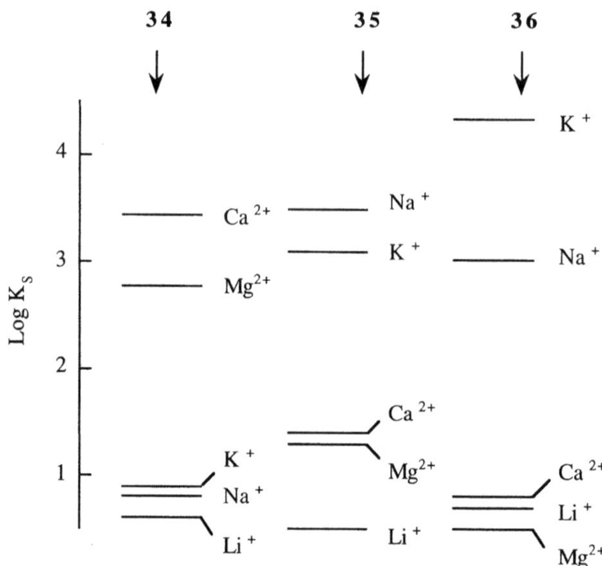

Fig. 10.12. Stability constants of complexes with crowned coumarins in ethanol

The stability constants in ethanol for the complexes with 34–36 are reported in Fig. 10.12. The complexes of 34 with alkaline-earth cations are much more stable than with alkali cations but the stability constants are smaller than in acetonitrile, as is to be expected. It is worth noting that the stability constants of the complexes of 35 and 36 with Na^+ and K^+ are much higher than with Li^+, Mg^{2+}, and Ca^{2+} because the latter cations have a higher charge density and are thus more solvated than complexed in contrast to the results in acetonitrile. The selectivity of 35 with respect to the other alkali and alkaline-earth cations is in good agreement with that of the same pendent ether lariat (dibenzo-16-crown – 5-oxyacetate) [58]. The K^+/Na^+ selectivity of 36 was found to be 20 (instead of 2 in acetonitrile) which was indeed expected from previously reported data [59]. This is encouraging in regard to the aim of selective detection of K^+ with respect to Na^+ which is of major interest in human serum where Na^+ is present at a larger concentration than Na^+ ($[Na^+]/[K^+] \sim 30$).

These examples show that the most important parameters responsible for selectivity in these crowned coumarins are (i) the rigidity of the link between the fluorophore and the crown, (ii) the rigidity the crown itself, and (iii) the size of the crown.

Additional photophysical effects can be observed when the fluoroionophore contains two fluorophores. In 37 [51] (Fig. 10.13), the carbonyl groups of the two coumarin moieties participate in the complexes: direct interaction between these groups and the cation explains the high stability constants and the photophysical changes. In addition to the shifts of the absorption and emission spectra, an interesting specific increase in the fluorescence quantum yield upon binding of K^+ and Ba^{2+} ions has been observed: in the complexes with these ions

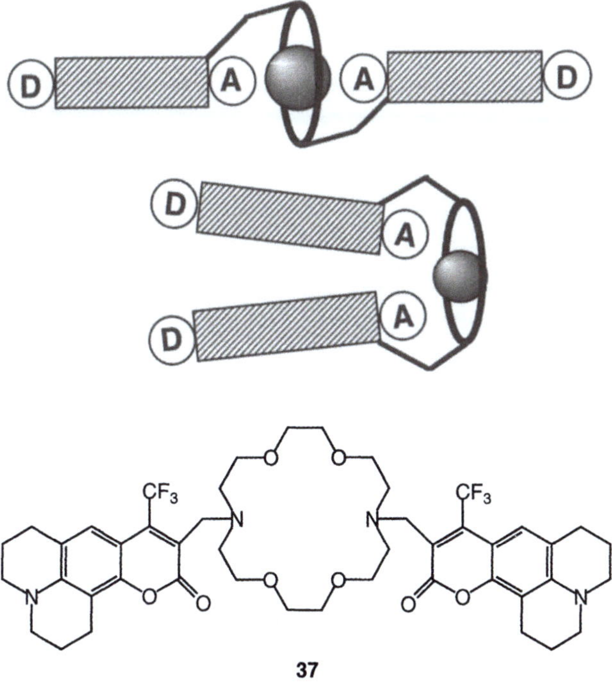

37

Fig. 10.13. Coumarin bifluorophoric PCT sensors in which the bound cation interacts with the acceptor group

that fit best into the crown cavity, the carbonyl groups of the two coumarins are preferentially on the opposite sides with respect to the cation and self quenching is thus partially or totally suppressed, whereas, with cations smaller than the cavity size of the crown (e.g., Li^+, Na^+, and Ca^{2+}), the preferred conformation of the relevant complexes may be such that the two carbonyl groups are on the same side and the close approach of the coumarin moieties accounts for static quenching.

10.4.2
Calixarene-Based PCT Sensors

Calixarene-based compound **38** (Fig. 10.14) [60] containing three ester groups and an appended naphthalenic fluorophore exhibits very interesting photophysical and complexing properties. Interaction of a cation with the carbonyl group of the fluorophore leads to a red shift of the absorption and emission spectra. A large enhancement of the fluorescence quantum yield has been observed upon cation binding (Fig. 10.15) which can be explained in terms of the relative locations of the singlet $\pi\pi^*$ and $n\pi^*$ states. In the absence of cation the lowest excited states has $n\pi^*$ character which results in an efficient intersystem crossing to the triplet state and consequently a low fluorescence quantum yield. In the pre-

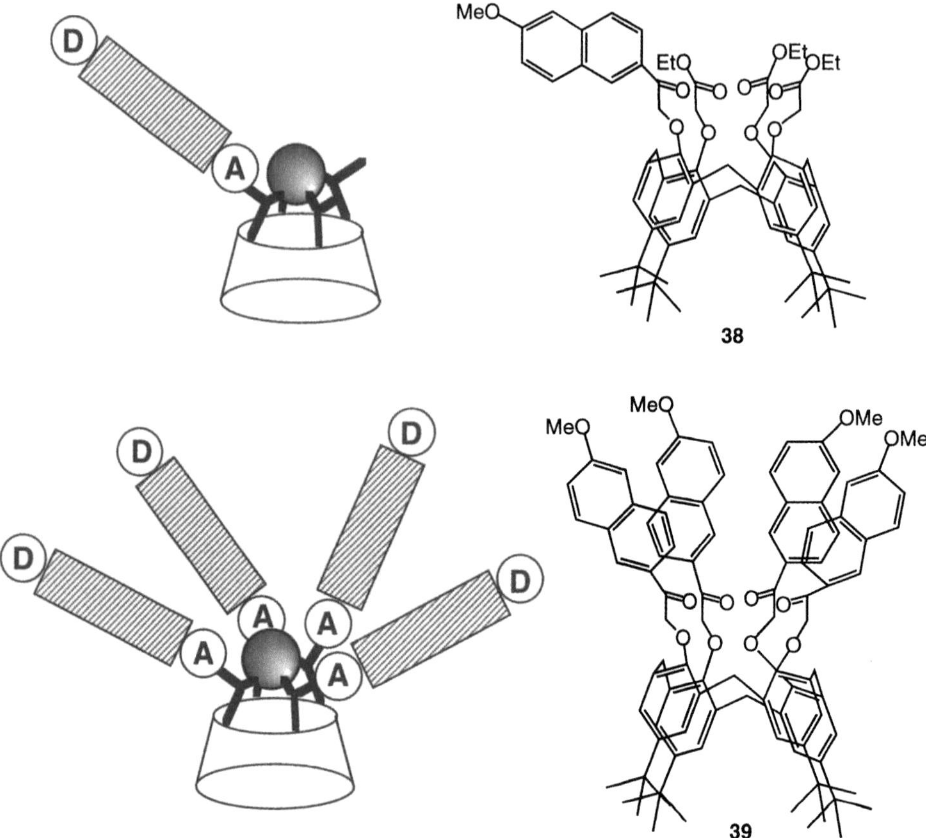

Fig. 10.14. Calixarene-based PCT sensors in which the bound cation interacts with the acceptor group

sence of cation which strongly interacts with the lone pair of the carbonyl group, the $n\pi^*$ state is likely to be shifted to higher energy so that the lowest excited state becomes $\pi\pi^*$. Transient absorption spectra in nanosecond scale have confirmed this phenomenon (to be published). The absorption of the triplet states observed in the absence of cation disappears in the complexes.

In the case of calixarene **39** with four appended naphthalenic fluorophores, similar effects were observed regarding the photophysical effects induced by complexation of alkali and alkaline-earth metal ions (i.e., red shift of the absorption and the emission spectra, and enhancement of the fluorescence quantum yield). However, the fluorescence quantum yields of the complexes of calixarene **39** are lower than those measured for calixarene **38**, which can be explained by self quenching of the chromophores. A broadening of the emission band or the appearance of a new band at higher wavelength were observed upon complexation with alkali metal ions. This is likely to be due to excimer formation between adjacent naphthalenic groups. The relative intensities of the mo-

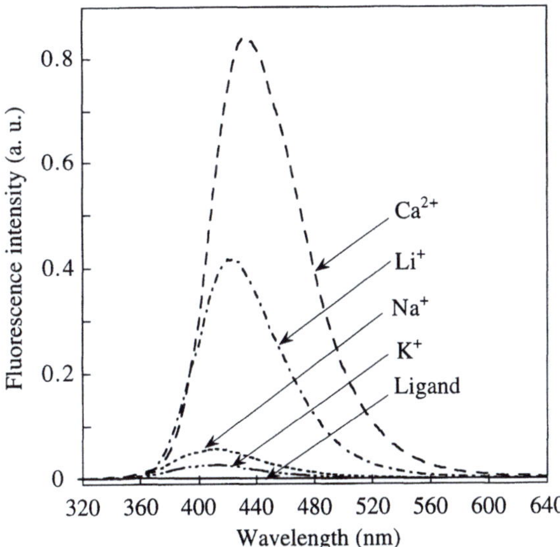

Fig. 10.15. Corrected fluorescence spectra of **38** and its complexes in acetonitrile (excitation wavelength: 320 nm)

nomer and excimer bands are related to the size of the complexed cation ($Li^+ > Na^+ > K^+$). The smaller the size of the cation, the smaller the average naphthalene-naphthalene distance, and the higher the probability of intramolecular excimer formation.

The selectivities Na^+/K^+ of **38** is 500 in ethanol and 1300 in a mixture of ethanol and water (60:40 v/v) [61]. Such an outstanding selectivity in the presence of water is very promising for practical applications in aqueous samples.

10.5
Conclusion

The fluorescent molecular sensors involving only crown ether receptors without the participation of external groups are of great interest for the understanding of cation-dipole interaction and photoinduced cation-steered charge transfer. However, these are unlikely to be of practical interest owing to the lack of selectivity. In contrast, the examples of crowned coumarins whose carbonyl group participates in the complexation of cations clearly show that an external group and the rigidity of the overall structure are of major importance for affinity and selectivity towards a given cation. Such a rigidity criterion is fulfilled in calixarene-based sensors which show great promise for the selective detection of sodium or potassium ions.

Regarding the photophysical effects, the interaction of a bound cation with an electron-donating group conjugated with an electron-withdrawing group leads in most cases to only a slight shift of the fluorescence spectrum due to photodisruption of the interaction with the cation because the donor group becomes positively polarized upon excitation as a result of charge transfer. In contrast, interaction of a cation with an electron-withdrawing group causes a redshift of the fluorescence spectrum due to enhancement of the photoinduced charge transfer. Moreover, large changes in fluorescence intensity can be observed upon cation binding when the cation induces inversion of the $n\pi^*$ and $\pi\pi^*$ states.

References

1. Czarnik AW (1993) (ed) Fluorescent chemosensors for ion and molecule recognition. ACS Symposium Series 358
2. Lakowicz JR (ed) (1994) Topics in fluorescence spectroscopy. Vol 4. Probe design and chemical sensing. Plenum Press, New-York
3. Valeur B, Bardez E (1995) Chem Br 31:216
4. Fabbrizzi L, Poggi A (1995) Chem Soc Rev 24:197
5. Desvergne JP, Czarnik AW (eds) (1997) Chemosensors of ion and molecule recognition. NATO ASI series, Kluwer Academic Publishers, Dordrecht
6. de Silva AP, Gunaratne HQN, Gunnlaugsson T, Huxley AJM, McCoy CP, Rademacher JT, Rice TE (1997) Chem Rev 97:1515
7. Valeur B (1994) In: Lakowicz JR (ed) Probe design and chemical sensing, topics in fluorescence spectroscopy, vol 4. Plenum, New York, p 21
8. Valeur B, Leray I (2000) Coord Chem Rev (in press)
9. Valeur B, Bourson J, Pouget J (1992) In: Czarnik AW (ed) Fluorescent chemosensors for ion and molecule recognition. A.C.S. Symposium series 538, American Chemical Society, Washington DC, p 25
10. Rettig W, Lapouyade R (1994) In: Lakowicz JR (ed) Topics in fluorescence spectroscopy. Vol 4. Probe design and chemical sensing. Plenum Press, New York, p 109
11. Valeur B, Badaoui F, Bardez E, Bourson J, Boutin P, Chatelain A, Devol I, Larrey B, LefÈvre JP, Soulet A (1997) In: Desvergne JP, Czarnik AW (eds) Chemosensors of ion and molecule recognition. NATO ASI series, Kluwer Academic Publishers, Dordrecht, p 195
12. Alfimov MV, Gromov SP (1999) In: Rettig W, Strehmel B, Schrader S, Seifert H (eds) Applied fluorescence in chemistry, biology and medicine. Springer, Berlin Heidelberg New York, p 161
13. Valeur B (1993) In: Schulman SG (ed) Molecular luminescence spectroscopy, part 3. Wiley, p 25
14. Löhr HG, Vögtle F (1985) Acc Chem Res 18:65 and references cited therein
15. Bourson J, Valeur B (1989) J Phys Chem 93:3871
16. Létard JF, Lapouyade R, Rettig W (1993) Pure Appl Chem 65:1705
17. Ushakov EN, Gromov SP, Fedorova OA, Alfimov MV (1997) Izv Akad Nauk Ser Khim 484 [Russ Chem Bull 46:463]
18. Druzhinin SL, Rusalov MV, Uzhinov BM, Gromov SP, Sergeev SA, Alfimov MV (1999) J Fluorescence 9:33
19. Rurack K, Bricks JL, Kachkovski A, Resch U (1997) J Fluoresc 7:63S
20. Mateeva N, Enchev V, Antonv L, Deligeorgiev T, Mitewa M (1995) J Incl Phenom 93:323
21. Martin MM, Plaza P, Dai Hung N, Meyer YH, Bourson J, Valeur B (1993) Chem Phys Lett 202:425
22. Martin MM, Bégin L, Bourson J, Valeur B (1994) J Fluorescence 4:271
23. Martin MM, Plaza P, Meyer YH, Badaoui F, Bourson J, Lefèvre JP, Valeur B (1994) J Phys Chem 100:6879
24. Dumon P, Jonusauskas G, Dupuy F, Pée P, Rullière C, Létard JF, Lapouyade R (1994) J Phys Chem 98:10,391

25. Mathevet R, Jonusauskas G, Létard JF, Lapouyade R (1995) J Phys Chem 99:15,709
26. Druzhinin SI, Rusalov MV, Uzhinov BM, Alfimov MV, Gromov SP, Fedorova OA (1995) Proc Indian Acad Sci 107:721
27. Rettig W (1986) Angew Chem Int Ed Eng 25:971
28. Létard JF, Delmond S, Lapouyade R, Braun D, Rettig W (1995) Rec Trav Chim Pays-Bas 114:517
29. Collins GE, Choi LS, Callahan JH (1998) J Am Chem Soc 120:1474
30. Jonker SA, Van Dijk SI, Goubitz K, Reiss CA, Schuddeboom W, Verhoeven JW (1990) Mol Cryst Liq Cryst 183:273
31. Kollmannsberger M, Rurack K, Resch-Genger U, Daub J (1998) J Phys Chem 102:10,211
32. Fery-Forgues S, Le Bris MT, Guetté JP, Valeur B (1988) J Chem Soc, Chem Commun 5:384
33. Fery-Forgues S, Le Bris MT, Guetté P, Valeur B (1988) J Phys Chem 92:6233
34. Fery-Forgues S, Bourson J, Dallery L, Valeur B (1990) New J Chem 14:617
35. Fery-Forgues S, Le Bris MT, Mialocq JC, Pouget J, Rettig W, Valeur B (1992) J Phys Chem 96:701
36. Addleman RS, Bennett J, Tweedy SH, Elshani S, Wai CM (1998) Talanta 46:573
37. Cazaux L, Faher M, Lopez A, Picard C, Tisnès P (1994) J Photochem Photobiol A: Chem 77:217
38. Delmond S, Létard JF, Lapouyade R, Mathevet R, Jonusauskas G, Rullière C (1996) New J Chem 20:861
39. Kasner SE, Ganz MB (1992) Am J Physiol 262:F462
40. Minta A, Tsien RY (1989) J Biol Chem 264:19,449
41. Crossley R, Goolamali Z, Gosper J, Sammes PG (1994) J Chem Soc Perkin Trans 2:513
42. Oguz U, Akkaya EU (1998) Tetrahedron Lett 39:5857
43. Grynkiewicz G, Poenie M, Tsien RY (1985) J Biol Chem 260:3440
44. Van den Bergh V, Boens N, De Schryver FC, Ameloot M, Steels P, Gallay J, Vincent M, Kowalczyk A (1995) Biophys J 68:1110
45. Haugland RP Handbook of fluorescent probes and research chemicals. Molecular Probes
46. Smith GA, Hesketh TR, Metcalfe JC (1988) Biochem J 250:227
47. Crossley R, Goolamali Z, Sammes PG (1994) J Chem Soc Perkin Trans 2:1615
48. Iwamoto K, Araki K, Fujishima H, Shinkai S (1992) J Chem Soc Perkin Trans 1:1885
49. Narita M, Higuchi Y, Hamada F, Kumagai H (1998) Tetrahedron Lett 39:8687
50. Bourson J, Borrel MN, Valeur B (1992) Anal Chim Acta 257:189
51. Bourson J, Pouget J, Valeur B (1993) J Phys Chem 97:4552
52. Bourson J, Badaoui F, Valeur B (1994) J Fluoresc 4:275
53. Habib Jiwan JL, Branger C, Soumillion JP, Valeur B (1998) J Photochem Photobiol A Chemistry 116:127
54. Leray I, Habib Jiwan JL, Branger C, Soumillion JP, Valeur B (2000) (submitted)
55. Yamauchi A, Hayashita T, Nishizawa S, Watanabe M, Teramae N (1999) J Am Chem Soc 121:2319
56. Xia WS, Schmehl RH, Li CJ (1999) J Am Chem Soc 121:5599
57. Leray I, Valeur B (unpublished results)
58. Ohki A, Lu JP, Hallman JL, Huang X, Bartsch RA (1995) Anal Chem 67:2405
59. Yang Y, Kou X, Kent Dalley N, Bartsch RA (1996) Supramol Chem 6:375
60. Leray I, O'Reilly F, Habib Jiwan JL, Soumillion JP, Valeur B (1999) Chem Commun 795
61. The solubility in pure water was not sufficient for the determination of the stability constants. It should be recalled that solubility in water is not recommended when the fluoroionophore is immobilized in the sensitive part of an optical chemical sensor (polymer or sol-gel matrix). Leaching is more likely with water soluble molecular sensors

CHAPTER 11

Fluorometric Detection of Anion Activity and Temperature Changes

L. Fabbrizzi, M. Licchelli, A. Poggi, G. Rabaioli, A. Taglietti

Molecular level fluorescent sensors can be built up following a two-component approach, i.e. by covalently linking a light-emitting fragment to the receptor suitable for the envisaged analyte. When the analyte is an anion, the recognition process can be based on the rather energetic and directional metal-ligand interaction. In this case, the receptor subunit must contain a coordinatively unsaturated metal centre, leaving one or more vacant coordination sites available for anion binding. In this perspective, the use of the $[Zn^{II}(tren)]^{2+}$ platform has been exploited with a special regard to the interaction with the -COO$^-$ group. Linking appropriate substituents to the tren framework allowed the design of Zn^{II}-based sensors which signalled carboxylate recognition through either fluorescence quenching (on/off) or revival (off/on). Finally, the same two-component approach has been utilised for the design of a fluorescent thermometric probe.

11.1
The Two-Component Approach to the Design of a Fluorescent Molecular Sensor

A molecular sensor is intended as a discrete molecular system capable of signalling the presence of a given analyte in a fluid and of monitoring its activity through the variation of a defined and well detectable property. If such a property is fluorescence, the sensor S, following the interaction with the analyte A, should have its emission properties (intensity and/or energy) drastically changed [1]. Important features of a molecular sensor are selectivity and efficiency [2]. Selectivity refers to the capability of S to establish an especially strong interaction with A, thus recognising it in the presence of other competing and less gifted analytes. Efficiency is related to the development of a sharp signal, to be instrumentally detected, which communicates to the outside the occurrence of the recognition process. The most direct way to build up a fluorescent molecular sensor is to put together, e.g. through a covalent linking, two distinct molecular subunits: (i) the receptor subunit, which displays selective affinity towards the envisaged analyte, and (ii) the signalling subunit, i.e. a fluorescent molecular fragment [3]. Sensing occurs if analyte complexation by the receptor subunit induces an intercomponent process with the nearby fluorophore, drastically modifying its emission. An example of a two-component molecular sensor is given by system 1, which has been designed to sense transition metal ions (see Fig. 11.1) [4].

In an MeCN/H$_2$O solution (4:1, v/v), 1 displays the typical fluorescent emission of the anthracene fragment, An, over the entire pH range (2–). At pH $\geq$ 5, the

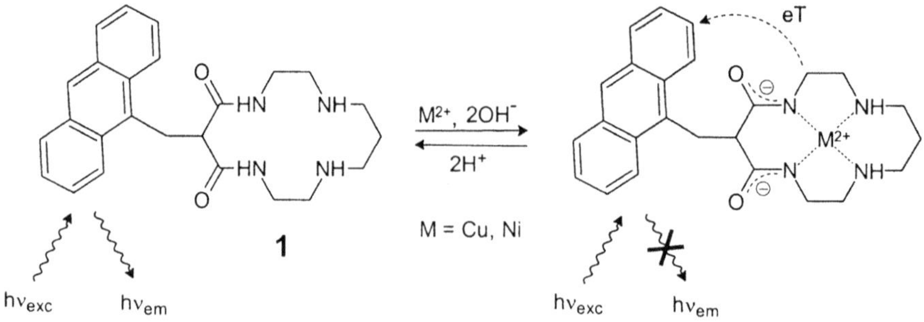

Fig. 11.1. A fluorescent sensor for Cu^{II} and Ni^{II} ions (on/off behaviour). The tetra-aza macrocyclic subunit of the two-component system **1** is able to incorporate an M^{II} metal ion (M = Cu, Ni), with simultaneous extrusion of two protons from the amide groups. The complexed metal is able to transfer an electron to the nearby excited anthracene fragment, thus quenching its fluorescent emission

Cu^{II} ion is quantitatively complexed by the diamine-diamide macrocyclic subunit (the so-called dioxocyclam macrocycle). Complexation involves the deprotonation of the two amide groups and formation of an especially stable metal-ligand complex of square stereochemistry. Interestingly for sensing purposes, Cu^{II} complexation induces full quenching of the anthracene emission (to less than 1 % of the original intensity). Quenching is to be ascribed to the occurrence of a Cu^{II}-to-An* electron transfer (eT) process, which is accounted for on a thermodynamic basis ($\Delta G^{\circ}_{eT} = -0.5$ eV) and ultimately depends on the fact that the strong in-plane interactions exerted by dioxocyclamato(2–) raise the energy of the metal centred HOMO level, thus favouring the electron release. Therefore, the transient species $Cu^{III} \sim An^{\bullet-}$ is easily achieved, which decays according to a non-radiative mode. Sensing of Cu^{II} by **1** is specific, as in a solution adjusted to pH = 4.7 only Cu^{II} is able to promote the endothermic deprotonation of the amide groups and complexation by the dioxocyclam subunit. At a higher pH (= 6) Ni^{II} can also be complexed by **1** and again metal binding induces the occurrence of a Ni^{II}-to-An* eT process, with full quenching of the anthracene emission ($\Delta G^{\circ}_{eT} = -0.35$ eV). No other 3d metal ions can be complexed by dioxocyclam even at a higher pH values. Thus, **1** is a specific fluorescent sensor for Cu^{II} and Ni^{II}, which can be discriminated by varying pH, and is the prototype of a class of fluorescent sensors which signal analyte recognition through the quenching of their emission: on/off behaviour.

In the on/off class of sensors, it is the analyte itself that is directly responsible for the signalling mechanism. In the examples illustrated before, the M^{II} cation transfers an electron to the nearby excited fluorophore, inducing fluorescence quenching. However, it may happen that the analyte is not capable of interfering directly with the nearby fluorophore, thus modifying its emission. This is the case of s block metal ions (e.g. Na^+, Mg^{2+}), which have a closed shell electronic configuration and cannot be involved in any electron exchange process. In such circumstances, the molecular framework of the sensor has to be engineered in such a way that metal complexation could alter a pre-existing mechanism affecting fluorophore activity.

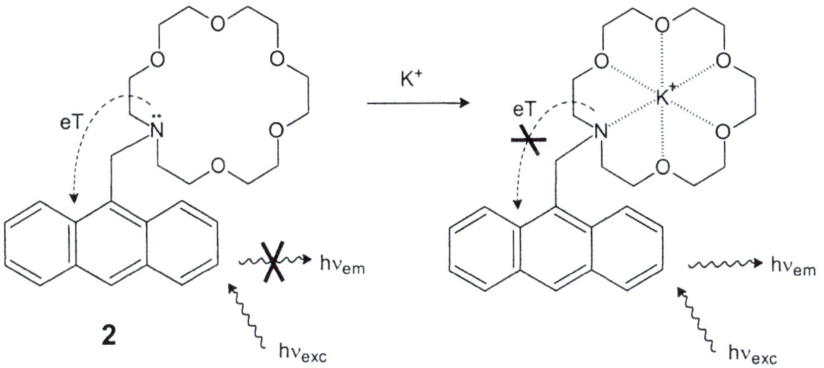

Fig. 11.2. A fluorescent sensor for K$^+$ (off/on behaviour). In absence of the metal ion, the sensing system **2** is not fluorescent, as an electron transfer process takes place from the tertiary amine nitrogen atom to the nearby excited anthracene fragment. On K$^+$ complexation by the macrocyclic subunit, the nitrogen lone pair is involved in metal binding and the electron transfer is suspended: fluorescence is restored

This case is nicely illustrated by system **2** (see Fig. 11.2), which consists of an anthracene fragment and of a polyether macrocycle [5]. The receptor subunit of **2** differs from the well known 18–crown–hexa-oxa ligand, which selectively binds K$^+$ over the other alkali cations, in that one of ethereal oxygen atoms has been replaced by a tertiary amine nitrogen atom. In particular, the latter atom acts as a bridge-head for the covalent linking to the signalling fragment. While this replacement does not change the affinity of the crown towards K$^+$, it has important consequences on the signalling mechanism. In fact, in the absence of metal ions, an eT process takes place from the tertiary amine nitrogen atom to the anthracene fragment, whose emission is substantially quenched. The signal is therefore off. On K$^+$ complexation, the lone pair of the nitrogen atom gets involved in the metal-ligand interaction and is no longer available for the photoinduced eT process. Thus, the anthracene fluorescence is fully restored: the signal is on.

It should be noted that one of the most popular and widely used fluorescent molecular sensors Fluo-3, **3**, belongs to the off/on class [6] (Fig. 11.3). Fluo-3 is used to monitor Ca^{2+} activity in biological fluids. In particular, cell biologists, using fluorescence microscopy, can detect intracellular Ca^{2+} with both spatial and temporal resolution. Fluo-3 consists of an EGTA type receptor, which gives a stable complex with Ca^{2+} at neutral pH, and of a xanthene fluorophore, covalently linked together. The metal free system is not fluorescent, due to the occurrence of an eT process from one of the tertiary amine nitrogen atoms of the EGTA fragment to the excited fluorophore. The eT process is arrested when the lone pair on each nitrogen atom is involved in metal binding, which makes fluorescence revive.

The two-component approach can be employed in the design of fluorescent molecular sensors for any kind of analyte, either electrically charged or neutral, by linking the appropriate receptor to a convenient fluorophore. The design is

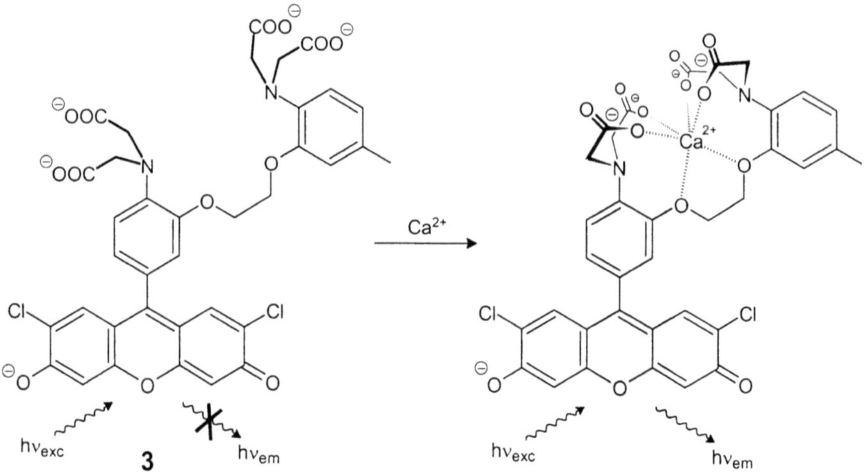

Fig. 11.3. The off/on behaviour of the Fluo-3 sensor, currently used for monitoring intracellular Ca^{2+} concentration [6]

especially simple when an on/off response is being sought. This is the case of a substrate which possesses electron donating or withdrawing tendencies pronounced enough to promote a photoinduced eT process with the fluorophore. In the following sections we will consider the design of luminescent molecular sensors for a negatively charged substrate: the carboxylate group of organic anions and of amino acids.

11.2
The Use of a [Zn^{II}(tren)]$^{2+}$ Platform for Anion Recognition and Fluorescent Sensing

Most of the receptors reported up to date for anion recognition operate through electrostatic interactions, which include hydrogen bonding [7]. In this context, the receptor molecular framework is equipped with positively charged groups, whose number and position cooperate to define the selectivity of the interaction. An example of a two-component fluorescent sensor for anions operating through electrostatic-hydrogen bonding interactions is illustrated in Fig. 11.4 [8].

The sensor **4** consists of two triamine compartments linked by two naphthyl spacers, which are also there to ensure the signalling function. In an aqueous solution buffered at pH = 6, the four amine groups adjacent to the naphthalene fragments are protonated. The tetrapositive cation, $4H_4^{4+}$, establishes especially strong interactions with ATP^{4-} (adenosine triphosphate). The $\log K$ for the $4H_4^{4+} + ATP^{4-} = [4H_4^{4+}ATP^{4-}]$ equilibrium, at pH = 6, is 5.1, a value distinctly higher than that observed for the binding equilibrium of AMP^{2-} (adenosine monophosphate), which possesses a "bite" too short to fit the distance between the two sides of the tetraammonium receptor ($\log K = 4.1$). On the other hand, ADP^{3-} (adenosine diphosphate) displays a binding ability ($\log K = 5.0$) compa-

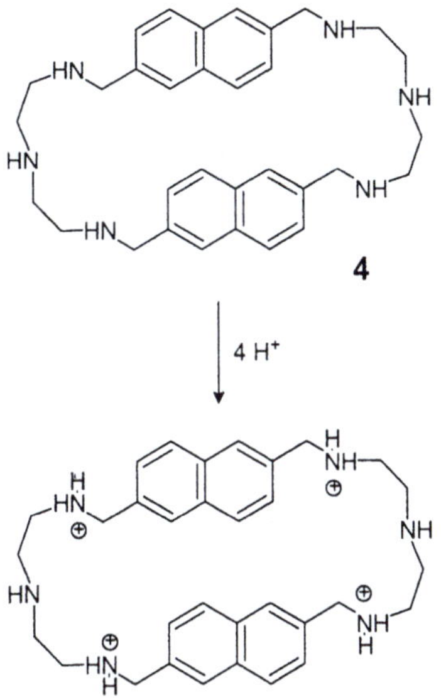

Fig. 11.4. The design of a fluorescent sensor for anions based on electrostatic interactions. At pH = 6, the four benzylic amine nitrogen atoms of azamacrocycle **4** are protonated. The $4H_4^{4+}$ cation establishes especially strong interactions with multicharged anions, in particular with ATP^{4-}

rable to that of ATP^{4-} towards $4H_4^{4+}$. Deficiency of linear discrimination is probably due to the rather flexible nature of the two receptor compartments, which can fold and adjust themselves to include the length of the anionic substrate. Selectivity towards ATP^{4-} may derive from both the favourable spatial fitting of $-NH_2^+$ and phosphate groups and from the correct matching of electrical charges of opposite sign. Moreover, ATP^{4-} binding induces quenching of the photoexcited bridges, so that recognition is announced through a distinct decrease of the fluorescent emission (another example of on/off sensor).

Besides electrostatic interactions, which include hydrogen bonding, there exists another type of interaction one could profit from for anion recognition: the metal-ligand interaction [9]. In this case, the receptor system should contain a metal centre which is coordinatively unsaturated, i.e. it should possess one (or even more) vacant coordination sites, to be occupied by donor atoms of the envisaged anion. Compared to electrostatic interactions, the metal-ligand interaction is in general more energetic and possesses a more defined directional character, two features which are expected to act in favour of a higher selectivity. In this perspective, we have developed in recent years a series of receptors and sensors based on metal complexes with the tripodal tetramine tris(2-aminoethyl)amine, tren (**5**). Tren forms stable complexes with 3d metal ions both in the solid state and in solution and tends to impose a trigonal bipyramidal stereochemistry. In particular the tetramine occupies four sites of the coordination polyhedron and leaves the remaining axial position available for either a solvent molecule or an anion.

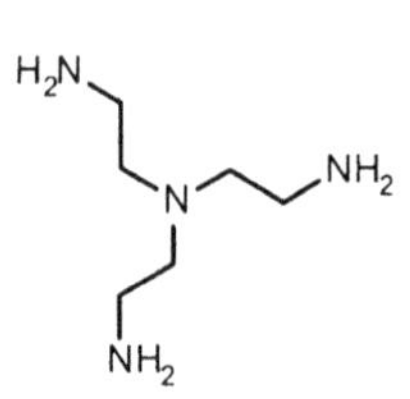

5

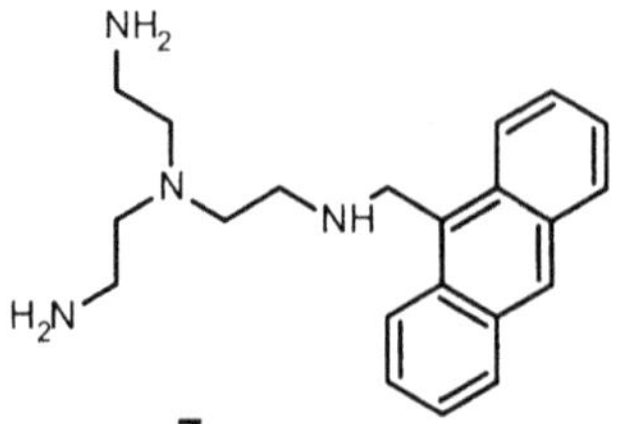

7

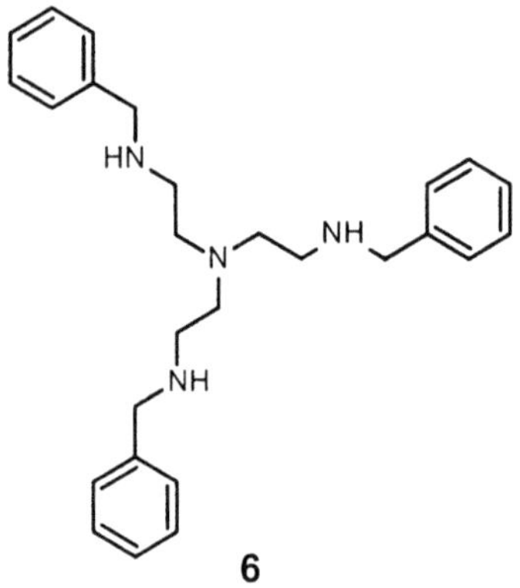

6

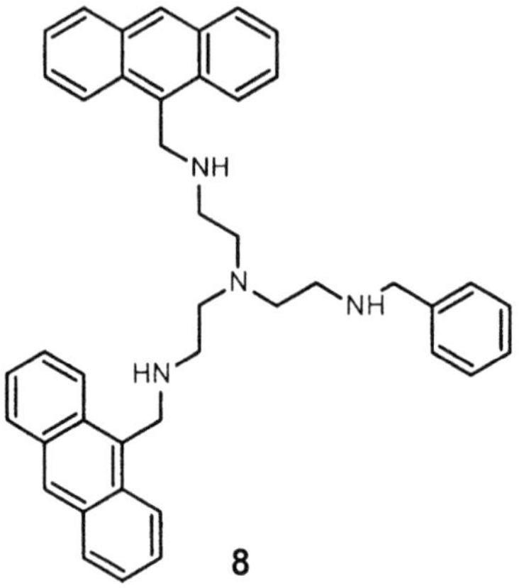

8

A variety of five-coordinate complexes of copper(II) with tren and its derivatives have been isolated and structurally characterised through X-ray diffraction studies. The structure of a Cu^{II} complex of the N,N',N''-tribenzyl tren derivative (**6**) is sketched in Fig. 11.5. An N_3^- anion is also bound to the metal, according to a regular bipyramidal coordination geometry [10]. The directionality of the metal-anion interaction is clearly indicated by the bent coordination of N_3^-(Cu^{II}-N-N angle: 127°). This reflects the fact that the metal-ligand interaction involves orbital interaction. In particular, the N_3^- anion offers to the metal a filled sp^2 hybridised orbital.

Following the two-component approach, an anthracene fragment has been appended to one of the terminal nitrogen atoms of tren, to give **7**, trenAn. Cu^{II} and other 3d cations, however, cannot be used as metal centres to design a fluorescent sensor based on the trenAn system. In fact, d block metal ions tend to quench any proximate excited fluorophore through either an electron transfer (eT) or an energy transfer (ET) process. This unpleasant behaviour results from the pronounced redox activity (for the eT mechanism, as discussed in Sect. 11.1) and from the availability of either empty or half-filled levels of low energy (for the energy transfer, ET mechanism, Dexter type [11]). We considered that the post-transition cation Zn^{II} could be an acceptable surrogate of genuine 3d metal ions for the following reasons: (i) it forms fairly stable complexes with tren in aqueous solution (M^{2+} + tren = $[M(tren)]^{2+}$, log K: Cu^{II}, 18.5; Ni^{II}, 14.6; Zn^{II}, 14.5; Co^{II}, 12.7; Fe^{II}, 8.8) [12]; (ii) it cannot be involved in any photoinduced eT pro-

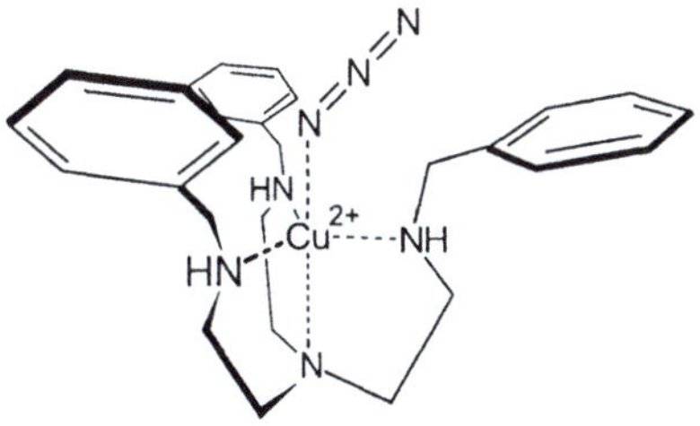

Fig. 11.5. The five-coordinated copper(II) complex of *N,N',N''*-tribenzyl-tren (**6**). An N_3^- anion completes the trigonal bipyramidal polyhedron. Azide bent coordination (Cu^{II}-N-N angle: 127°) stresses the directional nature of the metal-anion interaction

cess (as it is absolutely redox inactive) or ET process (due to its closed shell electronic configuration, d^{10}).

Since coordinatively unsaturated Zn^{II} polyamine complexes display a good affinity towards the COO^- group, the $[Zn^{II}(trenAn)]^{2+}$ platform was first tested for the fluorescent sensing of carboxylate anions. For instance, there is evidence from spectrophotometric titration experiments that $[Zn^{II}(trenAn)]^{2+}$ forms a stable adduct with the benzoate anion, in an ethanolic solution at 25 °C. However, when performing a spectrofluorimetric titration, even in large excess of benzoate, the typical fluorescent emission of the anthracene fragment remains unchanged: anthracene fails the reporter duty it was supposed to do. However, when an ethanolic solution of $[Zn^{II}(trenAn)]^{2+}$ is titrated with the 4-*N,N*-dimethylamine-benzoate, the anthracene emission is progressively quenched: the fluorescence intensity, I_F, vs anion equivalent profile corresponds to the formation of a 1:1 adduct and the $\log K$ value for the $[Zn^{II}(trenAn)]^{2+}$ + $RCOO^-$ = $[Zn^{II}(trenAn)(RCOO)]^+$ equilibrium is 5.45. Quenching has to be ascribed to the occurrence of an eT process from the *N,N*-dimethylaniline donor fragment, DMA, to the excited anthracene subunit, An*. The DMA-to-An* eT process is characterised by a negative value of the free energy change ($\Delta G^\circ_{eT} = -0.4$ eV); on the other hand, molecular modelling indicates that coordination to the Zn^{II} centre brings the 4-*N,N*-dimethylamine-benzoate donating fragment close enough to the anthracene subunit to ensure the occurrence of a fast and efficient electron transfer mechanism.

Fluorescence quenching is also observed for the titration of $[Zn^{II}(trenAn)]^{2+}$ with 4-nitro-benzoate and the $\log K$ value for the 1:1 adduct formation is 4.73. In this case, quenching is ensured by a thermodynamically favoured An*-to-nitrobenzoate eT process ($\Delta G^\circ_{eT} = -1.0$ eV). The intra-complex photoinduced electron transfer process responsible for the fluorescent sensing of 4-*N,N*-di-

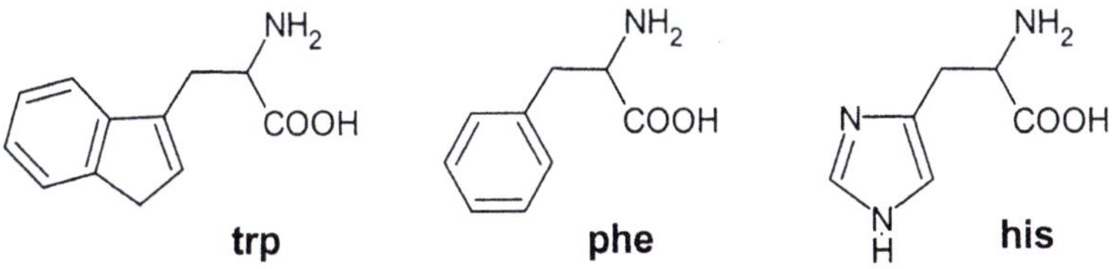

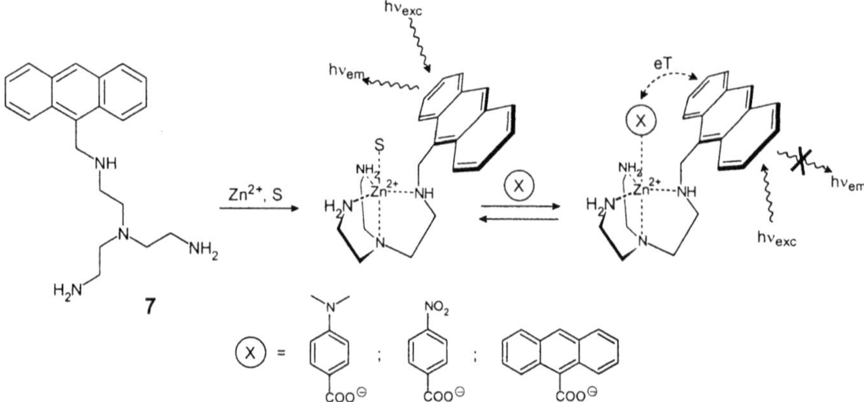

Fig. 11.6. Recognition of carboxylate anions, RCOO⁻, by the $[Zn^{II}(trenAn)]^{2+}$ platform. Recognition is announced through fluorescence quenching only when the R substituent is able to promote an electron transfer process involving the excited anthracene fragment (on/off fluorescent sensing)

methylamine-benzoate and 4-nitro-benzoate is schematically sketched in Fig. 11.6. In conclusion, the $[Zn^{II}(trenAn)]^{2+}$ coordinatively unsaturated complex represents an example of fluorescent sensor for anions of the on/off type, in which the analyte itself directly triggers the signalling mechanism, i.e. an electron transfer process which quenches the emission of the anthracene reporter [13].

The high affinity of the $[Zn^{II}(tren)]^{2+}$ platform for carboxylate anions group [14] would recommend its utilisation for building up sensors of an important class of analytes bearing the –COO⁻ group: natural amino acids. However, selectivity towards substrates of general formula $NH^{3+}–CH(\mathbf{R})–COO^-$ cannot rely on the –COO⁻ group, which is common to each member of the class, but on the individual **R** group of each amino acid. An example is given by the tren derivative **8**, trenAn₂Bz, in which the tren framework has been equipped with two anthracenyl subunits and a benzyl fragment. Among natural amino acids, only phenylalanine (phe) and tryptophane (trp) give stable 1:1 adducts with the $[Zn^{II}(trenAn_2Bz)]^{2+}$ receptor in an ethanolic solution. The corresponding $\log K$ values for the association equilibria are 4.48 (phe) and 4.21 (trp), as determined through absorbance measurements. In particular, in the spectrophotometric titration experiments, changes in the π-π^* region of the absorption spectrum were monitored. It has to be noted that other amino acids (e.g. glycine, proline, etc.), in the same conditions, interact very weakly with the $[Zn^{II}(trenAn_2Bz)]^{2+}$ receptor, with $\log K$ values of about 3. Low affinity seems to be attributed to the electrostatic repulsive effects between the Zn^{2+} centre and the $-NH_3^+$ portion of the amino acid. For phe and trp amino acids, such an unfavourable effect is more than compensated for by the additional π-stacking interaction by the aromatic substituents on tren framework and the R substituent of the amino acid: benzyl for phe and indole for trp.

Moreover, spectrofluorimetric titration experiments have shown that trp quenches the emission of the anthracene fragment of $[Zn^{II}(trenAn_2Bz)]^{2+}$ re-

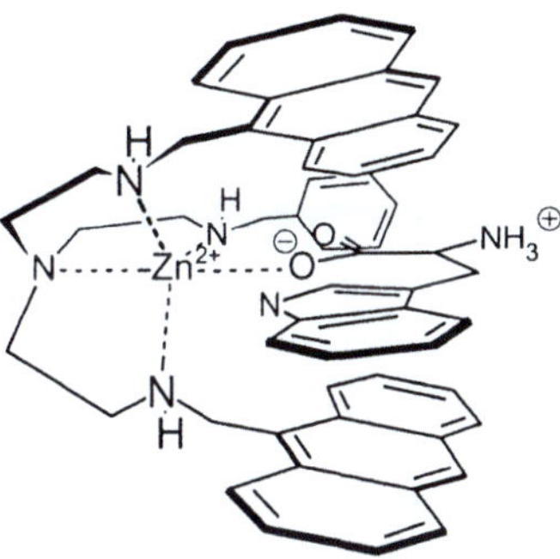

Fig. 11.7. Recognition of the tryptophane (trp) amino acid by the $[Zn^{II}(trenAn_2Bz)]^{2+}$ receptor. The indole moiety of trp and the aromatic substituents on the tren framework give π-stacking interaction. An indole-to-anthracene photoinduced electron transfer mechanism quenches the fluorescent emission, thus communicating the occurrence of the recognition process (on/off behaviour)

ceptor (log$K$ value from the I_F vs trp equivalent profile: 4.28), whereas phe does not. This may be due to the fact that only the indole moiety possesses distinct electron donating tendencies and is therefore able to transfer an electron to the facing excited An* fragment. The hypothesised structure of the $[Zn^{II}(trenAn_2Bz)(trp)]^{2+}$ adduct is sketched in Fig. 11.7. Such a stereochemical arrangement should account for the occurrence of a fast and efficient eT process. Thus, the $[Zn^{II}(trenAn_2Bz)]^{2+}$ system selectively recognises phe and trp, but signals only the interaction with the latter amino acid [15].

Absolutely specific recognition and sensing by a system based on the $[Zn^{II}(tren)]^{2+}$ platform is observed with histidine, his. In this case attention is centred on the **R** group of his: imidazole. Imidazole, imH, is a protic acid but, owing to its very high pK_A value (14.4) [16], it does not deprotonate in water. However, in the presence of two metal ions, imH may deprotonate and the imidazolate ion, im$^-$, which forms simultaneously bridges the two metals. Such a process was first observed in the case of two Cu^{II} ions coordinatively unsaturated and prepositioned within a hexaaza macrocycle L [17]. The $[Cu_2^{II}L]^{4+} + imH = [Cu_2^{II}(L)(im)]^{3+} + H^+$ equilibrium takes place at neutral pH. Quite interestingly, for sensing purposes, a similar equilibrium takes places also in the presence of two coordinatively unsaturated Zn^{II} ions, at a higher pH. On these premises, two tren subunits were linked together through the 9,10-dimethylanthracene spacer, to give the ditopic ligand **9**.

In an aqueous solution containing **9** and 2 equivalents of Zn^{II} and adjusted to pH = 9.6, the dizinc(II) complex is present in solution (see Fig. 11.8), which exhibits the typical anthracene fluorescent emission. On addition of imidazole, fluorescence is quenched substantially, while the I_F vs equivalent of imH profile points towards the formation of a 1:1 adduct ($[Zn^{II}_2(\mathbf{9})(im)]^{3+}$, $\log K = 3.65$), whose hypothesised structural arrangement is illustrated in Fig. 11.8. In particular, bridging of the two Zn^{II} centres would bring imidazolate close to the anthracene subunit, An, allowing the occurrence of an eT process from the electron rich im⁻ anion to the facing excited fluorophore An*. Histidine showed a similar behaviour, even if the $\log K$ value for the 1:1 adduct ($[Zn^{II}_2(\mathbf{9})(his)]^{4+}$, $\log K =$ 2.92) is lower than observed for plain imidazole, probably due to the unfavour-

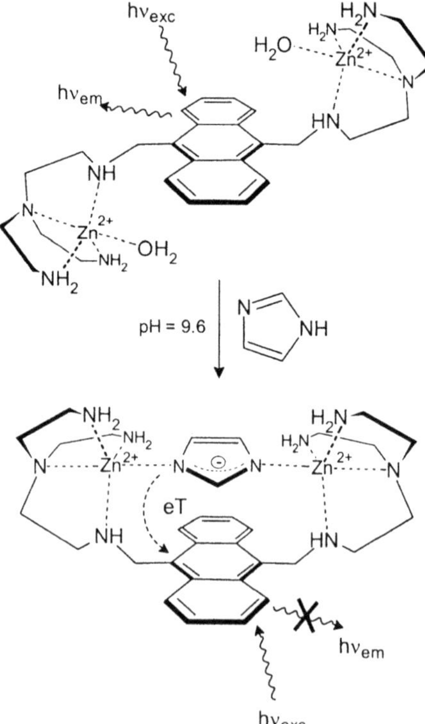

Fig. 11.8. Recognition of imidazole, imH, by a ditopic receptor containing two $[Zn^{II}(tren)]^{2+}$ subunits. At pH = 9.6, imH deprotonates and the im⁻ anion bridges the two Zn^{II} centres. The fluorescence of the anthracene spacer, An, is quenched through an im⁻–to–An* electron transfer process. A similar behaviour is observed with the amino acid histidine, which contains an imidazole residue (on/off fluorescent sensing)

able electrostatic repulsive interaction between the Zn^{2+} centre and the ammonium group of histidine.

Noticeably, the I_F vs equivalent of his profile did not change even if the solution contained a fivefold excess of any other natural amino acid, designating the $[Zn_2^{II}(9)]^{4+}$ system as a specific fluorescent molecular sensor for histidine. Other amino acids cannot compete for the $[Zn_2^{II}(9)]^{4+}$ receptor because they offer as a bridge the $-COO^-$ group, a fragment of much lower donating tendencies than the imidazolate subunit [18].

11.3
Carboxylate Recognition Signalled by Fluorescence Enhancement

In the previously discussed systems, anion recognition was in any case signalled through fluorescence quenching and was reserved to substrates capable to release/uptake an electron to/from the excited fluorogenic fragment covalently linked to the receptor [19]. Quenching of the light emission may not be the best

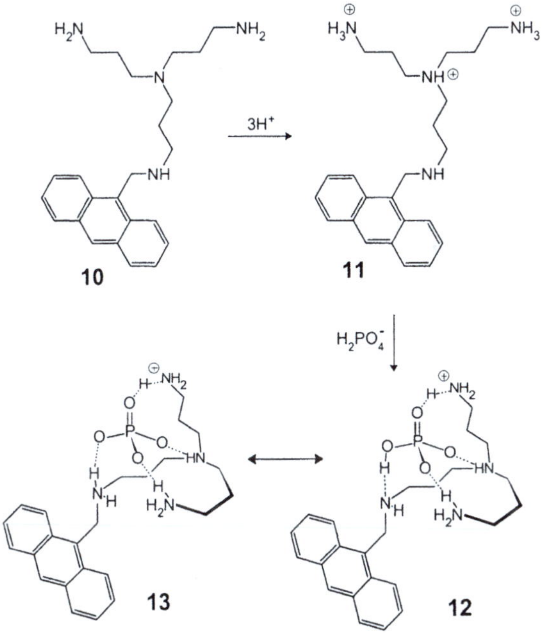

Fig. 11.9. Recognition and sensing of the HPO_4^{2-} anion. At pH = 6, the anthracenyl functionalised tetramine 10 is triply protonated (11). 11 acts as a receptor for $H_2PO_4^-$, through the formation of hydrogen bonding interactions (12). It is possible that an intra-complex proton transfer takes place and the receptor-substrate adduct is better represented by formula 13. Before anion recognition, 11 is weakly fluorescent, as an electron transfer (eT) process from the anthrylamine nitrogen atom quenches light emission. Following anion binding, the anthrylamine group is engaged in either hydrogen binding (12) or ammonium formation (13). In any case the eT process is suspended and recognition signalled through fluorescence revival (an example of off/on sensor)

way to signal the activity of a given substrate, in particular when using the fluorescence microscopy technique for intracellular investigations. Cell biologists say that they prefer to look at a lantern lighting up in the Black Forest, rather than to single out the light of an apartment switching off in the night at Manhattan.

Fluorescent molecular sensors that switch on their light emission on anion recognition have been obtained following the approach described for the design of sensors for s block metal ions (e.g. system **2**) [20].

An example of an off/on fluorescent molecular sensor for anions is illustrated in Fig. 11.9: in the two-component system **10**, an anthracene fragment is linked through a methylene group to a tripodal tetramine similar to tren (in this case, each aliphatic chain has an additional $-CH_2-$ group). In an aqueous solution at pH = 6, the emission intensity by the anthracene subunit is low. At this pH, all the amine groups are protonated, but the one linked to the anthracenyl fragment has a benzylic nature and is less basic than the other ones. This benzylamine group behaves as an electron donor and transfers an electron to the nearby photoexcited anthracene fragment, whose emission is substantially quenched. On addition of HPO_4^{2-}, fluorescence is enhanced (of about 150%). It is suggested that the polyammonium ion, **11**, acts as a receptor for hydrogenophosphate: in particular, hydrogen bonding interactions are established between the three ammonium groups and the three oxygen atoms of HPO_4^{2-}. Then a further hydrogen bond interaction is established between the benzylic amine group and the -OH fragment of the HPO_4^{2-} ion, as illustrated in formula **12** in Fig. 11.9; it is probable that an intramolecular proton transfer takes place from the HPO_4^{2-} -OH group to the benzylic amine group and a hydrogen bonding interaction is then established between the benzylammonium group and the fourth oxygen atom of the PO_4^{3-} ion, as shown in formula **13**. In any case, following either hydrogen bonding interaction with -OH or covalent bond formation with H^+, the benzylic nitrogen atom cannot make its lone electron pair available for the photoinduced

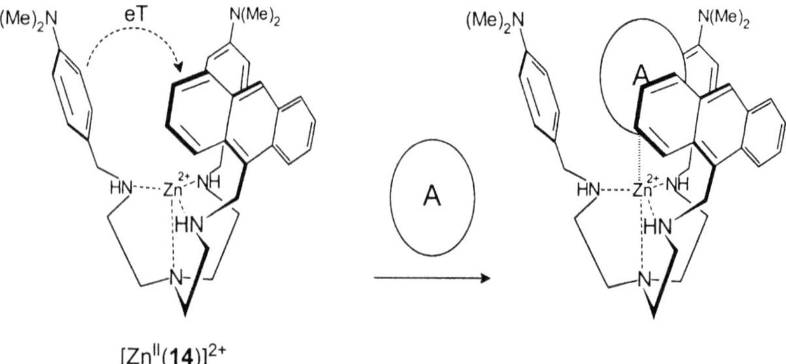

Fig. 11.10. The design of an off/on anion fluorescent sensor based on the $[Zn^{II}(tren)]^{2+}$ platform. A through-space intramolecular eT process from an N,N-dimethylamine substituent quenches the emission of the anthracene fragment. Following coordination by anion A, the eT process is interrupted and the fluorescent emission is restored

electron transfer process. Thus, the fluorophore can undergo radiative decay: fluorescence is fully restored and recognition is communicated through an off/on behaviour [21].

We looked at further fluorescent sensors of the off/on type, in which anion binding arrests an eT process taking place within the receptor cavity. In particular, we considered again our beloved [Zn^{II}(tren)]$^{2+}$ platform. This time the tripodal tetramine was equipped (i) with a fluorophore – anthracene – at one of the terminal amine nitrogen atoms, and (ii) with electron donor groups – N,N-dimethylaniline, DMA – at the other two, to give **14**. One would expect that in the [Zn^{II}(**14**)]$^{2+}$ system: (i) a through-space eT process from one of the DMA subunits to the excited anthracene fragment takes place, thus quenching fluorescence [22], and (ii) occupancy of the vacant axial position on the Zn^{II} centre by a coordinating anion would disturb the development of the through-space eT process, possibly inducing the recovery of the fluorescent emission.

To verify this hypothesis, we considered titration experiments on the [Zn^{II}(**14**)]$^{2+}$ complex, in presence of an especially bulky carboxylate anion: 1,1,1-triphenyl-acetate, Ph$_3$C-COO$^-$. In order to make a proper comparison, we looked first at the [Zn^{II}(**7**)]$^{2+}$ system, the precursor fluorescent platform not equipped with DMA donor substituents. In particular, an aqueous ethanol solution containing **7** (trenAn), 1 equivalent of Zn^{II}, plus excess acid, was titrated with standard base and its fluorescence was measured. The fluorescence intensity, I_F, vs pH profile is shown in Fig. 11a (filled triangles).

In the strongly acidic solution, pH = 2, highest fluorescence – full emission of the anthracene fragment – is observed. On increasing pH, I_F decreases, due to stepwise deprotonation of the amine groups and occurrence of amine-to-An* eT processes. At pH = 4 I_F stops decreasing and increases to reach a constant value until pH = 8. This effect is due to the formation of the [Zn^{II}(**7**)]$^{2+}$ complex and

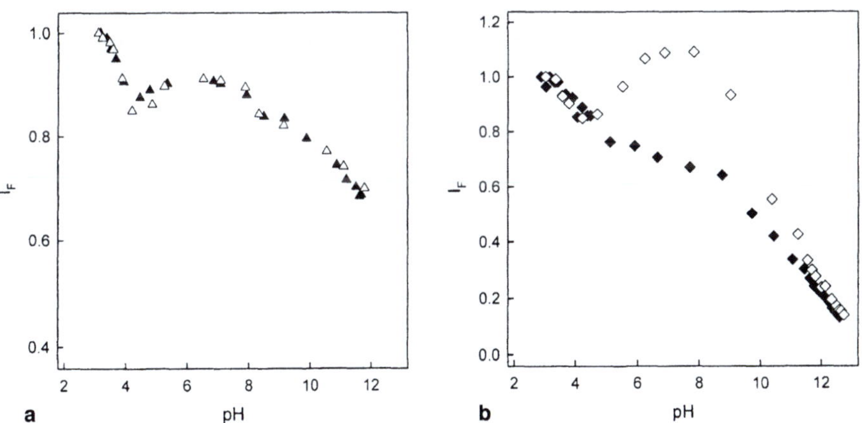

Fig. 11.11. **a** pH dependence of the fluorescence intensity, I_F, of an aqueous ethanol solution containing equimolar amounts of tetramine (**7**) and Zn^{II} (full triangles), plus 1 equivalent of 1,1,1–triphenyl-acetate, Ph$_3$C-COO$^-$ (*open triangles*); **b** The same as (a), but with tetramine **14**; in absence (*filled diamonds*) and in presence (*open diamonds*) of Ph$_3$C-COO$^-$

to the constitution of the metal-amine coordinative bonds, which stop the eT process. Then, at pH > 8, I_F decreases again; decrease has to be associated to the coordination of OH⁻ to the Zn^{II} centre: it is an eT mechanism from the electron rich hydroxide ion that quenches the nearby excited fluorophore. The same titration experiment has been carried out with a similar solution containing tetramine **14**: starting from the acidic region, I_F decreases, as observed with system 7 (Fig. 11b, filled diamonds). However, at pH = 4, I_F decrease is not arrested, as for 7, but keeps declining, even if with a less pronounced slope. This behaviour has to be ascribed to the fact that, on Zn^{II} complexation (at pH ≥ 4), a DMA substituent and the An* fragment are brought close enough to allow the occurrence of a through-space DMA-to-An* eT process, as anticipated in the hypothesised structural arrangement sketched in Fig. 11.10. Thus, with the Zn^{II}/**14** system, at pH = 4, the amine-to-fluorophore electron transfer mechanism is suspended and replaced by the slightly less efficient DMA-to-fluorophore eT process. Change of the quenching regime is indicated by a inflection point in the I_F vs pH plot.

Similar titration experiments were carried out on the same solutions, which also contained 1 equivalent of Ph_3C-COO^-. In the case of the Zn^{II}/trenAn system (Fig. 11a, open triangles), the titration profile superimposed perfectly on that obtained in absence of the carboxylate anion. Ph_3C-COO^- does not possess any defined electron donor/acceptor properties and, when coordinated to the Zn^{II} centre, cannot interfere with the facing anthracene subunit. In the case of the Zn^{II}/**14** system, the titration profile obtained in presence of 1 equivalent of Ph_3C-COO^- superimposes well on that obtained for the carboxylate free solution in the strongly acidic region. At pH = 4, I_F stops decreasing and begins to increase, to reach its maximum value at pH = 8 (Fig. 11b, open diamonds). It has to be noted that at pH ≥ 4 the $[Zn^{II}(4)]^{2+}$ complex forms and, simultaneously, coordinates the Ph_3C-COO^- anion. Within the $[Zn^{II}(4)(Ph_3C-COO)]^+$ adduct, the bulky anion prevents the occurrence of any occasional VdW contact between the DMA substituent and the photoexcited fragment, precluding the eT process and allowing the fluorophore to undergo its regular radiative decay. Thus, Ph_3C-COO^- recognition is signalled by the Zn^{II} based sensing system through a distinct enhancement of the fluorescent emission. Noticeably, the much smaller CH_3COO^- anion does not modify at all the Zn^{II}/**14** titration profile. A very modest I_F increase is observed in presence of benzoate. These preliminary results candidate the $[Zn^{II}(14)]^{2+}$ complex as a fluorescent sensor of bulky carboxylate anions [23].

11.4
The Design of a Molecular Fluorescent Thermometer

Molecular level fluorescent sensors have been designed to detect and monitor the activity of any kind of analytes (a chemical quantity), following the two-component approach. The same synthetic approach can be used to design a fluorescent sensor of a physical quantity: temperature. Monitoring of intracellular temperature is a long awaited opportunity in cell biology studies. Actually, most of the biochemical processes taking place within the cell are characterised by heat development and the determination of the associated temperature changes,

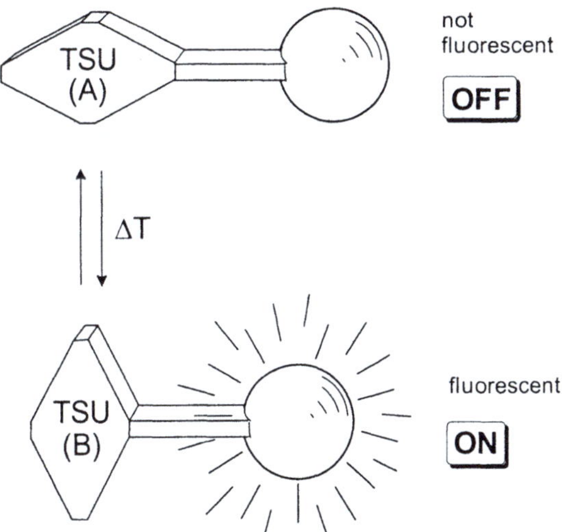

Fig. 11.12. The design of a molecular fluorescent thermometer. The light emission of the fluorescent fragment Fl* is controlled by the temperature sensitive unit TSU. TSU exists in two states, A and B, which have a different perturbing effect on Fl*. A and B are connected through a temperature dependent equilibrium

with both temporal and spatial resolution, may provide a useful additional tool to understand cell physiology mechanisms.

The two-component approach requires that the signalling unit (a fluorescent fragment) is covalently linked to a component sensitive to temperature changes (see the scheme illustrated in Fig. 11.12). In particular, the temperature sensitive unit (TSU) should exist in two different states, A and B, connected through a fast and reversible equilibrium. Essential prerequisites are: (i) the equilibrium must be temperature dependent, either endothermic or exothermic; (ii) A and B must interfere with the proximate fluorescent fragment to a distinctly different extent.

Let us consider, for instance, the case that: (1) the A = B equilibrium is endothermic, and (2) state A quenches the nearby photoactive fragment and state B does not. In such circumstances, a temperature increase will induce an increase of the concentration of B, which results in an enhancement of the fluorescence intensity. Thus, the two-component system will behave as a thermometer and the I_F vs temperature plot (which in the present case should have a positive slope) represents its calibration curve. Notice that, in principle, four types of fluorescent thermometer can be built up according to (i) whether A or B exerts the perturbing effect on the fluorophore, and (ii) whether the A-to-B equilibrium is endo- or exothermic, as illustrated in Table 11.1. In two cases, the I_F vs temperature plot has a positive slope, in the other two slope is negative.

A convenient TSU fragment is provided by coordination chemistry. In particular, the nickel(II) complex of the macrocyclic tetramine **15**, cyclam, when dissolved in a polar medium (water, acetonitrile) exists in two forms at the equilibrium: (a) a distorted octahedral species in which cyclam co-planarly chelates the

Table 11.1. The four possible types of fluorescent molecular probes based on the design illustrated in Fig. 11.12

Activity of A on Fl*	Activity of B on Fl*	Thermal nature of the A-to-B conversion	Slope of the I_F vs temperature plot
1 Quenches	Does not quench	Endothermic	Positive
2 Quenches	Does not quench	Exothermic	Negative
3 Does not quench	Quenches	Endothermic	Negative
4 Does not quench	Quenches	Exothermic	Positive

Ni^{II} centre and two solvent molecules, S, occupy the axial sites of the elongated octahedron; in the $[Ni^{II}(cyclam)S_2]^{2+}$ complex, the metal centre has two unpaired electrons and it is said to be *high-spin*; (b) a square-planar species, obtained from species (a) through the release of the two solvent molecules; in the square $[Ni^{II}(cyclam)]^{2+}$ complex, the stronger in-plane metal-ligand interactions induce the pairing of the two highest energy electrons, to give a diamagnetic, *low-spin* species [24].

15   **16**

The equilibrium is reversible, fast and temperature dependent. In particular, the (high-spin)-to-(low-spin) conversion is endothermic. This means that the high-spin form predominates at low temperatures, and that a temperature increase makes the concentration of the low-spin species to grow.

Having ascertained that the Ni^{II}/cyclam system can be a convenient TSU, we linked the macrocycle to a classic aromatic fluorophore, naphthalene, to give **15**, and we prepared the corresponding nickel(II) complexes with a variety of counterions, X^-: $[Ni^{II}(16)]X_2$ [25]. In the solid state, the X^- anions occupy the axial sites of a more or less elongated octahedron and their coordinating tendencies determine the spin-state of the Ni^{II} centre. When X^- is a strongly coordinating anion (Cl^-, NCS^-, NO_3^-), the complex is high-spin; when it is poorly coordinating (ClO_4^-, $CF_3SO_3^-$), the complex is low-spin. Then we studied the emitting behaviour of $CHCl_3$ solutions of the various $[Ni^{II}(16)]X_2$ salts. $CHCl_3$ molecules are not coordinating at all and, in a CH_3Cl solution, each complex maintains the same structure and spin multiplicity as in the solid state. We observed that in low-spin complexes the naphthalene emission was distinctly higher than for high-spin complexes (quantum yield: low-spin: ClO_4^-, 0.019; $CF_3SO_3^-$, 0.020; high-spin: anion Cl^-, 0.007; NCS^-, 0.005; NO_3^-, 0.009). Transition metal ions like Ni^{II}, when encircled by the cyclam ring, tend to quench N-linked fluorophore through an energy transfer (ET) mechanism, probably of the double electron exchange type [26]. The emission behaviour of $[Ni^{II}(16)]X_2$ complexes in $CHCl_3$ indicates that this ET mechanism is more efficient with the high-spin metal centre than with the spin paired cation, which exhibits a sort of "closed shell" electronic configuration.

Following the above-mentioned evidence, the $Ni^{II}/16$ system should behave as a type (1) thermometric probe, according to the classification illustrated in Table 11.1. In fact, a solution of $[Ni^{II}(16)](ClO_4)_2$ in MeCN at room temperature is poorly fluorescent. On increasing the temperature, the naphthalene emission sharply increases, as illustrated in Fig. 11.13. In the polar MeCN medium, the $[Ni^{II}(16)](ClO_4)_2$ salt dissociates and the complex cation undergoes the interconversion equilibrium, involving two solvent molecules, illustrated by Fig. 11.14.

At lower temperatures, the octahedral high-spin complex dominates, which has a pronounced quenching effect: fluorescence is low. On increasing temperature, the endothermic spin conversion (1) takes place and the relative concentration of the low-spin species increases: this species exerts a much lower perturbing effect on the fluorophore, whose emission is restored.

The response of the thermometer is practically instantaneous since the change of the emission intensity, both decrease and increase, is related to extremely fast events: (1) release/uptake of two solvent molecules by the labile Ni^{II} ion; (2) rearrangement of d electrons to the high-/low-spin configurations; (3) double electron exchange between the excited fluorophore and the metal centre. The three processes, on the whole, are well below the ms time scale.

The $[Ni^{II}(16)]^{2+}$ system can be used, at the 10^{-6} mol/l concentration level, to measure temperature of polar fluids, like MeCN, DMF, DMSO, and their aqueous

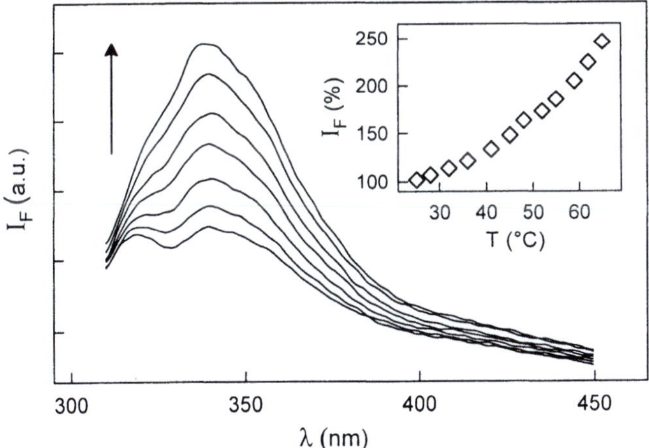

Fig. 11.13. The emission spectra of a 10^{-6} mol/l solution of $[Ni^{II}(16)](ClO_4)_2$ in MeCN, recorded over the 27–65 °C temperature range: on increasing temperature, the fluorescence intensity, I_F, grows. The I_F vs temperature plot (*inset*) gives the calibration curve of the molecular fluorescent thermometer

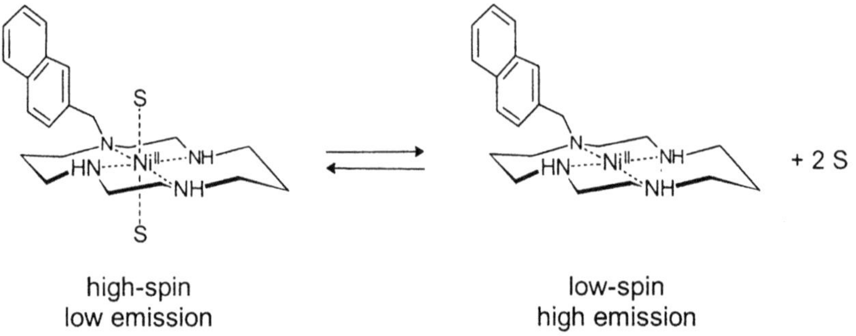

Fig. 11.14. The high-spin to low-spin interconversion equilibrium which takes places with the $[Ni^{II}(16)]^{2+}$ system in a polar solvent S. As the high-to-low-spin conversion is endothermic, a temperature increase makes fluorescence increase. The two forms reversibly interconvert in less than 1 ms and the thermometric response is practically instantaneous

mixtures. It cannot be used in pure water, where it is insoluble. Water solubility, and temperature testing of biological fluids, can be achieved either through the insertion of hydrophilic substituents on the macrocyclic framework, or by using a less lipophilic fluorogenic fragment than naphthalene.

Acknowledgement. This work has been supported by the Italian Council of Research – C.N.R. (progetto finalizzato "Biotecnologie" n. 99.00359.PF49, and "Progetto Chimica", n. 98.03733. PF28).

References

1. de Silva AP, Gunaratne HQN, Gunnlaugsson T, Huxley AJM, McCoy CP, Rademacher JT, Rice TE (1997) Chem Rev 97:1515
2. Czarnik AW (1994) Chem Biol 2:423
3. Fabbrizzi L, Poggi A (1995) Chem Soc Rev 24:197
4. Fabbrizzi L, Licchelli M, Pallavicini P, Perotti A, Taglietti A, Sacchi D (1996) Chem-Eur J 2:167
5. de Silva AP, de Silva SA (1986) J Chem Soc Chem Commun 1709
6. Minta A, Kao JPY, Tsien RY (1989) J Biol Chem 264:8171
7. Schmidtchen FP, Berger M (1997) Chem Rev 97:1609
8. Dhaenens M, Lehn JM, Vigneron JP (1993) J Chem Soc Perkin Trans 2:1379
9. (a)Fabbrizzi L, Licchelli M, Pallavicini P, Parodi L, Poggi A, Taglietti A (1996) Anion Sensing based on the metal ligand interactions. In: Echegoyen L, Kaifer A (eds) Nato ASI Series, Physical supramolecular chemistry. Kluwer Academic Publishers, Dordrecht, 433; (b) Fabbrizzi L, Pallavicini P, Perotti A, Parodi L, Taglietti A (1995) Inorg Chim Acta 238:5
10. Pallavicini P, Sardone N (unpublished results)
11. Suppan P (1994) Chemistry and light. The Royal Society of Chemistry, Cambridge, pp 66–68
12. (a) Moketakis RJ, Martell AE, Lehn JM, Watanabe EI (1982) Inorg Chem 21:4253; (b) Prue J, Schwarzenbach G (1950) Helv Chim Acta 33:963
13. De Santis G, Fabbrizzi L, Licchelli M, Poggi A, Taglietti A (1996) Angew Chem Int Ed Engl 35:202
14. Fabbrizzi L, Licchelli M, Parodi L, Poggi A, Taglietti A (1999) Eur J Inorg Chem 35
15. Fabbrizzi L, Licchelli M, Parodi L, Poggi A, Taglietti A (1998) J Fluoresc 8:263
16. George P, Hanani G, Irvine D (1964) J Chem Soc 5689

17. (a) Coughlin PK, Dewan JC, Lippard SJ, Watanabe EI, Lehn JM (1979) J Am Chem Soc
    101:265; (b) Strohkamp KG, Lippard SJ (1982) Acc Chem Res 10:318; (c) Coughlin PK,
    Martin AE, Dewan JC, Watanabe EI, Bulkowski JE, Lehn JM, Lippard SJ (1984) Inorg Chem
    23:1004; (d) Fabbrizzi L, Pallavicini P, Parodi P, Perotti P, Taglietti A (1995) J Chem Soc
    Chem Comm 2439
18. Fabbrizzi L, Francese G, Licchelli M, Perotti A, Taglietti A (1997) Chem Commun 581
19. Fabbrizzi L, Faravelli I, Francese G, Licchelli M, Perotti A, Taglietti A (1998) Chem Com-
    mun 971
20. Czarnik AW (1997) In: Czarnik AW (ed) Chemosensors of ion and molecule recognition.
    Kluwer Academic Publishers, Dordrecht, p 104
21. Huston ME, Akkaya EU, Czarnik AW (1989) J Am Chem Soc 111:8735
22. Fabbrizzi L, Licchelli M, Pallavicini P, Taglietti A (1996) Inorg Chem 35:1733
23. Fabbrizzi L, Licchelli M, Taglietti A (manuscript in preparation)
24. Fabbrizzi L, Sabatini L (1979) Inorg Chem 18:438
25. Engeser M, Fabbrizzi L, Licchelli M, Sacchi D (1999) Chem Commun 1191
26. De Santis G, Fabbrizzi L, Licchelli M, Sardone N, Velders AH (1996) Chem-Eur J 2:1243

CHAPTER 12

Oxygen Diffusion in Polymer Films for Luminescence Barometry Applications

X. Lu, I. Manners, M. A. Winnik

This chapter reviews recent results on the measurement of oxygen diffusion and oxygen permeability in thin polymer film coatings by luminescence quenching experiments. These coatings have the potential to serve as "pressure-sensitive paints" for luminescence barometry applications. We compare results of steady-state measurements with those of pulsed-laser experiments for polydimethylsiloxane (PDMS), a series of commercial silicone resins, and several poly (thionyl-phosphazene) homopolymers (C_nPATP) and copolymers. In some systems we compare the quenching properties of two different dyes and conclude that the efficiency of quenching by oxygen of platinum octaethylporphine triplets has twice the efficiency as quenching the excited state of tris(diphenylphenanthroline) ruthenium dichloride. Our major interest is the relationship between the polymer structure and oxygen permeation and diffusion in the polymer. When the polymer is crystalline, as in the case of poly(tetramethylene oxide) (PTHF), these properties evolve as the polymer undergoes slow crystallization. In PATP-PTHF block copolymers, microphase separation introduces another level of complexity, since the dye is likely partitioned into both microphases. The block copolymers have the very attractive feature that they exhibit the high permeability of PATP itself, without the surface tackiness of the homopolymer.

12.1
Introduction

This chapter reviews recent results on the measurement of oxygen diffusion and oxygen permeability in thin polymer film coatings that have the potential to serve as "pressure-sensitive paints" for luminescence barometry applications. In this technology, an object is coated with a thin film containing a fluorescent or phosphorescent dye and placed in a wind tunnel. The object is illuminated with light while exposed to flowing air. The dyes in the film are excited by the illumination, but the dye excited states are quenched by oxygen present in the film. Because the concentration of oxygen in the film is related to the partial oxygen pressure at the surface, the extent of quenching provides a measure of the surface air pressure profile across the object. One monitors the fluorescence of the coating with a digital camera. These intensities must then be converted to air pressure at each point on the surface of the object. The object can be an airplane, or airplane part such as a wing or propeller. It can be a turbine blade, or racing car, or even a building or bridge [1].

Engineering aspects of luminescence barometry have received a great deal of attention over the past ten years. Less attention has been paid to problems of design and selection of optimum materials for the pressure sensitive paint itself. The important issues involve the choice of dyes and polymer matrices. There are a variety of materials issues that need to be considered. The dye must resist photo-decomposition (photobleaching) during use. The polymer must exhibit abrasion resistance when exposed at high velocities to dust that is always present in large volumes of flowing air. Rapid response to pressure changes is essential. One must also have a sufficient understanding of the fundamental principles that connect the air pressure at the surface of the polymer film to the changes in intensity and decay time for the dyes in the matrix. This topic, the connection between changes in air pressure and changes in quenching parameters, particularly as they are affected by the choice of dye and polymer, is the major focus of this review. We begin by considering how the fundamental parameters that describe oxygen transport in polymers are determined. We then consider the use of luminescence quenching experiments on thin polymer films themselves to determine the permeability and diffusivity of oxygen in the films. Finally, we consider the behavior of a number of polymer films and film-dye combinations examined in our laboratory as candidates for pressure-sensitive paints (PSPs).

12.1.1
Measuring Oxygen Transport

The diffusion of oxygen (and other gases) in polymers is most commonly determined by measuring the rate of gas permeation across a membrane [2]. The parameters of interest are the diffusion coefficient D_{O_2}, oxygen solubility in the film S_{O_2}, and its permeability P_{O2}. These parameters are related through the expression $P_{O_2} = D_{O_2} \times S_{O_2}$. In a membrane diffusion experiment, one measures the gas flux as a function of time. D_{O_2} is determined from the kinetics of the approach of the flux to its steady-state value, and P_{O_2} is calculated from the steady-state flux. These membrane diffusion experiments require freestanding films, and are most convenient for glassy and semi-crystalline polymers. Viscoelastic polymers that flow cannot be examined directly in this way. One has to introduce cross-links to convert the polymer to an elastic film, and an uncertainty is introduced into the measurement. The cross-links have the potential to perturb both the diffusivity and solubility of a gas in the film.

Because oxygen is such a powerful quencher of fluorescence and phosphorescence, there has long been an interest in using oxygen quenching to determine oxygen transport parameters for polymer films [3–6]. In these experiments, one prepares a polymer film containing a luminescence dye, illuminates the film at a wavelength absorbed by the dye, and monitors the emission intensity or emission decay profile as a function of partial pressure of oxygen. To obtain diffusion coefficients, one has to carry out experiments in which the extent of oxygen penetration into the film or efflux from the film is somehow determined. A useful strategy for analyzing these experiments involves developing a model that couples Fick's Laws of diffusion to describe the oxygen transport with Stern-Volmer kinetics [3–7] to describe the quenching process. Early attempts at formulating a proper model did not take account of the spatial concentration profile created as a consequence of diffusion that satisfies Fick's Laws. Mills and Chang [7] were the first to take this concentration profile properly into account; they analyzed numerically the contribution of oxygen diffusion to oxygen quenching experiments in thin polymer film.

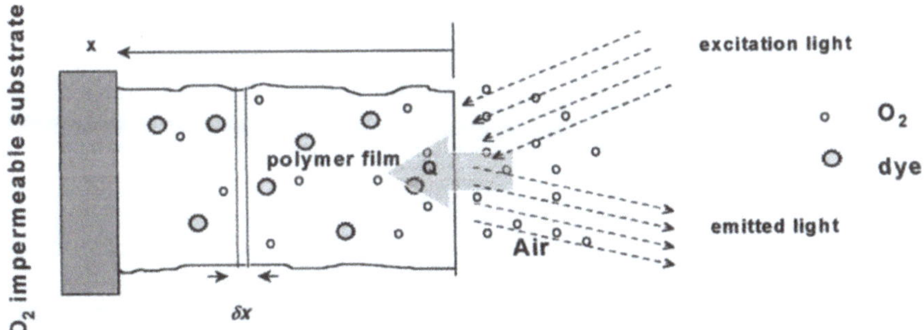

Fig. 12.1. A polymer film containing a uniform concentration of dye is supported on an oxygen-impermeable substrate. The film is illuminated with light as the film is exposed to an atmosphere of air. One monitors the luminescence intensity of the emitted light as a function of time

A full derivation of this model, based on fundamental physical principles, was developed by Yekta et al. [6]. This analysis begins by dividing the film into a series of layers. If these layers are sufficiently thin, one can assume that the *local concentration* of oxygen ($Q(x,t)$) in the layer located between x and $x + dx$ at time t is essentially uniform. Under these circumstances the instantaneous elemental luminescence intensity emanating from this layer is given by a Stern-Volmer-type equation. In the following section, we summarize this derivation, and then employ the equations derived to examine oxygen transport in a series of polymer films that serve as candidates for PSP applications. A presentation of this model is given in Fig. 12.1.

12.2
Oxygen Diffusion and Luminescence Quenching

In a luminescence quenching experiment, a luminescent dye is excited with either a pulse of light or with steady illumination. In the absence of oxygen or any other added quencher, the excited eye (dye*) will decay with its unquenched lifetime τ°. When oxygen or any other mobile quencher Q is present, interaction of dye* and Q leads to the deactivation of the excited state (Scheme 12.1).

$$\text{dye} \xleftarrow{\quad 1/\tau^\circ \quad} \text{dye*} + Q \xrightarrow{\quad k_q[Q] \quad} \text{dye} + Q^* \text{ or } Q$$

Scheme 12.1

From this simple mechanism, one derives the Stern-Volmer (SV) equation (Eq. 12.1) [8], which serves as the point of departure for all discussions of fluorescence and phosphorescence quenching involving mobile species in polymers:

$$\frac{I^\circ}{I} = \frac{\tau^\circ}{\tau} = 1 + k_q \tau^\circ [Q]_{eq} \tag{12.1}$$

In this expression, I is the emission intensity at some molar quencher concentration $[Q]$, and $I°$ is the corresponding intensity in the absence of quencher. The decay of the electronically excited dye follows first-order kinetics in the absence of Q and pseudo-first-order kinetics in the presence of Q. In other words, dye* exhibits an exponential decay profile, with lifetime $\tau°$ in the absence of Q, and τ in its presence.

In Eq. (12.1) we assumes a "well-stirred reactor," which implies that the relaxation of the solvent around the excited dye and the redistribution of reactants in the solution are faster than the reaction itself. As a consequence, the concentration of Q in the solution is effectively uniform, and the radial distribution function of Q surrounding each dye* is uniform. We emphasize this point because many sensor experiments employ dyes that exhibit non-exponential decay profiles even in the absence of oxygen. In other systems, where the excited dye itself undergoes simple first-order decay in the absence of quencher, the decays become non-exponential in the presence of oxygen. Non-exponential decays in the absence of quencher point to a distribution of emitting species in the system. Deviations from simple pseudo-first order kinetics in the presence of quencher are likely connected to spatial variations in quencher concentration which do not equilibrate on the time scale defined by $\tau°$. Oxygen is a very effective quencher of both excited singlet and triplet excited states.

12.2.1
Diffusion-Controlled Reactions

The rate coefficient k_q is related to the diffusivity of the quencher. For fluid media,

$$k_q = \alpha k_{diff} \tag{12.2}$$

where α is the probability of quenching in an encounter complex of the excited state/quencher, and k_{diff} is the diffusion-controlled rate constant for the formation of the encounter pair [3]. The rate constant k_{diff} depends upon the sum of the diffusion coefficients of the dye and the quencher ($D = D_{dye} + D_Q$), as well as the sum of the radii σ of the diffusing species, normally taken to be the interaction distance in the encounter complex (about 1 nm), and N'_A is $10^{-3} \times$ Avogadro's number.

$$k_{diff} = 4\pi\sigma N'_A D \,[1 + \sigma/(\pi D t)^{1/2}] \tag{12.3a}$$

At short times and high quencher concentrations, the transient ($t^{-1/2}$) term in Eq. (12.3a) contributes to the reaction rate, but under most circumstances, k_{diff} quickly evolves to its steady state value. For oxygen as the quencher, one commonly assumes that $D_{dye} \ll D_{O_2}$, and D in Eq. (12.3) can be replaced by D_{O_2}. Thus

$$k_{diff} = 4\pi\sigma N'_A D_{O_2} \tag{12.3b}$$

Substituting Eq. (12.3b) into Eq. (12.1) yields

$$\frac{I^\circ}{I} = \frac{\tau^\circ}{\tau} = 1 + (4\pi\sigma 2 N_A' \tau^\circ D_Q)\,[Q]_{eq} \tag{12.4a}$$

It is useful to define the parameter $B = (I^\circ/I_{eq} - 1)$ as a measure of the response of the system at equilibrium upon exposure to air. When the final equilibrium state is 1 atm air (0.21 atm oxygen),

$$B = (4\pi\sigma 2 N_A' \tau^\circ D_Q)\,[Q]_{0.21\,atm} \tag{12.5}$$

It is important to recognize that B contains the product $\sigma\alpha$, which can be thought of as an effective encounter radius. While much has been written about α, and how it differs for singlet quenching compared to triplet quenching [6, 11–13], it is common to assume that $\alpha = 1$ and $\sigma = 1.0$ nm. Deeper insights into the meaning of σ are possible through the theory of partially diffusion-controlled reactions [9, 10], originally formulated by Collins and Kimball [14]. In this treatment, one incorporates the quenching efficiency into k_{diff} and appreciates that σ is an effective interaction distance whose magnitude depends upon D: σ increases as D decreases. In the limit of small D, every encounter persists long enough for the reaction to occur.

In a typical experiment, a polymer film is mounted in the sample chamber of a fluorescence spectrometer in a cell that allows control of the surrounding atmosphere. One flushes the cell with nitrogen or argon until oxygen has been removed completely from the film, and the emission intensity holds steady at its most intense value. At this point, one has $[Q] = 0$, inside and outside of the sample, and $I = I^\circ$. At time zero, the sample is at once exposed to a fixed external partial pressure of oxygen (here 0.21 atm.). In the sample, $[Q](x, t)$ increases to its final uniform equilibrium value, $[Q]_{eq}$, and concurrently the measured intensity drops to its final value I_{eq}. The same sample is now ready for a diffusion-out (desorption) experiment. That is, the external concentration (pressure) of oxygen is at once brought to zero (e.g., flushing with N_2). There is a net diffusion of quenchers out of the sample, until $[Q] = 0$ everywhere in the sample. The measured intensity increases concurrently from I_{eq} to I°.

12.2.2
Quenching and Oxygen Diffusion

During the approach to equilibrium, the oxygen concentration $[Q]$ is not uniform through the polymer film. If the oxygen diffusion obeys Fick's second law, then the diffusion equation must be integrated in a way that takes proper account of the boundary conditions at the surfaces of the film. These boundary conditions will differ when the polymer film is suspended with both sides exposed to the atmosphere compared to a film coated on a gas-impermeable substrate. For a film on an impermeable substrate, the oxygen concentration in a layer a distance x away from the gas/film interface at time t is given by the expressions

- Oxygen sorption:

$$[Q](x,t) = [Q]_{0.21\,\text{atm}} \left[1 - \frac{4}{\pi} \sum_{n=\text{odd}}^{\infty} \frac{1}{n} \exp\left\{ -\frac{n^2 \pi^2 D_{O_2} t}{4L^2} \right\} \sin\left(\frac{n\pi x}{2L} \right) \right] \qquad (12.6\,\text{a})$$

- Oxygen desorption:

$$[Q](x,t) = [Q]_{0.21\,\text{atm}} \left[\frac{4}{\pi} \sum_{n=\text{odd}}^{\infty} \frac{1}{n} \exp\left\{ -\frac{n^2 \pi^2 D_{O_2} t}{4L^2} \right\} \sin\left(\frac{n\pi x}{2L} \right) \right] \qquad (12.6\,\text{b})$$

The intensity of the light emitted from this layer can be described by a Stern-Volmer-type expression

$$\frac{\delta I^\circ}{\delta I(x,t)} = 1 + \frac{B[Q](x,t)}{[Q]_{0.21\,\text{atm}}}; \quad \delta I^\circ = I^\circ\, dx/L \qquad (12.7\,\text{a})$$

The intensity of light emitted from the entire film at time t is given by the summation over the layers in the film:

$$I(t) = \frac{I^\circ}{L} \int_0^L \frac{dx}{1 + B[Q](x,t)/[Q]_{0.21\,\text{atm}}} \qquad (12.7\,\text{b})$$

If both D_{O_2} and the film thickness L are known, one can replace the $[Q](x,t)$ term in Eq. (12.7) by Eq. (12.6), and then integrate it to calculate the time profile of the emission intensity from the film. The result of this calculation is a *simulated* intensity vs time profile, calculated with measured values of B and L and an arbitrarily chosen value of D_{O_2}. The diffusion coefficient is treated as an adjustable parameter in comparing simulated $I(t)$ profiles with those measured experimentally. The best fit is obtained by a minimization of χ^2.

In terms of the external oxygen partial pressure p_{O_2} (0.21 atm), the equilibrium solubility $[Q]_{eq}$ is proportional to p_{O_2} in the range of low to moderate pressure, according to Henry's law, $[Q]_{eq} = S_{O_2} p_{O_2}$. In this equation S_{O_2} is the Henry's law constant of the gas solubility. When oxygen acts as quencher, Eq. (12.5) can be rewritten as

$$B = 4\pi N_A' \alpha \sigma \tau^\circ (S_{O_2} D_{O_2}) p_{O_2} = 0.84\, \pi N_A' \alpha \sigma \tau^\circ (S_{O_2} D_{O_2}) \qquad (12.8)$$

where the product of S_{O_2} and D_{O_2} is the oxygen permeability, P_{O_2}.

$$\frac{\tau^\circ}{\tau} = 1 + 4\pi \alpha \sigma N_A' \tau^\circ (D_{O_2} S_{O_2}) \qquad (12.4\,\text{b})$$

The permeability P_{O_2} can also be obtained from luminescence decay measurements carried out on the polymer film in equilibrium with various partial pressures of oxygen. We have rewritten Eq. (12.4a) as Eq. (12.4b) to emphasize the predicted linear dependence of the lifetime ratio τ°/τ vs p_{O_2}. The slope of that plot should be equal to $4\pi N_A' \alpha \sigma \tau^\circ P_{O_2}$. Implicit in this analysis is the assumption that the decay profile of the excited dye in the polymer matrix is exponential.

12.3
Silicone Polymers

Silicone resins represent a family of elastomers that usually show high oxygen permeability. Many commercial silicone resins contain functionality which allow them to be crosslinked (cured) in situ. The details of their composition and structure are often proprietary and confidential. We begin our examination of silicone polymers with polydimethylsiloxane (PDMS), a linear homopolymer. To minimize flow during a measurement, we examine a relatively high molecular weight sample ($M_w = 500,000$). PDMS has a very low glass transition temperature ($T_g = -120\,°C$). We then examine several commercial resins purchased from Genesee Resins Co.

12.3.1
PDMS

An example of a time-scan experiment for PDMS containing 100 ppm platinum octaethylporphine (PtOEP) are shown in Fig. 12.2. During oxygen sorption (curve 1), the emission intensity drops rapidly. In contrast, the intensity increases at a much slower rate during oxygen desorption experiment (curve 2).

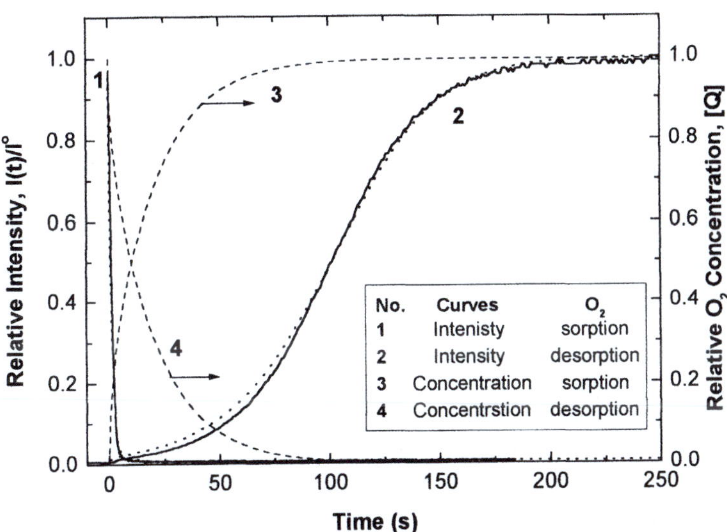

Fig. 12.2. Time scan of PtOEP (100 ppm) emission intensity in a film of PDMS ($L = 0.302$ mm) on glass. The emission is observed at 645 nm during steady excitation at 535 nm. Curve 1 is obtained from an oxygen sorption experiment; curve 2 is from an oxygen desorption experiment. The *solid lines* are the experimental data and the *dotted lines* are simulated data calculated according to Eqs. (12.6) and (12.7). The fits obtained are $D_{O_2} = 1.75 \times 10^{-5}$ cm^2 s^{-1} (curve 1) and 1.72×10^{-5} cm^2 s^{-1} (curve 2). The *dashed curves* 3 (sorption) and 4 (desorption) are normalized oxygen concentrations in the film calculated according to Eq. (12.6) using the values of $D_{O_2} = 1.74 \times 10^{-5}$ cm^2 s^{-1} and $L = 0.302$ mm

Table 12.1. Values of oxygen permeation properties for PDMS and Genesee silicone resins, calculated from the time-scan experiments

Polymer	B	$10^5 D_{O_2}$[a]	τ°, µs	$10^4 S_{O_2}$[a]	$10^{12} P_{O_2}$[a]
PDMS	170	2.0	70	12.5	16
GP-187[b]	50	1.1	60	5.3	5.8
GP-197[b]	3.6	0.057	99	4.2	0.24
GP-277[b]	195	1.1	44	26	28
EXP36-X-20[b]	8.2	0.81	70	1.2	1.1
EXP38-X-20[b]	1.2	0.1	98	0.8	0.08

[a] Units: D, cm^2 s^{-1}; S, M atm^{-1}; P, mol cm^{-1} s^{-1} atm^{-1}.
[b] Commercial silicone resins from Genesee Resins Co.

Although the intensity-time profiles are very different, they both fit well to the Fickian diffusion model as described by Eqs (12.6) and (12.7). The best-fit D_{O_2} values for both experiments are essentially identical: $D_{O_2} = 1.75 \times 10^{-5}$ cm^2 s^{-1} for oxygen-sorption and $D_{O_2} = 1.72 \times 10^{-5}$ cm^2 s^{-1} for oxygen-desorption.

For curves 3 and 4 in Fig. 12.2, we plot the total oxygen concentration in the film as a function of time, calculated with Eq. (12.6), using the average oxygen diffusivity $D_{O_2} = 1.74 \times 10^{-5}$ cm^2 s^{-1} and the measured film thickness ($L = 302$ µm). These curves show that the times for oxygen molecules to diffuse out of or into the PDMS film are the same, around 250 s. The intensity profiles differ because of the way in which the oxygen concentration profiles evolve. At late stages of the sorption experiment, there is a slow increase of oxygen concentration inside the film, and the corresponding changes in the extent of quenching (curve 1) are so small that they are hard to detect. The average D_{O_2} value obtained from 12 measurements on 4 PDMS samples with various thickness is $D_{O_2} = 2.0 \pm 0.4 \times 10^{-5}$ cm^2 s^{-1}. Values of P_{O_2} (1.6×10^{-11} mol cm^{-1} s^{-1} atm^{-1}) and S_{O_2} (1.25×10^{-3} M atm^{-1}) are calculated using the D_{O_2} and B values given in Table 12.1.

12.3.2
Genesee Resins

The GP series of silicone resins from Genesee are functional methyl silicone resins. GP-277 is sold as a dilute solution of resin that cures to form a relatively hard coating with high oxygen permeability. EXP36-X-20 and EXP38-X-20 are 20 % active epoxy-functional silicones. When exposed to atmospheric moisture in the absence of an added diamine-curing agent, they undergo slow hydrolysis and condensation to form a cross-linked resin film. The most significant characteristic of these resins is the large differences in B values for the different resins. In addition, it appears that in the resins with large B values the PtOEP dye has shorter τ°. Because the B is proportional to $\tau^\circ \times P_{O_2}$, the silicone samples with higher B values are much more permeable to oxygen. It is important to recognize for silicone resins that the permeability of oxygen depends sensitively on the composition and structure of the polymer. The second important feature is that for an individual resin sample, B values change with time, suggesting that oxygen permeability in these resins depends upon the extent of cure. In spite of these complications, individual samples are well described by Eqs. (12.6)– (12.8) [15].

In Fig. 12.3, we give the examples of EXP36-X-20, for which the B value is a factor of 20 smaller than that of PDMS. In time scan experiments, EXP36-X-20 shows emission intensity profiles similar to those of PDMS. The best fits to the data give a value of $D_{O_2} = 1.2 \times 10^{-5}\,\mathrm{cm^2\,s^{-1}}$ for the sorption curve and $D_{O_2} = 1.0 \times 10^{-5}\,\mathrm{cm^2\,s^{-1}}$ for the desorption curve [6].

When we examine the results in Table 12.1, we see that PDMS and GP-277 show very high permeability to oxygen and strong oxygen pressure dependence for oxygen quenching. The different permeabilities arise from differences in both oxygen solubilities and diffusivities. In the case of GP-277, GP-187, and the epoxy resin EXP36-X-20, the D_{O_2} values are remarkably similar, and the variation of permeabilities appear to come exclusively from large change in the solubility coefficient S_{O_2}. The value of D_{O_2} for GP-197 is much smaller. The oxygen permeability of PDMS is close to that of GP-277; however, the twofold higher D_{O_2} value in PDMS implies that S_{O_2} is only half that of GP-277. Usually the D_{O_2} for PDMS is higher than D_{O_2} for cross-linked silicone resins, which is a result of decreasing polymer chain flexibility by cross-links.

Under pulsed laser excitation, all silicone polymer film samples examined in the absence of oxygen or air exhibited exponential decays. This indicates the uniform microenvironments around the PtOEP dye molecules. When the PDMS

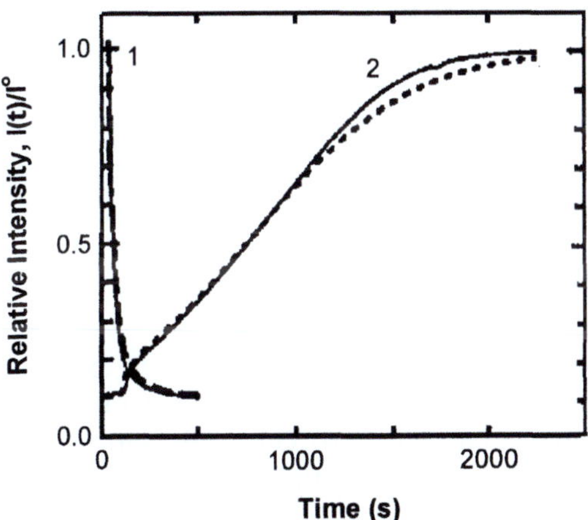

Fig. 12.3. Time scan of PtOEP (50 ppm) emission intensity in a film of EXP36-X-20 ($L = 0.745$ mm) on glass. The emission is observed at 645 nm during steady-state excitation at 535 nm. Curve 1, an oxygen sorption experiment; curve 2, an oxygen desorption experiment. The *solid lines* are experimental data and the *dashed lines* are simulated data calculated according to Eqs. (12.6) and (12.7). The fits are calculated with $D_{O_2} = 1.2 \times 10^{-5}\ \mathrm{cm^2\ s^{-1}}$ (curve 1) and $1.0 \times 10^{-5}\,\mathrm{cm^2\,s^{-1}}$ (curve 2). (Copyright 1995 National Research Council of Canada Research Press – Reprinted with permission from [6])

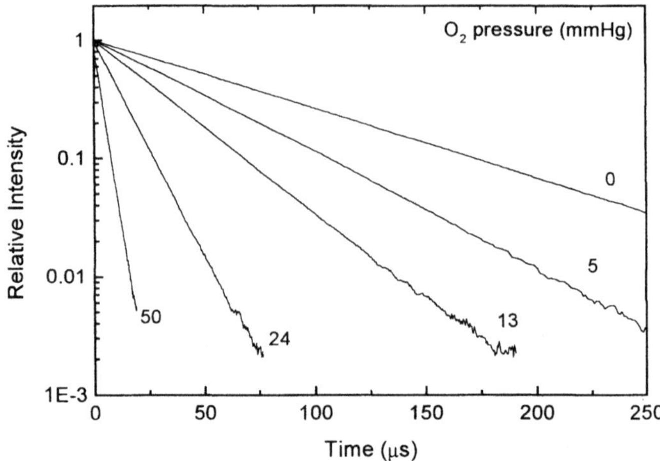

Fig. 12.4. Phosphorescence decays of PtOEP (10 ppm) in a film of PDMS equilibrated with various external pressures of oxygen. The emission is observed at 600 ± 25 nm following pulsed laser excitation at 532 nm

films were exposed to oxygen or air, the decay curves remained exponential until p_{O_2} became larger than 30 Torr (0.039 atm) (see Fig. 12.4). For the other silicone resins, the decays were all non-exponential in the presence of oxygen, and the extent of deviation from an exponential form increased with the external oxygen or air pressure. One normally interprets this type of curvature in semi-logarithmic intensity-decay plots to a distribution of dye sites in the film which have different accessibilities to quenching by oxygen. One could imagine a distribution of site solubilities and or of local diffusivities in the system. In order to analyze the data, each decay curve was fitted arbitrarily to a sum of exponential terms, and the mean decay time $\langle \tau \rangle$ was calculated from the fitting parameters τ_i and A_i:

$$\langle \tau \rangle = \frac{\sum_i A_i \tau_i^2}{\sum_i A_i \tau_i} \tag{12.9}$$

Rather surprisingly, plots of $(\tau^0/\langle \tau \rangle)$ are linear with external oxygen pressure over the entire range of $0 < p_{O_2} < 4.5$ atm for silicone resins (see Fig. 12.5) and $0 < P_{O_2} < 1$ atm for PDMS (see Fig. 12.6). One of the most striking features of the data in Fig. 12.5 is the large range of the slopes of the plots, which emphasizes the range of oxygen permeabilities in these silicone films. The values of P_{O_2} for these silicone films, calculated from the slopes of Stern-Volmer plots are listed in Table 12.2. The P_{O_2} values for the Genesee silicone resins obtained from B, are always larger, in some cases remarkably larger, than those determined from $\langle \tau \rangle$. For PDMS the P_{O_2} values from the two different experiments are very similar.

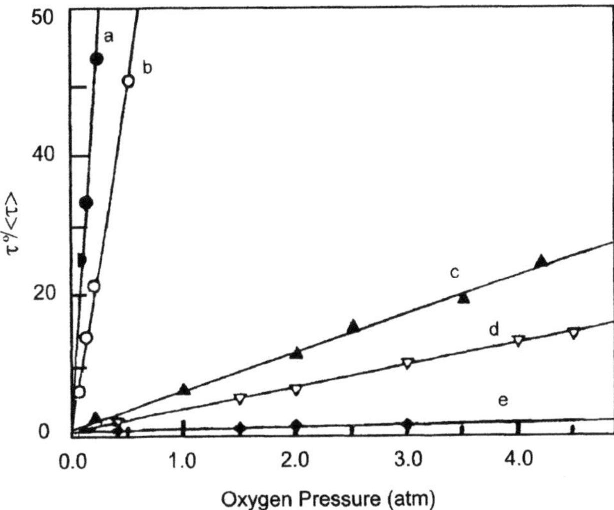

Fig. 12.5. Lifetime Stern-Volmer plots for PtOEP in a series of commercial silicone resins (Genesee Resin Co.). Each decay trace was fitted to a sum of exponentials and the mean decay time $\langle \tau \rangle$ was calculated with Eq. (12.9). From the top-to-bottom the silicone resins are (a) GP-277, (b) GP-187, (c) EXP36-X-20, (d) GP-197, (e) EXP38-X-20

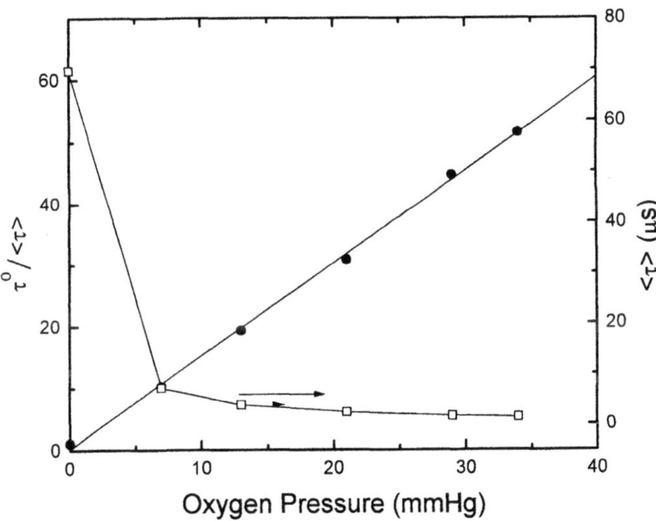

Fig. 12.6. Effect of oxygen pressure on the excited state lifetime of PtOEP (10 ppm) in a film of PDMS. The mean lifetimes (*open points*) are calculated according to Eq. (12.9). P_{O_2} obtained from the slope of lifetime Stern-Volmer plot (*filled points*) is equal to 19×10^{-12} mol cm^{-1} s^{-1} atm^{-1}

Table 12.2. Comparison of the P_{O_2} values calculated from steady-state and pulsed-laser experiments

Polymer	τ°, µs	$10^{12}\, P_{O_2}{}^a$ from lifetime SV plots	$10^{12}\, P_{O_2}{}^a$ from B
PDMS	70	19	16
GP-187[b]	60	2.4	5.9
GP-197[b]	99	0.042	0.24
GP-277[b]	44	9.4	28
EXP36-X-20[b]	70	0.10	1.1
EXP38-X-20	98	0.005	0.08

[a] Units: P, mol cm^{-1} s^{-1} atm^{-1}. [b] Commercial silicone resins from Genesee Resins Co.

12.4
Poly(aminothionylphosphazenes) (PATP)

This class of inorganic macromolecules contains a backbone made up of sulfur, nitrogen, and phosphorus atoms [16–18]. An attractive feature of this class of polymers is that backbone polarity and the large number of non-bonded electron pairs in the repeat unit assist dye solubility. We have used two different dyes, PtOEP and tris(diphenylphenanthroline) ruthenium dichloride ($Ru(dpp)_3Cl_2$) to examine oxygen permeability and oxygen diffusivity in a series of poly-(amino-thionylphosphazene) (PATP) homopolymers [19–21] and block copolymers [22].

The C_nPATP homopolymers, with linear aliphatic pendant groups of n carbons, are elastomers characterized by high oxygen permeability. We have carried out time-scan and pulsed laser experiments on several of these polymers, and the data are summarized in Table 12.3. The oxygen diffusion coefficients in C_nPATP, determined with PtOEP as the probe, increase as the size of the pendant chains increase. The increase of both D_{O_2} and P_{O_2} corresponds to the decrease in the T_g of the polymer, whereas S_{O_2} in this series of homopolymers remains almost constant. Since the polymer with lower T_g is considered to have larger

Table 12.3. Values of oxygen permeation properties for several C_nPATP homopolymers, calculated from the time scan experiments

Polymer	T_g, °C	B	$10^6 D_{O_2}{}^a$	$10^3 S_{O_2}{}^a$	$10^{12} P_{O_2}{}^a$	τ°, µs
C_2PATP	4	9.7 ± 0.2	0.50 ± 0.04	1.2 ± 0.1	0.62 ± 0.01	105
C_3PATP	6	28 ± 1	1.2 ± 0.2	1.5 ± 0.3	1.8 ± 0.1	103
C_4PATP	−16	74 ± 6	4.0 ± 0.6	1.2 ± 0.2	4.8 ± 0.4	103
C_6PATP	−18	80 ± 1	6.2 ± 0.8	0.87 ± 0.11	5.4 ± 0.1	98
C_7PATP	–	74 ± 4	6.0 ± 0.5	0.85 ± 0.08	5.1 ± 0.3	96

[a] Units: D, cm^2 s^{-1}; S, M atm^{-1}; P, mol cm^{-1} s^{-1} atm^{-1}.

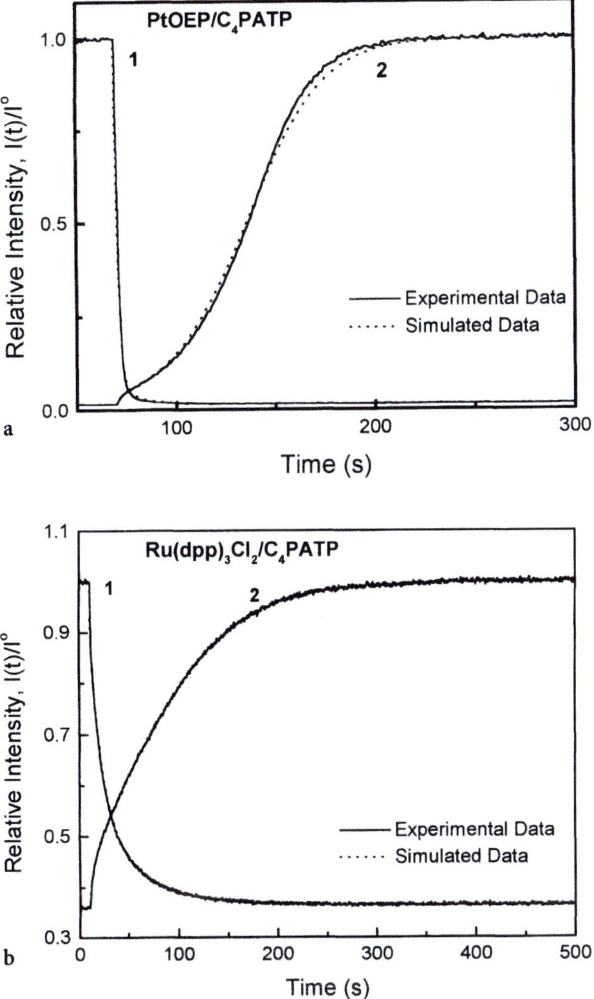

Fig. 12.7 a, b. Time scan experiments in C$_4$PATP using different luminescent dyes: **a** the emission of PtOEP (500 ppm) in a film with $L = 0.125$ mm is observed at 645 nm during steady excitation at 535 nm; **b** the emission of Ru(dpp)$_3$Cl$_2$ (500 ppm) in a film with $L = 0.135$ mm is observed at 620 nm during steady-state excitation at 452 nm. Curves 1 are obtained from oxygen sorption experiments; curves 2 are from oxygen desorption experiments. The *solid lines* are experimental data and the *dotted lines* are simulated data calculated according to Eqs. (12.6) and (12.7). In PtOEP/C$_4$PATP, the fits are calculated with $D_{O_2} = 3.06 \times 10^{-6}$ cm^2 s^{-1} (curve 1) and 3.58×10^{-6} cm^2 s^{-1} (curve 2); in Ru(dpp)$_3$Cl$_2$, the fits give the values D_{O_2} (4.76×10^{-6} cm^2 s^{-1}) for curve 1 and D_{O_2} (3.57×10^{-6} cm^2 s^{-1}) for curve 2

free volume at room temperature, these results are consistent with the idea that high free volume favors oxygen diffusion in the polymer matrix.

The most intensely studied polymer is poly(n-butylaminothionylphosphazene), C_4PATP. In this matrix, the two dyes Ru(dpp)$_3$Cl$_2$ and PtOEP were compared. Like many ionic dyes in polymer matrices, the decay profile of Ru(dpp)$_3$Cl$_2$ in C_4PATP is non-exponential. The curvature in the semi-logarithmic decay plot is small, and the mean lifetime is 6.31 µs. The excited state lifetime for PtOEP in C_4PATP (102 µs) is more than an order of magnitude larger. The longer lifetime makes PtOEP more sensitive to quenching by low concentrations of oxygen, and provides an interesting test to the validity of the equations derived in the previous section of this chapter.

The intensity profiles from the time-scan experiments for C_4PATP containing PtOEP and Ru(dpp)$_3$Cl$_2$ are shown in Fig 12.7 a, b, respectively. The good fit for both sets of data confirms that the oxygen diffusion in C_4PATP is Fickian. For several films containing PtOEP, we obtained an average of $D_{O_2} = 3.73 \times 10^{-6}$ cm^2 s^{-1} for both sorption and desorption experiments. For corresponding experiments with Ru(dpp)$_3$Cl$_2$ (e.g., Fig. 12.7b), the average diffusion coefficient ($D_{O_2} = 3.72 \times 10^{-6}$ cm^2 s^{-1}) is identical. To a high degree of precision, the D_{O_2} value, determined with two different probes, is found to be the same. The P_{O_2} values calculated from B and the probe lifetimes, however, differ by a factor of two for the two probes. This result leads to the questionable conclusion that S_{O_2} for oxygen in C_4PATP is different when determined with one probe than when determined with a different probe. In Table 12.4, we list D_{O_2}, P_{O_2}, and S_{O_2} calculated for C_4PATP films containing the two different dyes.

The resolution of this dilemma takes us back to the issue of the capture radius $\alpha\sigma$ in Eqs (12.4) and (12.5). P_{O2} values were calculated from B and τ° assuming that $\alpha = 1$ and $\sigma = 1.0$ nm. Steric effects, and the extent to which contact of the quencher with the ligands of the dye are effective at deactivating of the excited state, can play a role in determining the effective capture radius. The apparent difference in P_{O_2} and S_{O_2} values can be reconciled if we assume that $\alpha\sigma$ for O$_2$ quenching of Ru(dpp)$_3$Cl$_2$ is half the value of that for PtOEP; i.e., for Ru(dpp)$_3$Cl$_2$, $\alpha\sigma = 0.5$ nm.

In the presence of oxygen, the decays of both PtOEP and Ru(dpp)$_3$Cl$_2$ in C_4PATP are non-exponential. Nevertheless, the lifetime Stern-Volmer (SV)

Table 12.4. Comparison of oxygen permeation properties for C_4PATP determined from films containing different dyes, but assuming $\alpha\sigma = 1.0$ nm for oxygen quenching of both dyes

Experiment	Parameter	PtOEP	Ru(dpp)$_3$Cl$_2$
Time-scan	$10^6 \, D_{O_2}{}^a$	3.73	3.72
	$10^3 \, S_{O_2}{}^a$	1.03	0.49
	$10^{12} \, P_{O_2}{}^a$	3.85	1.82
Pulsed-laser	$10^6 \, D_{O_2}{}^{a,b}$	3.73	3.72
	$10^3 \, S_{O_2}{}^a$	1.08	0.54
	$10^{12} \, P_{O_2}{}^a$	4.03	2.01

[a] Units: D, cm^2 s^{-1}; S, M atm^{-1}; P, mol cm^{-1} s^{-1} atm^{-1}.
[b] The results are obtained from the time-scan experiments.

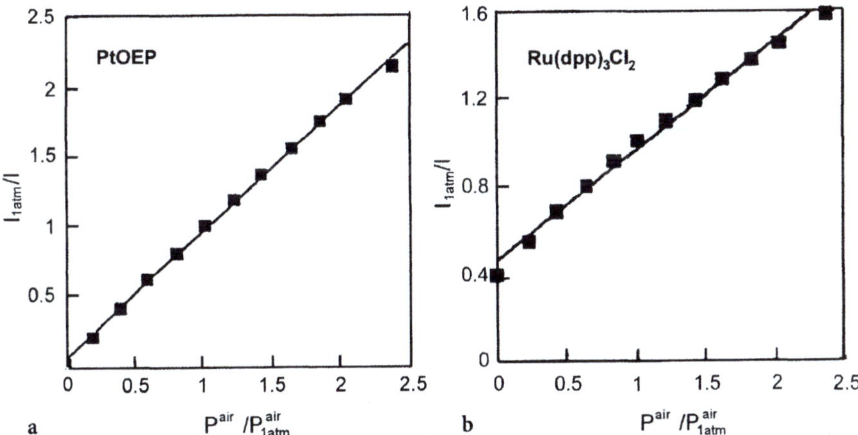

Fig. 12.8 a, b. Intensity Stern-Volmer-type plots for the two dyes in C_4PATP polymer films: a PtOEP; b $Ru(dpp)_3Cl_2$. The x-axis refers to the air pressure P^{air} in the sample chamber; the y-axis is the ratio of the intensity at P^{air} to that at 1 atm ($P^{air}_{1\,atm}$). The *solid line* represents the linear fit to the data. (Copyright 1996 Wiley-VCH – reprinted with permission from [20])

plot of $\tau^0/\langle\tau\rangle$ vs p_{O_2} for the films containing PtOEP is strictly linearity, and the corresponding plot for $Ru(dpp)_3Cl_2$ exhibits only a slight curvature. P_{O_2} can be calculated from the slopes of the lifetime SV plots (Eq. 12.4b). Here the P_{O_2} values for the two dyes appear to differ by a factor of two. This discrepancy can be removed by assuming that for $Ru(dpp)_3Cl_2$ in C_4PATP $\alpha\sigma = 0.5$ nm. The fact that the B values from the time scan experiments and the slopes of the lifetime Stern-Volmer plots are consistent with a twofold larger $\alpha\sigma$ value for PtOEP than for $Ru(dpp)_3Cl_2$ in C_4PATP, provides strong support that the kinetic parameters for the interaction of the two excited dyes with oxygen are different.

In a separate set of experiments, films of C_4PATP containing each of the two dyes were examined under steady illumination as a function of constant external air pressure. In Fig. 12.8 one sees that the intensity SV-type plot for PtOEP is linear, whereas that for $Ru(dpp)_3Cl_2$ exhibits a small but noticeable downward curvature. The non-exponential luminescence decays of $Ru(dpp)_3Cl_2$ in C_4PATP, and the curvature in the intensity Stern-Volmer plots, indicate that this system is more complex than that involving PtOEP. Not all the assumptions of the equations derived above apply to the ionic dye. The deviations may be due to the formation of dye aggregates in the matrix, or a non-uniform distribution of counterions around the dye in the matrix.

12.5
Modified Poly(aminothionylphosphazenes)

In this section we examine the properties of a series of polymers based upon the PATP backbone. These polymers were synthesized in the hopes that they might build on the useful properties of C_nPATP for luminescence barometry applications, but avoid some of the drawbacks. C_4PATP, in particular, provides an excellent polymer matrix for pressure sensitive paints because of its high oxygen permeability and the reproducibility of the oxygen quenching parameters. Because the polymer does not need to be cured in place, the sensing layer can be removed after use by dissolution in common organic solvents. Unfortunately C_4PATP forms rather tacky films, which make them sensitive to dust accumulation and mechanical deformation in wind tunnel tests. Tackiness is a consequence of the low T_g ($-16\,^\circ$C) of this amorphous polymer. In an attempt to explore possible options, we examined the properties of a variety of PATP-based polymers synthesized by the Manners group in Toronto [22]. Their structures are shown in Scheme 12.2. In the polymer denoted MSPTP the pendant groups are tris(methylsiloxyl)ethyl groups. Here one hopes to capture some of the oxygen transport features of a siloxane resin in a linear polymer with a PATP backbone. The other polymers are block copolymers, in which the crystalline poly(tetramethylene oxide) (PTHF) block is intended to improve film hardness.

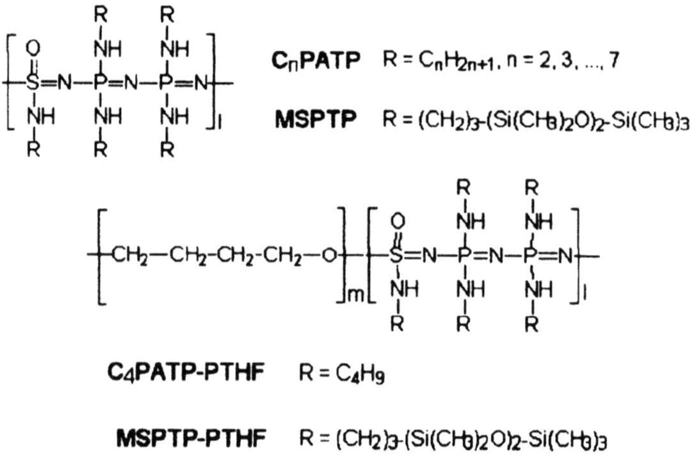

C_nPATP $R = C_nH_{2n+1}$, $n = 2, 3, ..., 7$

MSPTP $R = (CH_2)_3\text{-}(Si(CH_3)_2O)_2\text{-}Si(CH_3)_3$

C_4PATP-PTHF $R = C_4H_9$

Scheme 2   **MSPTP-PTHF** $R = (CH_2)_3\text{-}(Si(CH_3)_2O)_2\text{-}Si(CH_3)_3$

Many of these polymers undergo microphase separation. The existence of phase-separated domains introduces complications into the data interpretation. One has the possibility of oxygen transport at different rates in the different domains as well as partitioning of the dye between the domains.

MSPTP forms light yellow films, which are soft but not tacky. All C_4PATP-PTHF and MSPTP-PTHF block copolymer films that we have examined are hard to the touch and transparent when freshly annealed above the melting temperature (34–40 $^\circ$C) of the PTHF block [22]. These films become noticeably hazy on standing.

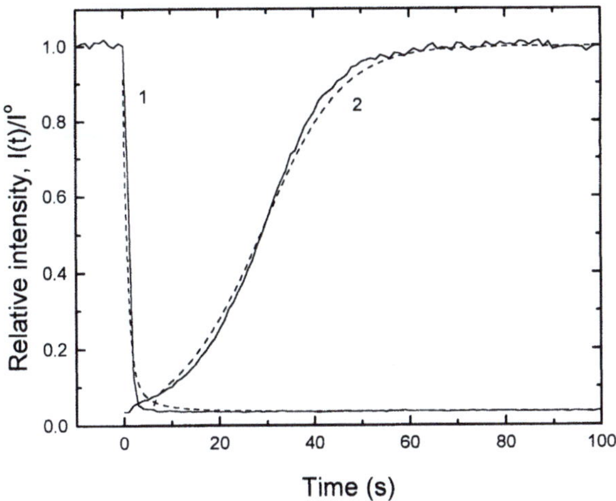

Fig. 12.9. Time scan of PtOEP (100 ppm) emission intensity in a film of MSPTP ($L = 0.139$ mm) on glass. The emission is observed at 645 nm during steady-state excitation at 535 nm. Curve 1, an oxygen sorption experiment; curve 2, an oxygen desorption experiment. The *solid lines* are experimental data and the *dashed lines* are simulated data. The fits are obtained with $D_{O_2} = 8.3 \times 10^{-6}$ cm^2 s^{-1} (curve 1) and 7.9×10^{-6} cm^2 s^{-1} (curve 2)

12.5.1
MSPTP

Time-scan experiments on PtOEP in MSPTP show small but consistent deviations between the experimental and simulated data. An example is given in Fig. 12.9. We have nevertheless chosen to obtain D_{O_2} values by choosing the simulated curves that fit the extrema and pass through the inflection point of the time-scan data. From 21 time-scan measurements on freshly-annealed films ranging in thickness from 0.158 mm to 0.340 mm, an average value of $D_{O_2} = (9.0 \pm 0.2) \times 10^{-6}$ cm^2 s^{-1} was obtained. This value is higher than that for C_4PATP. We see that the introduction of the more flexible silicone pendant chains into the PATP polymer increases the oxygen diffusivity.

While the unquenched decay of PtOEP in MSPTP is mono-exponential ($\tau^\circ = 90$ μs), the decays become increasingly non-exponential when the air pressure is increased. Mean lifetimes were calculated, and the plot of $\tau^\circ/\langle\tau\rangle$ as a function of p_{O_2} (Fig. 12.10) shows a pronounced downward curvature. From the "best-fit" D_{O_2} values, B, and τ°, values of S_{O_2} and P_{O_2} were calculated. These values are collected in Table 12.5. The curious feature of the data is that while oxygen diffusivity is larger for MSPTP than for C_4PATP, the permeability is lower. Although the τ° values are similar, B for MSPTP ($B = 25$) is significantly smaller than that ($B = 74$) for C_4PATP. This conclusion is at odds with the initial slope of the lifetime Stern-Volmer plot. In Fig. 12.10 we see that, from a lifetime perspective, PtOEP in this polymer has a higher sensitivity to quenching than in C_4PATP.

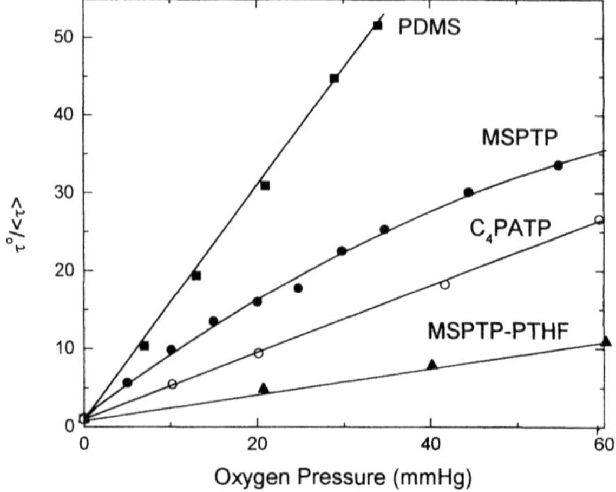

Fig. 12.10. Comparison of lifetime Stern-Volmer plots for PtOEP in different polymer materials. Each decay trace was fitted to a sum of exponentials and the mean decay time $\langle\tau\rangle$ was calculated according to Eq. (12.9)

Table 12.5. Permeation properties of oxygen in polymer films calculated from time-scan experiments

Polymer/dye	$10^6 D^a$	$10^4 S^a$	$10^{12} P^a$	$\tau^\circ, \mu s$	B
MSPTP/P2.C.OEP	9.0 ± 0.5	2.0 ± 0.05	1.8 ± 0.1	90.2 ± 1.0	25.0 ± 2.3
MSPTP-PTHF /PtOEP(1 h)[b]	6.0 ± 0.05	6.8 ± 0.9	4.1 ± 0.5	65.0 ± 1.0	39.8 ± 5.3
MSPTP-PTHF /PtOEP(3 d)[b]	3.3 ± 0.06	6.9 ± 0.7	2.30 ± 0.3	62.5 ± 0.8	22 ± 2.4
C₄PATP-PTHF(2) /PtOEP (1 h)[b]	10.9 ± 0.6	0.8 ± 0.1	0.86 ± 0.03	92.0 ± 1.0	12.0 ± 0.6
C₄PATP-PTHF(10) /PtOEP (1 h)[b]	1.55 ± 0.07	10.2 ± 0.5	1.58 ± 0.08	96.0 ± 0.5	23.0 ± 1.0
C₄PATP-PTHF(10) /PtOEP (3 d)[b]	1.01 ± 0.03	12.0 ± 0.7	1.19 ± 0.03	94.0 ± 0.5	17.4 ± 0.4
C₄PATP-PTHF(10) /[Ru(dpp)₃]Cl₂ (1 h)[b]	2.6 ± 0.7	3.7 ± 0.9	1.00 ± 0.06	6.9 ± 0.1	1.05 ± 0.03
C₄PATP-PTHF(10) /[Ru(dpp)₃]Cl₂ (3 d)[b]	1.4 ± 0.3	6.7 ± 0.4	0.95 ± 0.06	6.9 ± 0.1	1.00 ± 0.02
PTHF/PtOEP(1 h)[b]	3.2 ± 0.9	7.4 ± 2.1	2.4 ± 0.2	92.1 ± 0.6	33.5 ± 0.9
PTHF/PtOEP(10 d)[b]	1.0 ± 0.3	19 ± 5	1.9 ± 0.1	86.1 ± 0.7	24.8 ± 3.5
PTHF/PtOEP(30 d)[b]	0.68 ± 0.17	25 ± 6	1.7 ± 0.1	75.1 ± 0.5	18.9 ± 1.3

[a] Units: D, cm² s⁻¹; S, M atm⁻¹; P, mol cm⁻¹ s⁻¹ atm⁻¹.

[b] The time given in parentheses is the aging time (d, days; h, hours) at 22 °C in darkness after the samples were removed from the annealing oven.

12.5.2
PTHF

Before considering the properties of C_4PATP-PTHF block copolymers, we first describe oxygen transport properties in PTHF itself. PTHF is a semi-crystalline polymer with a melting point in the range of 34–40 °C [22]. Annealing the polymer at elevated temperatures followed by rapid cooling to room temperature generated transparent amorphous films. As these films undergo slow crystallization, they become increasingly cloudy. One imagines that the dyes in the polymer become excluded from the crystallites and remain in the amorphous domains. It is also likely that oxygen transport is dominated by diffusion through the amorphous domains.

Our observations are consistent with a strong influence of crystallization on oxygen transport: the intensity profiles from time-scan experiments change with time (Fig. 12.11) [22]. The data on freshly annealed films containing PtOEP fit quite well to Eqs. (12.6) and (12.7) (see Fig. 12.11a). As the films age, the time-scan profiles show increasing deviations from the "best-fit" simulations based upon Eqs. (12.6) and (12.7). For films aged 10 days and 30 days at 22 °C, we estimated "best-fit" D_{O_2} values from the simulated curves that pass through the mid-point of the time-scan plots, and used each value to calculate P_{O_2} and S_{O_2} from B and τ^0. With aging, there is nearly a factor of 2 decrease in B and an even larger decrease in D_{O_2}, from $3.2 \pm 0.9 \times 10^{-6}$ cm^2 s^{-1} for the freshly annealed sample to $0.68 \pm 0.17 \times 10^{-6}$ cm^2 s^{-1} for the sample aged 30 days. Since the P_{O_2} values decrease slightly, there is a dramatic difference in S_{O_2} values between two samples.

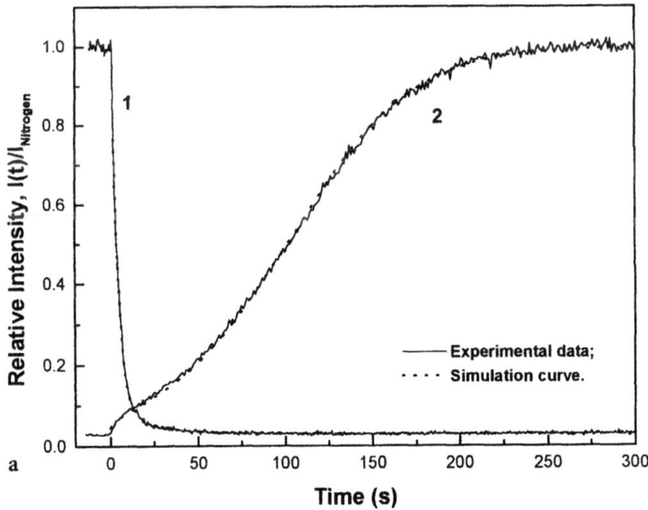

Fig. 12.11 a–c. Time scan of PtOEP (500 ppm) emission intensity in a film of PTHF ($L = 0.180$ mm) on glass: a freshly annealed film, $D_{O_2} = 3.4 \times 10^{-6}$ cm^2 s^{-1}.

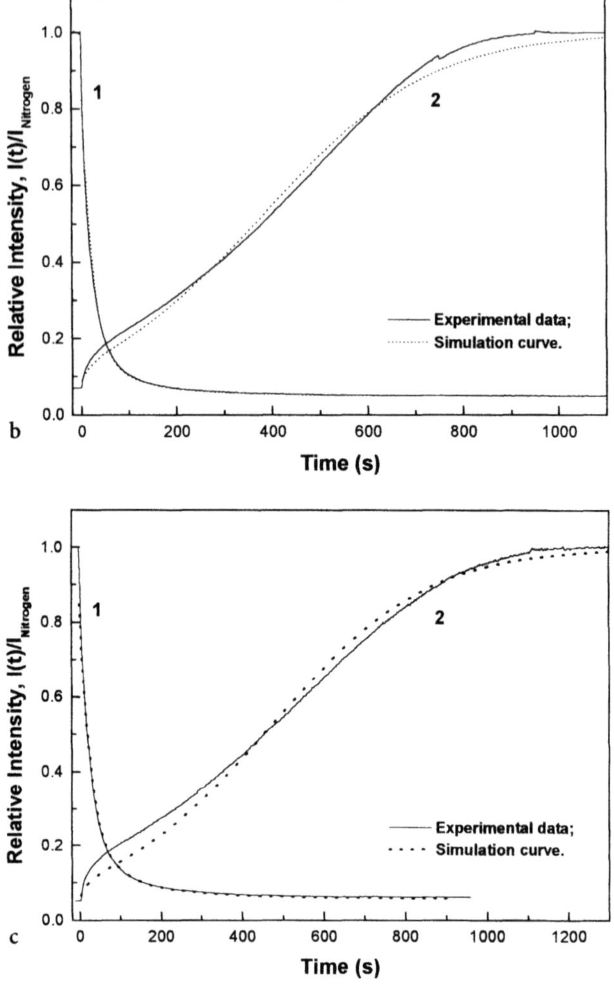

Fig. 12.11 (continued) **b** film allowed to age for 10 days, $D_{O_2} = 0.95 \times 10^{-6}$ cm^2 s^{-1}; **c** film allowed to age for 30 days, $D_{O_2} = 0.51 \times 10^{-6}$ cm^2 s^{-1}. The emission is observed at 645 nm during steady-state excitation at 535 nm. Curve 1 is obtained from an oxygen sorption experiment; curve 2, from an oxygen desorption experiment

12.5.3
C$_4$PATP-PTHF Block Copolymers

Our strategy in the synthesis of C$_4$PATP-PTHF block copolymers was to prepare samples in which the PTHF block was longer than the C$_4$PATP block in the hopes of obtaining hard and tack-free films. It was difficult to reproduce the individual and relative block lengths in the polymers we prepared. These polymers have very interesting properties. In our notation, the number in parenthesis in the polymer name is the ratio of the number of monomer units in the PTHF block to that of the C_4PATP block. One can appreciate that in C_4PATP-PTHF(30), the C_4PATP represents a small component of the system. Nevertheless, the presence of this small component has a large effect on oxygen transport. For C_4PATP-PTHF(2), only PtOEP was used as luminescence sensor. In case of C_4PATP-PTHF(30), we examined both PtOEP and $Ru(dpp)_3Cl_2$.

Figure 12.12 shows the time-scan intensity profiles for PtOEP in C_4PATP-PTHF(2) [22]. It is surprising to obtain such good data from a polymer system where one anticipates microphase separation. The experiment in Fig. 12.12 was carried out approximately 1 h after removing the sample from the annealing oven. The average D_{O_2} value obtained from 12 measurements on 3 film samples is $D_{O_2} = 10.9 \pm 0.6 \times 10^{-6}\,\mathrm{cm^2\,s^{-1}}$. Combining the D_{O_2} values with the value of $B(12.0 \pm 0.6)$, a value of $S_{O_2} = 0.8 \pm 0.1 \times 10^{-4}\,\mathrm{M\,atm^{-1}}$ is calculated (see Table 12.5).

C_4PATP-PTHF(30) is much richer in PTHF segments than C_4PATP-PTHF(2). For PtOEP, exponential decays ($\tau^\circ = 96\,\mu s$) were found in fresh C_4PATP-PTHF(30) films in the absence of oxygen. After three days the decay remained exponential, with a slight decrease of τ° (94 μs). Similarly, the lifetime of $Ru(dpp)_3Cl_2$ in this polymer changes little after three-day aging, although the unquenched decays are non-exponential. Recalling the large change of lifetime for PtOEP in PTHF after aging, it seems possible that the dye molecules partition into the C_4PATP domains, instead of PTHF domains.

C_4PATP-PTHF(30) forms films which are initially clear upon removal from the annealing oven. Upon cooling they become hazy and then turbid as the PTHF phase crystallizes. When film samples containing PtOEP are examined in time-scan experiments within 1 h or so of annealing, they give plots (see

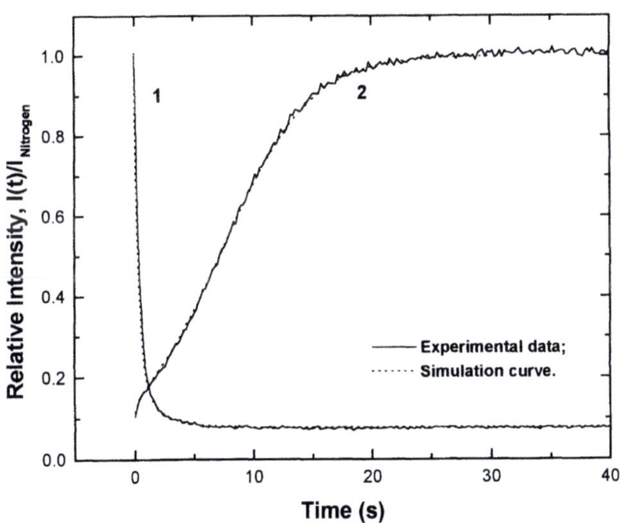

Fig. 12.12. Time scan of PtOEP (100 ppm) emission intensity in a film of C_4PATP-PTHF(2) block copolymer ($L = 0.092\,\mathrm{mm}$) on glass. Curve 1, an oxygen sorption experiment; curve 2, an oxygen desorption experiment. The simulated curves are calculated with $D_{O_2} = 9.80 \times 10^{-6}\,\mathrm{cm^2\,s^{-1}}$ (curve 1) and $D = 9.85 \times 10^{-6}\,\mathrm{cm^2\,s^{-1}}$ (curve 2)

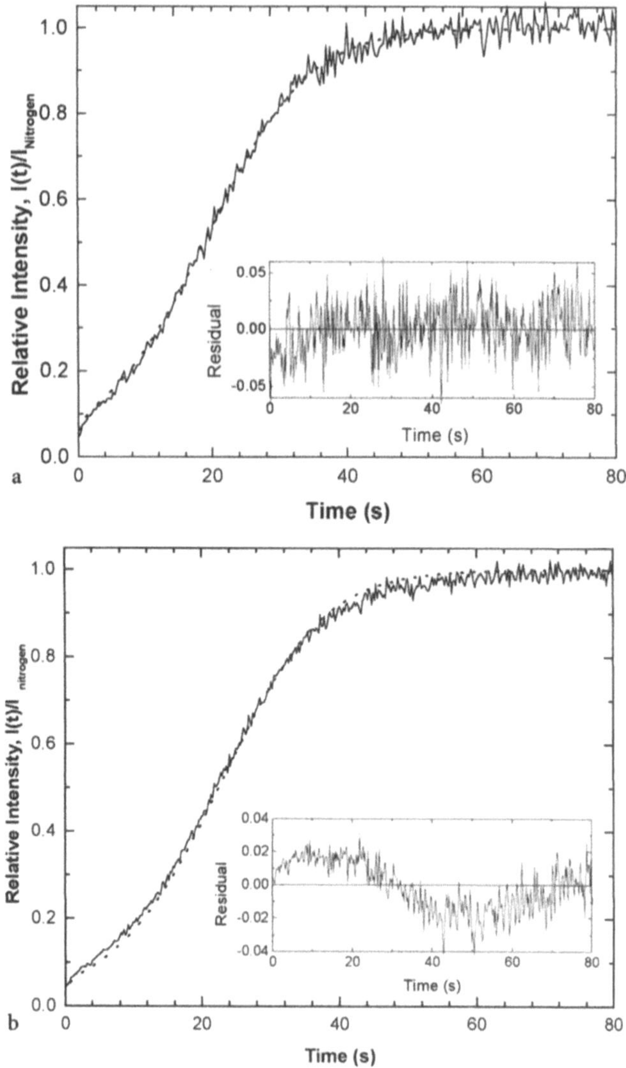

Fig. 12.13a, b. Time scan of PtOEP (100 ppm) emission intensity in a film of C_4PATP-PTHF(30) block copolymer ($L = 0.058$ mm) on glass for oxygen desorption experiments: **a** freshly annealed film, $D_{O_2} = 1.6 \pm 0.1 \times 10^{-6}\,\mathrm{cm^2\,s^{-1}}$; **b** film allowed to age for three days at 22 °C, $D_{O_2} = 1.0 \pm 0.1 \times 10^{-6}\,\mathrm{cm^2\,s^{-1}}$. The *solid curve* is for experimental data and the *dashed curve* is obtained from the theoretical fit. The *inset* is the plot of residuals ($I_{\mathrm{expt}} - I_{\mathrm{fit}}$)

Fig. 12.13a) strongly resembling those in Fig. 12.12. The simulated and experimental data overlap, and the D_{O_2} values from the sorption and desorption experiments agree to better than 5%. For three films, values of $B = 23.0 \pm 1.0$ and $D_{O_2} = 1.55 \pm 0.07 \times 10^{-6}\,\mathrm{cm^2\,s^{-1}}$ were obtained, from which we calculate a value of $S_{O_2} = 10.2 \pm 0.5 \times 10^{-4}\,\mathrm{M\,atm^{-1}}$. For C_4PATP-PTHF(30), the time-scan experiments were repeated after allowing the samples to age in the dark for three days at room temperature. The τ^o value of the PtOEP was essentially unchanged, whereas the value of B decreased to 17.4. The time-scan experiments (Fig. 12.13b) showed a small deviation between the experimental data and our model, which is likely related to the crystallization of the PTHF blocks. From repeated experiments, using simulated curves that gave the best fit to the midpoints of the sorption and desorption curves, we find $D_{O_2} = 1.01 \pm 0.03 \times 10^{-6}$ cm^2 s^{-1}. From the values of B in conjunction with measured values of τ^o, we calculated values of S_{O_2} and P_{O_2} for this polymer. All these values are collected in Table 12.5.

Time-scan experiments on this polymer containing 500 ppm [Ru(dpp)$_3$]Cl$_2$ are well behaved. For samples measured approximately 1h after annealing, $B = 1.05 \pm 0.03$ and similar values of D_{O_2} are obtained from the sorption and desorption experiments, $D_{O_2} = 2.6 \pm 0.7 \times 10^{-6}$ cm^{2}s^{-1}. This value is about 70% larger than the value determined for similar films containing PtOEP. From the B value, using $\tau^o = \langle \tau^o \rangle = 6.9$ µs, $S_{O_2} = 3.7 \pm 0.9 \times 10^{-4}$ M atm^{-1}. The non-exponential decay of the ruthenium dye, however, introduces uncertainty into the calculation of the oxygen solubility. Nevertheless, it appears that the [Ru(dpp)$_3$]Cl$_2$ dye senses an environment in C$_4$PATP-PTHF(30) with a lower oxygen solubility and somewhat higher oxygen diffusivity than PtOEP in this copolymer. After these films were allowed to age for three days, the time-scan experiments were repeated. Recall that these are now very turbid films. Deviations from a perfect fit of the data to Eqs (12.6) and (12.7) were found, but these were less pronounced than the deviations found for the PtOEP dye in this polymer (Fig. 12.13b). The B values remained identical ($B = 1.00 \pm 0.02$), and the best fit D values for the sorption and desorption experiments were very similar, with $D_{O_2} = 1.4 \pm 0.3 \times 10^{-6}$ cm^2 s^{-1}. Here, too, aging of the sample leads to a decrease in the measured oxygen diffusion coefficient.

12.5.4
MSPTP-PTHF

In comparison to the soft MSPTP films, MSPTP-PTHF films are hard and smooth. However, like C$_4$PATP-PTHF films, MSPTP-PTHF shows an aging effect. The freshly annealed samples are transparent, but become cloudy when they are allowed to stand in the dark at room temperature. Crystallization of PTHF block has a strong influence on the D_{O_2}, S_{O_2}, and P_{O_2} values in MSPTP-PTHF. For the freshly annealed films, an average of $D_{O_2} = 6.0 \times 10^{-6}$ cm^2 s^{-1} was found. During a three-day-aging period, D_{O_2} dropped by 20%, with the biggest decrease occurring during the first 24 h. Over the following 48 h, the change of D_{O_2} is very slow. D_{O_2} drops to 3.3×10^{-6} cm^{-2} s^{-1} after 72 h (see Fig. 12.14). A more pronounced change induced by sample aging can be observed in the time scan experiments themselves. Freshly annealed films give sorption and desorption intensity data that fit well to Eqs. (12.6) and (12.7), whereas in aged films, deviations between the model and the experiment become evident. During the oxygen-desorption experiments, the deviation is most pronounced in the latter stages when the increase of emission intensity is slow. It appears that the real rate of oxygen diffusion out of the film is slower than that predicted from the simulation, in which we assume that D_{O_2} is uniform throughout the polymer matrix.

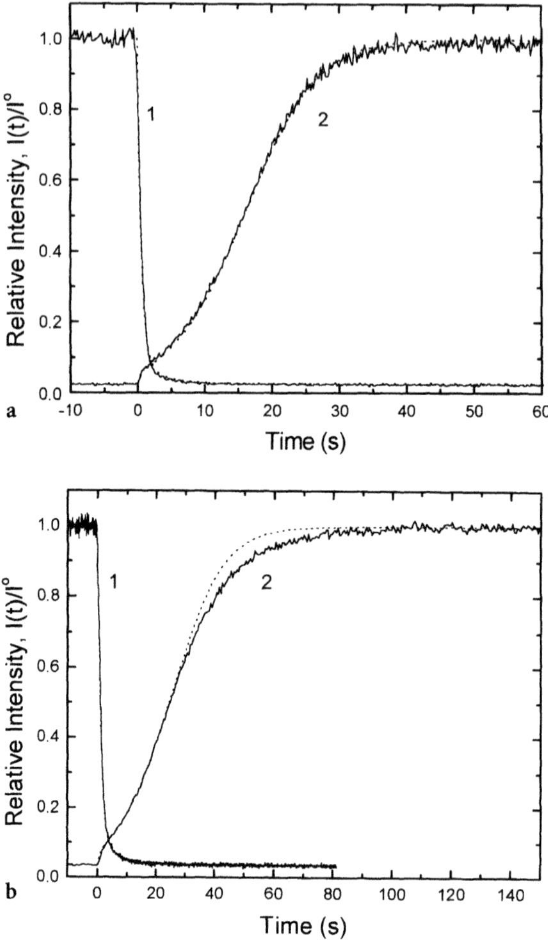

Fig. 12.14a, b. Time scan of PtOEP (100 ppm) emission intensity in a film of MSPTP-PTHF ($L = 0.076$ mm) on glass. From the fits, one obtains: **a** freshly annealed film, $D_{O_2} = 6.5 \times 10^{-6}$ cm^2 s^{-1} (O$_2$-desorption) and 6.0×10^{-6} cm^2 s^{-1} (O$_2$-sorption); **b** film allowed to age for 72 h, $D_{O_2} = 3.0 \times 10^{-6}$ cm^2 s^{-1} (O$_2$-desorption) and 3.8×10^{-6} cm^2 s^{-1} (O$_2$-sorption) Curve 1, an oxygen sorption experiment; curve 2, an oxygen desorption experiment. The *solid lines* are experimental data and the *dashed lines* are simulated

We do find concordance in the time-scan and lifetime analyses of oxygen quenching of PtOEP in MSPTP-PTHF. The unquenched decay of PtOEP in MSPTP-PTHF is mono-exponential with $\tau^\circ = 62.5$ µs. When the oxygen is introduced into the sample chamber, the decays become non-exponential, but the plot of $\tau^\circ/\langle\tau\rangle$ vs p_{O_2} for a film aged three days is linear (Fig. 12.10). From the slope of this Stern-Volmer plot, we find a value of $P_{O_2} = 2.52 \times 10^{-12}$ mol cm^{-1} s^{-1} atm^{-1} and $S_{O_2} = 7.6 \times 10^{-4}$ M atm^{-1}. These values are close to those calculated from the time scan experiments ($P_{O_2} = 2.30 \pm 0.3 \times 10^{-12}$ mol cm^{-1} s^{-1} atm^{-1} and $S_{O_2} = 6.9 \pm 0.7 \times 10^{-4}$ M atm^{-1}) (see Table 12.5). In addition, Fig. 12.10 shows that, in comparison with PDMS, C_4PATP, and MSPTP, MSPTP-PTHF has the smallest Stern-Volmer slope of the polymers depicted, indicating a smaller product of P_{O_2} and τ^0.

12.6
Summary

In luminescence barometry, one monitors air pressure profiles across the surface of an object through measurements of the intensity or decay time at different points on the surface. The response of the coating to various changes in air pressure depends upon the particular combination of dye and polymer in the coating. Important parameters, which determine the sensitivity and dynamic range of the system, include the lifetime of the dye and the oxygen permeability of the polymer matrix. The response time of the system depends upon the thickness of the coating and the diffusion coefficient of oxygen in the matrix. The proper design of an optimal pressure sensitive paint requires that one be able to determine these parameters for a wide variety of dyes and polymers.

In this review, we show that time-scan oxygen quenching experiments on thin polymer film samples is an effective methodology for determining D_{O_2} values. One compares measured intensity decays in oxygen sorption, and intensity growth profiles in desorption experiments, with simulated curves calculated from an appropriate model. For homogenous polymer films, this model couples Fickian diffusion with Stern-Volmer quenching kinetics, and takes full account of the concentration profile that develops as a consequence of quencher diffusion. Experiments with PtOEP in PDMS and in a series of C_nPATP homopolymers give precise values of D_{O_2}, P_{O_2}, and S_{O_2}. Similar values of oxygen permeability are obtained from time-scan experiments and lifetime Stern-Volmer plots. In some experiments, we compare PtOEP and Ru(dpp)$_3$Cl$_2$ as dyes. Time scan experiments give identical values of D_{O_2} in C_4PATP, but different values of P_{O_2}. The differences in the calculated permeability can be resolved if one assumes that the effective quenching radius ($\alpha\sigma$) for $O_2 + $Ru(dpp)$_3Cl_2$ is half of that for $O_2 + $PtOEP.

For PDMS and the series of C_nPATP homopolymers, there is a strong decrease of D_{O_2} with an increase in the T_g of the polymer matrix. For this series of polymers D_{O_2} and P_{O_2} vary by nearly a factor of 10, whereas S_{O_2} is almost constant. This result is consistent with an increase in free volume in the film favoring oxygen diffusion. In contrast to the experiments on linear homopolymers are the results on several commercial silicone resins that require cure to form tough films. These resins exhibit a large variation in oxygen permeability and diffusivity with variation in the largely unknown structure of the resin. More disturbing is the fact that these properties change as the resin cures.

Block copolymers and semi-crystalline polymers segregate spontaneously into microdomains. The types of experiments described above become more difficult to interpret when the dye can partition between different domains and oxygen transport in the domains is different. In freshly annealed samples such as PTHF, MSPTP, MSPTP-PTHF, and C_4PATP-PTHF, we observe excellent agreement between the experimental time-scan and theoretical curves calculated assuming a single diffusion coefficient in the system. In systems containing PTHF homopolymer or PTHF blocks, this simple model becomes less effective at describing the system as aging leads to increasing crystallization of the PTHF domains.

Acknowledgement. The authors thank NSERC Canada and Materials and Manufacturing Ontario (MMO) for their support of this work.

References

1. (a) Gouterman M (1997) Oxygen quenching of luminescence of pressure sensitive paints for wind tunnel research. J Chem Edu 74 (6): 697–702. (b) Puklin E, Carlson B, Gouin S, Green E, Ponomarev S, Tanji H, Gouterman M (2000) Ideality of pressure-sensitive paint. I. Platinum tetra(pentafluorophenyl)porphine in fluoroacrylic polymer. J Appl Polym Sci 77 (13): 2795–2804. (c) Gouin S, Gouterman M (2000) Ideality of pressure-sensitive paint. II. Effect of annealing on the temperature dependence of the luminescence. J Appl Polym Sci 77 (13): 2805–2814. (d) Gouin S, Gouterman M (2000) Ideality of pressure-sensitive paint. III. Effect of the base-coat permeability on the luminescence behavior of the sensing layer. J Appl Polym Sci 77 (13): 2815–2823. (e) Gouin S, Gouterman M (2000) Ideality of pressure-sensitive paint. IV. Improvement of luminescence behavior by addition of pigment. J Appl Polym Sci 77 (13): 2824–2831. (f) Bedlek-Anslow JM, Hubner JP, Carroll BF, Schanze KS (2000) Micro-heterogeneous oxygen response in luminescence sensor films. Langmuir 16 (24): 9137–9141
2. Wen WY (1993) Motion of sorbed gases in polymers. Chem Soc Rev 22: (2) 117–126
3. (a) Guillet JE (1986) Mass diffusion in solid polymers. In: Winnik WA (eds) Photophysical and photochemical tools in polymer science conformation, dynamics, morphology. D. Reidel, Holland, pp 467–494; (b) Guillet JE (1987) Polymer photophysics and photochemistry. Cambridge University Press, New York, pp 52–70
4. Wang B, Ogilby PR (1995) Activation barriers for oxygen diffusion in polystyrene and polycarbonate glasses. Can J Chem 73 (11): 1831–1840
5. Cox ME, Dunn B (1986) Oxygen diffusion in poly(dimethylsiloxane) using fluorescence quenching. J Polym Sci Part A: Polym Chem 24: 621–636
6. Yekta A, Masoumi Z, Winnik MA (1995) Luminescence measurements of oxygen permeation and oxygen diffusion in thin polymer films. Can J Chem 73: 2021–2029
7. Mills A, Chang Q (1992) Modeled diffusion-controlled response and recovery behavior of a naked optical film sensor with a hyperbolic-type response to analyte concentration. Analyst 117: 1461–1466
8. Lakowicz JR (1983) Principles of fluorescence spectroscopy. Plenum Press, New York
9. Rice SA (1985) Diffusion limited reactions. In: Bamford CH, Tipper CFH, Compton RG (eds) Comprehensive chemical kinetics, vol 25. Elsevier Science Publishers, New York
10. Martinbo JMG, Winnik MA (1987) Transient effect in pyrene monomer-excimer kinetics. J Phys Chem 91: 3640–3644
11. Jones PF (1968) On the use of phosphorescence quenching for determining permeabilities of polymeric films to gases. J Polym Sci Part B: Polym Lett 6: 487–491
12. Nowakowska PF, Najbar J, Waligora B (1976) Fluorescence quenching of polystyrene by oxygen. Eur Polym 12: 387–391
13. Guillet JE, Andrews M (1992) Studies of oxygen diffusion in poly(styrene-*co*-naphthylmethacrylate) by phosphorescence quenching. Macromolecules 25: 2752–2756
14. Collins FC, Kimball GE (1949) Diffusion-controlled reaction rates J Colloid Sci 4: 425–437
15. Masoumi Z, Stoeva V, Yekta A, Winnik MA, Manners I (1997) Studies of oxygen diffusion in polysiloxane resins with application to luminescence barometry. In: Shi L, Zhu D (eds) Polymers and organic solids. Science Press, Beijing, China, pp 157–168

16. Liang M, Manners I (1991) Poly(thionylphosphazenes): a new class of inorganic polymers with backbones of phosphorus, nitrogen, and sulfur(VI) atoms. J Am Chem Soc 113: 4044–4045
17. Ni Y, Stammer A, Liang M, Massey J, Vancso GJ, Manners I (1992) Synthesis, thermal transition behavior, and solution characterization of poly[(aryloxy)thionyl-phosphazenes] with halogen substituents at sulfur. Macromolecules 25:7119–7125
18. Ni Y, Park P, Liang M, Massey J, Waddling C, Manners I (1996) Polymers with sulfur(VI)-nitrogen-phosphorus backbones: synthesis, characterization, and properties of atactic poly[(amino)thionylphosphazenes]. Macromolecules 29:3401–3408
19. Masoumi Z, Stoeva V, Yekta A, Pang Z, Manners I, Winnik MA (1996) Luminescence quenching method for probing the diffusivity of molecular oxygen in highly permeable media. Chem Phys Lett 261:551–557
20. Pang Z, Gu X, Yekta A, Masoumi Z, Coll JB, Winnik MA, Manners I (1996) Phosphorescent oxygen sensors utilizing sulfur-nitrogen-phosphorus polymer matrices. Adv Mater 8 (9):768–771
21. (a) Jayarajah CN, Yekta A, Manners I, Winnik MA (2000) Oxygen diffusion and permeability in alkylaminothionylphosphazene films intended for phosphorescence barometry applications. Macromolecules 33 (15): 5693–5701. (b) Jayarajah CN (1998) Luminescence quenching studies of oxygen diffusion in highly permeable media. Msc thesis, University of Toronto
22. Ruffolo R, Evans C, Liu XH, Ni Y, Pang Z, Park P, McWilliams A, Gu X, Lu X, Yekta A, Winnik MA, Manners I (2000) Phosphorescent oxygen sensors utilizing sulfur-nitrogen-phosphorus polymer matrices: synthesis, characterization, and evaluation of poly(thionyl-phosphazene)-*b*-poly(tetrahydrofuran) block copolymers. Anal Chem 2:1894–1904

Dual Lifetime Referencing (DLR) – a New Scheme for Converting Fluorescence Intensity into a Frequency-Domain or Time-Domain Information

I. KLIMANT, C. HUBER, G. LIEBSCH, G. NEURAUTER,
A. STANGELMAYER, O. S. WOLFBEIS

Fluorescence spectroscopy and NMR spectroscopy are probably the most powerful spectroscopies at present albeit with very different (and highly complementary) fields of application. Fluorometry can be based on the intrinsic fluorescence of (bio)molecules or ions, or on the use of fluorescent probes, indicators, or labels. Numerous parameters can be measured which include intensity, decay time, polarization, radiative and non-radiative energy transfer, quenching efficiency, and combinations thereof. Fluorescence microscopy and imaging are other widely applied techniques, and multi-dimensional and synchronous fluorescence spectroscopy have gained some interest in recent years.

The determination of fluorescence intensity is still widely used in quantitative fluorometry. This, for one, is due to the linear relationship that exists between the concentration of a fluorophore and its fluorescence intensity, and also because of the availability of respective instrumentation. On the other hand, precise measurement of fluorescence intensity is compromised by adverse effects such as drifts of the opto-electronic system, variations in the optical properties of the sample including fluorophore concentration, turbidity, coloration and refractive index and photobleaching of the fluorophore.

We present a new and universally applicable scheme for converting fluorescence intensity into a phase shift, a signal that is hardly affected by the interferents discussed above. It is based on the addition of a luminescent reference dye having a decay time much longer than that of the fluorescent indicator. This scheme is called *Dual Lifetime Referencing* (DLR). When using a phosphorescent luminophore as the reference dye, the time domain is in the microsecond range, so that modulation frequencies in the lower kHz range are adequate. This enables the use of inexpensive optoelectronic devices and thus also provides cost advantages.

In this overview, the fundamentals of time-domain DLR and frequency-domain DLR are demonstrated and the potential of the new method is assessed with respect to alternative schemes. Furthermore, strategies are presented to design phosphorescent reference beads containing completely inert (non-quenchable) luminescent indicators since they are essential for an optimal performance of DLR. Examples are given for the application of DLR in optical chemosensing, and in 2-D fluorescence imaging to demonstrate its wide-spread applicability.

13.1
Introduction

Quantitative fluorometry is frequently performed in chemical and biochemical analysis [1–3]. Its extraordinarily high sensitivity (allowing the detection of single fluorescent molecules) and its versatility makes it the method of choice in immunoassay, in DNA analysis, and in fluorescent chemo- and biosensing. Fluorescence parameters useful for quantitative analysis include spectral changes, fluorescence anisotropy, decay time, and intensity. Anisotropy and decay time are highly attractive parameters since they are intrinsically referenced and hardly affected by fluctuations of the overall fluorescence intensity [3, 4].

Despite the advantages of intrinsically referenced parameters (i.e., decay time and anisotropy), fluorescence intensity is more commonly measured. This is due to an often linear correlation between the concentration of a fluorophore in a sample and the detectable intensity. This linearity, of course, deviates at high fluorophore concentration due to inner filter effect. Hence, fluorescence intensity is the most widely used parameter in bioanalytical assays and fluorescent sensing applications.

Notwithstanding this, fluorescence intensity is a crucial parameter which is strongly affected by numerous factors. These include drifts of the optoelectronic system, variable sample turbidity and color, and the effect of external quenchers. Such adverse effects can partially be overcome by making use of ratiometric measurement, i.e., by ratioing the intensities at two wavelengths. This approach is widely used, for example in calcium assays using fluorophores displaying two excitation bands or two emission bands [5]. Alternatively, an inert fluorophore may be added with spectral properties different from those of the indicator. Again, ratioing the intensities at two excitation or emission wavelengths results in a referenced parameter. The disadvantages of this method include the need for two separate optical channels, thus complicating the optical setup. For example, the drift in the sensitivity of both channels can be different, as can be the intensities at two excitation wavelengths. Light scatter and signal loss caused by fiber bending (e.g., in fiber optic sensors or certain microtiter plate readers) further contribute to effects not compensated for by 2-wavelength referencing.

We present an approach which is based on the use of two dyes (as in intensity-ratiometric schemes). However, we convert the ratio of the intensities into a phase shift that depends on the differences in the weighed decay times of the two fluorophores, namely that of the fluorescent probe (indicator) and that of an added reference dye, respectively [6–8]. The ratio can be determined in either the time domain or the frequency domain. Preferably, reference dyes are used that decay in the microsecond or millisecond time domain and so simplify the opto-electronic system.

In this overview we demonstrate the versatility of the new fluorometric scheme by giving specific examples on (a) fluorescence assays, (b) fluorescent sensing, and (c) fluorescence imaging.

13.2
Theoretical Background

13.2.1
Frequency Domain DLR Spectroscopy

Frequency domain spectroscopy is a well established technique to measure luminescence decay times. In case of single exponential decay, the phase angle of the luminescent signal modulated at a single frequency reflects the decay time of the luminophore. In the case of samples containing two luminophores of similar spectral properties, an overall luminescence signal will be measured. As a result, the phase angle obtained at a single frequency is determined by the ratio of the intensities of the two single signals and their respective decay times. Assuming a single exponential decay for both, the analytical signal of a 2-component system can be seen as the superposition of two single sine wave functions as shown in Fig. 13.1.

In a sample containing a reference luminophore of constant decay time and luminescence intensity, and in addition containing a fluorescence indicator of variable intensity, the following phase-dependent signals will be obtained:

$$A_m \cdot \cos \Phi_m = A_{ref} \cdot \cos \Phi_{ref} + A_{flu} \cdot \cos \Phi_{flu} \tag{13.1}$$

$$A_m \cdot \sin \Phi_m = A_{ref} \cdot \sin \Phi_{ref} + A_{flu} \cdot \sin \Phi_{flu} \tag{13.2}$$

where A_m represents the overall signal intensity and Φ_m the measurable phase shift. A_{flu} and A_{ref} are the amplitudes of the fluorescence indicator and the reference standard, respectively, and Φ_{flu} and Φ_{ref} are the respective phase shifts. In case of phosphorescent reference dyes, the decay time is longer by orders of magnitude compared to that of the fluorescent probe. At low modulation frequencies, e. g., in the kHz range, the fast fluorescence component causes no phase shift ($\Phi_{flu} = 0$) and therefore Eqs. (13.3) and (13.4) can be derived from Eqs. (13.1) and (13.2):

$$A_m \cdot \cos \Phi_m = A_{ref} \cdot \cos \Phi_{ref} + A_{flu} \tag{13.3}$$

$$A_m \cdot \sin \Phi_m = A_{ref} \cdot \sin \Phi_{ref} \tag{13.4}$$

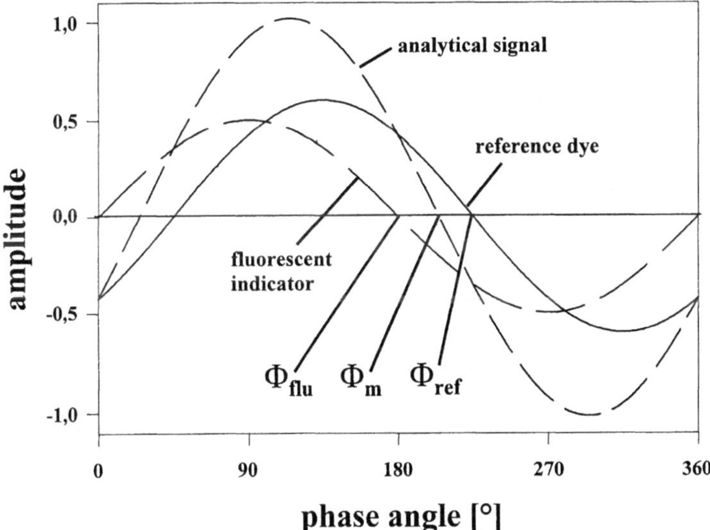

Fig. 13.1. Concept of frequency-domain dual-lifetime referencing (DLR). The overall in intensity ("analytical") signal is the total of the prompt fluorescence of the indicator (which is affected by the analyte) and the delayed phosphorescence of the reference dye. The phase angle Φ_m reflects the ratio of the amplitudes of the two components

Division of Eq. (13.3) by Eq. (13.4) results in a correlation of the phase signal (Φ_m) and the ratio of the intensities of the indicator/reference couple (Eq. 13.5):

$$\frac{A_m \cdot \cos \Phi_m}{A_m \cdot \sin \Phi_m} = \cot \Phi_m = \cot \Phi_{ref} + \frac{1}{\sin \Phi_{ref}} \cdot \frac{A_{flu}}{A_{ref}} \tag{13.5}$$

Thus, $\cot \Phi_m$ reflects the referenced intensity of the fluorescence indicator. In the case that Φ_{ref} is constant and known, there is a linear relationship between $\cot \Phi_m$ and A_{flu}/A_{ref}. It is obvious that $\cot \Phi_m$ does not relate to the overall signal intensity. This method is referred to as Dual Lifetime Referencing (DLR). If performed with frequency modulated light, we specifically refer to it as frequency-domain DLR.

13.2.2
Time-Domain DLR Spectroscopy

Fluorescence intensity may as well be referenced via DLR by adding a phosphorescent reference standard to the system and measuring in the *time domain*. This method is referred to as time-domain DLR sensing. In this case, luminescence is periodically excited by squared pulses of light. The resulting overall luminescence is measured in two time windows: in window 1 ("LED on") luminescence is measured during excitation, while in window 2 ("LED off") it is measured after excitation has been turned off (Fig. 13.2). The bandwidth of both windows is in the same order of magnitude as the decay time of the reference standard.

It is obvious that the signal of window 1 is the total luminescence. It arises from (a) the fluorescence indicator (squared) and (b) the reference standard (shark fin). In window 2 the phosphorescence of the reference standard is detected only. This clear separation of the two single components is due to the

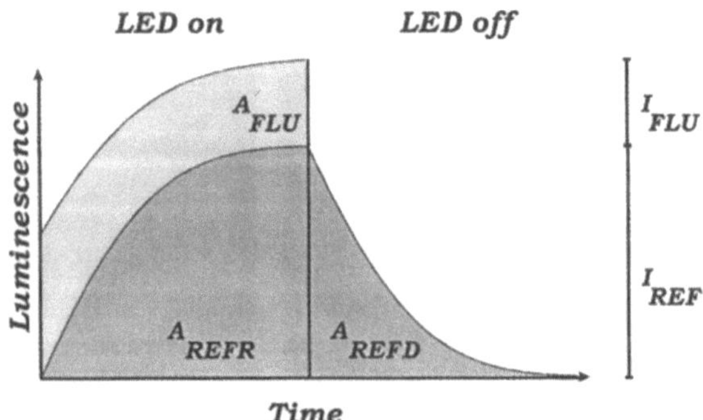

Fig. 13.2. Scheme of time-resolved DLR sensing. Fluorophore and luminophore are excited by squared pulses of light from an LED. During excitation ("LED on") the total signal is composed of the fluorescence indicator and the reference dye, whereas the slowly decaying phosphorescence is detected only when the LED is switched off

μs decay time of the reference standard. If the reference material has a mono-exponential decay, the following relationship (Eq. 13.6) applies:

$$\frac{A_{\text{REFR}} + A_{\text{FLU}}}{A_{\text{REFD}}} = R = (k_1 - 1) + k_2 \frac{I_{\text{FLU}}}{I_{\text{REF}}} \tag{13.6}$$

Here A_{REFR} and A_{FLU} are the signal intensities arising from the reference dye and the fluorescent indicator in the excitation window respectively. A_{REFD} is the signal intensity in the emission window arising from the reference dye with its long decay. k_1 and k_2 are constants. Thus, there is a linear correlation between R (the ratio between the excitation and the emission window) and the ratio of the fluorescence signal (I_{FLU}) and the reference signal (I_{REF}).

In the presence of ambient light it is, however, necessary to measure the signal intensity at a third window (where the light source is switched off) so to suppress any interfering effects.

Frequency-domain DLR and time-domain DLR are identical in terms of performance and both are capable of eliminating the same interferents. For practical applications frequency-domain DLR is the preferred technique due to the simple opto-electronic setup as described more detailed in the Instrumental Section. For fluorescence imaging applications, time-resolved techniques are more established. Nowadays, directly gateable CCD cameras allows the precise measurement of phosphorescence lifetimes without image intensifiers [9].

13.3
Phosphorescent Standards

The application of DLR in luminescence analysis requires the availability of phosphorescent reference standards. However, only a few classes of luminophores are useful. Metal ligand complexes of ruthenium, osmium, rhenium, or iridium containing polypyridyl ligands display promising properties. Their quantum yields of up to 50 % and decay times in the lower μs range make them very attractive. In addition, their photophysical and solubility properties can be tailored by proper selection of ligands and counter ions (in case of charged complexes). Other useful luminophores for use as reference standards include chelates of lanthanide ions, in particular Eu^{3+} and Tb^{3+}, as well as certain Pt^{2+} and Pd^{2+} porphyrins. Even solid state phosphorescent materials may be used.

An ideal reference standard features the following properties:

1. Luminescence quantum yield, decay time and spectrum are not affected by the analyte and any other substance in the system.
2. Its absorption matches that of the fluorescent indicator (Fig. 13.3), or at least there is strong overlap.
3. It can be mixed uniformly into the system and does not aggregate or sediment.

Ruthenium (polypyridyl) complexes are widely used, well-characterized, and easily accessible luminophores with decay times in the μs region. Unfortunately, their excited state can be quenched dynamically or statically by either oxidants

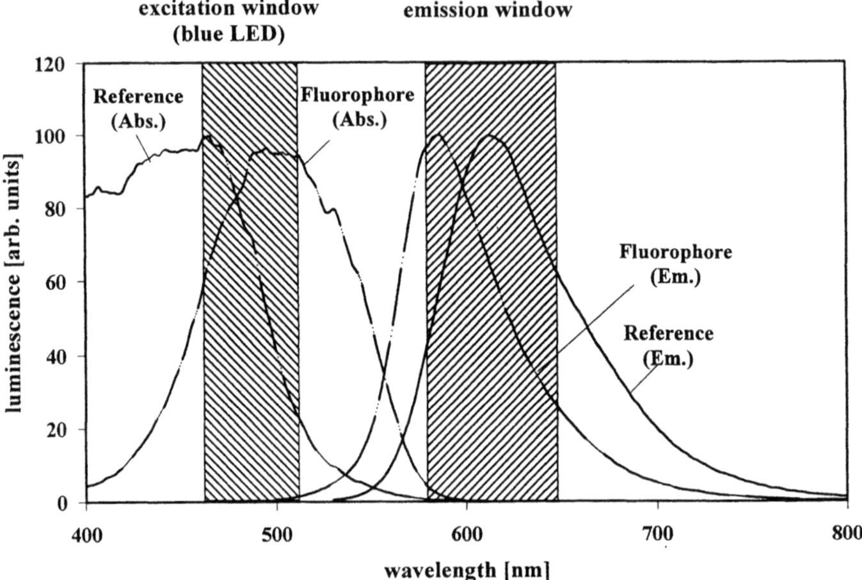

Fig. 13.3. Spectral properties of a fluorophore/reference couple suitable for DLR. Both luminophores can be excited by the same LED

or reductants. Oxygen is another notorious quencher of luminescence, and its partial pressure in a sample is often variable and unknown. To overcome the adverse effects of external quenchers, the phosphorescent dyes are preferably added in the form of a "solution" in a polymer, thus rendering it less accessible. Since dynamic quenching by oxygen is most critical, the matrix for encapsulation is preferably impermeable to gases. Promising materials include materials prepared by the sol-gel process recently presented in context with temperature-sensitive coatings [10]. Poly(acrylonitrile) (PAN) is another material of extraordinarily low oxygen permeability. Furthermore, PAN has a low water uptake. Metal ligand complexes such as the ruthenium(II)-tris-polypyridyl complexes are well soluble in PAN. The phosphorescence of thin films of such complexes dissolved in the PAN matrix are not at all quenched by oxygen and other species.

Another problem associated with the design of reference standards is the lack of phosphorescent dyes with spectral properties similar to those of commonly used fluorescent indicators. This is due to the fact that fluorescence usually displays smaller Stokes' shifts (with few exceptions) than phosphorescence.

The applicability of DLR to fluorometric assays depends on the availability of reference standards of adequate spectral properties. We find the ruthenium(II)-tris-4,7-diphenyl-1,10-phenanthroline complex [Ru(dpp)] dissolved in PAN to be useful as a reference dye to convert fluorescence intensities of either fluoresceins (which are excitable by blue light) or certain rhodamines (which are excitable by blue-green light of 490–520 nm. In addition, we find certain platinum(II)-porphyrins and certain osmium complexes (which absorb between 550 nm and 630 nm) to be suitable as reference dyes for use in combination with long-wavelength absorbing indicators.

13.4
Instrumentation

DLR, in contrast to 2-wavelength measurements, makes use of modulated light sources and gateable photodetectors. In case of using ruthenium(II)-poly-pyridyl complexes as reference standards, their spectra match those of several established decay time-based chemical sensors [11–19]. Thus, it is possible to convert such sensing systems into DLR type sensors. Bright blue or blue green LEDs are the light source of choice since they are inexpensive, small, and require little power. Modulation of these light sources in the upper kHz range (sine wave or square pulses) is state of the art and can be performed with commercially available oscillators. Electronic cross-talk is almost excluded at such low modulation frequencies. In case of sufficiently strong signals (> 100 pW), a PIN photodiode can be used as the photodetector. The phase shift of the phosphorescent signal is measured with a dual phase lock-in amplifier with a phase resolution of around 0.01°. If using phosphorescent standards with lifetimes in the milli-

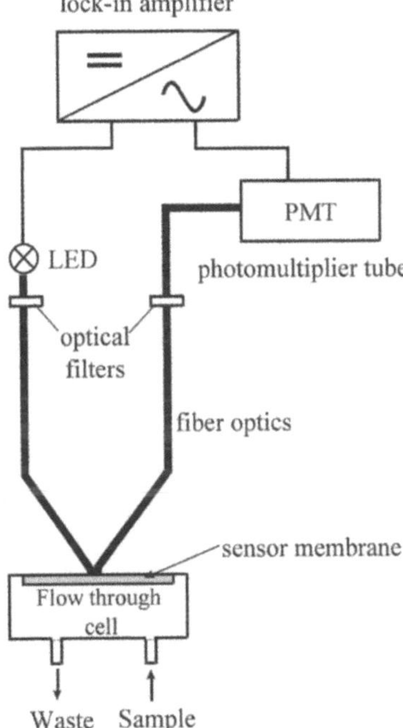

Fig. 13.4. Fiber optic setup for DLR sensing with microsensors consisting of light-emitting diodes (LED), photomultiplier tube (PMT), optical filters, and a fiber optic tip placed to a flow-through cell containing the sensor membrane. In case of sufficiently high signal intensities, the photomultiplier tube (PMT) can be replaced by a PIN photodiode. High resolution phase detection is performed with a lock-in amplifier board

second range, xenon flash lamps or electroluminescent lamps are adequate for use as modulated light sources. It is evident that the costs for an instrument designed for DLR luminescence sensing can be in the same order of magnitude as for an intensity-based instrument.

Two systems which are in use in our laboratory for DLR sensing are shown schematically in Figs. 13.4 and 13.5. Figure 13.4 gives a typical fiber-optic setup. Originally designed for decay time-based sensing using microsensors with a typical tip diameter of less than 30 µm [16, 17], it has a phase modulation-based system where the phase angle is measured using a dual phase lock-in amplifier. A phase resolution of 0.01° is achieved which corresponds to a change in intensity of 0.02% at the optimal modulation frequency which is 45 KHz in case of Ru(dpp) dissolved in PAN particles.

Figure 13.5 shows a system designed for time-resolved phosphorescence lifetime imaging of sensor layers placed on the bottom of wells of a microtiter plate. Here, a pulsed LED array is used as the light source and luminescence is recorded with a time-gated CCD camera. Such a setup is useful for 2-D sensing of the distribution of chemical parameters in complex samples, or in combination with quantitative fluorescence microscopy.

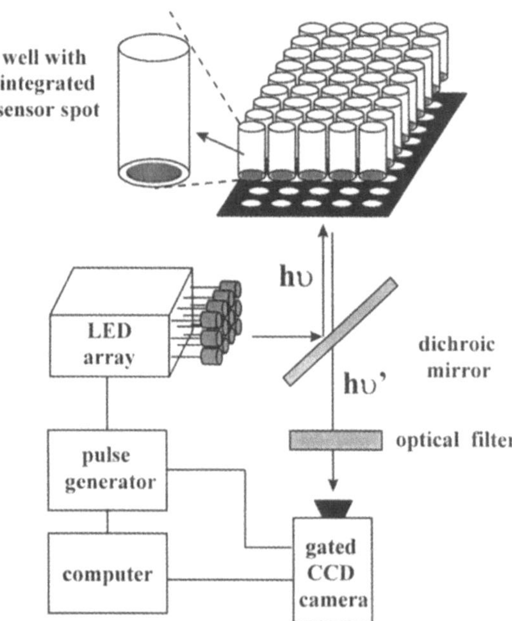

Fig. 13.5. Setup for time-resolved DLR imaging. A fast and directly gateable camera is synchronized with a pulsed LED array. The system is used to readout the signals arising from microtiter plates with integrated DLR chemical sensors for either pH, oxygen, or CO_2

13.5
DLR Applications

13.5.1
Homogeneous Assays

Any assay based on quantitative measurement of fluorescence intensity can be converted into a more precise DLR assay by addition of a reference standard to the sample and converting intensity changes into phase shifts. Provided that the ratio between fluorescent indicator and reference standard is known and constant, a calibration-free quantification of the fluorescence signal becomes possible by DLR. It is essential, however, to warrant a uniform distribution of indicator and reference standard in the sample. This is preferably achieved by adding the phosphorescent dyes in the form of PAN nanoparticles. These behave similar to dissolved dyes and do not sediment. The surface of such particles may even be chemically activated so as to enable coupling of fluorescent indicators, antibodies, or DNA fragments.

In a typical example of a DLR fluorescence pH-assay, 1-hydroxypyrene-3,6,8-trisulfonate (HPTS) is used as the pH indicator and the Ru(dpp) complex dissolved in PAN nanoparticles is added as a reference dye. The nanoparticles have an average diameter of 100 nm and are suspended in the sample. Figure 13.6 shows the differences in the plots of phase angle and intensity vs pH.

Figure 13.7 demonstrates the performance of a DLR calibration. A sample of defined pH, and the same indicator/reference couple, was used in this experiment. While keeping the pH constant, the optical properties of the sample (scattering, coloration, dilution) were varied. As expected, the signal intensity

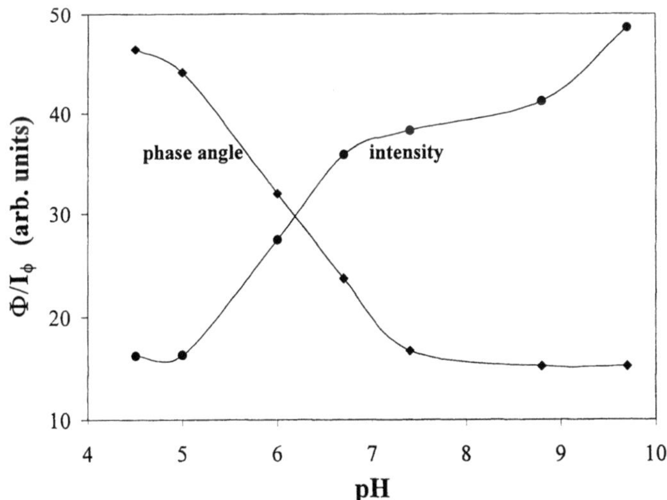

Fig. 13.6. pH calibration curves arising from intensity and phase shift data, respectively

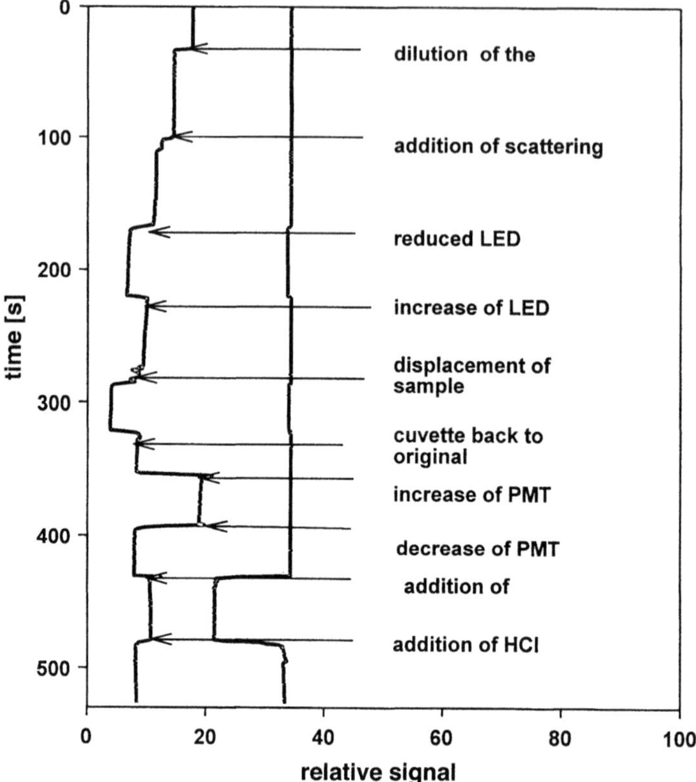

Fig. 13.7. Stability of the intensity (*left track*) and phase shift (*right track*) of a DLR-based pH measurement. Typical parameters that disturb fluorescence intensity are clearly canceled out in the DLR-based measurement

varies while the phase angle remains constant. Subsequently, the intensity of the light source, the voltage of the photomultiplier tube, and the position of the cuvette were varied. A stable phase shift was observed again, while intensity was strongly affected by the interferents.

This experiment demonstrates that most adverse effects are referenced out by DLR except for intrinsic background fluorescence. This was to be expected since the decay profile of background fluorescence cannot be distinguished from that of the fluorescent indicator.

Deviations in the linearity of intensity-based calibration plots may occur as a result of the so-called inner filter effect (IFE). DLR can even eliminate the IFE as shown in Fig. 13.8. On the other hand, any photodecomposition of the fluorescent indicator cannot be compensated for because it changes the ratio of fluorophore to reference. This has to be kept in mind when using DLR in combination with fluorescence microscopy, where high illumination intensities are common and photobleaching is frequently encountered.

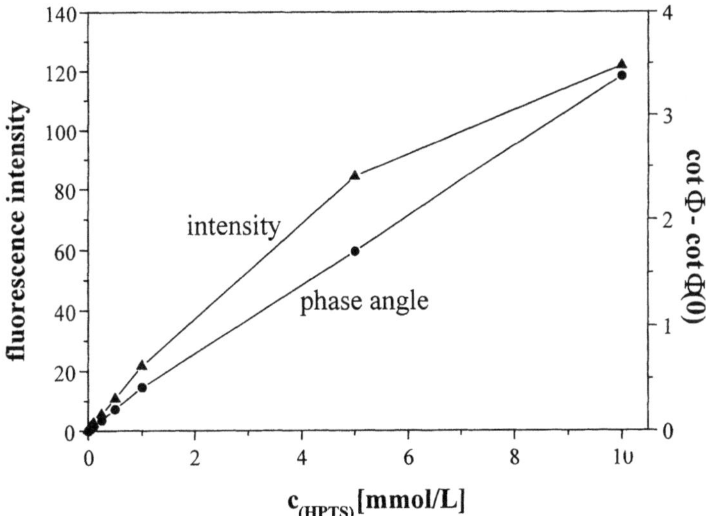

Fig. 13.8. Fluorescence intensity calibration plot and DLR calibration plot for the pH indicator dye HPTS. Non-linearities caused by the inner filter effect are seen in the intensity plot but are canceled out in the phase angle plot

The largest potential of DLR is in fluorescent assays of "real" samples containing scattering centers. Typical examples include cell suspensions, serum, urine, and environmental samples. Furthermore, intensity based measurements can be improved in systems of poorly defined geometry such as in flasks or microreactors for cell cultivation. Another application is in the quantification of algae bloom via the fluorescence of chlorophyll A in dirty or turbid samples.

13.5.2
DLR Based Optical Sensors

An attractive field of application of DLR is in optical sensor technology. Sensors are frequently applied to samples whose optical parameters vary considerably. In addition, a sensor may drift over time, as may the opto-electronic system. In contrast to instruments like flow cytometers or microtiter plate readers, sensors often cannot be recalibrated.

DLR provides a simple method for stabilizing the signal of many fluorescence intensity-based optical sensors. In fact, it can easily be adapted to many sensors described in the literature. To do so, the sensor material needs to be provided with a phosphorescent reference standard. Figure 13.9 shows typical designs for planar DLR sensor membranes.

The simplest approach exploits a multilayer design (1,3). Here, the reference standard forms a transparent ground layer which is covered by the second, i.e., chemically sensitive coating. The multilayer approach can be adapted to almost any existing fluorescence-based sensor by depositing the sensitive layer onto the

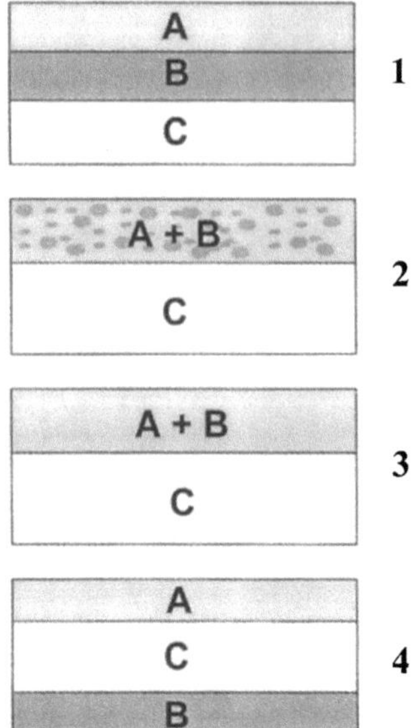

Fig. 13.9. Representative cross-sections of typical DLR sensors. (*A*) chemically responsive material; (*B*) reference material composed of polymer and the inert luminophore; (*C*) inert polymeric support

reference layer. Design 2 is most favorable in terms of production. Here, the reference standard is finely dispersed into the sensitive layer in the form of inert particles. Such materials are useful for single-layer optical sensors, and the ratio of reference dye and fluorophore can easily be adjusted. In addition, this design allows for the elimination of almost any disturbing effects because of the uniform distribution of both dyes. It thus paves the way to calibration-free optical sensors.

13.5.2.1
Optical Chloride Sensor Based on DLR

This sensor is based on the dynamic quenching of the fluorescence of lucigenin by halide ions [20]. Lucigenin is a cationic indicator whose fluorescence can be excited by the blue LED. It was electrostatically immobilized on Nafion membranes. Poly(acrylonitrile)-based nanoparticles containing Ru(dpp) were used as reference standard. The excellent match in the absorption spectra along with the highly different emission spectra makes this combination of dyes most useful. The two emissions were detected with a single photodetector by selection of a longpass filter with a cut-off wavelength of 530 nm.

Lucigenin on Nafion responds fully reversibly to chloride ions. From Fig. 13.10 it is evident that the sensitivity of the calibration plot can be tuned by selection of emission filters of different cut-off wavelengths. This is typical for an indicator couple showing little overlap in the emission spectra. Not-

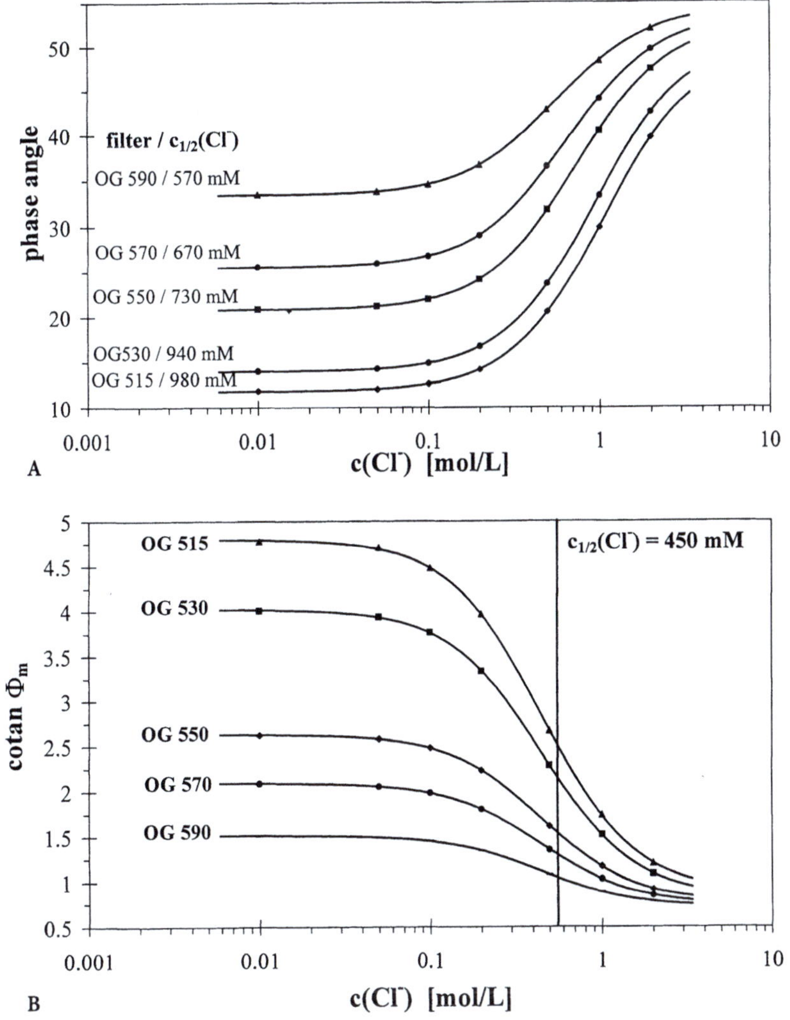

Fig. 13.10. A Phase calibration plots of the chloride sensor when using different long-pass filters. **B** Plot of cotan Φ_m vs the chloride concentration using different long-pass filters

withstanding the use of different filters, the plots of the cotangents vs the respective chloride concentration result in calibration curves of identical slope. The chloride sensor presented here has the highest sensitivity at 440 mmol/l which is the optimal concentration range to measure seawater salinity [21]. If lucigenin and Ru(dpp)/PAN beads are immobilized in a hydrogel matrix, the dynamic range is shifted to physiological concentrations [22].

13.5.2.2
Fiber Optic pCO$_2$ Microsensor Based on DLR

This sensor is of the fiber-optic type, the diameter of the tip being < 30 μm. In optical microsensing it is difficult to control the geometry of the sensing tip. In particular, microbending of the tip can result in significant losses of signal intensity. Therefore, an effective referencing scheme is essential. Two types of DLR-based pCO$_2$ microsensor are presented here, one based on a double layer design, the other on a single layer design (Fig. 13.11). The double layer sensor has the advantage of being insensitive to quenching by oxygen. This makes it – in principle – the sensor of choice for determination of pCO$_2$ gradients in samples whose oxygenation varies a lot, e.g., in marine samples. It is, however, difficult to coat reproducibly the sensor tip and this results in non-identical calibration curves for each individual sensor.

Again, the best sensor is of the monolayer type, with finely dispersed particles of the reference standard contained in the pCO$_2$ sensitive layer. In the case of microsensors, the reference standard has to be present in the form of particles of nanoscale dimensions, because otherwise it is difficult to accomplish a thin and uniform coating on the tip of the fiber. The conventional pCO$_2$ sensor is based on the pH indicator HPTS (see Sect. 13.5.1.) dissolved in an ethylcellulose matrix as described before [9, 23]. Finely dispersed nanoparticles of Ru(dpp) encapsulated in poly(acrylonitrile) were added to this sensor to act as a

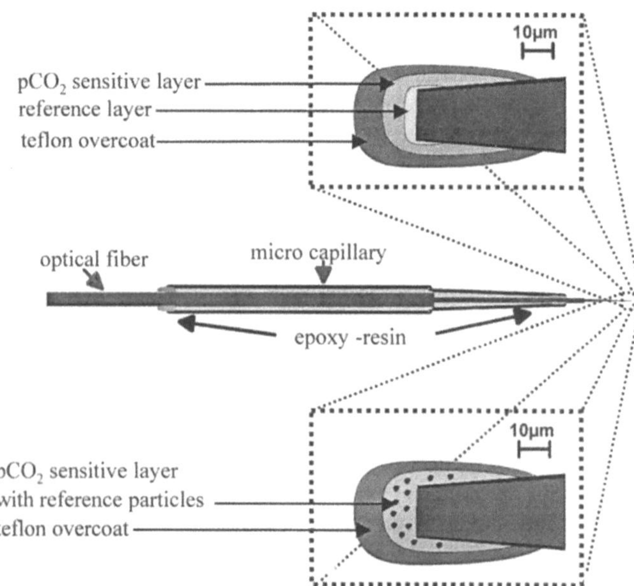

Fig. 13.11. Cross section through the tip of DLR-type pCO$_2$ fiber optic microsensors. *Top:* double-layer design. *Bottom:* single-layer design. The Teflon overcoat is permeable to CO$_2$, but not to ions and larger molecules

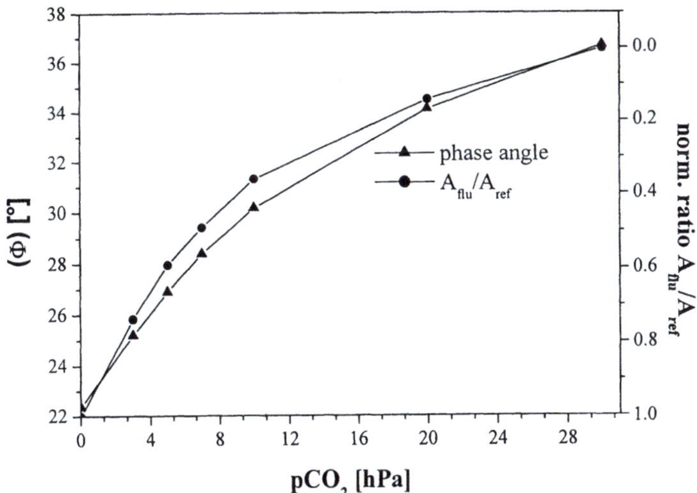

Fig. 13.12. pCO_2 calibration plots of a DLR-based microsensor with phase angles measured at a modulation frequency of 45 kHz. The normalized fluorescence intensity (A_{flu}/A_{ref}) was calculated from the phase angles

reference standard. The phase shift calibration plots of a single layer DLR pCO_2 microsensor is given in Fig. 13.12. Measurements were performed with the opto-electronic setup used for phosphorescence lifetime sensing (Fig. 13.4).

13.5.3
DLR Imaging Using Planar Optical pH Sensors

DLR sensors are ideally suited for combination with time-resolved luminescence imaging. Lifetime-based sensing of oxygen [11, 14, 18], pH [15, 17], CO_2 [23] or ions [24, 25] has been demonstrated and extended to imaging of pH, oxygen, and CO_2 [26]. Here, we demonstrate DLR-based imaging of pH using a pH-sensitive layer [27]. It consists of a lipophilic fluorescein dissolved in a hydrogel layer. Ru(dpp)-stained nanoparticles were used again as reference standard. Conventional images suffer from inhomogeneities due to non-uniform illumination of the areas. However, by forming the ratio of the excitation and the emission windows, this effect can be eliminated to a wide extent. Figure 13.13 shows the pH images of a microfilter soaked with buffer solutions of pH 5 and 9, respectively. The imaging scheme was also applied to measure the pH of solutions contained in microtiter plates with a pH-sensitive spot on their bottom (Fig. 13.14). It is evident that the reference particles are not homogeneously distributed in the sensing layers placed in the wells. This has to be improved, particularly if pH is to be measured with high spatial resolution.

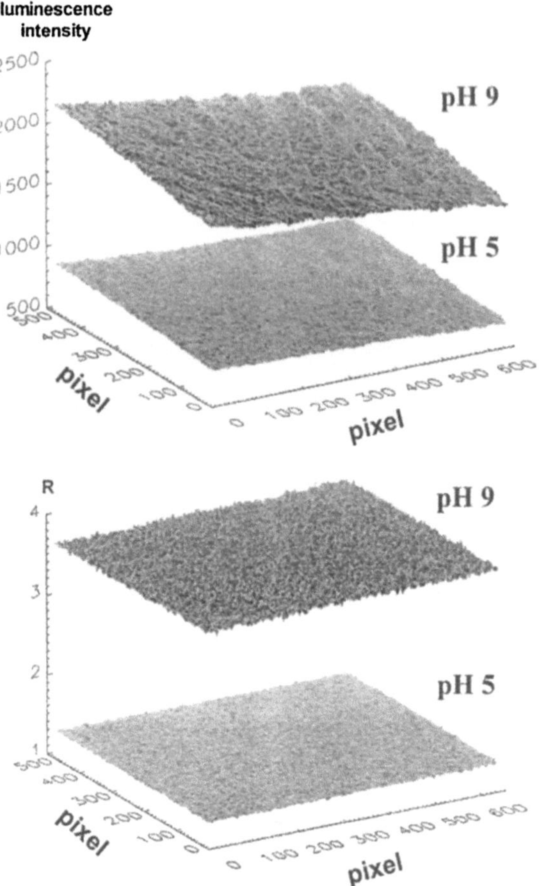

Fig. 13.13. *Top:* fluorescence intensity image of a planar optical pH-sensor at pH 5 and 9, respectively. *Bottom:* same area imaged via DLR (100 pixels correspond to 10 mm)

13.5.4
Outlook

The results presented here clearly demonstrate the usefulness of DLR as a new concept for quantitative fluorescence assays. The major task in optimizing the performance of DLR assays is the design indicator/reference couples with matched spectral properties. This is either possible by using phosphorescent dyes other than ruthenium(II)-polypyridyl complexes, or to look for chemically sensitive dyes with larger Stokes' shifts.

If performed at two different modulation frequencies, two phase angles are obtained as the analytical information. While the first can be used for DLR referencing, the second allows the calculation of the decay time of the reference standard. Hence, any interferences by dynamic luminescence quenchers (such

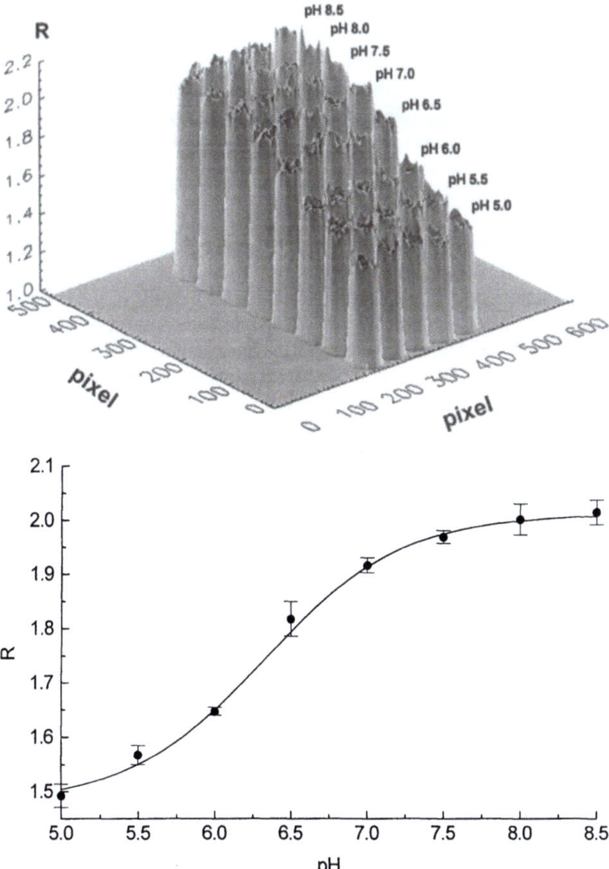

Fig. 13.14. *Top:* pH image of a microtiter plate with sensor spots into the wells. *Bottom:* time-domain DLR calibration plot for the pH sensors integrated into the wells

as oxygen) may be compensated for. In addition, parameters which effect the lifetime of the reference standard can be determined as well. A typical example is temperature which therefore can be monitored via 2-frequency DLR. Thus, its effect can be compensated for. In conclusion, DLR is a powerful concept to design a new generation of intrinsically referenced optical assays, sensors and imaging methods.

References

1. Rettig W, Strehmel B, Schrader S, Seifert H (eds) (1999) Applied fluorescence in chemistry biology and medicine. Springer, Berlin Heidelberg New York
2. Daehne S, Resch-Genger U, Wolfbeis OS (eds) (1998) Near infrared dyes for high technology applications. Kluwer Academic Publishers, Dordrecht

3. Lakowicz JR (1999) Principles of fluorescence spectroscopy. Plenum Press, New York
4. Wolfbeis OS (ed) (1992) Fluorescence spectroscopy: new methods and applications. Springer, Berlin Heidelberg New York
5. Haugland RP (ed) (1999) Handbook of fluorescent probes. Eugene, chap 1
6. Klimant I (1997) Ger Pat Appl DE 198.29.657 (1 Aug 1997)
7. Klimant I, Wolfbeis OS (1998) Book of abstracts, 6th European Conference on Optical Chemical Sensors & Biosensors (Europt(r)ode) (Ger), p 125
8. Lakowicz FN, Castellano JD, Dattelbaum L, Tolosa L, Rao G, Gryszynski I (1998) Anal Chem 70:5115
9. Holst G, Kohls O, Klimant I, Kühl M, König B, Richter T (1998) Sensors Actuat 51:163
10. Liebsch G, Klimant I, Wolfbeis OS (1999) Adv Mat 15:1296
11. Lippitsch ME, Pusterhofer J, Leiner MJP, Wolfbeis OS (1988) Anal Chim Acta 205:1
12. Holst AG, Köster T, Voges E, Lübbers DW (1995) Sensors Actuat B29:231
13. Trettnak W, Kolle C, Reininger F, Dolezal C, O'Leary (1996) Sensors Actuat B35/36:506
14. Klimant I, Kühl M, Glud RN, Holst G (1997) Sensors Actuat B 38/39:29
15. Kosch U, Klimant I, Wolfbeis OS (1999) Fresenius' J Anal Chem 364:48
16. Klimant I, Meyer V, Kühl M (1995) Limnol Oceanogr 40:1159
17. Kosch U, Klimant I, Werner T, Wolfbeis OS (1998) Anal Chem 70:3892
18. Wolfbeis OS, Klimant I, Werner T, Huber C, Kosch U, Krause C, Neurauter G, Dürkop A (1998) Sensors Actuat 51:17
19. Hartmann P, Ziegler W (1996) Anal Chem 68:4512
20. Huber C, Fähnrich K, Krause C, Werner T (1999) J Photochem Photobiol 128:111
21. Huber C, Klimant I, Mayr T, Krause C, Werner T, Wolfbeis OS – Fres J Anal Chem
22. Huber C, Klimant I, Krause C, Wolfbeis OS (2000) (submitted)
23. Neurauter G, Klimant I, Wolfbeis OS (1999) Anal Chim Acta 382:67
24. Krause C, Werner T, Huber C, Klimant I, Wolfbeis OS (1998) Anal Chem 70:3983
25. Werner T, Klimant I, Huber C, Krause C, Wolfbeis OS (1999) Mikrochim Acta 131:25
26. Liebsch G, Klimant I, Frank B, Holst G, Wolfbeis OS (2000) Appl Spectrosc 54:548
27. Liebsch G, Klimont I, Krause C, Wolfbeis OS (2001) Anal Chem (submitted)

CHAPTER 14

Two-Photon Fluorescence Fluctuation Spectroscopy

Y. Chen, J. D. Müller, J. S. Eid, E. Gratton

Two-photon excitation spectroscopy has inherent 3-D resolution with excitation volumes as small as 0.1 fl. Compared to conventional fluorometers a reduction by a factor of 10^{10} of the excitation volume can be achieved. The fluorescence fluctuations within the small excitation volume provide, via fluorescence correlation spectroscopy (FCS), a unique way to study biological phenomena. In this contribution, we outline the instrumentation of two-photon fluorescence correlation spectroscopy and highlight technical details of our experimental setup. We discuss the autocorrelation analysis for single and multiple species with emphasis on the normalized fluctuation amplitude $G(0)$. Furthermore, we revisit another data analysis technique called moment analysis. This method was originally introduced 10 years ago and provides a very fast and convenient characterization of the fluctuation amplitude. We compare experimental results from moment analysis with that of another data analysis method, the photon counting histogram (PCH), and discussed differences between these two methods.

14.1
Introduction

Typically, macroscopic systems do not allow the observation of fluctuations, since they are masked by the enormous number of independent contributions that make up the observable. In the microscopic world, however, fluctuations are abundant. Fluctuations of fluorescent light emerging from a small region of the sample were first treated more than two decades ago [1, 2]. The technique is known as fluorescence correlation spectroscopy (FCS) and was originally used to measure translational and rotational diffusion and chemical reactions [1–3].

The time-dependent decay of the fluorescence intensity fluctuations is characterized by the autocorrelation function $G(\tau)$, which is directly obtained from FCS experiments. Theoretical models for a number of physical processes, such as diffusion or chemical reactions, exist [2–4]. FCS has been applied to study translational and rotational diffusion [5, 6], flow [7], chemical reactions [8], triplet state kinetics [9], and hybridization reactions [10, 11]. It has been applied to study processes in bulk solution, on surfaces, and in cells [6, 12, 13]. In the case of pure translational diffusion of a single species, two parameters can be recovered from the autocorrelation function: the average number of molecules $\bar{N}$ in the observation volume, which is inversely proportional to $G(0)$, and the diffusion coefficient D of the particles [14, 15].

While the temporal behavior of fluctuations is best described by the autocorrelation function, the amplitude of the fluctuations is characterized by its probability distribution. The determination of the autocorrelation function from the raw data is a data reduction technique. Hence, some information encoded in the time series of the photon counts is lost. While the autocorrelation approach is the method of choice for characterizing kinetic processes embedded in the stochastic signal, it lacks information regarding the amplitude distribution of the intensities [16]. To overcome this limitation of the autocorrelation approach we developed a new technique, the photon counting histogram (PCH), which uses additional information not harnessed by the autocorrelation function [17]. We demonstrated that it is possible to resolve biological molecules with either one or two dye labels by a single PCH measurement [18].

The theory and some applications of PCH analysis have been shown elsewhere [17, 18]. In this contribution, we discuss another data analysis technique, fluctuation moment analysis, which was introduced 10 years ago by Qian and Elson [19, 20]. Moments provide a model-independent way to characterize certain aspects of probability distributions. The first moment marks the average of the distribution, while the second central moment, the variance, describes the width of the distribution. These are the two most important moments for most applications. There are also higher order moments, like the third moment, which represents the skewness of a distribution [21]. Besides the ordinary moments other types of moments exist, such as the factorial moments, the cumulants, and the factorial cumulants, which are advantageous in certain circumstances [22]. Here we demonstrate the use of moment analysis to calculate directly the normalized fluctuation amplitude $G(0)$ in a model independent way.

14.2
Instrumentation

The primary technique in our laboratory is two-photon microscopy and FCS. Fluorescence fluctuation experiments are typically performed on samples at very low concentrations (in the range from pmol/l to µmol/l). It is important to maximize the detected photon counts per molecule to achieve good signal to noise ratios. Therefore, to carry out successful fluorescence fluctuation measurements it is of the utmost importance to have an efficient two-photon setup. Each parameter that determines the two-photon absorption has to be optimized. The instrumental setup for two-photon fluorescence fluctuation spectroscopy is shown in Fig. 14.1. This setup is based on the instrumentation described by Berland with some modifications [23].

14.2.1
Laser

Two-photon excitation efficiency is largely determined by the pulse width and repetition frequency of the laser. At the same average power, a short pulse width ensures higher instantaneous power and a more effective two-photon excitation compared to a longer pulse width. We use a Ti:Sapphire laser (Mira 900, Coherent Inc., Santa Clara, CA) in this work. The pulse width of the laser is ~100 fs at optimal conditions. This ultra short laser pulse is susceptible to pulse broadening caused by chromatic dispersion of glass materials, such as objectives, lens elements, or polarizers. The final pulse width of the laser at the

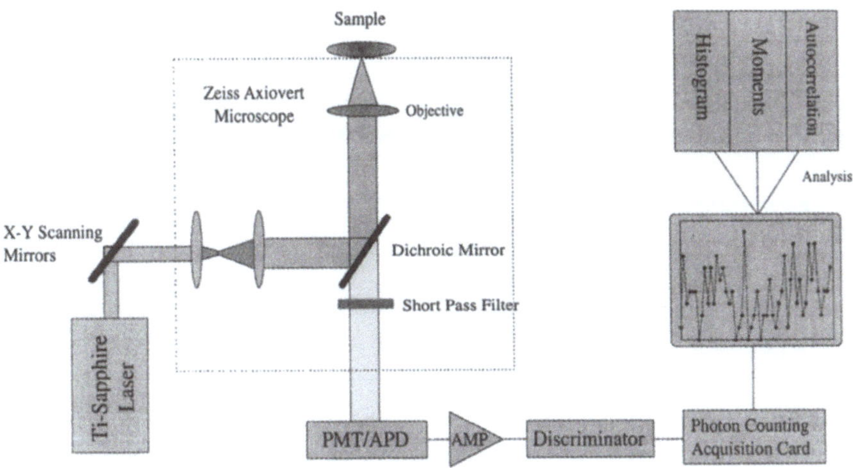

Fig. 14.1. Schematic of the two-photon instrument used for FCS. The modelocked Ti-Sapphire laser provides efficient two-photon excitation of the sample. The excitation light is send to a Zeiss microscope. Either a PMT or an APD detects the fluorescence generated by the two-photon excitation. The photon counts from the photon counting module are saved on the computer hard disk, and analyzed by different algorithms

sample is estimated to be on the order of 150 fs [24, 25]. The repetition rate of the laser is 80 MHz. The gain medium of the system, the titanium:sapphire crystal will lase over the wavelength range from 720 nm to 990 nm.

14.2.2
Microscope Objectives

Another crucial element in two-photon experiments is the microscope objective. Aberration and color corrected objective are not only crucial for two-photon excitation, but also important for obtaining a diffraction limited excitation volume for fluorescence fluctuation experiments. Most experiments were performed with a Zeiss $63\times$ Plan Neofluar ($N.A.$ 1.25), a $40\times$ Fluar ($N.A.$ 1.3) or a $63\times$ Plan Apochromat ($N.A.$ 1.4) infinity corrected oil immersion objective (Zeiss, Thornwood, NY). One requirement for objectives, that is particular to the two-photon technique, is a stringent color correction over a broad wavelength range, from near UV to near IR. Otherwise, chromatic aberrations will occur, which lead to compromises of the excitation volume in fluorescence fluctuation measurements [26]. The $63\times$ Plan Apochromat and the $63\times$ Plan Neofluar objective have good chromatic corrections and a high transmission in the visible region. The $40\times$ Fluar objective has a high transmission in the near UV (approximately 70% transmittance at 340 nm), but poor chromatic corrections for the two-photon excitation wavelengths in the near IR.

Table 14.1 gives a more detailed comparison of fluorescence fluctuation measurements using these three objectives. A sample of rhodamine 110 in water was used for this purpose. The excitation wavelength was 780 nm with a power

Table 14.1. Comparison of three commonly used objectives for two-photon fluorescence fluctuation experiments

Objectives	Power at the sample (mW)	Diameter of back aperture (mm)	cpsm	$\langle N \rangle$	w_0 (μm)
Plan neofluar (63×) N.A. = 1.25	7	7	8,300	1.85	0.304
Fluar (40×) N.A. = 1.3	20	11	22,000	4.52	0.295
Plan apochromat (63×) N.A. = 1.4	7	7	13,000	1.95	0.310

before the objectives of 38 mW. The fluorescence signal was detected with a PMT. The back aperture of the Plan Neofluar and the Plan Apochromat were 7 mm, and that of the Fluar was 11 mm. The excitation beam with a diameter of 13 mm overfilled the back aperture of the objectives. The power measured after the objectives were 7 mW (Plan Neofluar), 7 mW (Plan Apochromat) and 20 mW (Fluar).

Figure 14.2A–C displays the experimental data analyzed by the autocorrelation function, and Fig. 14.2D shows the same experiments analyzed with PCH. The important experimental parameter for fluorescence fluctuation measurements is not the average photon counts of the sample, but the photon counts per second and per molecule (*cpsm*). The experimental setup determines the *cpsm* for a given fluorophore; thus *cpsm* is not a universal parameter. The powers at the sample were identical for the Plan Neofluar and Plan Apochromat; therefore the difference in the performance of these two objectives can only originate from their different numerical aperture. An average of 1.9 molecules inside the excitation volume was recovered for the Plan Neofluar objective, an average of 2.0 molecules for the Plan Apochromat objective. Assuming a diffusion coefficient for rhodamine 110 of 272 μm^2/s, we recovered the Gaussian-Lorentizan beam waist w_0 of the Plan Neofluar objective as 0.304 μm, and of the Plan Apochromat objective as 0.310 μm. The $\langle N \rangle$ ratio between these two objectives is 1.05, which is very close to 1.06 determined by taking the ratio of their beam waist w_0 to the third power. The *cpsm* of these two objectives differs by a factor of 1.57; this is almost identical to the value of 1.5 given by the ratio of their numerical apertures to the fourth power. Therefore, these two objectives demonstrated that the number of molecules in the observation volume $\langle N \rangle$ scale with the cube of the beam waist of the diffraction limited excitation volume, and the *cpsm* scale with the fourth power of the N.A. of the objectives. The situation is more complicated for the Fluar 40× objective. The differences in *cpsm* and $\langle N \rangle$ between the 40× Fluar objective with any of the two 63× objectives cannot be explained solely by the differences in their N.A. and the power at the sample. The discrepancy can be contributed to the imperfect chromatic correction of the Fluar objective, which leads to an apparent shift of the fluorescence emission focal point relative to the focal point of the excitation light.

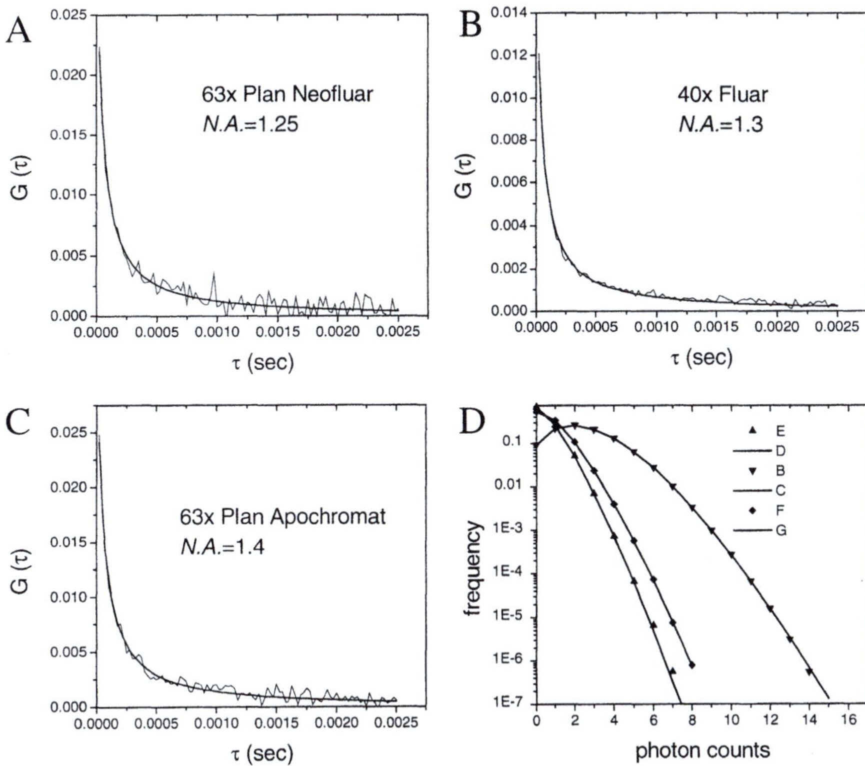

Fig. 14.2 A–D. Comparison of three commonly used objectives. A sample of 10 nmol/l rhodamine 110 in water was measured. The excitation power before the objectives was 38 mW (see Table 14.1). The fluorescence was detected by a PMT. $\langle N \rangle$ and *cpsm* were recovered from PCH analysis; $\langle N \rangle$ was recovered from fitting the autocorrelation function. The diffusion coefficient used for rhodamine 110 was 272 $\mu m^2/s$

We can summarize the results as follows:

1. For a specified power before the objectives, the power at the sample is almost three times higher for the Fluar objective than it is for the Plan Neofluar and the Plan Apochromat objective, which is due to the different back aperture of the objectives. Consequently, the Fluar objective yields the highest *cpsm* for a given power *outside* the microscope.
2. The Plan Apochromat objective gives the highest *cpsm* for a given power *at* the sample.
3. For a given sample concentration, the recovered number of molecules $\langle N \rangle$ for either the Plan Neofluar or Plan Apochromat is ~2 times lower than for the Fluar objective. This means that fluctuation amplitudes will be twice as strong when measured with the Plan Neofluar or Plan Apochromat compared to the Fluar objective.

14.2.3
Microscope, Filters, and Electronics

The experiments were carried out with either a Zeiss Axiovert 35 or a Zeiss Axiovert 135 TV microscope. The Zeiss Axiovert 135 TV has a detector port directly underneath the emission tube lens, which results in a twofold increase in the photon collection efficiency compared to the Zeiss Axiovert 35. The dichroic mirror (650DCSP, Chroma Technology, Brattleboro, VT) reflects the laser excitation from 650 nm to 850 nm, and transmits 90% of the fluorescence from 400 nm to 625 nm. Glass filters, either BG-39, BGG-22, or CM-500 (Chroma Technology, Brattleboro, VT) were used as emission filters to eliminate the remaining excitation light. Among the three filters, the CM-500 has the broadest transmission band and passes more fluorescence above 600 nm than either BG-39 or BGG-22. The broad wavelength range provided by CM-500 makes it an ideal choice for most fluorescence fluctuation measurements. All three filters block light very well above 750 nm (6 O.D. with a 1 mm thick filter), and two filters coupled together with immersion oil are sufficient to eliminate all contributions from scattered laser light.

The photon counting module consists of a photon detector, preamplifier, and discriminator. Two-types of detectors have been used in this laboratory, a photo multiplier tube (PMT) R-5600 (Hamamatsu, Japan) and an avalanche photodiode (APD) (EG&G, Canada). Both detectors have low dark counts, high gain, and a fast response time, and are optimized for single photon counting. The detailed comparisons of these two detectors can be found in [27]. For the PMT, the signal is first amplified (Phillips Scientific, Model 6931, Ramsey, NJ), and then converted to TTL pulses with a discriminator (Phillips Scientific, Model 6930). The APD detector has the preamplifier and discriminator built into a single unit. The detected photon counts are fed into a home-build acquisition card and sampled with a variable time-resolution from 0.1 µs to 1 ms [28]. The recorded photon counts are saved to a data file in the acquisition computer and analyzed with either PV-WAVE version 6.10 (Visual Numerics, Inc.) or LFD Globals Unlimited software (Champaign, IL).

14.3
Autocorrelation

Two fundamental physical properties can be determined from the autocorrelation function, $G(\tau)$ (for a review, see [29]):

1. Kinetic information, which is characterized by the decay of the autocorrelation function [1, 6, 30–32].
2. The fluctuation amplitude of $G(\tau)$ at $\tau = 0$, $G(0)$, which characterizes the strength of the fluctuation signal and represents the normalized variance of the fluorescence intensity, $\langle \Delta F^2 \rangle / \langle F \rangle^2$ [15, 19, 33].

14.3.1
Single Species

For a single fluorescent species, the time-zero value of the autocorrelation function, $G(0)$ depends on the average number of molecules inside the excitation volume:

$$G(0) = \frac{\langle F^2 \rangle - \langle F \rangle^2}{\langle F \rangle^2} = \frac{\gamma}{\langle N \rangle} \qquad (14.1)$$

where $\langle N \rangle$ is average number of molecules inside the excitation volume, and γ is a geometric factor and depends only on the shape of the volume,

$$\gamma = w_2/w_1 \qquad (14.2)$$

with the constant w_n defined as

$$w_n = \int [I^2(r)/I^2(0)]^n \, d^3r \qquad (14.3)$$

The geometric factor γ is equal to $3/4\,\pi^2$ (0.07599) for the Gaussian-Lorentzian PSF and equal to $\sqrt{2}$ (0.3535) for the 3D Gaussian PSF.

The experimental observable in a fluorescence fluctuation experiment is not the fluorescence intensity, but the photon counts. The $G(0)$ value contains the shot noise contribution of photon counts. A direct calculation of $G(0)$ based on the photon counts according to Eq. (14.1) leads to an overestimation of $G(0)$. Two approaches have been applied to recover the "true" $G(0)$ value without the shot noise contribution of the photon counts.

The first, a straightforward way, is to extrapolate $G(0)$ from $G(\tau)$ at $\tau \neq 0$ [15, 34]. Detectors have no memory of the photons detected in the past because the absorption process is almost instantaneously. Thus, there is no shot noise contribution at any other channels except at $\tau = 0$. The extrapolation of $G(0)$ obtained by fitting to the appropriate model relies on the knowledge of the underlying kinetic processes that contribute to the intensity autocorrelation function. Hence, it is necessary to develop physical models to describe and account for these kinetic processes.

The second approach is called Scanning FCS (S-FCS) and was introduced by Weissman et al. [35] to determine molecular weight of DNA samples labeled with ethidium bromide. This method requires a periodic spatial sampling of the probe with a time period T. If the diffusion is negligible within the first time period T, then $G(T)$ will be equivalent to the ideal $G(0)$ without shot noise contributions. The merits of this method are to measure simultaneously multiple independent excitation volumes and to eliminate background contributions to the autocorrelation function. The signal to noise ratio increases significantly when measuring molecules with slow diffusion. S-FCS has been successfully implemented to investigate several types of oligomer dissociation [36], and it has been used to determine the diffusion coefficients of biopolymers and the number of particles in biological systems [37, 38].

The temporal decay of the autocorrelation function, $G(\tau)$ contains information about the dynamics of the system. Depending on the time scale of interest, one can observe fluorescence lifetimes, triplet state reactions, rotational and translational diffusion [1, 39–41]. The dynamic information contained in FCS is often overwhelming. Here, we limit our discussion only to translational diffusion.

For a system governed by Brownian motion, the relation between the diffusion coefficient and the concentration fluctuation is given by Fick's second law:

$$\partial c_i(r,t)/\partial t = D_i \nabla^2 c_i(r,t) \tag{14.4}$$

where $c(r,t)$ is the concentration of the sample at position r, and at time t. For spherical particles, the diffusion coefficient D is given by the Stokes-Einstein relationship:

$$D = k_B T/6\pi\eta R \tag{14.5}$$

where R is the radius of the particle, η is the viscosity of the solution, k_B is Boltzmann's constant, and T is the temperature. For diffusion in three dimensions, the autocorrelation function of the particle concentration is related to the diffusion coefficient by [2]

$$\langle \delta c(r,t) \cdot \delta c(r,t+\tau) \rangle = \langle c \rangle (4\pi D\tau)^{-3/2} \exp\left(\frac{-|r-r'|^2}{4D\tau}\right) \tag{14.6}$$

and $\langle \delta c(r,t) \cdot \delta c(r,t+\tau) \rangle$ is related to $\langle \delta F(r,t) \cdot \delta F(r,t+\tau) \rangle$ by

$$\langle \delta F(r,t) \cdot \delta F(r,t+\tau) \rangle = \kappa^2 Q^2 \int dr \int dr' I^4(r,t)\, I^4(r',t) \tag{14.7}$$
$$\langle \delta c(r,t) \cdot \delta c(r,t+\tau) \rangle$$

where $I^2(r,t)$ is the laser intensity profile for two-photon excitation, κ is a constant, and Q is a function of the absorption coefficient, the fluorescence quantum yield, and the detection efficiency of the optical system. By substituting either the 2 D-Gaussian, the 3 D-Gaussian, or Gaussian-Lorentzian PSF into Eq. (14.7), the functional form of $G_{PSF}(\tau)$ is given by [23]

$$G_{PSF}(\tau) = G(0) \cdot g_{PSF}(D,\tau) \tag{14.8}$$

where $g_{PSF}(D,\tau)$ is defined for the 2-D Gaussian case as

$$g_{2DG}(D,\tau) = \left(1 + \frac{8D\tau}{w_{2DG}^2}\right)^{-1} \tag{14.9}$$

in the 3-D Gaussian as

$$g_{3DG}(D,\tau) = \left(1 + \frac{8D\tau}{w_{3DG}^2}\right)^{-1} \left(1 + \frac{8D\tau}{z_{3DG}^2}\right)^{-1/2} \tag{14.10}$$

and in the Gaussian-Lorentzian case as

$$g_{GL}(D,\tau) = \frac{4\,w_{GL}^2}{3\lambda}\left(\frac{\pi}{D\tau}\right)^{1/2} \int db \, \frac{\exp\left(-z_R^2\,b^2/2D\tau\right)}{b^4 + \left(2 + \dfrac{16\,D\tau}{w_{GL}^2}\right)\cdot b^2 + \dfrac{1}{4}\left(\dfrac{16\,D\tau}{w_{GL}^2}\right)^2}$$

$$\times \left[\frac{2b^2 + \dfrac{16\,D\tau}{w_{GL}^2} - 4}{\sqrt{2}\,(2+b^2)} + \frac{2}{\left(2 + b^2 + \dfrac{16\,D\tau}{w_{GL}^2}\right)^{1/2}} \right] \tag{14.11}$$

The parameters w_{2DG}, w_{3DG}, and w_{GL} are the beam waist of the corresponding PSF.

To extrapolate the $G(0)$ value and to recover kinetic coefficients from the autocorrelation function, it is important to have an accurate and robust model. The expression of $g_{GL}(D,\tau)$ in Eq. (14.11) has no analytical solution and needs to be solved by numerical integration. Compared to the expression of $g_{GL}(D,\tau)$, both the $g_{2GL}(D,\tau)$ and $g_{3GL}(D,\tau)$ are mathematically easier to handle and analytically more convenient to calculate.

14.3.2
Calibration of the Excitation Volume

To convert the experimentally recovered $G(0)$ into a concentration for the chemical species of interest, the size of the excitation volume must be determined for a given instrumental setup. The size of the excitation volume is influenced by the beam profile of the laser, the alignment of the microscope optics, and the maintenance of exterior optics. Variations in the excitation volume by a factor of 2 can occur [23]. To calibrate the excitation volume, we often perform an experiment with a sample of known concentration and of known diffusion coefficient. Experimentally, we do not measure diffusion coefficient directly, but we measure the residence time of a molecule inside the excitation volume given by $w_o^2/8D$. To recover the excitation volume from the residence time of the molecule, the diffusion coefficient must be obtained by other methods. In the past, fluorescent spheres of known diameter were used for calibration. However, spheres tend to aggregate as a function of time. Depending on the size and the concentration of the spheres, they aggregate on the time scale of minutes to hours. Furthermore, spheres also adsorb to many other materials, such as test tubes, glass slides, and biological cells [32]. Frequently, the fluorescent intensity from the spheres varies during the cause of hours due to the adsorption of spheres to the sample holder.

All the calibration work presented in this thesis is based on specific fluorescent dyes. Fluorescent dyes are stable and less susceptible to sample preparations. The most commonly used dye is fluorescein in high pH buffer (50 mmol/l tris(hydroxymethyl)amino-methane (tris) buffer (Sigma, MO), pH = 10). Fluorescein is a pH sensitive dye, and its spectroscopic properties vary drastically from pH 7.5 to pH 2. At pH >7.5, fluorescein has a constant quantum yield (greater than 0.93) and very good water solubility [42]. The major benefit of using fluorescein at high pH buffer (pH >8) is its lack of interactions with surfaces; consistent results are always obtained regardless of the type of sample holder used. Another frequently used dye is rhodamine 110; rhodamine 110 has lower water solubility than fluorescein, but is, under identical instrumental conditions, almost by a factor of two brighter than fluorescein. Figure 14.3 A shows the autocorrelation curve of 4 nmol/l fluorescein in tris buffer. Figure 14.3 B shows the autocorrelation curve of rhodamine 110 at 3 nmol/l in water. A global analysis was performed on the data sets. The diffusion coefficient of fluorescein was fixed at 300 μm^2/s. The Gaussian-Lorentizan beam waist, w_{GL}, was a link parameter between the two data sets and a value of 0.319 μm was recovered. Consequently, the diffusion coefficient of rhodamine 110 was recovered as 272 μm^2/s.

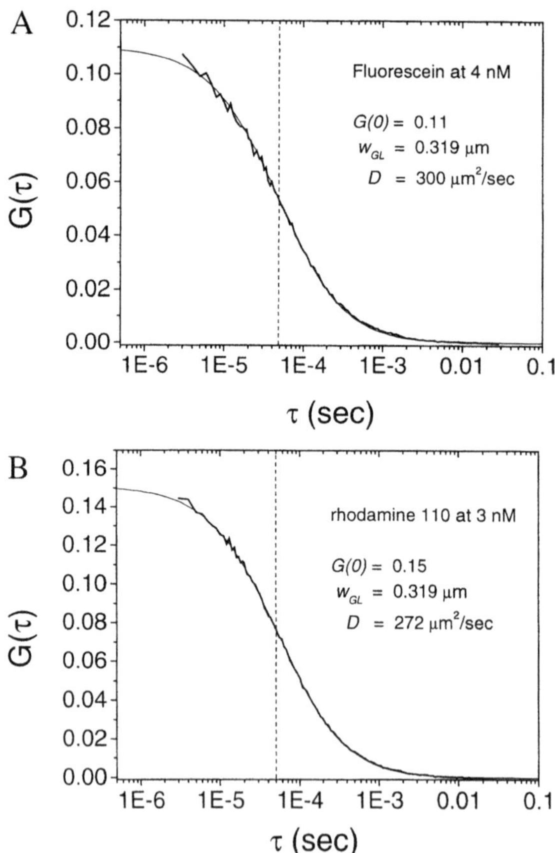

Fig. 14.3 A, B. Autocorrelation curve $G(\tau)$ of two fluorescent dyes: **A** fluorescein; **B** rhodamine 110. A 40× fluar objective was used for these measurements and the fluorescence was detected by an APD. The data were taken in the photon mode

14.3.3
Comparison of Models

Autocorrelation functions based on either 2D-Gaussian, 3D-Gaussian, or Gaussian-Lorentzian PSF have been used extensively in FCS analysis [32, 43, 44]. In this section, we first demonstrate the differences when we approximate a simulated $G_{GL}(D, \tau)$ by either a $G_{2DG}(D, \tau)$ or a $G_{3DG}(D, \tau)$. Second, we will compare the fitting results of an experimental autocorrelation function using the three different models.

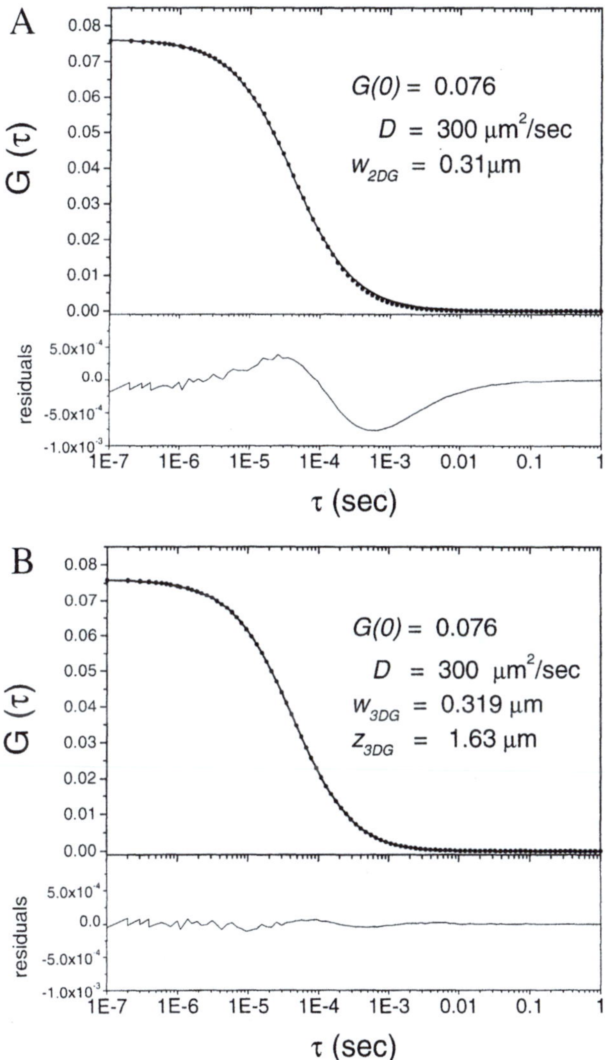

Fig. 14.4A, B. Approximating a simulated autocorrelation function, $G_{GL}(\tau)$ by: **A** $G_{2DG}(\tau)$; **B** $G_{3DG}(\tau)$. The simulation parameters were chosen as, $G(0) = 0.076$, $D = 300$ μm²/s, and $w_{GL} = 0.3$ μm

Figure 14.4 compares the simulated $G_{GL}(\tau)$ using Eq. (14.11) with the best approximated $G_{3DG}(\tau)$ and $G_{2DG}(\tau)$. The parameters used in the simulation were $G(0) = 0.076$, $D = 300\ \mu m/s$, and $w_{GL} = 0.3\ \mu m$. Figure 14.4A displays the result when the simulated data were fitted with the $G_{2DG}(\tau)$ model. We recovered a $G(0)$ of 0.076 and a beam waist w_{2DG} of 0.31 $\mu m^2/s$. Figure 14.4B shows the result when the simulated data fitted with the $G_{3DG}(\tau)$ model. We recovered a $G(0)$ of 0.076, a beam waist w_{3DG} of 0.319 μm and z_{3DG} of 1.6 μm. The residuals of the two model fits are also shown for comparison. The systematic deviations by approximating $G_{GL}(\tau)$ by a 2D-Gaussian autocorrelation function are more severe than with a 3D-Gaussian model. When the residuals in Fig. 14.3A are inspected more carefully, the deviation is especially strong at the tail of the autocorrelation function. This is caused by the lack of z dependence in the $G_{2DG}(\tau)$ model. The intensity of the focused laser beam decreases in the z direction less than in the radial direction. Thus, the characteristic z dependence appears in the tail of the autocorrelation function.

The intensity distribution in the x or y direction is Gaussian for all three models. The beam waist in the z direction is infinity for the 2D-Gaussian PSF, Gaussian distributed for the 3D-Gaussian PSF, and Lorentzian distributed for the Gaussian-Lorentzian PSF. The 2D-Gaussian PSF is a very good approximation for an extended beam waist w_{DG} [43]. If w_{DG} is small, as in this simulation, the 2D-Gaussian model fails to give a reasonable fit. Conversely, the small w_{DG} in the Gaussian-Lorentzian PSF can be better approximated by the 3D-Gaussian PSF.

The $G(\tau)$ based on the simulation contains no experimental uncertainties, except for residuals caused by the numerical calculations. Next, we will compare the residuals from fitting an experimental autocorrelation function to the $G(\tau)$ of three models (Fig. 14.5). The sample measured was rhodamine 110, and the actual experimental autocorrelation curve and the corresponding fit to a Gaussian-Lorentzian model are shown in Fig. 14.3B. The recovered $G(0)$ and beam waist in axial and radial direction for the three fits are summarized in Table 14.2. The residuals obtained by fitting to either the Gaussian-Lorentzian or 3D-Gaussian model are almost flat, while the residuals by the 2D-Gaussian model fit deviate to a greater extent. A closer examination of the residuals reveals that the residuals for the 3D-Gaussian model display less systematic deviation than the residuals for the Gaussian-Lorentzian model. However, the experimental data do not allow one to distinguish clearly between the 3D-Gaussian and Gaussian-Lorentzian model. Additional experiments, such as determining the experimental PSF directly, are needed to address this issue. Nevertheless, considering that the integration time for this measurement was

Table 14.2. Comparison of the recovered parameters by fitting an autocorrelation function from Rhodamine 110 in water using three different models

	$G_{2DG}(\tau)$	$G_{3DG}(\tau)$	$G_{GL}(\tau)$
$G(0)$	0.153	0.152	0.151
$w\ (\mu m)$	0.328	0.335	0.319
$z\ (\mu m)$	–	2.4	–

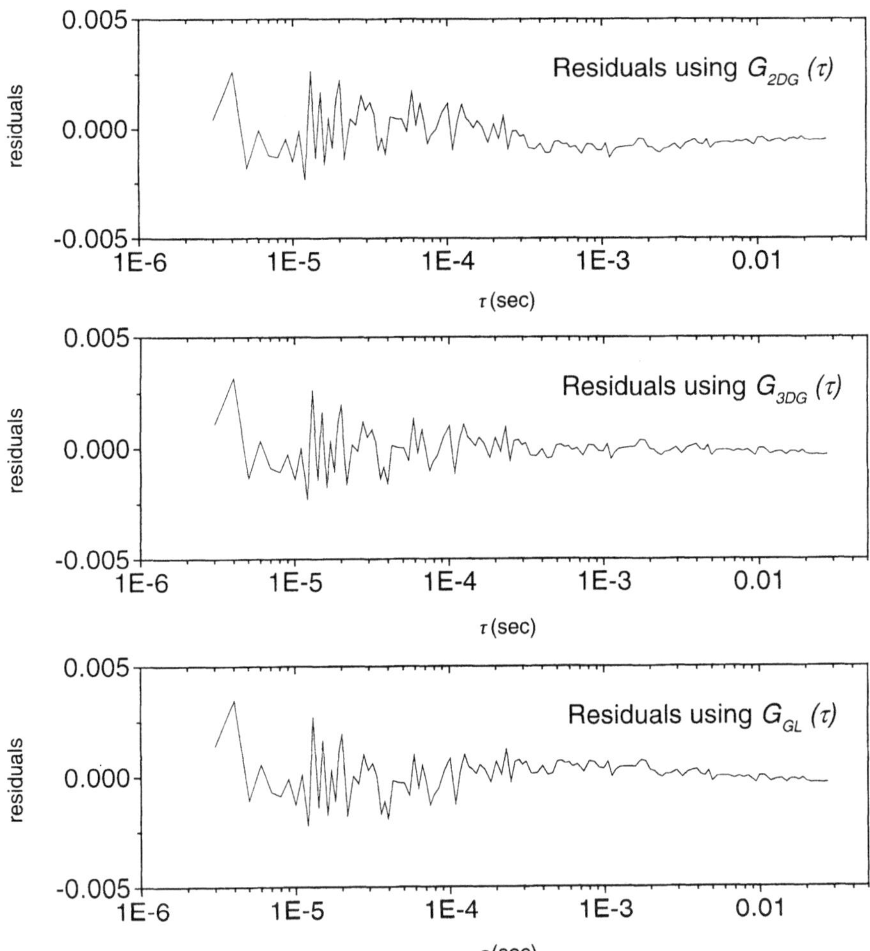

Fig. 14.5. Comparisons of the residuals from fitting experimental data with three different models. The experimental autocorrelation function was determined for rhodamine 110 at 3 nmol/l in water (see Fig. 14.3 B). The recovered parameters are listed in Table 14.2. Both $G_{3DG}(\tau)$ and $G_{GL}(\tau)$ represent the experimental data well, while the $G_{2DG}(\tau)$ model leads to slightly stronger deviations

more than 10 min using one of the brightest samples, most experiments will have a lower signal to noise ratio, since the integration time is typically much shorter and molecules are less bright than rhodamine 110. Thus, we can conclude that both models can fit experimental data accurately.

We can summarize our observations as follows:

1. The $G(0)$ values can be extrapolated by any one of the three autocorrelation models: the differences in the recovered $G(0)$ among the three different models are less than one percent.

2. For our experimental conditions, both autocorrelation functions $G_{GL}(\tau)$ and $G_{3DG}(\tau)$ represent the experimental autocorrelation function well. The auto-correlation function $G_{2DG}(\tau)$, on the other hand, cannot fit the tail of the experimental autocorrelation function.

14.3.4
Multiple Species

The autocorrelation function, $G(\tau)$ of multiple species is a superposition of the autocorrelation functions of individual species weighted by their corresponding fractional intensity squared:

$$G(\tau) = \sum_{m=1}^{M} \left(\frac{F_m}{F_T}\right)^2 \cdot G_m(D_m, \tau) = \sum_{m=1}^{M} \left(\frac{\varepsilon_m \cdot \langle N_m \rangle}{F_T}\right)^2 \cdot G_m(D_m, \tau) \qquad (14.12)$$

where F is the average fluorescence intensity in cps, and M is the total number of species. The average fluorescence intensity of a mixture, F_T is given by $F_T = \sum_{m=1}^{M} F_m$. The intensity F_m of the m-th species is given by:

$$F_m = \varepsilon_m \cdot \langle N_m \rangle \qquad (14.13)$$

where ε_m is the molecular brightness in $cpsm$ of the m-th species for a given experimental setup. At $\tau = 0$, Eq. (14.12) can be written as

$$G(0) = \sum_{m=1}^{M} \left(\frac{\varepsilon_m \langle N_m \rangle}{\sum_{m=1}^{M} \varepsilon_m \langle N_m \rangle}\right)^2 \cdot \frac{\gamma}{\langle N_m \rangle} = \gamma \cdot \sum_{m=1}^{M} \left(\frac{\varepsilon_m}{F_T}\right)^2 \cdot \langle N_m \rangle \qquad (14.14)$$

For a two-species mixture A and B, $G(0)$ is given by

$$G(0) = \gamma \cdot \left[\left(\frac{\varepsilon_A}{F_T}\right)^2 \cdot \langle N_A \rangle + \left(\frac{\varepsilon_B}{F_T}\right)^2 \cdot \langle N_B \rangle\right] \qquad (14.15)$$

The $G(0)$ of a mixture is a non-linear combination of the number of molecules $\langle N \rangle$ of each individual species. When the brightness of the two species is not identical, the direct relation between $G(0)$ and $\langle N \rangle$ is no longer valid and additional information is needed to recover the concentration of each species.

If the hydrodynamic size of the two species is substantially different, which is, e.g., the case for a small fluorescent dye binding to a large protein, the difference in their diffusion coefficient can be used to recover the number of molecules of each species. The autocorrelation $G(\tau)$ for the two species mixture is given by

$$G(\tau) = \left[\left(\frac{F_A}{F_T}\right)^2 \cdot G_A(0) \cdot g_A(D_A, \tau) + \left(\frac{F_B}{F_T}\right)^2 \cdot G_B(0) \cdot g_B(D_B, \tau)\right] \qquad (14.16)$$

Equation (14.16) can be simplified:

$$G(\tau) = [G'_A(0) \cdot g_A(D_A, \tau) + G'_B(0) \cdot g_B(D_B, \tau)] \tag{14.17}$$

where $G'_A(0)$ and $G'_B(0)$ are $G_A(0)$ and $G_B(0)$ multiplied by the square of their fractional intensity squared.

Equation (14.17) contains four variables, and an unequivocal fit of experimental data to this equation can be difficult depending on the quality of the experimental autocorrelation function. The extraction of $G'_A(0)$ and $G'_B(0)$ from an experimental measurement of $G(\tau)$ is often not reliable without using some additional knowledge about the systems. If the diffusion coefficient of each species is known by some other experiment, then the determination of the two $G'(0)$ becomes straightforward. To calculate the average number of molecules for each individual species from the $G'(0)$ values, the brightness of the two species, ε_A and ε_B must be determined by other experiments.

The hydrodynamic size of species does not necessarily differ substantially for biological applications. For example, the difference in the translational diffusion coefficient between dimer and monomer is about $\sqrt[3]{2}$ (1.26), and it becomes challenging to recover N of the individual species relying solely on their diffusion coefficients [45]. Figure 14.6 displays the residuals of simulated autocorrelation curves for mixtures fitted by a single species model. The autocorrelation func-

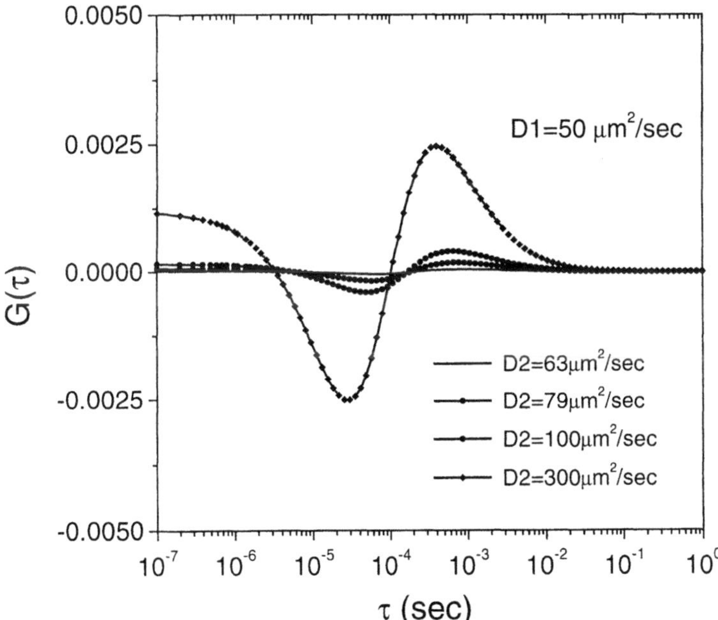

Fig. 14.6. Residuals of simulated autocorrelation functions fitted to a single species model. The autocorrelation functions of binary mixtures with different diffusion coefficient ratios were fitted to a single species model. The residuals of the fits are plotted

tions are simulated according to Eq. (14.17): the two $G'(0)$ are set to the same value, 0.038. The diffusion coefficient of species one, D_1, is kept constant at 50 $\mu m^2/s$, while the diffusion coefficient of species two, D_2, is varied as indicated in Fig. 14.6. The ratios between D_2/D_1 are chosen so that they represent molecular weight ratios of 2, 4, 8, and 216. Based on the absolute values of the residuals obtained from fitting experimental data (see Fig. 14.6), it will be exceedingly difficult to identify mixtures of monomer and dimer, or monomer and tetramer by judging the residuals of data fits to a single species model. When the diffusion coefficient ratio between two species becomes 6, similar to the value obtained for a small ligand, such as fluorescein, binding to an antibody, the residuals become large enough to identify two species conveniently.

For real biological applications, the situation becomes even more complicated. Not only the diffusion coefficients of the species differ, but also their $G'(0)$ values, which depend on the brightness and the number of molecule of each species. In additional, other kinetic processes, such as chemical reactions, can further influence the accuracy of the recovered diffusion coefficients [44]. Thus, to recover reliably the number of molecules from a mixture, the two species either must have very different diffusion coefficients or the diffusion coefficients of the individual species have to be determined independently.

14.4
Moment Analysis

Fluctuation moment analysis was introduced in the early 1990s by Qian and Elson to characterize the fluctuation amplitude $G(0)$ [19, 20]. The ordinary moments of a probability distribution $p(x)$ are given by

$$\langle x^n \rangle = \int x^n p(x)\, dx\,, \quad \text{for} \quad n = 0, 1, 2, \ldots \tag{14.18}$$

The first moment is the mean or the average, $\langle x \rangle$, of the random variable x. It can be shown mathematically that the knowledge of all the moments $\langle x^n \rangle$ is equivalent to the information content of the probability distribution $p(x)$ [22]. If one is interested mainly in the deviations from the mean, it is more convenient to define the central moments,

$$\langle \Delta x^n \rangle = \langle (x - \langle x \rangle)^n \rangle \tag{14.19}$$

The second central moment is given by

$$\langle \Delta x^2 \rangle = \langle x^2 \rangle - \langle x \rangle^2 \tag{14.20}$$

and is called the variance of the variable x. The variance $\langle \Delta x^2 \rangle$ and the average $\langle x \rangle$ are the two most commonly used moments in statistical analysis.

The normalized intensity fluctuation amplitude $G(0)$ is defined by the normalized variance of the fluorescence:

$$G(0) = \frac{\langle \Delta I^2 \rangle}{\langle I \rangle^2} = \frac{\langle I^2 \rangle - \langle I \rangle^2}{\langle I \rangle^2} \tag{14.21}$$

which is equivalent to the asymptotic value of the intensity autocorrelation function $G(\tau)$ for $\tau \to 0$. However, photon counts k rather than fluorescence intensities I are measured. The $G(0)$ value calculated from experimental photon counts is always larger than the $G(0)$ value based on Eq. (14.21), due to the shot noise contribution of the photon detection process.

The statistical relationship between the intensity at the detector and the photon counts is given by Mandel's formula [46]:

$$\Pi(k) = \int_0^\infty \frac{(\eta I)^k \exp(-\eta I)}{k!} \Pi(I)\, dI \tag{14.22}$$

The experimentally observed photon count distribution $\Pi(k)$ is the Poisson transformation of the probability distribution $\Pi(I)$ of the fluorescence intensity. For the following discussion, it is more convenient to rescale the fluorescence intensity, so that the detection efficiency η is equal to one. The rescaled intensity allows a more convenient comparison with the photon counts without loss of generality.

Using Mandel's formula, it can be shown that the ordinary moments of the fluorescence intensity I are equal to the factorial moments of the photon counts k [22]:

$$\langle I^m \rangle = \left\langle \frac{k!}{(k-m)!} \right\rangle = \langle k(k-1)(k-2)\cdots(k-m+1)\rangle \tag{14.23}$$

The first and second moment of the fluorescence intensity are therefore related to the first two moments of the photon counts by

$$\langle I \rangle = \langle k \rangle , \tag{14.24}$$

$$\langle I^2 \rangle = \langle k(k-1)\rangle = \langle k^2 \rangle - \langle k \rangle \tag{14.25}$$

The average of the photon counts $\langle k \rangle$ is equal to the average of the photon intensity $\langle I \rangle$, while the second moment of the photon counts $\langle k^2 \rangle$ is larger than the second moment of the intensity $\langle I^2 \rangle$ due to the shot noise contribution. The variance of the fluorescence intensity $\langle \Delta I^2 \rangle$ can now be expressed in terms of the variance and the average of the photon counts by combining Eqs. (14.20), (14.24) and (14.25):

$$\langle \Delta I^2 \rangle = \langle \Delta k^2 \rangle - \langle k \rangle \tag{14.26}$$

where the shot noise contribution $\langle k \rangle$ is subtracted from the variance of the photon counts $\langle \Delta k^2 \rangle$.

Now we can express the normalized fluctuation amplitude $G(0)$ in terms of the average and the variance of the photon counts [19, 20]:

$$G(0) = \frac{\langle \Delta I^2 \rangle}{\langle I \rangle^2} = \frac{\langle \Delta k^2 \rangle - \langle k \rangle}{\langle k \rangle^2} \tag{14.27}$$

The autocorrelation or PCH method requires the knowledge of the molecular kinetics or the shape of the PSF to determine $G(0)$. Moment analysis, however, computes $G(0)$ directly and model-independently according to Eq. (14.27), and is computationally extremely fast. Hence, we incorporated moment analysis in our data acquisition software to monitor the fluctuation amplitude $G(0)$ and the average photon counts online.

Moment analysis allows the model independent calculation of $G(0)$. If we want to determine, for the single species case, the average number of molecules $\langle N \rangle$ and their molecular brightness ε, we have to know the geometric factor γ. This means that the calculation of $\langle N \rangle$ and ε is not model independent any more. The shape of the PSF determines γ, which is given by $1/2\sqrt{2}$ for a three-dimensional Gaussian beam profile and by $3/4\pi^2$ for a Gaussian-Lorentzian beam profile. The average number of molecules $\langle N \rangle$ is then given by

$$\langle N \rangle = \gamma / G(0) \tag{14.28}$$

and the molecular brightness ε can be calculated as

$$\varepsilon = \langle k \rangle / \langle N \rangle \tag{14.29}$$

14.4.1
Comparison Between PCH and Moment Analysis

Moment analysis was introduced by Qian and Elson [19, 20] almost a decade ago, but has not been applied in any other studies we are aware of. This is mostly due to the commercial hardware autocorrelators that are used by most researchers, which do not allow direct access to the record of photon counts. Consequently, moment analysis has not been characterized in practical applications. In Table 14.3 we compare the results of several dyes analyzed by moment and PCH analysis.

The recovered values of $\langle N \rangle$ and ε from both analysis methods are very close, and are actually almost identical in two cases. PCH characterizes the photon counts by probability distribution, while moment analysis characterizes the same data by the fluctuation moments. Thus, both PCH and moment analysis recover $G(0)$ of a single species reliably. However, one should keep in mind that moment analysis only provides the individual moments, while the PCH analysis contains more detailed information and characterizes the complete photon count distribution. A careful examination of the residuals of the PCH analysis allows the detection of multiple species or sample contamination.

Table 14.3. Comparison of the recovered molecular brightness ε and average number of molecules $\langle N \rangle$ by using either PCH or moment analysis

	$\langle k \rangle$	ε(PCH)	$\langle N \rangle$ (PCH)	ε(moments)	$\langle N \rangle$(moments)
Rhodamine 6G (water)	9.97	2.83	3.518	2.83	3.518
Rhodamine 6G (ethanol)	6.11	6.05	1.011	6.03	1.013
Tetrafluoro-fluorescein	5.64	1.73	3.271	1.72	3.274
Fluorescein	4.38	1.75	2.498	1.76	2.497
Rhodamine 110	4.95	2.93	1.692	2.93	1.692

As demonstrated in Table 14.3, moment analysis is a very robust technique to characterize the intensity fluctuations, and to recover $\langle N \rangle$ for a single species measurement. In fact, from a pure mathematical point of view, histogram and moment analysis are equivalent, since the knowledge of all moments is equivalent to the knowledge of the distribution function of the photons. However, the statistical error affects the intensity moments and the histogram differently. A simple, analytical form describes the statistical error of PCH, and this error is taken into account in the PCH analysis. This allows us to fit data to models and judge the quality of the fit from the residuals and the reduced χ^2 in a quantitative manner, which is an important factor for identifying and resolving multiple species [18]. The transformation of the simple expression for the error of PCH analysis to moments is quite complex. In fact, the statistical uncertainty has not been determined analytically.

Both moment analysis and PCH can be applied to dim samples. Low excitation power, low $N.A.$ objectives, or intrinsically dim fluorophores yield low molecular brightness values. Autocorrelation analysis is often not feasible under such conditions because of the experimental uncertainty in the measured autocorrelation function. But for the same conditions, both PCH and moment analysis still return accurate fluctuation amplitudes. Furthermore, PCH and moment analysis can be used to determine $G(0)$ of heterogeneous samples that consist of distinct species with different diffusion times [47]. This is very different from the autocorrelation method, where the different decay of each component of the autocorrelation function has to be considered explicitly in order to evaluate the $G(0)$ value.

14.5
Conclusions

After more than 20 years of development, fluorescence fluctuation spectroscopy is reemerging with exciting new applications in biology and chemistry. The sensitivity provided by fluorescence techniques combined with the extremely small observation volumes achieved by modern microscopy techniques allows the study of biological processes even on the single molecule level. In this contribution, we focused on two-photon fluorescence fluctuation spectroscopy. The experimental setup is relatively straightforward and all components are commercially available. We compared objectives and optical filters that have been used in our lab and discussed their relevance in the context of fluctuation measurements. Furthermore, we outlined the autocorrelation analysis with emphasis on practical issues. We revisited another analysis method, fluctuation moment analysis, and compared it with the PCH approach.

References

1. Magde D, Elson E, Webb WW (1972) Phys Rev Lett 29:705
2. Elson EL, Magde D (1974) Biopolymers 13:1
3. Ehrenberg M, Rigler R (1974) Chem Phys 4:390
4. Aragon SR, Pecora R (1975) Biopolymers 14:119
5. Kask P, Piksarv P, Pooga M, Mets Ü (1989) Biophys J 55:213

6. Koppel DE, Axelrod D, Schlessinger J, Elson EL, Webb WW (1976) Biophys J 16:1315
7. Magde D, Webb WW, Elson EL (1978) Biopolymers 17:361
8. Magde D, Elson EL, Webb WW (1974) Biopolymers 13:29
9. Widengren J, Mets U, Rigler R (1995) J Phys Chem 99:13, 368
10. Kinjo M, Rigler R (1995) Nucleic Acids Res 23:1795
11. Schwille P, Oehlenschlager F, Walter NG (1996) Biochemistry 35:10, 182
12. Borejdo J (1979) Biopolymers 18:2807
13. Brock R, Hink MA, Jovin TM (1998) Biophys J 75:2547
14. Magde D (1976) Q Rev Biophys 9:35
15. Palmer AG, Thompson NL (1987) Biophys J 52:257
16. Bendat JS, Piersol AG (1971) Random data: analysis and measurement procedures. Wiley-Interscience, New York
17. Chen Y, Muller JD, So PT, Gratton E (1999) Biophys J 77:553
18. Müller JD, Chen Y, Gratton E (1999) Biophys J (in press)
19. Qian H, Elson EL (1990) Biophys J 57:375
20. Qian H, Elson EL (1990) Proc Natl Acad Sci USA 87:5479
21. Press WH, Teukolsky SA, Vetterling WT, Flannery BP (1993) Numerical recipes in C: the art of scientific computing. Cambridge University Press
22. van Kampen NG (1981) Stochastic processes in physics and chemistry. Elsevier Science Publishing
23. Berland KM (1995) Two-photon fluctuation correlation spectroscopy: method and applications to protein aggregation and intracellular diffusion. University of Illinois at Urbana-Champaign
24. Guild JB, Xu C, Webb WW (1997) Appl Opt 36:397
25. Müller M, Squier J, Brakenhoff GJ (1995) Optics Letters 20:1038
26. Keller HE (1995) In: Pawley JB (ed) Handbook of biological confocal microscopy. Plenum Press, New York, p 111
27. Chen Y, Muller JD, Berland KM, Gratton E (1999) Methods 19:234
28. Eid JS, Müller JD, Gratton E (1999) Rev Sci Instrum (in press)
29. Thompson NL (1991) In: Lakowicz JR (ed) Topics in fluorescence spectroscopy, vol 1. Plenum, New York, p 337
30. Fahey PF, Barak LS, Elson EL, Koppel DE, Wolf DE, Webb WW (1977) Science 195:305
31. Rigler R, Mets U, Widengren J, Kask P (1993) Eur Biophys J 22:169
32. Berland KM, So PTC, Gratton E (1995) Biophys J 68:694
33. Palmer AG, Thompson NL (1989) Proc Natl Acad Sci USA 86:6148
34. Icenogle RD, Elson EL (1983) Biopolymers 22:1949
35. Weissman M, Schindler H, Feher G (1976) Proc Natl Acad Sci USA 73:2776
36. Berland KM, So PTC, Chen Y, Mantulin WW, Gratton E (1996) Biophys J 71:410
37. Meyer T, Schindler H (1988) Biophys J 54:983
38. Koppel DE, Morgan F, Cowan AE, Carson JH (1994) Biophys J 66:502
39. Aragon SR, Pecora R (1976) J Chem Phys 64:1791
40. Mets U, Widengren J, Rigler R (1997) Chem Phys 218:191
41. Widengren J, Dapprich J, Rigler R (1997) Chem Phys 216:417
42. Sjoback R, Nygren J, Kubista M (1995) Spectrochim Acta Part A 51:L7
43. Qian H, Elson EL (1991) Appl Opt 30:1185
44. Schwille P, Bieschke J, Oehlenschlager F (1997) Biophys Chem 66:211
45. Meseth U, Wohland T, Rigler R, Vogel H (1999) Biophys J 76:1619
46. Mandel L (1958) Proc Phys Soc 72:1037
47. Chen Y, Müller JD, Gratton E, Tetin SY, Tyner JD (1999) Biophys J (submitted)

CHAPTER 15

Fluorescence Lifetime Imaging Microscopy of Signal Transduction Protein Reactions in Cells

P. I. H. Bastiaens, P. J. Verveer, A. Squire, F. Wouters

The cell is a highly compartmentalised parallel processing device whose response to the extracellular environment is dependent on the spatio-temporal patterns of activity of its intrinsic biochemical machinery. This machinery mediates information transfer by interaction between – and covalent modification of – proteins. Thus, insight into the operation of signal transduction in response to cell surface stimuli could be obtained by from a 4-D image of a cell (x, y, z, t) overlaid with a map of the population evolution of proteins in one of their specific interactive or covalent states. Advances in fluorescence lifetime imaging microscopy (FLIM) in combination with the use of genetically encodable biosensors based on fluorescent protein fusions provides access to these "protein state" maps. Fluorescence resonance energy transfer (FRET) measured by donor lifetime reduction allows the detection of interacting proteins in live cells in excess of acceptor probe. Acceptor photobleaching on live cells following a FLIM time-lapse sequence provides a reference donor lifetime in the absence of an acceptor. With this reference value the pixel-by-pixel FRET efficiency for each image in the time-lapse series can be calculated. This type of reference measurement in time-lapse experiments is only possible with photophysical observables that are independent of concentration or light path length such as fluorescence lifetimes. Multi-frequency fluorescence lifetime imaging microscopy allows more complex fluorescence decay kinetics to be sampled and therefore the mapping of populations in multi-protein or multi-state processes. The encoding of a priori knowledge about the states of proteins in a global analysis of the FLIM data is essential for successfully recovering the population images.

15.1
Imaging Protein States by FRET

Cellular signal transduction is mediated by proteins through interactions, covalent modifications, conformational change and proteolytic processing. In order to observe these processes in live cells, fusion proteins can be engineered such that donor/acceptor fluorescent protein pairs from *Aequorea victoria* [1] or from the *Anthozoa* species [2] are in key structural positions that change their relative spatial configuration in one of the processes listed above [3–5]. These changes in the spatial configuration of the two fluorophores are detected via changes in the FRET efficiency. Specific antibodies labelled with acceptor probes to green fluorescent proteins such as the indocyanine dye Cy3 can be used to detect covalent modification of fusion proteins via FRET upon binding of the antibody to the modified protein site [6, 7]. On the molecular level the occurrence of FRET is discretely dependent upon the protein state due to the acute R^{-6} distance dependence of its efficiency. In other words, the fluorescent molecules are either in close proximity exhibiting efficient FRET or too far apart for detectable FRET. Consequently, on the microscopic scale, the measured FRET efficiency at each pixel in an image is a linear function of the population of protein states. These populations can thus be approximately calculated from FRET efficiency images of a two-component interacting protein system when the true FRET efficiency in the protein complex is known.

15.2
FRET Imaging by Donor Fluorescence Lifetime

The fluorescence lifetime is an intrinsic property of the fluorophore independent of probe concentration or light path length but dependent on processes depopulating the excited state such as FRET. By imaging donor fluorescence lifetimes with FLIM one can measure the FRET efficiency at each pixel through the donor photophysical properties exclusively [7, 8]. This implies that the acceptor labelled antibody need not be absolutely specific for its epitope and can be present in large excess over the donor chromophore. The latter has been exploited, for example, in live cell experiments where antibodies against a protein kinase C (PKC) phosphorylation site were microinjected in cells expressing a fusion construct of green fluorescent protein (GFP) with PKCα. Binding of the labelled antibody to PKCα, and thus phosphorylation, was detected through the donor GFP lifetime exclusively and enabled imaging of protein phosphorylation in live cells [6]. Donor lifetime FRET measurements have the additional advantage that the acceptor dye need not be fluorescent, obliterating the need for stringent optical filtering of the donor emission, resulting in improved detection sensitivity. In donor lifetime imaging, the FRET efficiency map E_i for each pixel i can be calculated from the lifetime image τ_i of the donor/acceptor system divided by the reference lifetime of the donor exclusively:

$$E_i = 1 - \frac{\tau_i}{\langle \tau^r \rangle} \tag{15.1}$$

The scalar reference lifetime $\langle \tau^r \rangle$ is calculated from a pixel averaged donor lifetime image obtained from an independent experiment with donor fluorophores exclusively. In contrast, a pixel-by-pixel donor reference lifetime measurement can be obtained by photobleaching the acceptor dye in the specimen as outlined below.

15.3
Acceptor Photobleaching in FRET Imaging

An internal reference donor lifetime measurement in the absence of acceptor can be made on the same specimen by photobleaching the acceptor dye by excitation in its absorption maximum [9]. Now the FRET efficiency E_i can be calculated with a reference lifetime τ_i^r for each pixel i:

$$E_i = 1 - \frac{\tau_i}{\tau_i^r} \tag{15.2}$$

Acceptor photobleaching can be performed with high selectivity since absorption spectra tend to tail in the blue part of the spectrum but are steep at their red-edge. Photobleaching the acceptor can take considerable time (minutes) as compared to cellular diffusion processes. A pixel-by-pixel donor reference lifetime from which a FRET map can be calculated by Eq. (15.2) can therefore, in most practical cases, only be obtained in immobilised fixed cell specimens. This type of FRET imaging is also possible with fluorescence intensity measurements where an image of the donor fluorescence is recorded prior and after acceptor bleaching [10–12]. The FRET efficiency can be easily calculated from the relative increase in the donor fluorescence after acceptor photobleaching. Care should be taken with acceptor photobleaching FRET measurements that the photochemical product of the bleached acceptor does not have residual absorption at the donor emission, and more importantly, that it does not fluoresce in the donor spectral region. In live cell experiments, the reference lifetimes τ_i^r cannot be determined at each pixel, but the average reference donor lifetime $\langle \tau^r \rangle$ can be obtained at the end of a time-lapse FLIM series by acceptor photobleaching. The pixel-by-pixel FRET efficiency in each time-lapse image can then be calculated with Eq. (15.1). This type of measurement is only possible by using fluorescence lifetime imaging since lifetimes are independent of local probe concentration and light path length, variables that change in the cell during a time-lapse experiment.

15.4
Fluorescence Lifetime Imaging Microscopy

FRET detection by frequency domain fluorescence lifetime imaging microscopy is made possible by the use of image intensifiers operating as frequency mixing devices for homodyne/heterodyne detection [13–17]. Classically, the phase shift ($\Delta\phi_i$) and demodulation (M_i) of sinusoidally modulated fluorescence are recorded at a single excitation frequency (ω), from which both the phase (τ_i^ϕ) and modulation (τ_i^M) fluorescence lifetimes are calculated at each pixel i:

$$\tau_i^\phi = \frac{\tan(\Delta\phi_i)}{\omega} \tag{15.3}$$

$$\tau_i^M = \frac{\sqrt{\dfrac{1}{M_i^2} - 1}}{\omega} \tag{15.4}$$

These are only equal to the true fluorescence lifetime for mono-exponential homogeneous lifetime samples. In donor lifetime FRET measurements the sample contains various quantities of lifetime species, the composition of which differs at every location in the image dependent on the relative populations of interacting molecules. In order to determine the true lifetime composition at each pixel within the sample, the phase and modulation must be recorded at multiple frequencies, where the reciprocal of the frequencies are in general chosen so as to span the full lifetime range in the sample. This can be achieved by exploiting the higher harmonic content in the gain characteristics of an image intensifier modulated at its photocathode in multiple frequency fluorescence lifetime imaging microscopy (mfFLIM) [18]. These frequencies are then available for simultaneous homodyne frequency mixing with matched harmonics in the fluorescent signal. Higher harmonic content in the excitation field can be achieved by using matched pairs of acousto-optic modulators [19] or pulsed light sources such as mode-locked lasers [20, 21]. By taking a series of phase dependent images over a full period (0–360°) of the fundamental, all the harmonics in the signal can be sampled at once. This requires that the Nyquist sampling criterion is satisfied for the highest harmonic component present. A Fourier analysis of the resulting phase dependent images gives the phase and modulation at each of the matching harmonic frequencies ($n\omega$). These can then be fitted to dispersion relationships for the phase shift ($\Delta\phi_i$) and demodulation (M_i) in order to determine the fluorescence lifetime composition of the sample at each pixel (i):

$$\Delta\phi_i(n;\alpha_{i,q},\tau_{i,q}) = \tan^{-1}\left(\frac{\sum\limits_{q=1}^{Q}\dfrac{\alpha_{i,q}n\omega\tau_{i,q}}{1+(n\omega\tau_{i,q})^2}}{\sum\limits_{q=1}^{Q}\dfrac{\alpha_{i,q}}{1+(n\omega\tau_{i,q})^2}}\right) \tag{15.5}$$

$$M_i(n;\alpha_{i,q},\tau_{i,q}) = \left(\left(\sum\limits_{q=1}^{Q}\frac{\alpha_{i,q}n\omega\tau_{i,q}}{1+(n\omega\tau_{i,q})^2}\right)^2 + \left(\sum\limits_{q=1}^{Q}\frac{\alpha_{i,q}}{1+(n\omega\tau_{i,q})^2}\right)^2\right)^{1/2} \tag{15.6}$$

where $\alpha_{i,q}$ is the fractional contribution to the steady state fluorescence from the q-th emitting species with corresponding lifetime $\tau_{i,q}$ at pixel i. Because the Nyquist criterion requires two images for each frequency, we limit the number of harmonics (typically $n=4$) in our excitation to minimise exposure to the cells [18]. With so few harmonics, the accuracy of population estimates ($\alpha_{i,q}$) is dramatically improved when a global analysis is used for fitting FLIM images to the dispersion relationships.

15.5
Global Analysis and the Population of States

The populations of proteins that are in a specific biochemical state at each pixel of an image can be estimated by exploiting prior knowledge about the biochemical system under study in a global analysis [22] of the FLIM data [23]. A typical application is a protein interaction system where it can be assumed that only two states are present: bound and unbound. Tagging of this binary protein system with a suitable donor/acceptor chromophore pair results in efficient FRET only in the bound state reducing the donor fluorescence lifetime for that state to a discrete value. Mixing of the bound/unbound protein states results in double exponential fluorescence decays for the donor at each pixel with spatially invariant lifetimes (global parameters) and spatially varying amplitudes (local parameters). The assumption of distinct discrete lifetimes is valid since it is likely that the spatial configuration of the donor and acceptor dyes is consistent for each protein complex. The global analysis is implemented by simultaneously analysing all pixels in the image subject to the constraint that the lifetime values of each molecular species are equal in all pixels. The results of the calculation are the two spatially invariant lifetime values and an image of the bound/unbound protein populations. This implies that the reference donor lifetime measurement (for example by acceptor photobleaching) is not necessary, a distinct advantage in live cell imaging. This analysis can be simultaneously performed on several independent FLIM data sets as long as the assumption holds that the lifetime values of each molecular species are invariant. The result of this approach is a dramatic improvement in accuracy of the estimated lifetimes and the corresponding fractional populations of each fluorescent species. A further advantage of this type of analysis is the possibility of estimating populations of two molecular species with single frequency FLIM without prior knowledge of the two associated lifetimes [23]. The true lifetimes of the donor in the presence and absence of the acceptor are also obtained in a single analysis allowing the calculation of the true FRET efficiency in the protein complex.

15.6
Conclusions

Mapping of population of protein reactants in living cells is accessible by multiple frequency fluorescence lifetime imaging microscopy on cells expressing fluorescent protein tagged fusion constructs. Interactions, proteolytic processing, covalent modifications or conformational change can be followed in a live cell by detecting FRET between tagged proteins. FLIM-based FRET imaging has the obvious advantage that donor fluorescence lifetimes are independent of probe concentration and light path length, variables difficult to correct or control in live cells. Also, a reference donor lifetime determination in the absence of an acceptor can be performed on the same cell in a time-lapse experiment by photobleaching the acceptor fluorophore at the end of the time-series. More importantly, by exploiting a priori knowledge about the biological system, population of protein states can be calculated for each pixel by using global analysis methods on FLIM data without further reference measurements.

References

1. Tsien RY (1998) Ann Rev Biochem 76:509–538
2. Matz MV, Fradkov AF, Labas YA, Savitsky AP, Zaraisky AG, Markelov ML, Lukyanov SA (1999) Nat Biotechnol 17:969–973
3. Tsien RY, Bacskai BJ, Adams SR (1993) Trends Cell Biol 3:242–245
4. Mahajan NP, Linder K, Berry G, Gordon GW, Heim R, Herman B (1998) Nat Biotechnol 16:547–552
5. Miyawaki A, Llopis J, Heim R, McCaffery JM, Adams JA, Ikura M, Tsien RY (1997) Nature 388:882–887

6. Ng T et al. (1999) Science 283:2085–2089
7. Bastiaens PIH, Squire A (1999) Trends Cell Biol 9:48–52
8. Bastiaens PIH, Jovin TM (1996) Proc Natl Acad Sci USA 93:8407–8412
9. Wouters FS, Bastiaens PIH (1999) Curr Biol 9:1127–1130
10. Bastiaens PIH, Majoul IV, Verveer PJ, Soling HD, Jovin TM (1996) EMBO J 15:4246–4253
11. Bastiaens PIH, Jovin TM (1998) In: Celis JE (ed) Cell biology: a laboratory handbook, vol 3. Academic Press, New York, pp 136–146
12. Wouters FS, Bastiaens PIH, Wirtz KWA, Jovin TM (1998) EMBO J 17:7179–7189
13. Clegg RM, Feddersen BA, Gratton, E, Jovin TM (1992) Proc SPIE 1640:448–460
14. Gadella TWJ, Jovin TM, Clegg RM (1993) Biophys Chem 48:221–239
15. Schneider PC, Clegg RM (1997) Rev Sci Instrum 68:4107–4119
16. Lakowicz JR, Berndt K (1991) Rev Sci Instrum 62:1727–1734
17. Squire A, Bastiaens PIH (1999) J Microsc 193:36–49
18. Squire A, Verveer PJ, Bastiaens PIH (1999) J Microsc (in press)
19. Piston, DW, Marriott G, Radivoyevich T, Clegg RM, Jovin TM, Gratton E (1989) Rev Sci Instrum 60:2596–2600
20. Alcala JR, Gratton E, Jameson DM (1985) Anal Instrument 14:225–250
21. Watkins AN, Ingersoll CM, Baker GA, Bright FV (1998) Anal Chem 70:3384–3396
22. Beechem JM (1992) Meth Enzym 210:37–54
23. Verveer PJ, Squire A, Bastiaens PIH (1999) Biophys J (submitted)

CHAPTER 16

New Techniques for DNA Sequencing Based on Diode Laser Excitation and Time-Resolved Fluorescence Detection

M. Neumann, D.-P. Herten, M. Sauer

16.1
Introduction

In the foreseeable future the complete sequencing of all 3×10^9 base pairs of the human genome will be finished using Sanger's enzymatic chain termination method [1] in combination with various automated DNA sequencing machines [2–10]. However, to understand the function of each gene and the corresponding health implications, genetic variations in different cell types, individuals, and organisms have to be investigated. Hence, alternative methods have to be developed that are even faster, more efficient, more accurate, and more cost-effective. Among these new methods are capillary array electrophoresis [11–17], mass spectrometry [18], sequencing by hybridization [19–23], and single-molecule sequencing [24–31]. Probably, the human genome will be sequenced before one of these techniques is widely used. However, they will provide a powerful tool for biological scientists in the twenty-first century. In the following sections of this chapter we restrict our discussion on new DNA sequencing methods with fluorescence-based detection and identification schemes.

16.1.1
The Multiplex Dye Principle and Pattern Recognition

In most established methods, identification of DNA fragments is based on difference in emission spectra of the dyes used to label the fragments. The selectivity is achieved by suitable bandpass filters in front of a photomultiplier tube detector or by a spectrometer combined with a charge-coupled device (CCD) camera. However, as the emission spectra of different fluorescent labels overlap, the identification accuracy decreases. In addition, in routine diagnostics there is growing need for multiparameter analysis, i.e., more than four labels have to be identified with high fidelity. To increase the number of discernable labels the "Multiplex Dye Principle" (Fig. 16.1) might be a useful alternative. The principle takes advantages from the fact that each fluorescent dye exhibits a characteristic fluorescence decay time (lifetime) that can be used for its unequivocal identification. Hence, not only can the spectral information of the label be utilized, but so can the fluorescence lifetime.

In order to excite different labels efficiently with a single monochromatic light source we developed fluorescent dyes whose absorption and emission

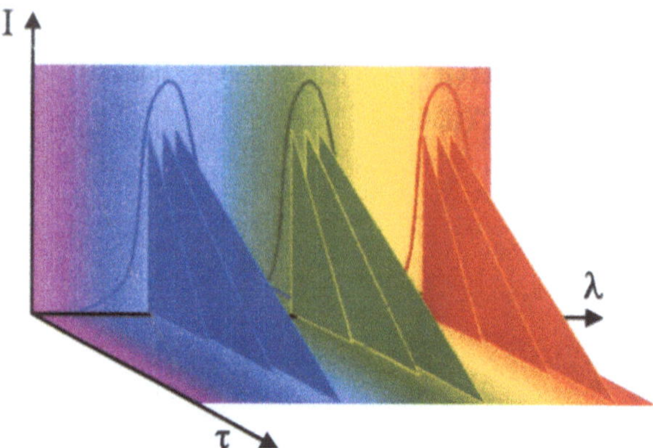

Fig. 16.1. The multiplex dye principle. Each label can be identified by both the spectral signature and the characteristic fluorescence lifetime

wavelengths are nearly identical, which differ, however, in fluorescence lifetime. Fluorescence dyes with these characteristics are called multiplex dyes [32–35].

In practice, more than 10,000 photon counts have to be acquired for the exact determination of an unknown fluorescence lifetime using the time-correlated single-photon counting (TCSPC) technique. However, from the experimental point of view, in ultrasensitive fluorescence measurements only a restricted number of photon counts are available for the correct identification of each label. One method for fluorescence lifetime determination in experiments with low photon count statistics is the use of a monoexponential maximum likelihood estimator (MLE)-algorithm [36, 37]:

$$1 + (e^{T/\tau} - 1)^{-1} - m(e^{mT/\tau} - 1)^{-1} = N^{-1} \sum_{i=1}^{m} i N_i , \qquad (16.1)$$

where T is the width of each time channel, m the number of utilized time channels, N the number of photon counts taken into account, and N_i the number of photon counts in time channel i. MLE-algorithms have been successfully used to distinguish between different fluorescent labels on the single molecule level with high accuracy [38–42]. In these experiments efficient lifetime determination was performed even with less than 100 collected photon counts per dye.

Besides the determination of the measured fluorescence decays, the discrimination can also be achieved by comparing the raw data with the fluorescence decay expected for the used labels, i.e., by use of a pattern recognition technique. Here the fluorescence decay patterns of the labels have to be recorded under the same experimental conditions as the test data but with high photon count statistics. Hence, from a statistical point of view, one is confronted in the experiment with a problem of classification, i.e., which pattern describes the measured decay at best. An optimal way of classification has been developed in the framework of information theory and is described in detail elsewhere [43, 44]. The application of this statistically optimal classification algorithm to the identification of fluorescence patterns of different labels leads to an extraordinary reliability even in case of very weak signals. Recently, Köllner et al. [45] demonstrated that three dyes can be distinguished with an identification accuracy of ~99.9% with only 300 detected photon counts, i.e., only one misclassification in 10,000 measurements.

16.2
DNA Sequencing in Capillary Gel Electrophoresis by Diode Laser-Based Time-Resolved Fluorescence Detection

16.2.1
Semiconductor Lasers as Efficient Excitation Source in the Red Spectral Region

Due to the limited number of compounds which show intrinsic absorption and emission above 600 nm, the use of near-infrared (NIR) fluorescence detection is a desirable alternative. Therefore we aimed for the development of a DNA sequencing technique which is based on semiconductor technology and time-resolved identification of single-stranded DNA fragments due to the characteristic fluorescence lifetime of the labels. Semiconductor lasers as consumer electronic device offer the advantages of low cost, small size, low power consumption, and long service lifetime. Meanwhile time-correlated single-photon counting (TCSPC) acquisition can be realized by new PC plug-in cards [46, 47]. In addition, pulsing of semiconductor lasers is easily obtained by current modulation at repetition rates of up to some hundreds of MHz. The avalanche photodiode detectors (APDs) used generate standard digital pulses of arriving photon counts with high quantum efficiencies and count rates of more than 10 MHz. Hence, both devices are ideally suited for a rugged, reliable, and miniaturized instrument.

16.2.2
Design of Multiplex DNA Sequencing Primers

Fluorescent dyes used for a one-lane four-dye time-resolved DNA sequencing technique with pulsed diode lasers should fulfill several requirements:

1. The dyes should exhibit similar absorption and emission spectra with high extinction coefficients at the emission wavelength of the used diode laser and high fluorescence quantum yields.
2. Their fluorescence lifetimes should be distinguishable under sequencing conditions.
3. They should be easily coupled to modified oligonucleotides.
4. The mobility shifts of labeled DNA fragments in capillary gel electrophoresis (CGE) should be uniform.

Figure 16.2 shows the molecular structure of some new developed fluorescent dyes together with the commercial available carbocyanine dye Cy5. Details of dye preparation are published in the literature [48]. The dyes have carboxyl group for mild covalent coupling to amino groups of modified oligonucleotides.

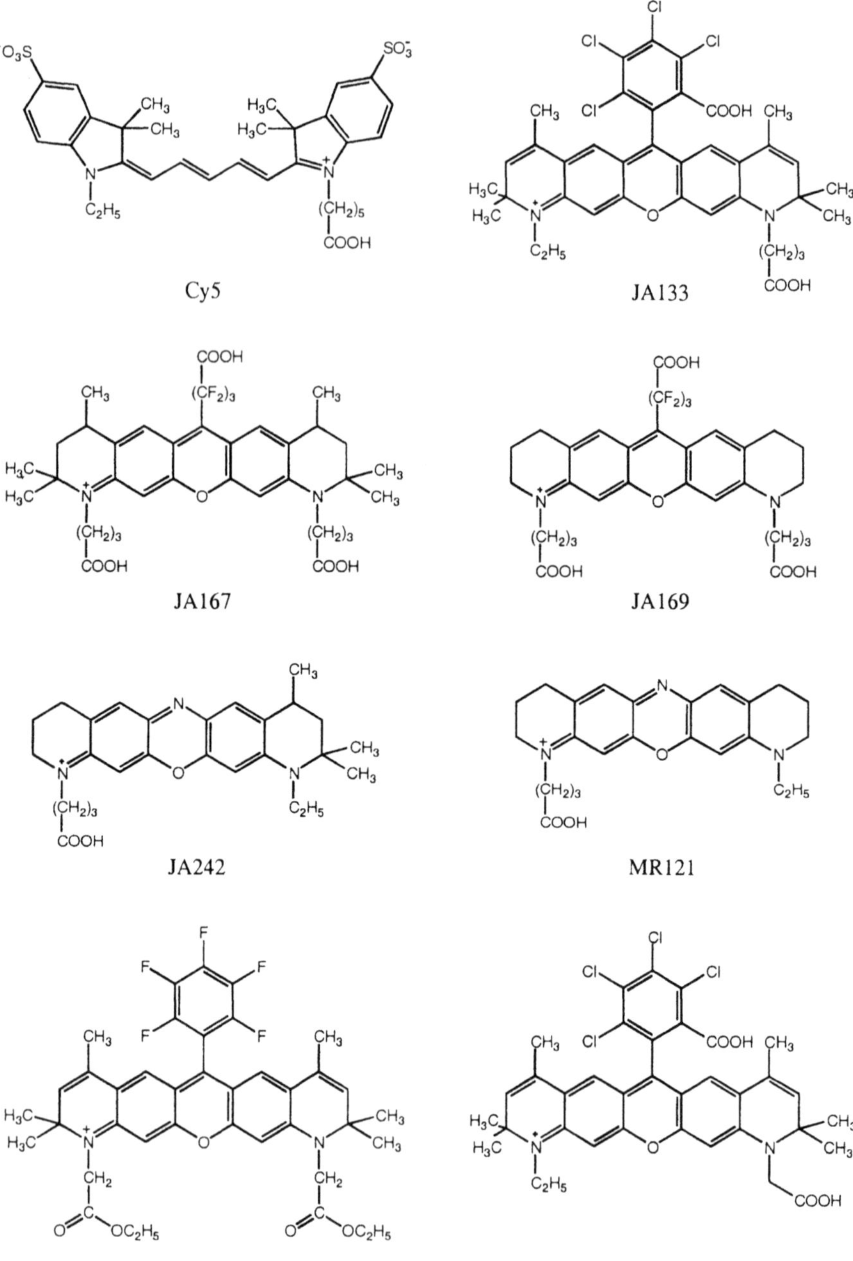

Fig. 16.2. Molecular structures of the used multiplex dyes

Table 16.1. Spectroscopic characteristics (fluorescence lifetime τ, quantum yield Φ_f, and absorption $\lambda_{abs,max}$ and emission maxima $\lambda_{em,max}$) and relative mobilities μ_{rel} of several investigated dye-primer conjugates

Primer	τ (ns)	μ_{rel}	$\lambda_{abs,max}$ (nm)	$\lambda_{em,max}$ (nm)	Φ_f
MR200–1-M13–21	3.6	1009	624	643	0.77
JA169-M13–21	2.9[a]	1006	632	656	0.36
JA242-o-M13–21	2.4[a]	1015	672	686	0.21
CY5-M13–21	1.6	1000	649	666	0.44

The spectroscopic data were measured in double-distilled water. Relative mobilities μ_{rel} were obtained in a linear polyacrylamide gel with 8% T at 25 °C and 200 V/cm.
[a] Average fluorescence lifetimes τ_{av} are given.

Their spectroscopic properties are listed in Table 16.1. As can be seen, the fluorescence lifetimes of the dyes are adequately long (ns range) to allow their detection and identification with standard detectors. While MR200–1, and JA169 are rhodamine dyes, JA242 is an oxazine derivative. The commercially available carbocyanine dye CY5 is used because of its suitable absorption and emission properties and its short fluorescence lifetime. After covalent attachment to oligonucleotides the spectroscopic properties of the dyes change slightly. A bathochromic shift in the absorption and emission maxima occurs, except in case of Cy5, which is not affected. All dye primer conjugates exhibit a longer average fluorescence lifetime than the free dye under the same experimental conditions. The multiexponential fluorescence kinetics of some conjugates can be described by a biexponential model with a longer lifetime and a shorter lifetime compared to the monoexponential fluorescence lifetimes of the free dyes [35, 49, 50]. While an interaction of rhodamine and oxazine dyes with adenosine, cytosine, or thymidine leads to the prolongation of the measured decay, an aggregation with guanosine residues results in fluorescence quenching (shorter fluorescence lifetime).

The activated dyes were coupled to the standard oligonucleotide primer M13–21, which carried an amino group attached either directly at the 5′-terminus or at a modified thymidine-base [49] and were purified by reversed phase high performance liquid chromatography (HPLC). Dependent on the dye and the oligonucleotide, respectively, the achieved reaction yields are 10–80%. In order to attain similar peak heights (due to different fluorescence quantum yields and extinction coefficients) the dye labeled primers are used at different concentrations in the four sequencing reactions.

As expected, the mobility of labeled DNA fragments is influenced by the dye structure, the coupling position and the length of the spacer. Table 16.1 shows the relative mobilities μ_{rel} of the optimized dye-labeled-primers. The relative mobility μ_{rel} of the sample (μ_{sample}) was calculated corresponding to the equation

$$\mu_{rel} = \frac{\mu_{sample}}{\mu_{standard}} = \frac{L_e L_t / U t_{R,sample}}{L_e L_t / U t_{R,standard}} = \frac{t_{R,standard}}{t_{R,sample}} \tag{16.2}$$

where L_e and L_t are the effective and the total length of the capillary, U is the applied voltage and t_R is the migration time of the dye labeled oligonucleotide. As standard we used a labeled DNA fragment with a very low mobility.

As can be seen in Table 16.1, the mobilities of the differently labeled primers are well distinct in a linear polyacrylamide gel $(8\% T)$ under denaturing conditions. Therefore problems in sequence determination might arise. Poor resolution of adjacent bases and miscalling of bases might occur. To circumvent this problem, the mobility shifts of multiplex DNA sequencing primers were matched by varying the spacer length between the dye and the oligonucleotide and the coupling position. In addition, the dye structure strongly affects the mobility of the conjugate. We have experimental evidence that both the hydrophobicity and the steric requirements of the dye structure play an important role in the mobility of dye labeled DNA fragments in CGE. A higher hydrophobicity of the dye enhances the aggregation tendency to the oligonucleotide. Using a relatively hydrophobic dye (MR200–1), the shorter the linker between the chromophore and the oligonucleotide, the better the mobility in capillary gel electrophoresis. Although the mass-charge-ratio of the conjugate with a small linker is greater, it migrates faster during gel electrophoresis. On the other hand, the mobility of hydrophilic dyes is controlled by the mass-charge-ratio of the conjugate (JA242, JA169, Cy5). However, the mobilities of the differently labeled sequencing primers can be matched. Hence, we developed four dye-primer-conjugates with similar mobility shifts in $5\% T$ polyacrylamide gels [49, 51].

The unique electrophoretic properties in CGE of DNA fragments generated by our four different labeled sequencing primers is confirmed by single-base extension experiments. The extended (ddTTP, dNTPs) DNA fragments exhibit almost the same mobilities. DNA fragments generated with Cy5 labeled primer move slightly slower than the JA169, MR200–1 and JA242-OCTA fragments (about 0.2 bases), but the base sequence is never changed during electrophoresis. A correction of the mobility shift is not necessary throughout a sequencing run.

16.2.3
4-Dye-1-Lane Multiplex DNA Sequencing

The schematic confocal set-up for multiplex dye DNA sequencing in capillary electrophoresis is shown in Fig. 16.3. A semiconductor laser at 630 nm serves as the excitation source. Repetition rates of up to 60 MHz are achieved using an electronic pulse generator.

Laser-light pulses of less than 500 ps (FWHM) at an average power of up to 3 mW are generated. The instrument response function (IRF) of the entire system was measured to be about 600 ps (FWHM). The laser beam is collimated, passes a bandpass filter, and is directed to a microscope objective by a dichroic beam splitter. The capillary is mounted on top of an xy-translational stage in front of the objectives aperture. The fluorescence signal is collected by the same objective, filtered by a bandpass filter, and imaged by an achromatic lens onto

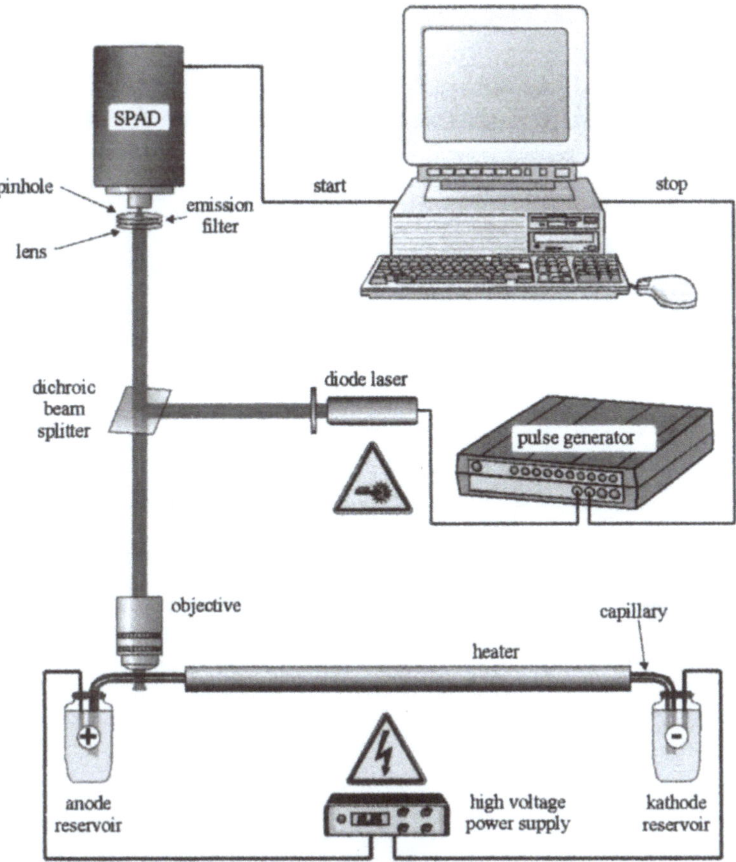

Fig. 16.3. Schematic diagram of the optical and electronic set-up

the active area of an avalanche-photodiode (SPAD: single photon avalanche diode). A 200 µm pinhole located directly in front of the SPAD limits the confocal volume. The signal of the SPAD is directly fed into a time-correlated single-photon counting (TCSPC) PC plug-in card. The fluorescence signal is collected in histograms with an integration time of 1 s each. The data are buffered in the memory and saved onto the computers hard disk drive. To get an electropherogram out of the histograms all photon counts of each decay are accumulated into one second time intervals and plotted vs the migration time. The background signal of the multiplex DNA sequencer is measured to be about 10 kHz with a stable baseline. The DNA-fragments are detected with an average signal-to-background (S/B) ratio of 6.5. Even for the low intensity bands of long DNA-fragments (longer than 500 bp) signal-to-background ratios of about 4 were determined. Figure 16.4 shows a section of a typical sequencing run. An overall accuracy of correct classification of better than 90 % for up to 660 bp was achieved using the pattern recognition technique described above [49, 51].

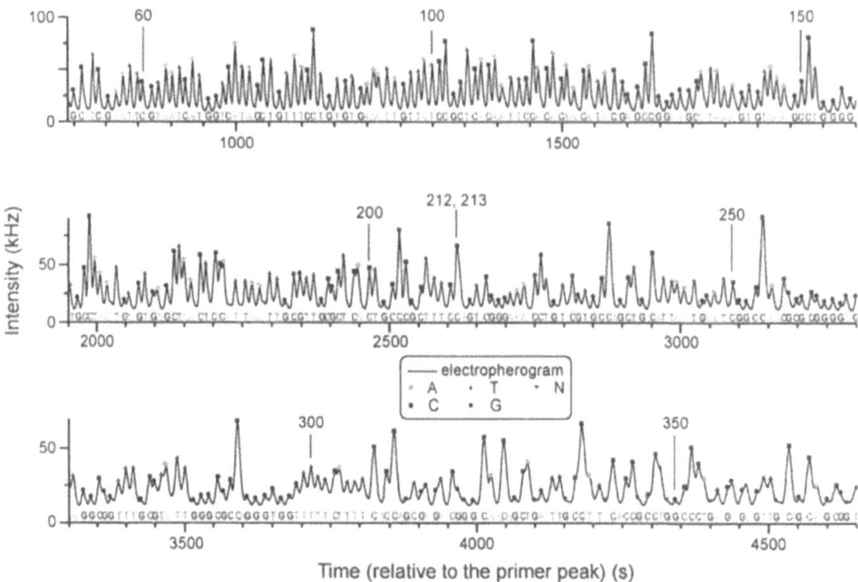

Fig. 16.4. Section of an electropherogram of a time-resolved DNA sequencing run. The raw data are shown. The electrophoresis was performed in a 5% PAA gel at 160 V/cm and 52 °C. N denotes missing bases

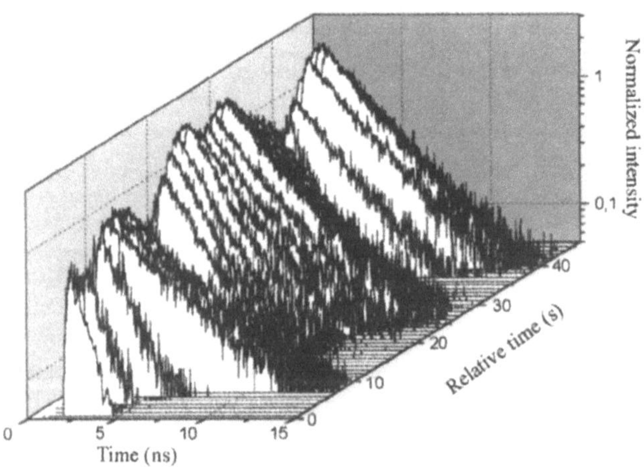

Fig. 16.5. 3-D view of a small section of an electropherogram (bases 135–138)

Figure 16.5 gives a 3-D view of a small section of an electropherogram (bases 135–138). The different fluorescence decays of the four multiplex dyes are shown. In fact, we detect four parameters simultaneously in our time-resolved DNA sequencer to identify a fragment in a given sample: (a) the fluorescence intensity (count rate), (b) the spectral property, (c) the fluorescence decay profile, and (d) the migration time of the compound. The results indicate the superior performance of pattern recognition compared to the fluorescence lifetime determination by a maximum-likelihood estimation assuming monoexponential fluorescence decays. A correct distinction between the two dyes JA169 and JA242 is difficult using the latter approach because JA169 and JA242-OCTA DNA fragments show a biexponential fluorescence decay. In contrast to the maximum-likelihood-estimation the pattern recognition method considers this multiexponential behavior. Hence, an unequivocal identification of JA169- and JA242-labeled fragments is possible, even during electrophoresis. The base identification due to the different fluorescence lifetimes of the attached dyes is nearly independent of the fragment length.

The results clearly demonstrate that one-lane four-dye DNA sequencing with time-resolved identification represents an elegant alternative to conventional sequencing methods. Dyes in the red region of the visible spectrum are promising in DNA sequencing. They have reasonably high fluorescence quantum yields and extinction coefficients. The reduced Rayleigh scattering and fluorescence background improves the obtainable signal-to-background ratio. Another advantage is the use of low cost and compact diode lasers for excitation and avalanche photodiodes for detection. The entire system is compact and rugged compared to the conventional instruments with argon ion lasers and photomultipliers or CCD cameras.

16.3
High-Throughput DNA Analysis

Electrophoretic separations are a bottleneck in DNA sequencing since the sample throughput of a single capillary DNA sequencer is poor compared to a slab gel instrument with parallel lanes. The main advantage of CE is automation. Hence, the labor intensive key steps in slab gel electrophoresis like gel pouring and sample loading can be easily replaced in CE by using low-viscosity polymer solutions as sieving matrix and autosampling. Besides the importance of large scale sequencing and its need for long read sequencing runs, there is an increasing demand for even more rapid DNA sequencing techniques, especially in cDNA mapping.

16.3.1
Increasing the Speed of Electrophoresis

Due to the onset of biased reptation [52, 53] of DNA fragments at higher electric fields it is not possible to reduce the duration of electrophoresis without significant loss of resolution by increasing the electric field strength. The effective length of the capillary is another parameter that determines the resolution of adjacent bases in DNA sequencing. The separation efficiency increases with capillary length at the same electric field, but the separation affords more time in longer capillaries. Hence fast sequencing is a compromise of two competing strategies: (a) separation in short capillaries, or (b) separation at higher field strength. For both a loss of read length is expected.

The electrophoretic mobility μ of a DNA fragment depends on the separation distance l, the electric field E and the migration time t:

$$\mu = \frac{l}{Et} \tag{16.3}$$

Different models have been developed to predict the mobility of DNA in gel electrophoresis. The Ogston model [52, 54] treats the gel as a molecular sieve with a distribution of different pore sizes. It is assumed that the DNA fragments move as a random coil smaller than the average pore size, which is determined by the gel concentration. Ogston derived an expression for the mobility of a DNA fragment of length N (in bases):

$$\ln \mu = \ln \mu_0 - \alpha N \tag{16.4}$$

where μ_0 is the mobility in pure solvent and a is a constant, which is characteristic for the concentration of the gel. A plot of log mobility vs fragment length would be linear in this separation regime.

If the random coil of DNA is too large to pass through the gel pores, it is assumed, that the fragment moves forward like a reptile. The biased reptation model [52–55] describes the behavior of DNA fragments larger than the average pores. The fragment mobility under a low electric field is expressed by the following equation:

$$\mu = \mu_0 \left(\frac{1}{3N} + \frac{\varepsilon^2}{27} + \frac{2N\varepsilon^4}{1215} + \dots \right) \tag{16.5}$$

where N and e are reduced variables. $N = M/M_a$ with M the length of DNA and M_a corresponds to the size of an unperturbed random coil of DNA that fits exactly into the average pore size a. $\varepsilon = E/E_a$ where E is the applied electric field and E_a the electric field, for which the drop in potential energy of a molecule of size M_a across pore size a is equal to the thermal energy:

$$E_a = \frac{2kT}{qM_a} \tag{16.6}$$

Here q is the net electric charge per base, k is the Boltzmann constant, and T is the temperature in degrees Kelvin. The plot of log μ vs log N is linear if reptation occurs. The high field saturation, for which there is no DNA size dependence of mobility, limits the maximum read length in DNA sequencing.

Figure 6a shows the mobility vs fragment length relation in a 7-cm short capillary at different electric fields. The slope of the reptation regime decreases if a higher electric field is applied. Thus the difference in the mobility of adjacent bases is smaller at high field strengths and the onset of oriented reptation due to stretching of fragments is shifted to shorter fragment lengths. Figure 6b indicates that the onset of oriented reptation is nearly independent of the capillary length. There is only a slight difference between the measured mobilities at different capillary lengths.

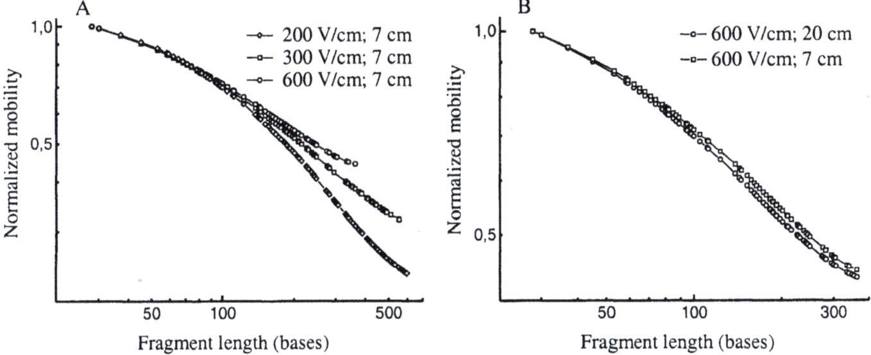

Fig. 16.6 A, B. Normalized mobility of DNA fragments vs fragment length: **A** 7 cm capillary length at different electrical fields; **B** influence of the capillary length on the onset of oriented reptation

It is remarkable that the onset of oriented reptation coincides with a decrease of the peak to peak distances. Although the total separation times for the 20-cm capillary at 600 V/cm and the 7-cm capillary at 300 V/cm are quite similar, it is remarkable that the shorter capillary separates even more fragments under these conditions. Table 16.2 summarizes the migration times for the primer, the end peak and some fragments at different electrophoresis conditions.

The electropherogram of a typical DNA sequencing run in a 7-cm capillary is shown in Fig. 16.7. Besides the good resolution of single-base extension fragments of up to 400 bases in length, the electropherogram shows the excellent quality of raw data that could be achieved using diode laser excitation of red fluorescent dyes. The onset of oriented reptation around 400 bases coincides with the loss of single base resolution in the electropherogram. The resolution of up to 400 bases would be sufficient for cDNA mapping. In cDNA mapping a relative short sequence of up to 200–300 bases of a PCR-amplified DNA and its size has to be determined [56].

Table 16.2. Migration times of DNA fragments N of different length at different field strengths E_{elec} and different effective capillary lengths L_{eff}

E_{elec} [V/cm]	L_{eff} [cm]	Migration time [min]				
		Primer	End peak	N 100	N 201	N 295
200	7	3.7	19.8	6.2	9.2	12.1
300	7	2.7	11.5	4.7	6.4	7.8
600	7	1.2	3.7	2.1	2.8	3.2
600	20	3.5	10.4	6.3	8.4	9.6

Experimental conditions: capillary i.d. 50 µm, o.d. 375 µm, total length 27 cm; 6% T, 0% C polyacrylamide gel; buffer 50 mmol/l TAPS; 50 mmol/l tris; 2 mmol/l EDTA; 7 mol/l urea; temperature 50 °C.

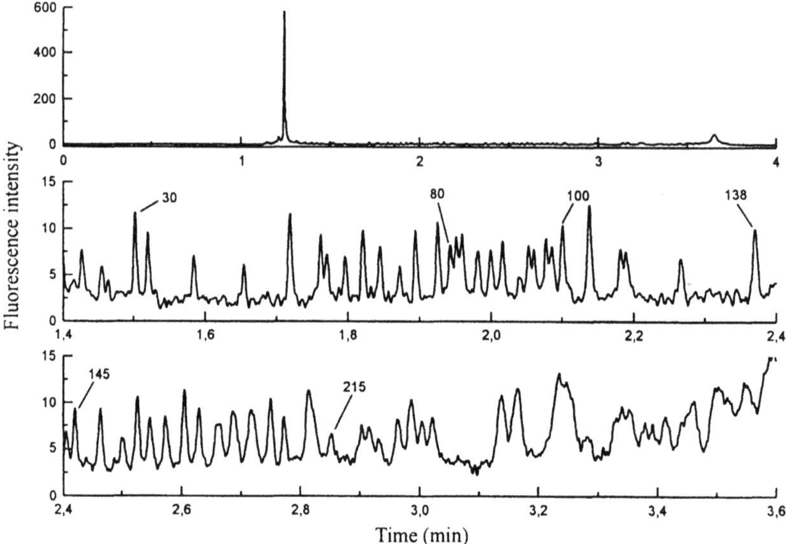

Fig. 16.7. Electropherogram of a typical DNA sequencing run in a 7-cm capillary at 200 V/cm and 50 °C (capillary i.d. 50 μm, o.d. 375 μm, 6% T polyacrylamide gel, buffer 50 mmol/l TAPS; 50 mmol/l tris; 2 mmol/l EDTA; 7 mol/l urea)

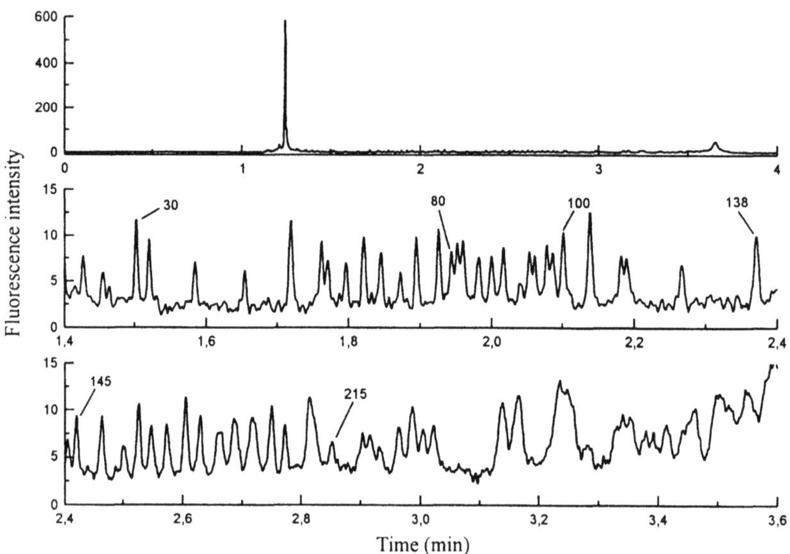

Fig. 16.8. Electropherogram of a DNA sequencing run in a 7-cm capillary at 600 V/cm and 50 °C (same conditions as Fig. 16.7)

If sequencing of up to 200 bases suits a distinct application, the speed of separation can be further enhanced. Figure 16.8 shows a very rapid separation of DNA fragments in a 7-cm capillary at 600 V/cm. The end peak is observed at 230 s and fragments up to 200 bases in length are resolved. The average peak to peak distance of adjacent bases amounts to 1 s (standard deviation $\sigma = \pm 0.17$). Hence, the minimum data rate that should be handled by the detection system is 2 Hz. In addition, a capillary array for high throughput rapid DNA sequencing would afford multiple rates for data acquisition.

The quotient of read length and run time demonstrates the increased throughput of rapid separations in short capillaries. In standard CE approximately 600 bases are separated in 2 h resulting in 5 b/min (bases per minute). Separating 400 bases in 20 min gives an increase by a factor of four (20 b/min). For the rapid separation of 200 bases in less than 4 min the sequencing performance reaches 50 b/min, which is about 20 times the speed of a single slab gel lane. This result indicates the dramatic enhancement of separation speed in short capillaries.

16.3.2
Construction of an Ideal Capillary Array Electrophoresis Instrument (CAE)

An ideal high-throughput DNA analyzer should offer high sensitivity for minimum sample and reagent consumption, a compact size to save lab space, a rugged design, an easy set-up procedure, and a long period of "walk away operation". At least the instrument should work automatically for 24 h without any human interference (sample loading, injection, etc.). An automated CAE system has to carry out different tasks like sample preparation, capillary pre-rinse, autosampling, injection, electrophoresis, detection, and data analysis. Therefore many mechanical, optical, and electrical components have to be integrated, e.g., a high voltage supply, a LIF detection unit, a capillary holder, microtiter plate carriers, robotic arms, a pump, and a heater.

There are two main concepts for fluorescence detection in a capillary array: the imaging and the scanning approach. Imaging systems [11,57–60] collect the fluorescence light of all capillaries simultaneously, thus providing a 100% duty cycle without any moving parts. In general, the laser enters the array from the side or the laser beam is shaped into a line in order to illuminate the capillaries at the same time. Due to the Gaussian laser beam profile of the used gas lasers the excitation light intensity varies from capillary to capillary. A large-diameter lens with a low numerical aperture has to be used to image the fluorescence light with a CCD camera. Imaging systems are more susceptible to laser scatter from the capillary walls compared with confocal scanners because of limited spatial filtering. Multiple-sheath flow systems offer higher sensitivity. The detection takes part outside the capillary in a gel-free optical cell. The DNA fragments are eluted from the capillary into the sheath flow. Compared to on-column detection the scatter is drastically reduced.

In scanning systems [13–17] either the capillary array or the detection system has to be moved along the scan axis. The sampling rate and the scan velocity should allow the detection of DNA fragment bands without loss of resolution. The background fluorescence from the capillary and scattering at its surface can be reduced using a high numerical aperture microscope objective and a pinhole in front of the detection device for confocal sectioning. A confocal capillary array scanner offers high sensitivity. Additionally, there is no demand for excessive laser power. A laser power of less than 1 mW is sufficient for efficient excitation.

The maximum number of capillaries is limited in both types of systems. In imaging systems the maximum size depends on the number of optically resolved detection channels, i.e., by the pixel number of the CCD camera and the optical resolution of the collection lens. In scanning systems the motion of the translational stage and the resulting sampling rate are the limiting factors. In addition, the fluorescence intensity of each data point decreases drastically at higher scan velocities.

16.3.3
Capillary Array Scanner for Time-Resolved Fluorescence Detection

For time-resolved detection in a capillary array the basic multiplex DNA sequencer is equipped with a motion controller driven x, y-microscope scanning stage. For synchronization of the scanning motion with the detection device Windows 32-based software was developed that controls the TCSPC-card and the motion controller simultaneously. The motion controller is connected to the serial port of the PC. The acquired electropherograms are displayed online. Other software allows data analysis on the basis of pattern recognition algorithms and other mass data procedures.

16.3.3.1
Discontinuous Bidirectional Scanning

A capillary array scanner must exhibit proper alignment and scanning of the capillaries along with sensitive detection. In order to obtain high scan rates and sufficient duty cycles the motion of the scanning stage should be as quick as possible. On the other hand, in time-resolved fluorescence detection the time to acquire a fluorescence decay profile (collection time) in each capillary should be as long as possible to collect enough photon counts. In continuous scanning mode the collection time per capillary depends on the scan speed, the diameter of the capillaries, and their distances, i.e., the collection time decreases with increasing scan speed. The capillaries should be mounted close to each other. The ratios of the outer and inner diameter of the used capillaries should be small to obtain sufficient collection times and scan rates. Nevertheless such capillaries are susceptible to mechanical breakage. Furthermore, a data overhead is generated by unsynchronized continuous scanning. For example, in capillaries with an outer to inner diameter ratio of 2 the same amount of data points is collected inside and outside of the capillaries. Of the collected data, 50% do not coincide with fluorescence light from the capillaries. In fact, the actual amount of information is even smaller due to strong scattering at the inner capillary walls.

A discontinuous, bidirectional scanning mode can circumvent this problem. The synchronization and the control of the alignment and the measurement process by software on a personal computer offers two main advantages: (a) the collection time is independent from the scan speed (determined by the user), and (b) there is no data overhead generated during the measurement and the lanes can be separated into individual data sets.

In practice, the collection of photon counts starts as the scanning stage stops at the desired capillary. After the desired collection time a signal is sent to the motion controller to align the next capillary. Once the capillary is aligned properly (settling time) data acquisition is restarted. After each capillary is sampled once, the scan direction is reversed. The next scan line starts at the last capillary of the previous line. Hence, the total time (t_{total}) for a scan line with n capillaries is given by $t_{\text{total}} = t_{\text{collection}} n + t_{\text{settling}} (n-1)$ with an average scan rate $f = 1/t_{\text{total}}$. The sampling intervals for the capillaries vary due to their positions in the array.

The settling time of the entire system is 80–150 ms for capillaries with an outer diameter of 375 µm at a collection time of 50 ms (the settling time is not completely independent of the collection time due to hardware problems, see [61] for further details). For example a continuous scanning mode with a constant velocity of 10 mm/s and a settling time of 40 ms passes the inner diameter of the capillary in only 7.5 ms (capillary i.d. 75 µm). To reach a collection time of 50 ms in each capillary the speed of motion has to be reduced more than six times, resulting in a poor settling time of 267 ms. This demonstrates the superior performance of the discontinuous scanning mode which allows faster settling with longer collection times. The settling process in discontinuous scanning is 2–3 times faster compared to the continuous approach. We obtained scan rates of up to 0.5 Hz on 16 parallel capillaries.

16.3.3.2
Time-Resolved Detection in Parallel Capillaries

The detection limits of the confocal scanner with time-resolved fluorescence detection were calculated from measurements of diluted solutions of multiplex dyes in the capillary. The detection volume is estimated from the magnification of the used microscope objective (40×) and the spatial filter (200 µm pinhole) to approximately 500 fl. The detection limits were determined at the point of deviation from linearity in log intensity vs log concentration plots. The achieved detection limits in the subzeptomol range (corresponding to about 20–150 dye molecules) clearly demonstrate the incredible sensitivity of the developed system.

The experimental setup for capillary array electrophoresis with time-resolved fluorescence detection is similar to Fig. 16.3. Figure 16.9 shows the raw data of a multiplex DNA sequencing run (same primer set as in single capillary experiments) in eight parallel capillaries. The sequencing fragments are well resolved although the electropherograms depict raw data without any further processing. As expected from the high spatial resolution of confocal microscopy, no cross-talks between different capillaries are observed. Using a collection time

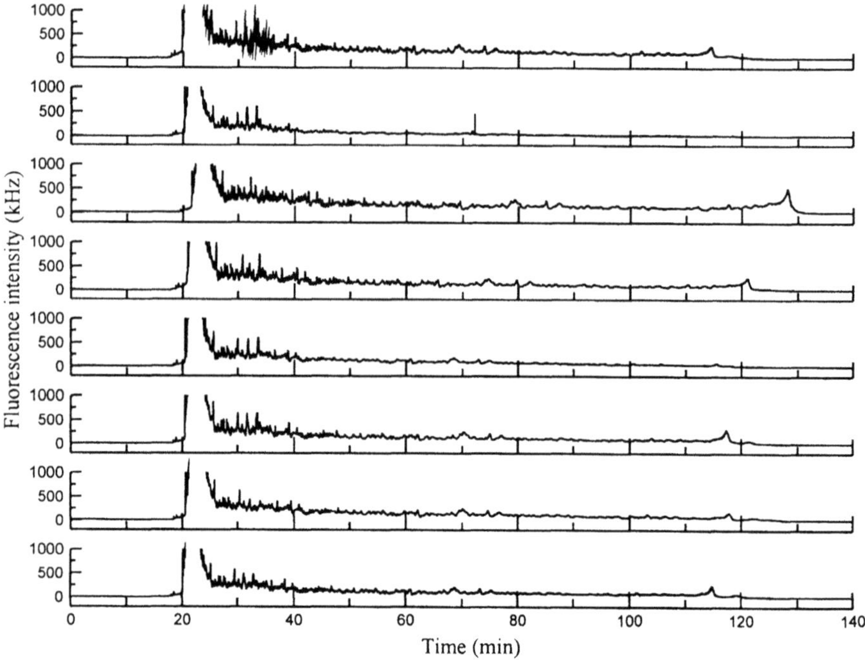

Fig. 16.9. Electropherograms of a time-resolved multiplex DNA sequencing run in eight parallel capillaries

of 100 ms, the system reached a scan rate of 0.45 Hz. As a peak crosses the detection region at least four sweeps were made under separation conditions which is sufficient to render the peaks. With a background rate of 5 kHz, signal-to-background ratios varying between 10 and 120 were achieved during electrophoresis, which corresponds to signal intensities between 50 kHz and 600 kHz for different DNA fragments. Up to 400 bases can be classified to the corresponding region of the M13mp18 sequence directly from raw data. It should be noted that data processing with peak identification algorithms would further extend the read length.

Figure 16.10 gives an insight into the time-resolved dimension of the data. The capillary array scanner utilized for time-resolved fluorescence detection and identification uses an unique discontinuous, bidirectional scanning mode. This mode minimizes the settling time in relation to the fluorescence collection time, allowing longer sampling intervals compared to continuous scanning devices. In addition, the synchronization of the alignment and measurement process does not produce any data overhead and is capable of online displaying of results. Besides high detection sensitivity the confocal detection setup offers spatial separation of the individual channels without any cross-talk. Detection limits for different dyes in the subzeptomol range corresponding to 20–150 detected dye molecules have been achieved. The identification of different fluorescently labeled probes in parallel capillaries by the characteristic fluores-

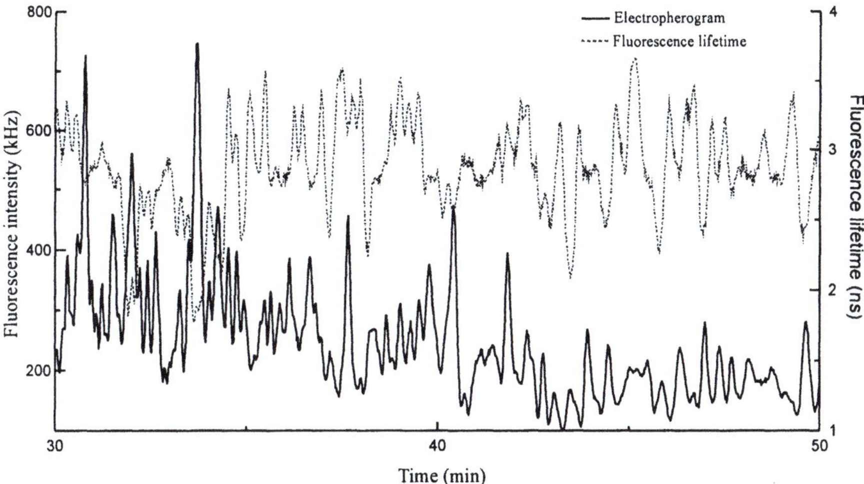

Fig. 16.10. Expanded section of an electropherogram recorded in one capillary. In addition the fluorescence lifetime determined by use of a MLE algorithm are given

cence decay time of the attached dyes is possible. Thus the entire system can be applied to parallel rapid sizing of labeled DNA fragments (e.g., PCR fragments or dsDNA) and for multiplex DNA sequencing in a capillary array. To increase the number of capillaries, i.e., the throughput of the system, a more sophisticated scanning hardware has to be developed to decrease the settling time per capillary. With collection times of 10 ms the settling time should be in the order of a few milliseconds to obtain suitable scan rates. Another very promising technique is the use of time-gated CCD cameras for time-resolved fluorescence detection. In combination with low-cost excitation sources like diode lasers, such cameras would provide a 100% duty cycle in capillary array electrophoresis.

16.4
Sequencing by Hybridization (SBH)

Besides enzymatic DNA sequencing, other important applications in the field of fluorescence techniques related to DNA-sequencing are DNA-chips. Their potential to simplify, accelerate, and even increase the sensitivity of clinical tests for almost all diseases give rise to increasing commercial interest. Those advantages are met due to a process of molecular recognition that has been evolved within the last two billion years to recognize complementary strands of DNA with an incredible high accuracy: the hybridization of oligonucleotides from 11 to 20 bases in length can be performed mismatch free. Even hybrids containing a single mismatch can be eliminated due to their decreased melting point, which is about 5–10 °C below the perfect hybrid [62]. Hybridization on a DNA-chip is utilized for recognition of a gene of interest (e.g., a specific region of a germ or a pathogenous human gene) by specific binding to complementary strands of DNA or oligonucleotides of 20 to some 100 bases length, that were immobilized on the surface of a silicon wafer or a glass chip. Although the potential of DNA-chips as specific and sensitive assays for genomic analysis is favored currently, it must be mentioned that the principle idea of DNA-chips was developed for their application in DNA-sequencing. On a theoretical basis Drmanac et al. showed in 1988 that DNA-chips could be used as an universal library of DNA-sequences [21]. The different number of sequences, that may be expressed by oligonucleotides of eight bases in length is $4^8 = 65536$ (65 K). Hence any existing DNA-sequence should completely hybridize on a DNA-Chip consisting of all 65 K possible sequences of octa-nucleotides, e.g., by forming a 256×256 array. Since hybridization on all 65 K spots yields only a meaningless pattern containing no information, the analysis of a genome has to be done partially and stepwise with approximately 2120 clones per million base pairs of DNA with inserts of between 500 and up to 7000 bases length each. Due to the differences of genomes of different species, each genome yields other patterns of hybridization on the chips. Even point mutations within a single base yield eight different spots due to the redundancy of this method.

Nevertheless this challenging target of sequencing by hybridization (SBH) has not yet been achieved. The method is currently limited by the fact that different sequences result in different melting temperatures of their corresponding double strands due to the inhomogeneous distribution of the four bases. The reason for the dependency of the melting temperature from the distribution of the bases in the sequence is simply explained by the different numbers of hydrogen-bonds between the base pairs A–T and G–C. Approaches towards homogeneous melting temperatures are the usage of mixtures of DNA and RNA, modified mononucleotides or as a recent effort peptide nucleic acid (PNA) [63].

The immobilized oligonucleotides are usually organized in arrays of spots each spot containing a single specific DNA-sequence. Generally, two methods of arraying are favored:

1. *Spotting* is very similar to the well known printing methods of needle or inkjet printers: small droplets with a volume of some picoliters (pl) are spotted on the surface of the chip by small capillaries or needles. The array is formed by moving the chip with a scanning stage. This method requires each oligonucleotide to be synthesized in its own specific process. In addition, the spotting device has to be designed in such a way that each oligonucleotide reaches its destiny spot without being contaminated by any other DNA-sequence. This method is preferred if only a few DNA-sequences of some hundreds bases length are required to be recognized.

2. *Photolithographic on-chip synthesis* of oligonucleotides utilizes recent advances in photochemistry. In contrast to the above method, all oligonucleotides are synthesized directly and simultaneously on the surface of the chip within a single process consisting of multiple successive steps. This method simplifies the generation of DNA-chips due to its parallel elongation steps and avoids contamination of spots carrying oligonucleotides of different sequences. Each step consists of two reactions: the photolysis of a chemical protection group, e.g., with a flash of light activating different spots or areas

for the following elongation. In the next step a specific mononucleotide is bound only to the activated sites. The mononucleotides carry a photo-chemical protection group by themselves in order to provide a homogenous protection for the successive cycle. Currently, with this method a length of about 20-bases per oligonucleotide can be synthesized. Hence, this method is ideally suited for chips containing hundreds or thousands of different short oligonucleotides which are used for instance in expression analysis [64].

Before hybridization of the DNA on the chip, the sample has to be cut into smaller pieces using specific DNA-endonucleases. After denaturing of the sample on the DNA-chip at 70–90 °C, hybridization takes place at lower temperatures of about 30–40 °C; strands containing complementary regions to those presented on the chip are immobilized by binding to their counterpart. In the following washing step any other unspecific material of the sample is swept away. Hence, unwanted genes and any other organic molecules, like proteins, are removed and cannot interfere the following steps of detection and identification of the hybridized parts of the sample.

For the detection of hybridized DNA fluorescence techniques are usually used. The selective labeling of hybridized DNA with fluorescent dyes can be achieved by the following methods:

- Long DNA-sequences can be labeled by addition of intercalating dyes.
- Other methods are either to label the sample covalently with fluorescence dyes or to synthesize complementary oligonucleotides with incorporated labels by use of the polymerase chain reaction (PCR). Both methods are applied before hybridization on the chip occurs.
- A method recently developed [65] places the process of labeling into the synthesis of the DNA-chip, which significantly simplifies the preparation of the sample. The intelligent probes used consist of a stem-loop structure that is labeled at one end with a multiplex dye. Upon specific hybridization to the target DNA a fluorescence quenching process is prevented and fluorescence released. While these hairpin oligonucleotides exhibit almost no fluorescence signal the hybridized counterparts exhibit a ~ 10-fold increase in fluorescence intensity and lifetime.

For all methods mentioned above, intensity measurements are, theoretically, sufficient to discriminate between hybridized and non-hybridized spots. However, a discrimination between fluorescence and scattered light from the surface or residing salts is hindered. To overcome these difficulties, time-resolved fluorescence detection techniques are the method of choice. As shown in Fig. 16.11, the time-profiles of scattered light and fluorescence exhibit significant differences. In addition, time-resolved fluorescence detection provides a powerful tool to discriminate efficiently between false-positive signals since each dye exhibits its characteristic fluorescence decay, even on the surface.

Another advantage of the application of time-resolved fluorescence measurements in SBH is the possibility of discriminating between differently labeled samples simultaneously on the same chip.

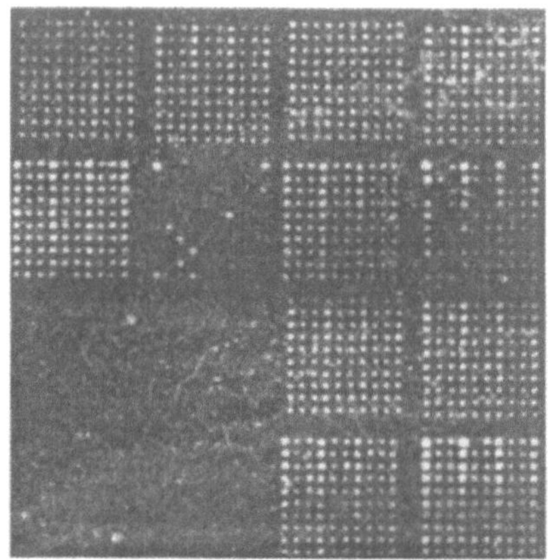

Fig. 16.11. Fluorescence intensity image of a DNA-chip (*left image*). Cy5 labeled oligonucleotides were spotted onto a modified glass chip. By using time-resolved fluorescence detection in combination with a pattern recognition technique an improved discrimination between background signal (scattered light) and fluorescence from the probes is easily achieved (*right image*)

16.5
Single Molecule DNA Sequencing in Submicrometer Channels

Since the first detection of single fluorescent molecules in solution by laser-induced fluorescence detection [66], several groups have developed the capability to detect and identify single fluorescent molecules in solution as they flow through a focused laser beam (for reviews see [26, 67–72]). As early as 1989 Keller and coworkers [24, 25, 31] proposed a new method for high-speed DNA sequencing based upon fluorescence detection of single molecules. In contrast to current DNA sequencing schemes where only relatively short DNA fragments of up to 1000 base pairs are sequenced, the method would allow one to sequence a single fragment of DNA, several tens of kilobases (kb) or more in length, at the rate of several hundred bases per second. As originally proposed, the technique is based on the detection and identification of fluorescently labeled nucleotides in flowing sample streams as they are released sequentially from a DNA strand by an exonuclease enzyme. The DNA to be sequenced should be copied using a biotinylated primer, a DNA polymerase, and the four nucleotide triphosphates (dNTPs), each containing a different fluorescent label which exhibits a characteristic laser-induced fluorescence. Recently [73], the use of native DNA as substrate for single molecule sequencing has been proposed as an alternative technique. Although this technique offers many advantages on the enzymatic side, the low fluorescence quantum yield and photostability of native nucleotides renders their detection on the single molecule level. Hence, the use of fluorescently labeled nucleotides seems to be the method of choice. As a single DNA fragment is bound to a microsphere or another solid support coated with avidin or streptavidin by the biotinylated primer, the microsphere or solid support is transferred into a flowing sample stream by mechanical micromanipulation or optical trapping. Upon addition of a $3' \rightarrow 5'$ exonuclease fluorescent nucleotide monophosphate, molecules (dNMPs) will be cleaved and transported to the detection area down stream. Each fluorescent dNMP molecule is identified by its characteristic fluorescence decay time [38, 39, 42, 74] or spectral property. Recently [75], it was shown that single fluorescent molecules can also be identified by their characteristic time-resolved fluorescence anisotropy in solution. However, besides the problems associated with the complete enzymatic substitution of the native nucleotides by dye-labeled nucleotides and the definite selection of a single DNA strand, the detection and characterization of each released nucleotide with high accuracy is also full of problems. Furthermore, since the DNA sequence is determined by the order in which labeled dNMPs are detected, diffusional misordering of sequentially cleaved dNMPs has to be prevented, i.e., the enzymatic cutting rates, flow velocities, and the distance to the detection volume have to be optimized [76]. To ensure the efficient detection of each dNMP, hydrodynamic focusing of the sample stream in a sheath flow cuvette down to $< 10\ \mu m$ was applied. Using an excitation laser beam focused to $\sim 10\ \mu m$, and a spatial filter in the detection path, a detection volume of approximately 1 picoliter (pl) is attained. To remove Rayleigh- and Raman scattering, pulsed excitation in combination with time-gated detection has to be used [66]. Especially Raman scattering, which is proportional to the number of molecules in the detection volume, prevents the definite detection of fluorescence bursts from individual molecules in pl-volumes if no time-gated detection is used. To reduce scattered light from the sample injection capillary tip, the capillary was removed and a Nd:YAG laser was used to optically trap $1\ \mu m$ microspheres about $20\ \mu m$ upstream of the detection volume. Under these conditions, fluorescent molecules released from an optically trapped microsphere have been detected with high efficiencies [31, 77]. On the other hand, single molecule fluorescence signal-to-background ratios (SBRs) can be drastically improved by using confocal excitation/detection techniques with detection volumes in the femtoliter (fl) range [78–81]. Using confocal fluorescence microscopy individual fluorescent analyte molecules can be detected without time-gated detection with SBRs > 100. Unfortunately, there is a drawback associated with the use of such small volumes in applications requiring efficient detection of all analyte molecules, e.g., in single molecule DNA sequencing. To constrain all analyte molecules to flow through a volume with linear dimensions of $< 1\ \mu m$ the detection volume must be confined by walls, i.e., a detection channel has to be used. While the refractive index differences at the outer walls of such channels can be matched by the use of the appropriate index-matching oil, the refractive index differences at the inner wall and deviations of the beam profile generally result in higher background rates and smaller photon bursts. In addition, the use of channels with such small volume-to-surface ratios results in strong adsorption of analyte molecules on the walls. Due to this dynamic adsorption process, burst durations of up to 60 ms have been measured for single rhodamine 6G molecules in aqueous buffer in submicrometer channels [82]. To circumvent this problem, polymethylmethacrylate (PMMA) microchannels with channel diameters of about $10\ \mu m$ in the detection area were employed [28]. The laser beam is shaped by a cylindrical lens and focused by the microscope objective over the whole channel. To reduce Raman scattering the volume element is imaged onto a glass fiber bundle where seven fibers are aligned. Each fiber is connected to its own separate detector, thus producing seven overlapping smaller volume elements. On the other hand, the sequential counting of single rhodamine 6G molecules dissolved in ethylene glycol in a 1-µm capillary was successfully demonstrated. No problems were observed with adsorption of analyte molecules to the walls [83]. Hence, if the adsorption can be efficiently suppressed by addition of detergents or other additives, the use of submicrometer channels should allow a precise control of the movement of single molecules by electrokinetic or electroosmotic forces. Recently [46, 84], we were able to demonstrate the time-resolved identification of individual fluorescent dyes as the they flow through a microcapillary with an inner diameter of 500 ± 200 nm applying an electrical tension of a few volts. Addition of a nonionic detergent (Tween 20) efficiently suppressed adsorption of molecules on the glass surface of the capillary and reduced the electroosmotic flow. Using a 3% polyvinyl pyrrolidone (PVP) matrix, single fluorescently labeled nucleotides were detected and identified with comparable high efficiency in the microcapillary as they were enzymatically cleaved from DNA strands bound to a fiber. In these experiments, about 60,000 DNA molecules were attached via biotin/streptavidin binding on a 3-µm optical fiber [29]. Hence, this technique seems to be a valuable alternative for single molecule DNA sequencing. However, in view of a routine application of the proposed sequencing method, background bursts which originate from impurities in the used buffer and enzyme solutions must be drastically reduced [85]. This problem is of minor importance for high counting rates of > 100 nucleotides/s (i.e., cleavage rate of the exonuclease > 100 Hz on labeled DNA) but will distort the sequence information for low counting rates. This fact, as well as the availability of low-cost diode lasers and suited fluorescent dyes in this spectral region, has prompted current efforts to use red and far-red dyes for single molecule detection methods [35, 42]. Recently [30] we described and discussed a new method for single molecule DNA sequencing which is based upon detection and identification of single fluorescently labeled mononucleotide molecules degraded from DNA-strands in a cone shaped microcapillary with an inner diameter of 0.5 µm. As illustrated in Fig. 16.12. the DNA is attached at an optical fiber via streptavidin/biotin binding and placed about 50 µm in front of the detection area inside of the microcapillary. The 5′-biotinylated 218-mer model sequence used in the experiments contains six fluorescently labeled cytosine and uridine residues, respectively, at well defined positions.

The negatively charged mononucleotide molecules were released by addition of exonuclease I and moved towards the detection area by electrokinetic forces. Adsorption of mononucleotide molecules onto the capillary walls as well as the electroosmotic (EOF) flow could be prevented efficiently by the use of a 3% polyvinyl pyrrolidone (PVP) matrix. For efficient excitation of the labeled mononucleotide molecules a short-pulse diode laser emitting at 638 nm with a repetition rate of 57 MHz in combination with a confocal fluorescence microscope was applied. In one experiment 5–6 single-stranded model DNA molecules each containing $6 \times$ Cy5-dCTP and $6 \times$ MR121-dUTP residues were attached at

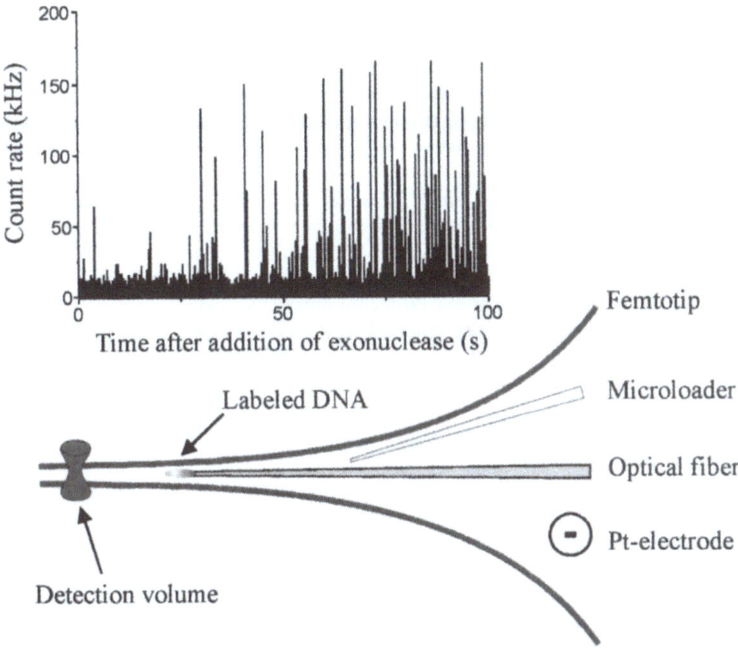

Fig. 16.12. Schematic diagram of the set-up used for single molecule DNA sequencing in submicrometer channels. The DNA is attached at the end of an optical fiber via biotin/ straptavidin. Upon addition of exonuclease with a microloader the DNA is degraded. Released mononucleotides are moved towards the detection volume by electrokinetic forces (for details see [29, 30])

the tip of a fiber, transferred into the microcapillary and degraded by addition of exonuclease I solution. In this experiment 86 photon bursts were detected (43 Cy5-dCMP and 43 MR121-dUMP) over 400 s and identified due to the characteristic fluorescence decay time of the labels of 1.43 ± 0.19 ns (Cy5-dCMP), and 2.35 ± 0.29 ns (MR121-dUMP). In these experiments we obtained the first sequence information from single DNA strands in a submicrometer channel. Under the applied experimental conditions, cleavage rates of exonuclease I on single-stranded labeled DNA molecules between $3 - 24$ Hz were observed. These results clearly demonstrate the feasibility of single molecule DNA sequencing in submicrometer channels in the near future.

Acknowledgements. We thank K. H. Drexhage for the generous disposal of the multiplex dyes, J. Hoheisel for preparation of the DNA chip, and J. Wolfrum for fruitful discussions. Financial support by the Deutsche Forschungsgemeinschaft (Grant WO 175/30 – 1), Roche Diagnostics GmbH, Volkswagen-Stiftung (Grant I/74 443), and the Bundesministerium für Bildung, Wissenschaft, Forschung und Technologie (Grants 0310158 A, 0311663 A, 11864BFA082) is gratefully acknowledged.

References

1. Sanger F, Niklen S, Coulson AR (1977) DNA sequencing with chain-terminating inhibitors. Proc Natl Acad Sci USA 74:5463–5467
2. Smith LM, Sanders JZ, Kaiser RJ, Hughes P, Dodd C, Connell CR, Heiner C, Kent SBH, Hood LE (1986) Fluorescence detection in automated DNA sequence analysis. Nature 321:674–679
3. Ansorge W, Sproat BS, Stegemann J, Schwager C, (1986) A non-radioactive automated method for DNA sequence determination. Biochem Biophys Meth 13:315–323
4. Connell CR, Fung S, Heiner C, Bridgham J, Chakerian V, Heron B, Jones B, Menchen S, Mordan W, Raff M, Recknor M, Smith L, Springer J, Woo S, Hunkapillar M (1987) Automated DNA sequence analysis. Biotechniques 5:342–348
5. Prober JM, Trainor GL, Dam RJ, Hobbs FW, Robertson CW, Zagursky RJ, Cocuzza AJ, Jensen MA, Baumeister K (1987) A system for rapid DNA sequencing with fluorescent chain-terminating dideoxynucleotides. Science 238:336–341
6. Brumbaugh JA, Middendorf LR, Grone DL, Ruth JL (1988) Continous, on-line DNA sequencing using oligodeoxynucleotide primers with multiple fluorophores. Proc Natl Acad Sci USA 85:5610–5614
7. Middendorf LR, Bruce JC, Bruce RC, Eckles RD, Grone DL, Roemer SC, Sloniker GD, Steffens DL, Sutter SL, Brumbaugh JA, Patonay G (1992) Continous on-line DNA sequencing using a versatile infrared laser scanner/electrophoresis apparatus. Electrophoresis 13:487–494
8. Ju J, Ruan C, Fuller CW, Glazer AN, Mathies RA (1995) Fluorescence energy transfer dye-labeled primers for DNA sequencing and analysis. Proc Natl Acad Sci USA 92:4347–4351
9. Metzker ML, Lu J, Gibbs RA (1996) Electrophoretically uniform fluorescent dyes for automated DNA sequencing. Science 271:1420–1422
10. Lee LG, Spurgeon SL, Heiner CR, Benson SC, Rosenblum BB, Menchen SM, Graham RJ, Constantinescu A, Upadhya KG, Cassel JM (1997) Nucleic Acids Res 25:2816–2822
11. Anazawa T, Takahashi S, Kambara H (1996) A capillary gel electrophoresis system using multiple laser focusing for DNA sequencing. Anal Chem 68:2699–2704
12. Trost P, Guttman A (1998) Fiber bundle based scanning detection system for auto mated DNA sequencing. Anal Chem 70:3930–3935
13. Huang XC, Quesada MA, Mathies RA (1992) DNA sequencing using capillary array electrophoresis. Anal Chem 64:2149–54
14. Mathies RA, Huang XC (1992) Capillary array electrophoresis: an approach to high-speed, high-throughput DNA sequencing. Nature 359:167–169
15. Kheterpal I, Scherer JR, Clark SM, Radhakrishnan A, Ju, Ginther CL, Sensabaugh GF, Mathies RA (1996) DNA sequencing using a four-color confocal fluorescence capillary array scanner. Electrophoresis 17:1852–1859
16. Kheterpal I, Mathies RA (1999) Capillary array electrophoresis DNA sequencing. Anal Chem 71:31A-37A
17. Scherer JR, Kheterpal I, Radhakrishnan A, Ja WW, Mathies RA (1999) Ultra-high throughput rotary capillary array electrophoresis scanner for fluorescent DNA sequencing and analysis. Electrophoresis 20:1508–1517
18. Murray KK (1996) DNA sequencing by mass spectrometry. J Mass Spectr 31:1203–1215
19. Bains W, Smith GC (1988) A novel method for nucleic acid sequence determination. J Theor Biol 135:303–307
20. Lysov YP, Florentev VL, Khorlin AA, Khrapko KR, Shik VV, Mirzabekov AD (1988) Determining the nucleotide sequence of DNA by hybridization with oligonucleotides: a new method. Dokl Akad Nauk SSSR 303:1508–1511
21. Drmanac R, Labat I, Brukner I, Crkvenjakov R (1989) Sequencing of megabase plus DNA by hybridization: theory of the method. Genomics 4:114–128
22. Southern EM, Maskos U, Elder JK (1992) Analyzing and comparing nucleic acids sequences by hybridization to arrays of oligonucleotides: evaluation using experimental methods. Genomics 13:1008–1017

23. Fodor SPA (1997) Massively parallel genomics. Science 277:393–395
24. Jett JH, Keller RA, Martin JC, Marrone BL, Moyis RK, Ratliff RL, Seitzinger NK, Shera B, Stewart CC (1989) High-speed DNA sequencing: an approach based upon fluorescence detection of single molecules. J Biomol Struc&Dyn 7:301–309
25. Harding JD, Keller RA (1992) Single molecule detection as an approach to rapid DNA sequencing. TIBTECH 10:55–57
26. Goodwin PM, Ambrose WP, Keller RA (1996) Single-molecule detection in liquids by laser-induced fluorescence. Acc Chem Res 29:607–613
27. Goodwin PM, Cai H, Jett JH, Ishaug-Riley SL, Machara NP, Semin DJ, Van Orden A, Keller RA (1997) Application of single molecule detection to DNA sequencing. Nucleosides &Nucleotides 16:543–550
28. Dörre K, Brakmann S, Brinkmeier M, Han KT, Riebeseel K, Schwille P, Stephan J, Wetzel T, Lapczyna M, Stuke M, Bader R, Hinz M, Seliger H, Holm J, Eigen M, Rigler R (1997) Techniques for single molecule sequencing. Bioimaging 6:139–152
29. Sauer M, Angerer B, Han KT, Zander C (1999a) Detection and identification of single dye labeled mononucleotide molecules released from an optical fiber in a microcapillary: first steps towards a new single molecule DNA sequencing technique. Phys Chem Chem Phys 1:2471–2477
30. Sauer M, Angerer B, Ankenbauer W, Földes-Papp Z, Göbel F, Han KT, Rigler R, Schulz A, Wolfrum J, Zander C (1999b) Single molecule DNA sequencing in submicrometer channels: state of the art and future prospects. J Biotechnology, in press
31. Werner JH, Cai H, Goodwin PM, Keller RA (1999) Current status of DNA sequencing by single molecule detection. SPIE Vol 3602:355–365
32. Sauer M, Schulz A, Seeger S, Wolfrum J, Arden-Jacob J, Deltau G, Drexhage KH (1993) Design of multiplex dyes. Ber Bunsenges Phys Chem 97:1734–1738
33. Bachteler G, Drexhage KH, Arden-Jacob J, Han KT, Köllner M, Müller R, Sauer M, Seeger S, Wolfrum J (1994) Sensitive Fluorescence Detection in Capillary Gel Electrophoresis using Laser Diodes and Multiplex Dyes. J Luminesc 62:101–108
34. Sauer M, Han KT, Ebert V, Müller R, Schulz A, Seeger S, Wolfrum J, Arden-Jacob J, Deltau G, Marx NJ, Drexhage KH (1994) Design of multiplex dyes for the detection of different biomolecules. Proc SPIE 2137:762–773
35. Sauer M, Han KT, Müller R, Nord S, Schulz A, Seeger S, Wolfrum J, Arden-Jacob J, Deltau G, Marx NJ, Zander C, Drexhage KH (1995) New fluorescent dyes in the red region for biodiagnsotics. J Fluores 5:247–261
36. Tellinghuisen J, Wilkerson CW Jr (1993) Bias and precision in the estimation of exponential decay parameters from sparse data. Anal Chem 65:1240–1246
37. Tellinghuisen J, Goodwin PM, Ambrose WP, Martin JC, Keller RA (1994) Analysis of fluorescence lifetime data for single rhodamine molecules in flowing sample streams. Anal Chem 66:64–72
38. Zander C, Sauer M, Drexhage KH, Ko DS, Schulz A, Wolfrum J, Brand L, Eggeling C, Seidel CAM (1996) Detection and identification of single molecules in aqueous solution. Appl Phys B 63:517–523
39. Müller R, Zander C, Sauer M, Deimel M, Ko DS, Siebert S, Arden-Jacob J, Deltau G, Marx NJ, Drexhage KH, Wolfrum J (1996) Time-resolved identification of single molecules in solution with a pulsed semiconductor laser. Chem Phys Lett 262:716–722
40. Enderlein J, Goodwin PM, Van Orden A, Ambrose WP, Erdmann R, Keller RA (1997) A maximum likelihood estimator to distinguish single molecules by their fluorescence decays. Chem Phys Lett 270:464–470
41. Sauer M, Zander C, Müller R, Ullrich B, Kaul S, Drexhage KH, Wolfrum J (1997) Detection and identification of individual antigen molecules in human serum with pulsed semiconductor lasers. Appl Phys B 65:427–431
42. Sauer M, Arden-Jacob J, Drexhage KH, Göbel F, Lieberwirth U, Mühlegger K, Müller R, Wolfrum J, Zander C (1998) Time-resolved identification of individual mononucleotide molecules in aqueous solution with pulsed semiconductor lasers. Bioimaging 6:14–24

43. Köllner M, Wolfrum J (1992) How many photons are necessary for fluorescence-lifetime measurements. Chem Phys Lett 200:199–203
44. Köllner M (1993) How to find the sensitivity limit for DNA sequencing based on laser-induced fluorescence. Appl Opt 32:806–820
45. Köllner M, Fischer A, Arden-Jacob J, Drexhage KH, Müller R, Seeger S, Wolfrum J (1996) Fluorescence pattern recognition for ultrasensitve molecule identification: comparison of experimental data and theoretical approximations. Chem Phys Lett 250:355–360
46. Becker W, Hickl H, Zander C, Drexhage KH, Sauer M, Siebert S, Wolfrum J (1999) Time-resolved detection and identification of single analyte molecules in microcapillaries by time-correlated single-photon counting (TCSPC). Rev Sci Instr 70:1835–1841
47. TimeHarp100 (1999) Picoquant GmbH, Berlin, Germany (TimeHarp100 is a PC plug-in card from the company Picoquant)
48. Arden-Jacob J (1992) Neue Fluoreszenzsonden und Laserfarbstoffe. PhD Thesis, Universität Siegen
49. Lieberwirth U, Arden-Jacob J, Drexhage KH, Herten DP, Müller R, Neumann M, Schulz A, Siebert S, Sagner G, Klingel S, Sauer M, Wolfrum J (1998) Multiplex dye DNA sequencing in capillary gel electrophoresis by didoe laser-based time-resolved fluorescence detection. Anal Chem 70:4771–4779
50. Nord S, Sauer M, Arden-Jacob J, Drexhage KH, Lieberwirth U, Seeger S, Wolfrum J (1997) Ground and excited state reactions of new red fluorescent dyes with the DNA base guanosine. J Fluoresc 7:15–18
51. Müller R, Herten DP, Lieberwirth U, Neumann M, Sauer M, Schulz A, Siebert S, Drexhage KH, Wolfrum J (1997) Efficient DNA sequencing with a pulsed semiconductor laser and a new fluorescent dye set. Chem Phys Let 279:282–288
52. Slater GW, Mayer P, Hubert SJ, Drouin G (1994) The biased reptation model of DNA gel electrophoresis: a user guide for constant field mobilities. Appl Theor Electrophor 4:71–79
53. Slater GW (1997) In: Heller C (ed) Analysis of Nucleic Acids by Capillary Electrophoresis. Vieweg, Wiesbaden, pp 24–65
54. Slater GW, Mayer P, Drouin G (1996) In: Karger BL, Hancock WS (eds) Methods in Enzymology. Academic Press, San Diego, p 272
55. Lumpkin OJ, Déjardin P, Zimm BH (1985) Theory of gel electrophoresis of DNA. Biopolymers 24:1573
56. Okubo K, Hori N, Matoba R, Niiyama T, Fukushima A, Kojima Y, Matsubara K (1992) Large scale cDNA sequencing for analysis of quantitative and qualitative aspects of gene expression. Nature Genet 2:173–179
57. Takahashi S, Murakami K, Anazawa T, Kambara H (1994) Multiple sheath-flow gel capillary-array electrophoresis for multicolor fluorescent DNA detection. Anal Chem 66:1021–1026
58. Ueno K, Yeung ES (1994) Simultaneous monitoring of DNA fragments separated by electrophoresis in a multiplexed array of 100 capillaries. Anal Chem 66:1424–31
59. Quesada MA, Zhang S (1996) Multiple capillary DNA sequencer that uses fiber-optic illumination and detection. Electrophoresis 17:1841–51
60. Behr S, Mätzig M, Levin A, Eickhoff H, Heller C (1999) A fully automated multicapillary electrophoresis device for DNA analysis. Electrophoresis 20:1492–1507
61. Neumann M, Herten DP, Dietrich A, Wolfrum J, Sauer M (1999) Capillary array scanner for time-resolved detection and identification of fluorescently labeled DNA fragments. J Chromat A, in press
62. Wallace RB, Shaffer J, Murphy RF, Hirose T, Itakura K (1979) Hybridization of synthetic oligonucleotides to $\phi\chi$ 174 DNA: The effect of single base pair mismatch. Nucleic Acids Res 6:3543–3557
63. Weiler J, Gausepohl H, Hauser N, Jensen ON, Hoheisel JD (1997) Hybridization based DNA screening on peptide nucleic acid (PNS) oligomer arrays. Nucleic Acids Research 25:2792–2799

64. Weiler J, Hoheisel JD (1996) Combining the Preparation of Oligonucleotide Arrays and Synthesis of High Quality Primers. Analytical Biochemistry 243:218–227
65. Knemeyer J, Marmé N, Sauer M (2000) Probes for the detection of specific DNA sequences at the single-molecule level. Anal Chem 72:3717–3724
66. Shera EB, Seitzinger NK, Davis LM, Keller RA, Soper SA (1990) Detection of single fluorescent molecules. Chem Phys Lett 174:553–557
67. Eigen M, Rigler R (1994) Sorting single molecules: Application to diagnostics and evolutionary biotechnology. Proc Natl Acad Sci USA 91:5740–5747
68. Barnes MD, Whitten WB, Ramsey JM (1995) Detecting single molecules in liquids. Anal Chem 67:A418-A423
69. Rigler R (1995) Fluorescence correlations, single molecule detection and large number screening Applications to biotechnology. J Biotechnology 41:177–186
70. Keller RA, Ambrose WP, Goodwin, PM, Jett JH, Martin JC, Wu M (1996) Single-molecule fluorescence analysis in solution. Appl Spectrosc 50:12A-32A
71. Nie S, Zare RN (1997) Optical detection of single molecules. Annu Rev Biophys Biomol Struct 26:567–596
72. Weiss S (1999) Fluorescence Spectroscopy of Single Biomolecules. Science 283:1676–1683
73. Dapprich J, Nicklaus N (1998) DNA attachment to optically trapped beads in microstructures monitored by bead displacement. Bioimaging 6:25–32
74. Wilkerson CW jr, Goodwin PM, Ambrose WP, Martin JC, Keller RA (1993) Detection and lifetime measurement of single molecules in flowing sample streams by laser-induced fluorescence. Appl Phys Lett 62:2030–2032
75. Schaffer J, Volkmer A, Eggeling C, Subramaniam V, Striker G, Seidel CAM (1999) Identification of Single Molecules in Aqueous Solution by Time-Resolved Fluorescence Anisotropy. J Phys Chem A 103:331–336
76. Pratt LR, Keller RA (1993) Estimate of the probability of diffusional misordering in high-speed DNA sequencing. J Phys Chem 97:10254–10255
77. Machara NP, Goodwin PM, Enderlein J, Semin DJ, Keller RA (1998) Efficient detection of single molecules eluting off an optically trapped microsphere. Bioimaging 6:33–42
78. Rigler R, Mets Ü, Widengren J, Kask P (1993) Fluorescence correlation spectroscopy with high count rate and low background: analysis of translational diffusion. Eur Biophys J 22:169–175
79. Mets Ü, Rigler R (1994) Submillisecond detection of single rhodamine molecules in water. J Fluoresc 4:259–264
80. Nie S, Chiu DT, Zare RN (1994) Probing individual molecules with confocal fluorescence microscopy. Science 266:1018–1021
81. Nie S, Chiu DT, Zare RN (1995) Real-time detection of single molecules in solution by confocal fluorescence microscopy. Anal Chem 67:2848–2857
82. Lyon WA, Nie S (1997) Confinement and detection of single molecules in submicrometer channels. Anal. Chem. 69:3400–3405
83. Zander C, Drexhage KH (1997) Sequential counting of single molecules in a capillary. J Fluoresc 7:37S-39 S
84. Zander C, Drexhage KH, Han KT, Wolfrum J, Sauer M (1998) Single-molecule counting and identification in a microcapillary. Chem Phys Lett 286:457–465
85. Affleck RL, Ambrose WP, Demas JN, Goodwin PM, Schecker JA, Wu M, Keller RA (1996) Reduction of luminescent background in ultrasensitive fluorescence detection by photobleaching. Anal Chem 68:2270–2276

The Integration of Single Molecule Detection Technologies into Miniaturized Drug Screening: Current Status and Future Perspectives

C. BUEHLER, K. STOECKLI, M. AUER

17.1
Introduction

Optical single molecule detection is an emerging field which currently impacts various scientific disciplines. In particular, laser induced fluorescence detection and spectroscopy of individual fluorescent molecules open new avenues for the development of fluorescent based assays in biophysical research but also in biotechnology and pharmaceutical industries. The ability to detect and identify either tagged or autofluorescent single molecules allows for the separation of sub-populations from heterogeneous ensembles which therefore renders single molecule detection (SMD) an ideal tool for selecting, trapping, sorting, tracking, picking, and even manipulating biological (macro)molecules. Moreover, since the physical observables in SMD experiments are unmasked by ensemble averaging, time-dependent pathways of chemical reactions can be resolved without the need to synchronize all the molecules of the ensemble. Currently, optical SMD has been successfully applied in separation science [1–7] near field [8–10] and far field microscopy [11–14], and total internal reflection (TIR) microscopy on quartz-liquid interfaces [15–17].

The need for detecting sparse quantities is critical in pharmaceutical high-throughput screening (HTS). Often, bioactive molecules are only available in minute quantities but need to be discriminated against a large number of compounds. With the extension to use wavelength ranges of 450–800 nm for excitation and emission, multi-color SMD becomes an ideal tool to investigate molecular cascades like inter- and intra-cellular signaling pathways [18]. Although still at the very beginning, the combination of individual assay modules to cellular assay systems is likely to become a future concept in drug discovery and validation.

While SMD has become routine in many research laboratories, the high throughput drug screening (> 100,00 samples per day), handling of minute liquid quantities (nano-to-microliter range), real-time analysis, data management, and the precarious industrial environment (impure substances) still require additional effort to achieve the necessary robustness of SMD screening platforms. To our knowledge, two drug companies are currently implementing SMD-based screening machines (Evotec Biosystems, Hamburg, Germany, [19]); Molecular Machines & Industries Heidelberg, Germany [20]).

Albeit SMD represents the ultimate level of sensitivity, the ability to detect a single molecule is primarily a matter of background reduction rather than a question of ultra-sensitive detectors [21]. Typical background sources are Raman or Rayleigh scattered laser light, fluorescent impurities from the solvent, optical components, and the sample carrier's cover glass, and also the inherent electronic noise of the detector [22]. The major SMD strategy focuses on (1) the reduction of the excitation and observation volume since the signal-to-background ratio is inversely related to the detection volume, (2) the use of bright and photostable dyes (3) low-noise and high quantum efficiency photodetectors, and (4) high collection efficiency optics and low-fluorescing materials. Additional options include pre-bleaching of impurities in the solvent, or time gated detection schemes to discriminate the fluorescence signal against the instantaneous light scatter [23, 24]. Moreover, the sample molecules might be excited by two photons (TPE). This strategy is based on a non-linear molecular absorption process. TPE involues the simultaneous absorption of two 'red' photons each carrying only half of the energy of the corresponding 'blue' photon commonly causing the electronic transition [25]. The large spectral shift between the (infra)-red laser light and the two-photon induced fluorescence allows for an efficient rejection of Raman and Rayleigh scatter by spectral filtering. Since the non-linear nature of two-photon absorption limits the region of photo-interaction to a sub-femtoliter volume at the focal spot, virtually no out-of-focus background generation and photo-bleaching occurs [26–28]. Correspondingly, the signal amplitude can easily exceed the background level by three orders of magnitude [29]. In order to attain an efficient two-photon excitation process, pulsed laser sources such as mode-locked Titanium:Sapphire lasers are used. Of course, the purchase of such an expensive light source has to be balanced carefully against the experimental needs. Currently, the corresponding costs for a low-cost Ti:Sapphire laser are still about one order of magnitude higher than those of a conventional continuous wave (CW) laser, e.g., an air-cooled argon ion laser. In comparison to a confocal setup the detection efficiency of a TPE microscope is slightly reduced, possibly due to non-linear competitive optical processes [29].

The passage of a single molecule through the observation volume becomes evident by a correspondingly emitted burst of photons characterized by its duration and number (burst size). In principle, single molecule transits can be monitored in real-time provided that the instrument is sufficiently sensitive [12, 30]. By using high quantum yield single photon detectors (PMT, avalanche photodiode) along with high numerical aperture objectives, the detection efficiency of a confocal or two-photon microscope can be as high as 5%. Assuming further a moderate fluorophore excitation rate of 10 MHz, and a typical molecular transit time through the observation volume of 200 μs, approximately 100 photons can be recorded from a single transit event. This is already sufficient for some approximate data analysis. For a more accurate analysis, multiple transit events have to be averaged over a longer time scale. The appropriate data acquisition time depends on the photophysical properties of molecules, chemical assay parameters, particular instrumental properties and settings, but also on the actual experimental question to be answered. If an assay only demands simple yes or no decisions, data acquisition times on the order of 1 s are already sufficient. This type of problem is typically encountered in primary HTS runs with their primary focus on the discrimination between target-bound vs target-unbound molecules. In a scientific research environment, on the other hand, SMD primarily addresses the detailed study of fundamental physico-chemical processes and reaction pathways. This inherently requires statistical accuracy of the data and consequently longer measurement times. In addition, often, no a priori information about the system under study is available. Therefore, the data fitting procedure has to involve the full dimensionality of the model (fitting) function which further increases the demands on statistical accuracy. Research-like data acquisition times are therefore on the order of tens of minutes.

Each detected photon generates an electric pulse which is then recorded by the data acquisition system. Correspondingly, single-molecule detection is a digital process. The 'analogue' fluorescence intensity impinging onto the detector is converted into photon counts. Unpredictable fluctuations in the laser light and the photo detection process change the statistics of measured intensities by adding shot noise [31]. Hence, the time of arrival and registration of a single photon is governed by probability laws. Therefore, single molecule experiments need to be analyzed in terms of statistical data analysis methods [32].

A widely used approach in single molecule analysis (SMA) is fluorescence correlation spectroscopy (FCS). Autocorrelation analysis is a powerful mathematical tool for noise-reduction and data processing [33]. Its full potential has become evident over the last few years, particularly due to advances in laser technology and microscopy. The temporal autocorrelation of the time series of detected photons yields as its most important readout parameters the average transit time of the molecules through, and their average number density within the volume element. Clearly, FCS is ideally suited to assay bimolecular reactions by means of the change in the diffusion time upon complex formation. FCS has been extensively applied to measure translational and rotational diffusion rates [34–37], flow rates [38], molecular aggregation [39, 40], chemical reaction rates [41], and protein-ligand interactions [42], triplet state kinetics [43], surface processes [36, 44, 45], hybridization reactions [46, 47], intra-cellular reactions [48–50], and numerous miscellaneous interactions [39, 51–53]. On the other hand, binding interactions can only be resolved by FCS measurements if the reaction partners' diffusion times differ quite substantially. This is a serious limitation of FCS. Further, at sample concentrations less than 10^{-9} mol/l the background signal becomes significant and requires careful background correction procedures [54]. Conversely, for concentrations exceeding approximately 10^{-8} mol/l the sample volume becomes overpopulated which results in a large mean fluorescence intensity superimposed by a minute (immeasurable) fluctuation signal.

Two basic directions have historically been followed to expand the autocorrelation analysis towards a molecular weight independent detection technology: (1) methods primarily based on temporal signal analysis such as higher order autocorrelation analysis [40, 55], higher order moment analysis [56, 57], two-color fluorescence cross-correlation spectroscopy [19, 58–60], and coincidence analysis [61], and (2) intensity (amplitude) dependent analysis methods which will be discussed in the following paragraphs.

While FCS is ideally suited to retrieve temporal information of dynamic processes buried in the stochastic signal, it is mostly deficient concerning the amplitude distribution of intensities, i.e., the probability to detect k light quanta within a given time interval [62]. It is intuitively clear that two molecules which differ in their specific fluorescence intensity (but not necessarily in their diffusion time) might be distinguishable via their characteristic brightness amplitudes. Since the molecules randomly diffuse in all three dimensions through an inhomogeneously illuminated sampling volume, and since the photon detection process is an additional source of randomness, the statistical distribution of the measured fluorescence amplitude is not accurately described by a simple Poisson statistic. Recently, three related but specifically distinct detection technologies in terms of instrumental equipment and data analysis procedures have been invented. One of these methods has been introduced on the basis of conventional economical continuous wave excitation (Fluorescence Intensity Distribution Analysis (FIDA)) whereas the other two methods use pulsed laser excitation for either two-photon excitation (Photon Counting Histogram (PCH)) or time-resolved analysis (Burst Integrated Fluorescence Lifetime (BIFL)). The Fluorescence Intensity Distribution Analysis [63, 64]) and the Photon Counting Histogram method [65] are two very similar approaches which both analyze the frequency distribution of recorded photons in terms of 'super-Poissonian' [66] model functions. However, the two methods basically differ in the actual treatment of the observation volume and the numeric calculation of the fitting function. The essential readout parameters are the molecular concentrations and brightnesses (fluorescence intensity per molecule) of different fluorescent species. While FCS cannot distinguish different species of similar diffusion times, both methods overcome this shortcoming provided that the individual fluorescence intensities of the species differ. Heterogeneous fluorescent populations are ubiquitous in biochemistry, life science, and medicine which renders FIDA and PCH as widely applicable methods for measuring the concentrations of various fluorescent species in heterogeneous samples. In particular, the important cases of molecular multimerization, oligomerization, or aggregation can be tackled since both methods are able to resolve a brightness ratio of two or less [64, 65]. While FCS is able to reject uncorrelated noise sources by the autocorrelation procedure, uncorrelated and correlated background signals both erroneously contribute to the FIDA and PCH analysis. Particularly in one-photon excitation setups using uncooled APD photodetectors, the background rate can easily exceed 3 kHz. Therefore, additional optical measures are required to minimize particularly the out-of-focus fluorescence of the molecules accumulated on the bottom of the sample carriers (Evotec Biosystems GmbH, Germany). FIDA and PCH hold great promise for increasing the efficiency and applicability of miniaturized HTS.

In the methods discussed so far, different single molecules can be distinguished on the basis of adequate changes in their diffusion times and molecular intensities. However, these signal parameters are relative quantities due to their dependence on experimental settings (laser power, detection efficiency, instrument alignment). Since fluorescence emission is multi-dimensional in its nature, other readout modalities can be exploited for a more effective molecular identification. By using two chromatically distinct detection channels, single molecules can be wavelength discriminated provided that their emission bands are sufficiently separated [67]. Alternatively, time resolved detection schemes allow for an accurate identification of molecular species in terms of the substance-specific fluorescence lifetime, i.e., the characteristic mean time a molecule spends in its electronically excited state before emitting a fluorescence photon. The fluorescence lifetime is an absolute parameter due to its insensitivity to variations in the light intensity and sampling concentrations [68]. On the other hand, fluorescence lifetime is most sensitive to structural changes in biological macromolecules including ligand/substrate binding or protein folding, as well as to particular environmental variables such as pH, divalent metal concentrations, or environmental polarity [69, 70]. Besides the numerous applications encountered in macroscopic ensemble measurements, also individual solute molecules have been lifetime-identified [6, 71].

BIFL or Burst Integrated Fluorescence Lifetime is a recently introduced multiplexed spectroscopic method using pulsed excitation and time-correlated single photon counting (TCSPC) technique to record simultaneously the size of fluorescence bursts and their intra-burst fluorescence lifetime [2, 53, 72]. A further extension of BIFL also yields intra-burst anisotropy [73]. The data analysis procedure in a BIFL experiment first selects the well-defined bursts by an appropriate burst search algorithm and subjects them individually to a fluorescence decay pattern recognition algorithm for the retrieval of the intra-burst fluorescence lifetime. By correlating the burst size and the intra-burst fluorescence lifetime of each observation volume crossing molecule in a two-dimensional histogram (like a scatter plot in Fluorescence Activated Cell Sorting or FACS [74]), individual species are clearly identified as distinct islands in the 2-D plot. Undoubtedly, this technique has tremendous potential for basic research. However, the poor statistics of the small numbers of photons collected during a single transit event does not allow more than one lifetime component to fit. Thus, two different molecules concurrently passing through the sampling volume cannot be positively distinguished on the bases of their lifetimes. Consequently, the sample concentration has to be sufficiently low (picomolar range) such that the probability of a double-transit event becomes insignificant. This restricts the technique's applicability to molecular binding reactions with equilibrium constants as low as the sample concentration.

Within the scope of this chapter we shall critically discuss and compare currently available SMA methods covering both the underlying mathematical algorithms and the practical applicability for the HTS.

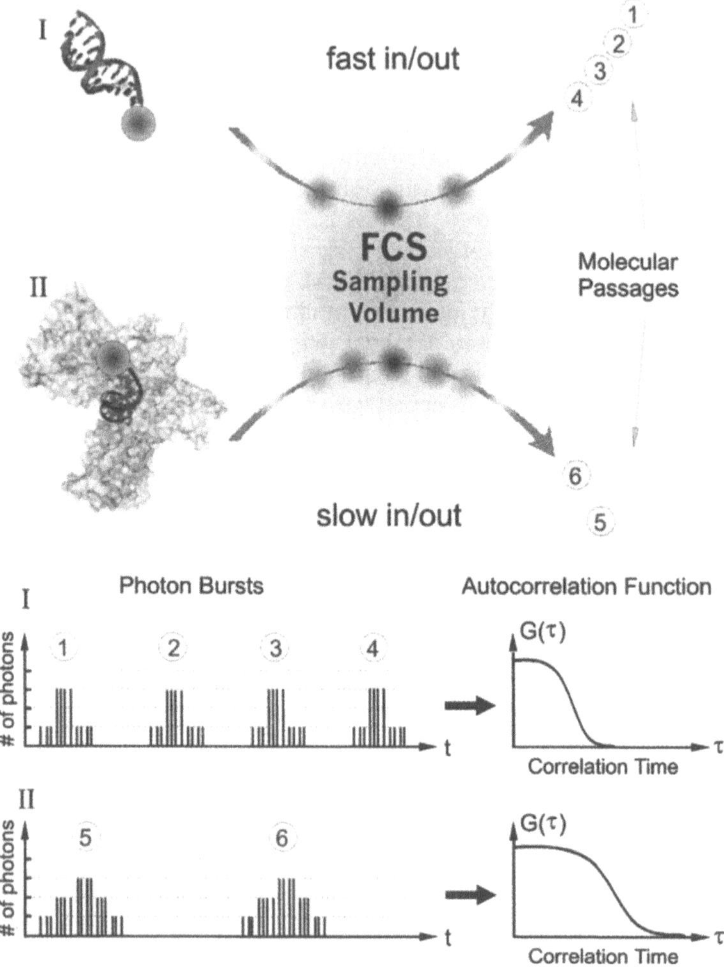

Fig. 17.1. Concept of fluorescence correlation spectroscopy. Molecules are analyzed based on their differences in translational diffusion times and particle numbers. The top part of the figure schematically depicts four sequential passages or transit events (❶❷❸❹) of a small fluorescently tagged (*gray dot*) oligonucleotide (ODN) through the laser illuminated observation volume (*shaded ellipsoid*). For the sake of clarity, the actual Brownian motion diffusion pathways of the ODN through the volume element are represented by the *up-pointing curvature arrow*. The *blurred dots along the arrow* symbolize the stroboscopically sampled fluorescence emission from a single molecule during a transit event. The *gray-level of the dots* accounts for the spatial dependent intensity of emitted fluorescence as a result of the nonuniform intensity profile of the laser light. The number of sampled fluorescence points depends on the sampling frequency and the molecular diffusion time. In this simplified representation the fast diffusing ODN is assumed to be sampled three times (3 dots). If the small ODN binds to a transcription factor (TRF) the complex diffuses slower. The bottom part of the figure depicts two sequential (❺❻) transit events of the ODN-TRF complex occurring during the same detection time frame. With respect to unbound ODN, the slow diffusion time of bound ODN increases the number of fluorescence sampling points (5 dots compared to 3 in the *top trace*) but correspondingly reduces the number of transit events (2 passages com-

17.2
Theoretical Background of Common Approaches in Single Molecule Analysis (SMA)

17.2.1
Principles of Fluorescence Correlation Spectroscopy (FCS)

Fluorescence correlation spectroscopy (FCS) measures the temporal intensity fluctuations of fluorescent light emitted by a small number of fluorescent molecules diffusing in and out of a very small open volume element. At equilibrium, the variations in the fluorescence signal reflect spontaneous fluctuations in the local concentration of a fluorescent species within the small sampling cavity. The fluctuation signal is analyzed in terms of the temporal autocorrelation function $G(\tau)$. While the magnitude of $G(\tau)$ is related to the particle density of different sampling volume crossing species, the shape and the characteristic decay time of $G(\tau)$ are a measure for the mean temporal duration of a fluorescence fluctuation. The principle of FCS is shown in Fig. 17.1.

17.2.2
Autocorrelation Analysis

The correlation method analyzes the measured fluorescence intensity $F(t)$ by autocorrelating the fluctuations around the average fluorescence value $\langle F(t) \rangle$ within a finite time period. Mathematically, the normalized autocorrelation function $G(\tau)$ may be written as

$$G(\tau) = \frac{\langle \delta F(t)\, \delta F(t+\tau) \rangle}{\langle F(t) \rangle^2} \tag{17.1}$$

where the brackets denote time averaging, and the fluctuation signal δF is the difference between the actual and the mean fluorescence intensity:

Fig. 17.1 (continued)
pared to 4 in the *top trace*). For the low and high molecular weight particles ODN (I) and ODN-TRF (II) the resulting time series of recorded fluorescence photons are separately shown in the *photon burst section of the graph*. Each transit event produces a burst of fluorescence photons. Within the simplified model chosen for explanation, the package of photons detected within the sampling time is represented by 3 – 4 *adjacent vertical lines* and their height (# of photons) corresponds to the fluorescence intensity (*gray-level*) sampled from the particle at a distinct spot (*blurred dot*) along its trajectory through the observation volume. The three and five sampled points of ODN and ODN-TRF respectively correspond to two and three levels of photon count rate in the photon burst diagram. Assuming equal intensities of the label in the unbound and bound state, the size of the photon burst increases with the diffusion time of a molecule while the number of registered bursts per measuring time correspondingly decreases. Since the total number of photons detected in a single burst is typically not sufficient to determine accurately the underlying diffusion time, the data analysis is performed over many photon bursts using numerical autocorrelation as an efficient mathematical tool for noise-rejection. A relative increase in the diffusion time of the labeled ligand is revealed by the calculated autocorrelation function $G(\tau)$ as a temporal shift towards longer correlation times

$\delta F = F(t) - \langle F(t) \rangle$. $G(\tau)$ decays with correlation time τ since the fluctuations become more and more uncorrelated at increasing temporal separation. At infinite temporal separation $(\tau \to \infty)$ the autocorrelation function approaches the limiting value of zero, i.e., $G(\tau \to \infty) = 0$. The decay rate of $G(\tau)$ is a direct measure for the average duration of the fluctuation signal, or equally correct, the characteristic molecular transit time through the sampling volume. Further, the fluctuation signal δF becomes more pronounced at lower particles concentrations. Hence, the fluctuation amplitude is inversely related to the mean number of observation volume crossing particles. According to Eq. (17.1), the autocorrelation function at time zero, $G_0 = G(\tau \to 0)$, is a direct measure for the (squared) magnitude of the fluctuation signal. Mathematically, G_0 is the normalized variance of $F(t)$ and inversely proportional to the average number of fluorescent molecules diffusing through the open volume element. For an infinite dilute homogeneous solution of non-interacting fluorophores G_0 is given by [75]

$$G_0 = \frac{\langle \delta F(t)^2 \rangle}{\langle F(t) \rangle^2} = \frac{\gamma}{\langle N \rangle} \tag{17.2}$$

where $\langle N \rangle$ is the average number of fluorescent particles in the observation volume, and the constant γ is determined by the spatial intensity profile of the excitation light, or more accurately, by the point-spread function (PSF) of the instrument. The PSF describes the spatial intensity distribution of an imaged point light source around the focal plane of a lens [76]. Due to the wave nature of the light and the limited resolution of the lens, the point is imaged into a blurred spot. The spot size depends on the wavelength of the light λ and the numerical aperture NA of the lens (objective). Typical NAs for microscope objectives are around 1.3. The intensity pattern in the focal plane of the imaging system is referred to as the Airy disc and its radius is given by $r_{\text{Airy}} = 0.61 \, \lambda/\text{NA}$. For a PSF with a three-dimensional (3-D) Gaussian intensity profile as encountered in a confocal microscope (see below) γ is 1/2. For the Gaussian-Lorenzian PSF characteristic for a two-photon microscope, γ is $3/(4 \pi^2)$ [48].

The measured mean fluorescence intensity $\langle F(t) \rangle$ is proportional to $\langle N \rangle$. The proportionality constant κ is a function of the position vector $\vec{r} = (x, y, z)$ and determined by both the species' fluorescent characteristics and the optical system. The actual detectable fluorescence intensity $F(t, \vec{r})$ emitted from a position $\vec{r}$ at time t can be expressed in terms of the actual local molecular concentration $C(t, \vec{r})$, i.e., $F(t, \vec{r}) = \kappa(\vec{r}) N(t, \vec{r}) = \kappa(\vec{r}) C(t, \vec{r}) d\vec{r}$. For a temporal constant light source with a spatial intensity profile $I(\vec{r})$ the total measured average fluorescence intensity can be written as

$$\langle F(t) \rangle = \kappa' \langle C \rangle \int I(\vec{r}) d\vec{r} \tag{17.3}$$

where κ' is the product of the molecular absorptivity, the fluorescent quantum yield, and the instrumental detection efficiency. The actually measured temporal fluctuations $\delta F(t)$ in the fluorescence intensity reflect the spontaneous local fluctuations of the particles around their equilibrium concentration

$$\delta F(t) = F(t) - \langle F(t) \rangle = \kappa' \int I(\vec{r}) \delta C(t, \vec{r}) d\vec{r} \tag{17.4}$$

where $\delta C(t,\vec{r})$ is the time and position dependent fluctuation in the number density of fluorescent particles. Inserting Eqs. (17.3) and (17.4) into Eq. (17.1) leads to

$$G(\tau)\frac{\iint I(\vec{r})I(\vec{r}')\langle\delta C(t,\vec{r})\,\delta C(t+\tau,\vec{r}')\rangle\,d\vec{r}\,d\vec{r}'}{\langle C\rangle^2\left(\int I(\vec{r}'')\,d\vec{r}''\right)^2}\tag{17.5}$$

The concentration fluctuation term $\langle\delta C(t,\vec{r})\,\delta C(t+\tau,\vec{r}')\rangle$ denotes the correlation between a fluorescence fluctuation at spatial position $\vec{r}$ and time t with a fluorescence fluctuation at spatial position $\vec{r}'$ and time $t+\tau$. The evaluation of the concentration fluctuation term involves the solution of the translational diffusion equation for $\delta C(t,\vec{r})$ [36, 77]:

$$\frac{\partial\delta C(t,\vec{r})}{\partial t}=D\nabla^2\delta C(t,\vec{r})\tag{17.6}$$

where D denotes the simple translational diffusion coefficient (m²/s). Equation (17.6) can be solved in Fourier space, and the closed expression for the concentration correlation term is given by [75]

$$\langle\delta C(t,\vec{r})\,\delta C(t+\vec{r})\rangle=\langle C\rangle(4\pi D\tau)^{-3/2}\exp\left(-\frac{|\vec{r}-\vec{r}'|}{4D\tau}\right)\tag{17.7}$$

The remaining terms to be solved in Eq. (17.5) require the knowledge of the geometrical shape of the laser beam profile in 3-D which is determined by the point spread function of the actual optical system. For a confocal microscope using one-photon excitation, the intensity normalized point spread function can be approximated by a three-dimensional Gaussian [78]

$$\overline{\mathrm{PSF}}_{3\mathrm{DG}}(x,y,z)=\frac{I(x,y,z)}{I_0}=\exp\left[-\frac{2(x^2+y^2)}{\omega_0^2}-\frac{2z^2}{z_0^2}\right]\tag{17.8}$$

where ω_0 and z_0 are the effective beam waists in the radial and axial directions, respectively. I_0 is the maximum light intensity in the center of the beam waist. Alternatively, the sample molecules might also be excited by a two-photon absorption process. The radial and the axial intensity distribution of the two-photon focal spot can then be approximated by a 2-D Gaussian and a 1-D Lorentzian function [48], respectively:

$$\overline{\mathrm{PSF}}_{\mathrm{GGL}}(x,y,z)=\frac{I(x,y,z)}{I_0}=\exp\left[-\frac{2(x^2+y^2+z^2)}{\omega_0^2\,(1+(z/z_0)^2)}\right]\tag{17.9}$$

The axial extent is given by $z_0=\pi\omega_0^2/\lambda$, and λ is the wavelength of the laser light.

The autocorrelation function for the Gaussian-Lorentzian intensity profile can only be solved numerically which is in general of minor importance with today's computer power. The autocorrelation function for a molecule diffusing through the triple-Gaussian PSF has been solved analytically [75, 78]:

$$G(\tau) = \frac{1}{\langle N \rangle} \frac{1}{1 + \dfrac{\tau}{t_{\mathrm{diff}}}} \frac{1}{\sqrt{1 + \left(\dfrac{w_0}{z_0}\right)^2 \dfrac{\tau}{t_{\mathrm{diff}}}}} \tag{17.10}$$

where the characteristic molecular diffusion time through the probe cavity, t_{diff}, is

$$t_{\mathrm{diff}} = \frac{\omega_0^2}{4D} \tag{17.11}$$

The diffusion coefficient can be predicted by the Stokes-Einstein Equation assuming spherically shaped molecules of hydrodynamic radius r', in solution of viscosity $\eta\,(\mathrm{g\,cm^{-1}\,s^{-1}})$, at a temperature T [79]:

$$D = \frac{kT}{6\pi\eta r'} \tag{17.12}$$

where T is the absolute temperature (e.g., 293 K) and k is the Boltzmann constant ($1.381 \times 10^{-23}\,\mathrm{JK^{-1}}$). The hydrodynamic radius r' is proportional to the cube root of the molecular mass m (Da):

$$r' = \sqrt[3]{\frac{3\,m/N_A}{4\pi\rho}} \tag{17.13}$$

where N_A is the Avogadro number ($6.022 \times 10^{23}\,\mathrm{mol^{-1}}$) and ρ is the mean density of the molecule (g/cm^3). Although the majority of the molecules are not adequately described as spheres, Eq. (17.12) is useful as a qualitative predictor of the diffusion coefficient for a variety of molecules. For illustration purposes, the diffusion coefficients of distinct molecules along with their corresponding mean times to diffuse through three differently sized observation volume elements are shown in Table 17.1.

Equation (17.10) is the fundamental single species model function for FCS data analysis with the mean particle number $\langle N \rangle$, the diffusion time t_{diff}, and the structure parameter K as fitting parameters. The characteristic effects of these parameters on the correlation amplitude $G(\tau)$ are demonstrated in the first three panels of Fig. 17.2. The dependence of the G_0 magnitude on the mean occupation number of the sample volume $\langle N \rangle$ is illustrated by the simulation shown in Fig. 17.2 A. The inverse proportionality between G_0 and $\langle N \rangle$ reveals the rather surprising fact that the sensitivity of FCS increases at decreasing fluorophore concentration. The overall shape and the decay rates of $G(\tau)$ are inherently determined by the mass of the fluorescent molecule m (via t_{diff}) and the geometry of the sample volume. The sample volume is described by its radial extent ω_0 and the structure parameter K, i.e., the ratio between the volume's axial and radial dimension $K = z_0/\omega_0$. Note, the observed diffusion time t_{diff} is a relative measure since it depends on the molecular mass m but also on the size of the sampling

Table 17.1. The values assume globular or in the case of DNA rod-like molecules and are taken from the Application Handbook for the ConfoCor and FCS ACCESS, Evotec Biosystems GmbH, Germany

Substance	MW [g/Mol]	$D\,[m^2/s]$	t_{diff} ($\omega_0 = 0.2\ \mu m$)	t_{diff} ($\omega_0 = 0.3\ \mu m$)	t_{diff} ($\omega_0 = 0.4\ \mu m$)
Dye	550	$3.6\ \times 10^{-10}$	28 µs	63 µs	111 µs
EGF – ligand	6,800	$1.5\ \times 10^{-10}$	67 µs	150 µs	267 µs
Unspec. (spherical)	10,000	1.36×10^{-10}	74 µs	165 µs	294 µs
Antibody	140,000	5.62×10^{-11}	178 µs	400 µs	712 µs
Bead, →10 nm	N/A	$4.3\ \times 10^{-11}$	233 µs	523 µs	930 µs
Bead, →100 nm	N/A	$4.3\ \times 10^{-12}$	2,330 µs	5,230 µs	9,300 µs
ds DNA 10 bp	6,600	$1.3\ \times 10^{-10}$	77 µs	173 µs	308 µs
ds DNA 20 bp	13,200	$9.8\ \times 10^{-11}$	102 µs	230 µs	408 µs
ds DNA 30 bp	19,800	$8.0\ \times 10^{-11}$	125 µs	281 µs	500 µs
ds DNA 40 bp	26,400	$6.8\ \times 10^{-11}$	147 µs	331 µs	588 µs
ds DNA 50 bp	33,000	$6.0\ \times 10^{-11}$	167 µs	375 µs	667 µs
ds DNA 100 bp	66,000	$3.8\ \times 10^{-11}$	263 µs	592 µs	1,053 µs
ds DNA 500 bp	330,000	$1.2\ \times 10^{-11}$	833 µs	1,875 µs	3,333 µs
ds DNA 1000 bp	660,000	$6.7\ \times 10^{-12}$	1,493 µs	3,358 µs	5,970 µs

volume (c.f. Eqs. 17.11–17.13). Figure 17.2B shows the shift of $G(\tau)$ towards longer correlation times for a series of DNA strands with increasing molecular weight. For comparison, the autocorrelation curve of a freely diffusion dye (rhodamine) is also depicted. The selected DNA strands differ substantially in their molar mass in order to visualize the mass-dependent shift between individual $G(\tau)$ curves. Next, the effect of the structure parameter K on the molecular diffusion time t_{diff} and the mean particle density $\langle N \rangle$ are investigated. A change in the size of the sampling volume affects the G_0 magnitude and the temporal decay rate of $G(\tau)$ since $\langle N \rangle$ and t_{diff} are both sampling volume related parameters. Thus, an inaccurate determined structure parameter results in erroneously retrieved number densities and diffusion times. Typically, K is determined by means of a standard dye at the beginning of a measurement series and then kept fixed for all subsequent measurements. The experimentally determined value for K depends on the alignment and actual setting of the microscope. Typical K-values are between 3 and 10, thus, slightly larger than the value predicted by the theoretical PSF (for instance, using the 1.25 NA oil objective and 488 nm light in an aberration-free confocal microscope, K is 2.9 (page 433 of [80]). For a series of 11 structure parameters, the single species model function (Eq. 17.10) was fitted against an analogously simulated reference curve with representative parameter values of $K = 5$, $\langle N \rangle = 1$, and $t_{\mathrm{diff}} = 63\ \mu s$. Figure 17.2C depicts the fitted diffusion times and their relative errors with respect to the reference diffusion time. For structure parameters larger than the reference value the relative error in t_{diff} is less than 5 % while it becomes much more significant for small K-values (maximum of – 17 % at $K = 2$). Further, the fitted mean particle numbers $\langle N \rangle$ are much less affected (data not shown) and reach a maximum deviation of – 0.2 % at $K = 2$.

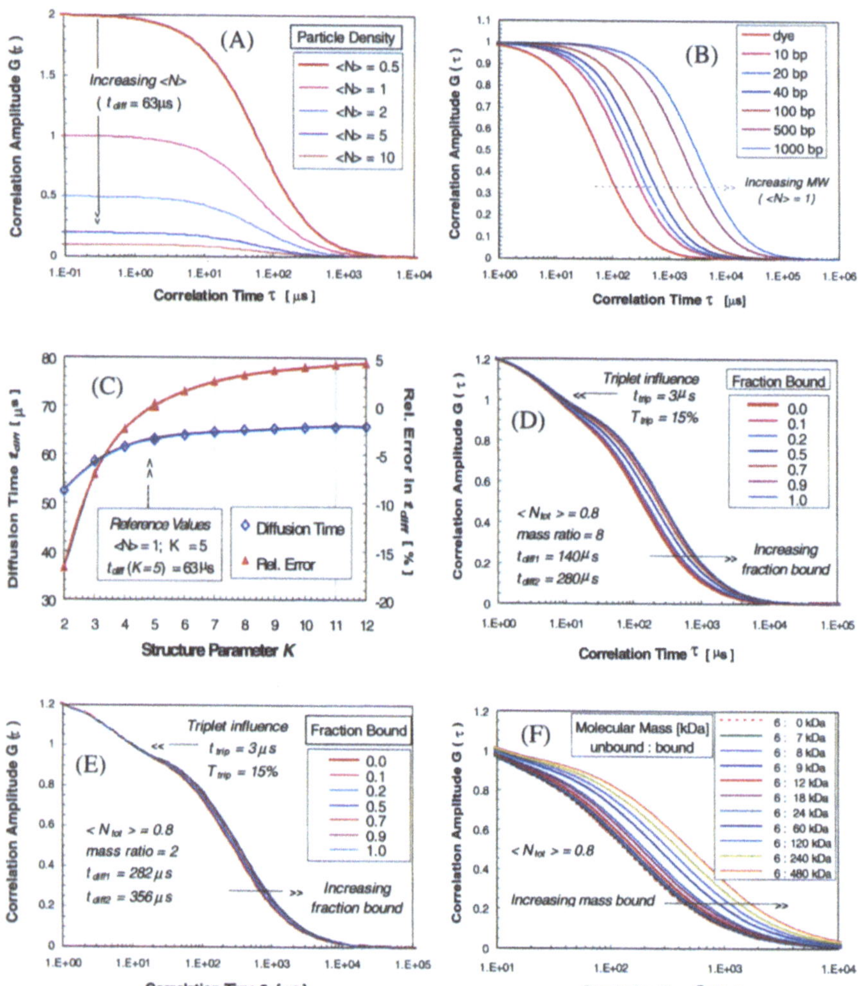

Fig. 17.2 A–F. Examples of simulated FCS autocorrelation functions. The single species and the two-component curves are calculated by Eq. (17.10) and Eq. (17.16), respectively. The diffusion times are calculated for a 0.3 μm radial extent of the sampling volume by Eq. (17.11) or taken from Table 17.1: **A** dependence of the correlation amplitude on the mean occupation numbers $\langle N \rangle$ of the sampling volume of a typical dye (rhodamine). Parameters: $K = 5$, $t_{\text{diff}}(\text{dye}) = 63$ μs; **B** effect of the molecular weight on the correlation time shown by a series of double stranded DNA with increasing numbers of base pairs (bp). Parameters: $\langle N \rangle = 1$, $K = 5$, $t_{\text{diff}}(\text{dye}) = 63$ μs; **C** influence of an erroneously determined structure parameter K on the fitted diffusion time t_{diff}. For a series of (fixed) K-values the diffusion times and the particle numbers were fitted to a reference curve ($K = 5$, $\langle N \rangle = 1$, $t_{\text{diff}} = 63$ μs). While the fitted mean particle number $\langle N \rangle$ remains approximately constant across the K-range (<0.2%, data not shown), the diffusion times significantly deviate from the 'true' 63 μs value particularly towards smaller structure parameters (–17% at $K = 2$); **D** FCS binding curves of a bimolecular reaction between a small (6 kDa) and a large protein (42 kDa). The small protein is fluorescently labeled (e.g., Cy5) and binds at increasing fractions to the 7 times larger receptor protein. The large mass difference between unbound and bound protein (factor of 8) results in

In order to obtain absolute values for the molecular concentration $\langle C \rangle$ or the diffusion coefficient D, it is a prerequisite to determine the shape and the PSF volume (V_{PSF}) since $\langle C \rangle = \langle N \rangle / V_{PSF}$ and $D = \omega_0^2 / 4 t_{diff}$. For optical systems with raster-scanning capabilities such as the confocal or two-photon laser-scanning microscope [26, 80], the actual point-spread function can be measured by a 3-D volume scan of a fluorescent bead which is sufficiently smaller ($\leq 20\%$) than the optical resolution [81]. Alternatively, by using a standard fluorophore with known diffusion coefficient D, the radial ω_0 and axial extent z_0 of the observation volume can be obtained via $\omega_0^2 = 4 D t_{diff}$ and $z_0 = \omega_0 K$. However, if a non-standard dye is used, the corresponding diffusion coefficient can be derived from the fluorophore's rotational correlation time ϕ by means of the Debey-Stokes-Einstein Equation [79]:

$$\phi = \frac{1}{6D} = \frac{\eta V}{RT} \tag{17.14}$$

R is the gas constant ($8.314 \, \mathrm{J \, mol^{-1} \, K^{-1}}$) and V is the hydrodynamic volume [82]. The ϕ parameter is commonly determined in a separate time-resolved anisotropy measurement by means of a high-quality fluorometer with pico-second time resolution. The rotational correlation time can also be on-line acquired by an extended FCS setup [34, 35, 37]. However, only high-precision measurements of rotational correlation times lead to diffusion coefficients accurate enough to provide for absolute concentration determinations. As recently shown, the diffusion coefficient can also be directly and rapidly determined from the time trace of recorded photons by analyzing the width distribution of individual fluorescence bursts [83].

If the calibration of the instrument has to be performed by means of a dye with different spectral properties than the label which will actually be used for screening, concentration measurements are inaccurate since the PSF depends on the wavelength of the light. A simple estimation assuming an elliptically shaped PSF yields a volume difference and correspondingly a concentration difference of about 48% for a spectral shift of 160 nm. Therefore, it is certainly a good policy to calibrate the instrument with the dye which further on will be used for the assay. Within the framework of HTS, the primary focus is on the discrimination between an unbound and a bound reaction state. For these ap-

<hr>

Fig. 17.2A–F (continued)
easily discernable binding curves. Further, the $G(\tau)$ series includes triplet state crossing which affects the curves in the low correlation time region. Parameters: $K = 5$, $\langle N_{tot} \rangle = 1$, $T = 0.15$, $t_{tr} = 10 \, \mu s$, $Q_r = 1$, t_{diff1} (ligand) $= 140 \, \mu s$, t_{diff2} (complex) $= 280 \, \mu s$; **E** aggregation between two identical 50 kDa molecules. The small mass difference between monomer and dimer (factor of two) correspondingly generates closely spaced binding curves which are difficult to resolve by eye. Same parameters as in D except for t_{diff1} (monomer) $= 282 \, \mu s$, t_{diff2} (dimer) $= 356 \, \mu s$; **F** two-component FCS curves of a small tagged protein (6 kDa) which binds at close saturation (90%) to a series of molecules covering the mass range from 1 kDa to 480 kDa. For binders with a molecular mass less than about 6 kDa, the corresponding curves are difficult to distinguish by eye

plications it is sufficient to calibrate the instrument with respect to the fast diffusion time of the unbound reaction partner. The accurate knowledge of the geometry of the observation volume is only a prerequisite for the determination of absolute molecular concentrations.

In FCS it is crucial to correct for a variety of experimental artifacts which degenerate the measured autocorrelation function. For very diluted samples, or correspondingly low fluorescence count rates, the background signal I_B contributes significantly to the autocorrelation function. As a result, the experimental value of G_0 decreases and therefore yields erroneous particle densities. If the mean background and signal fluctuations are not correlated, the actual G_0 value in Eq. (17.10) can be restored with a multiplicative correction factor $f_c = (1 - I_B/I_S)^2$. I_B is the background signal and I_S denotes the total signal amplitude [54]. Typically the background rate is much less than 1 kHz in standard equipment.

When higher laser intensities are used, molecules are prone to turn dark due to singlet-triplet transitions. As a result, the determined diffusion time is shorter then the actual one. If T denotes the mole fraction of molecules in the triplet state with its characteristic lifetime t_{tr}, the appropriate model function accounting for triplet state crossing is given by [43, 84]

$$G_{Tr}(t) = G(\tau) \frac{1 - T + T \exp(-t/t_{tr})}{1 - T} \tag{17.15}$$

Prior to the straightforward expansion of the single-species $G(\tau)$ to a two-component system, we also include the correction terms accounting for a possible difference in the quantum yield between the two species. For instance, different molecular brightnesses are encountered in bimolecular binding reactions in which the tagged molecule changes its fluorescence intensity upon binding. As a result, the FCS analysis would yield erroneous results in terms of the species' molecular concentrations. It is intuitively clear that the brighter species is overestimated in its concentration. If the ratio between the bound and unbound fluorescence intensities is denoted as Q_r, and if the correction factors accounting for background noise and triplet state crossing are further included, the general autocorrelation function for two-species $G_2(\tau)$ now becomes [75]

$$G_2(\tau) = \frac{1}{\langle N \rangle} \left(1 - \frac{I_B}{I_S}\right)^2 \frac{(1 - T + T \exp(-t/t_{tr}))}{(1 - T)} \frac{1}{(1 - Y + YQ_r)^2} \tag{17.16}$$

$$\left(\frac{1}{1 + \dfrac{\tau}{t_{\text{diff}1}}} \frac{1 - Y}{\sqrt{1 + \left(\dfrac{w_0}{z_0}\right)^2 \dfrac{\tau}{t_{\text{diff}1}}}} + \frac{1}{1 + \dfrac{\tau}{t_{\text{diff}2}}} \frac{YQ_r^2}{\sqrt{1 + \left(\dfrac{w_0}{z_0}\right)^2 \dfrac{\tau}{t_{\text{diff}2}}}} \right)$$

The fluorescent species are assumed to have comparable triplet decay times t_{tr}. The diffusion times of the unbound and bound species are $t_{\text{diff}1}$ and $t_{\text{diff}2}$ and their molecular fractions are $1 - Y$ and Y, respectively. Note, $\langle N \rangle$ denotes the total number of sampling volume molecules also including the (dark) triplet state population. The corresponding expansion of $G_2(\tau)$ to a multi-component system is straightforward.

Equation (17.16) is the basic fitting function in the analysis of the ubiquitous bimolecular binding reactions between a receptor protein R and its associated (fluorescently tagged) ligand L^*

$$L^* + R \underset{k_{\mathrm{off}}}{\overset{k_{\mathrm{on}}}{\rightleftharpoons}} RL^*$$

where the dissociation constant K_d is defined as the ratio between the off- and on-rate constants, i.e., $K_d = k_{\mathrm{off}}/k_{\mathrm{on}}$. The fluorescent receptor-ligand complex RL^* is slowed down in its diffusion time with respect to the unbound fluorophore linked ligand L^*. Provided that the diffusion times of the two molecular species differ sufficiently [85], their corresponding molecular fractions $1 - Y$ and Y can be mathematically retrieved by least square fitting procedures, e.g., [68, 86]. In FCS the fluorescent tag should always be conjugated to the smaller molecular species in order to achieve the largest possible shift in the diffusion time.

The equilibrium constant K_d of the bimolecular reaction can be obtained by successively increasing the receptor concentration at an appropriate constant ligand concentration (typically ~ 1 nmol/l), and analyzing the resulting series of autocorrelation curves in terms of a two-component model (c.f. Eq. (17.16). In conventional fluorescence experiments the intensity change ΔF is proportional to the concentration of the formed receptor-ligand complex $[RL^*]$:

$$\frac{\Delta F}{\Delta F_{\mathrm{max}}} = \frac{[RL^*]}{[L_0^*]} \tag{17.17}$$

The total ligand concentration L_0^* corresponds to the maximum fluorescence change ΔF_{max}. Depending on the affinity between ligand and receptor it can be a very difficult experimental task to determine the maximal saturation point. This, however, is a prerequisite for an accurate calculation of the free receptor concentration $[R^*]_{\mathrm{free}}$ using the law of mass action for curve fitting. During recent years the majority of interaction data found in the literature have therefore been theoretically fitted using the equation based on the total ligand and receptor concentrations $[L_0^*]$, $[R_0]$. For a simple competition study this already leads to a cubic equation for an algebraic exact description. In contrast, the fractional parameters $1 - Y$ and Y in FCS directly denote the percentage of the free ligand and the complex in the equilibrium state. The direct determination of the free and bound molecular concentrations in FCS results in a considerable simplification of the mass law description of equilibrium binding. For HTS it is important to keep in mind that the uncertainties of the $1 - Y$ and Y parameters, i.e., their probability distributions, are not directly reflected by the variance of the observed (raw) FCS signal. Actually, the molecular fractions of the (two) species are retrieved by a multi-parameter fitting algorithm from the calculated autocorrelation curve. Hence, FCS requires additional data processing steps to obtain the statistical distributions of the analysis parameters Y and $1 - Y$. Representative examples of simulated two-component FCS curves are shown in Fig. 17.2 D – 2 F. The primary aim of these panels is to illustrate graphically the increasing difficulties to resolve (visually) adjacent autocorrelation curves at decreasing mass difference between the two reaction partners.

In order to achieve an optimal performance of the FCS reader as preparation for an HT screening run, the optical system has to be carefully aligned and calibrated. In the following paragraph we outline the major steps in the instrumental setup procedure of a confocal microscope. First, the optical system is aligned with an aqueous (~ 1 nmol/l) dye solution. The laser power is set to generate an average fluorescent count rate well below the saturation limit of the detector, and the dye is preferentially the same as used for the actual screening process. The alignment strategy primarily maximizes the number of photon counts detected per molecule (cpm). The spatial position of the pinhole is adjusted to the position yielding the highest cpm values. The objective's correction ring which accounts for the different cover glass thicknesses allows an additional fine-tuning. Occasionally we then check for the optimal pinhole size by increasing its diameter until the cpm rate reaches saturation phase. Next, the laser power is successively increased until either the fluorescence rate per molecule does not considerably increase or the triplet population becomes unacceptably high (preferentially, the triplet state should be less than 15%). After finishing the (iterative) adjustment procedure, a series of 5–10 autocorrelation curves is acquired at high statistical precision (acquisition times on the order of tens of seconds). Using a one-component fitting model on the basis of Eq. (17.16), the structure parameter K is retrieved and kept fixed in all subsequent experiments. Further, if the background rate of the assay buffer solution is comparable to the dark noise of the photodetector, no background correction procedure is applied. Next, the autocorrelation functions of the unbound and the target-bound fluorescent tracer are measured and analyzed in terms of a one- and two-component fit, respectively. The one-component fit yields the unbound diffusion time t_{diff1} which is then kept fixed in the two-component model function correspondingly yielding the diffusion time t_{diff2} of the complex along with its molecular fraction Y. As a control, Y should be close to saturation level (> 0.9) for a reliable t_{diff2} readout. For compound screening, the measured FCS data are also fitted by the two-component model function. At least the previously retrieved diffusion times are fixed since a reduction in the number of fitting parameters increases the accuracy of remaining determinants [87].

17.2.3
Features and Issues of FCS-Based Screening

Within the framework of HTS, the statistical accuracy of correlation experiments is a crucial factor in order to assess reliably the binding affinity of potential drug compounds to target molecules. In a screening setup with a throughput rate of 80,000 samples per day, the resulting data acquisition time per sample is on the order of 1 s. The rough screening environment also introduces additional sources of noise which can further compromise a reliable readout. The following discussion qualitatively outlines the major issues and limitations of (auto)correlation-based SMA methods.

Common macroscopic fluorescence methods, whether based on intensity, polarization, or lifetime detection, are ensemble measurements, i.e., the recorded fluorescence emission is the average signal of all excited molecules, hence, only yielding information on average molecular properties. Further, a reduction of the common 50–200 µl assay volume conventionally used in microtiter plates (MTP) typically results in a decrease of the macroscopic (ensemble) fluorescence signal due to interference from surface interactions, from fluorescent compounds, from light-scattering particles, and inner-filter effects. The resolution limit is eventually set by the background noise. A practical detection limit for ensemble methods is around 10^{-9} mol/l, still comprising the vast number of approximately one billion (10^9) particles per microliter assay volume. The current limit for the miniaturization of many types of assays is approximately 10–20 µl using commercially available readers. With the establishment of single molecule spectroscopy the common cuvette is replaced by an open detection volume of the size of an *E. coli* cell ($\sim 0.2 \times 10^{-15}$ l) which only contains an average number of about 1–2 labeled molecules at a standard assay concentration of about 10 nmol/l. Such small sampling volumes are ideal for NTP-based assay platforms and correspondingly reduce both the total consumption of chemical substances and the running assay costs. Thus, in a simple approximation, the cost reduction is proportional to the sample carrier's well size. For instance, an assay performed with NTPs is about 100 times less expensive than assays in MTPs.

According to Eq. (17.2), small mean occupation numbers $\langle N \rangle$ produce large fluctuations in the sampling volume (cf. Fig. 17.2 A). Thus, the sensitivity of FCS actually increases with decreasing fluorophore concentrations. Typical working concentrations for a confocal (or two-photon) microscope are between 10^{-9} to 10^{-15} M. At very low dye concentrations, FCS measurements become impractical due to the time required to record sufficient photon bursts to generate a statistically accurate autocorrelation function. (For a 1 fmol/l solution, a particle crosses a confocal-like observation volume every 15 min). Therefore, FCS is best suited to measure chemical equilibrium with dissociation constants (K_d) in the pico- to nanomolar range. Of course, the 'sampling problem' is also encountered with very slowly diffusing particles such as membrane-bound receptors. Scanning or imaging correlation spectroscopy can overcome these limits by measuring the intensity fluctuation as a function of position rather than time [88–90]. Scanning of the specimen improves the signal-to-noise ratio since data are collected from different regions of the sample. This technique has been implemented in a commercially available screening platform (Mark II, Evotec Biosystems, Germany).

The ability to distinguish two different particles in solution is fundamental in molecular binding studies. Specifically, the binding of ligands to receptors can only be resolved by FCS measurements if their diffusion times differ quite substantially. This is a serious limitation of FCS, taking into account that the diffusion constant is approximately inversely proportional to the cube root of the molecular weight (cf. Eqs. 17.12 and 17.13). For example, the diffusion times of a common dye (~ 500 Da) and BSA (66 kDa) only differ by a factor of 5; their molecular mass ratio, however, is about 130. The statistical accuracy and the resolution limit of FCS has been examined theoretically and practically [54, 85, 91, 92]. As expected, the statistical uncertainties of FCS readout parameters inherently depend on experimental variables such as the data acquisition time, the brightness of the fluorophore, and the geometry of sample volume. For instance, experimental and simulation data both demonstrated that two reasonably bright species both at equal mole fractions can be distinguished within 300 s (!) of data acquisition time if their diffusion times differ by at least a factor of 1.6 [85]. In order to resolve the two species over a larger fractional range a factor of 4–8 was required. Such data acquisition times on the minute time scale are intolerable for HTS. Detection times around 1 s pose stringent conditions on the difference in molecular weight between free and bound ligand.

Under HTS conditions, attempts to fit experimentally obtained autocorrelation curves of simple two-component systems frequently fail when more parameters than only the two molar fractions $(1-Y, Y)$ are fitted. Moreover, the statistical accuracy is often not sufficient the determine the underlying physical model, i.e., whether a solution contains one or two species. The corresponding one or two component fitting procedures might both yield chi-squares close to one. Hence, to retrieve reliably the two molar fractions of a ligand-receptor binding assay under HTS acquisition times < 3 s, appropriate fitting parameters have to be substituted by their actual (previously determined) values and only a minimum number of fitting parameters is allowed to vary in the fitting procedure. Typically, the fixed parameter set includes the structure parameter, the ligand's diffusion times in the bound and unbound state, the factors correcting for background and quenching (Q_r), and occasionally the triplet state parameters. Such in-advance-knowledge can be obtained either off-line or during an actual screen. If fitting parameters can be connected between individual experiments such as encountered in binding studies, global analysis schemes can be applied. This further enhances the accuracy of the fitting results [87].

The variance of the diffusion time $\sigma(t_{\mathrm{diff}})$ was mathematically derived by Koppel (Eq. 51 in [54]). It is based on low photon counting rates ($\varepsilon \ll 1$, see section PCH) and assumes an exponential decay behavior of $G(\tau)$. Koppel's formula allows us to estimate the resolution limit for a diffusion time based discrimination between two chemical species under HTS conditions. The normalized variance of the diffusion time

$$\frac{\sigma(t_{\mathrm{diff}})}{t_{\mathrm{diff}}} = 2^{3/2} \frac{t_s/t_{\mathrm{diff}}}{\varepsilon} \frac{1}{\sqrt{t_{\mathrm{acq}}/t_{\mathrm{diff}}}} \tag{17.18}$$

with t_{acq} denoting the data acquisition time, and Δt_s the photon counting interval (bin time), also enables the identification of the critical experimental parameters. Within this context ε is not the molar extinction coefficient but corresponds to the molecular brightness. It is defined as the number of photons detected per molecule per time Δt_s. To use Eq. (17.18) the bin time has to be sufficiently short with respect to the diffusion time. Typically, Δt_s is around 10–50 µs. Since the average number of detected photons scales linearly with the sampling time, ε can be normalized to $\varepsilon^* = \varepsilon/\Delta t_s$ (counts per molecule per second). Common (one-photon excited) dyes have an ε^* within the range

0–100 kHz. Clearly, the statistical accuracy in the determination of t_{diff} improves with increasing molecular brightness ε. Therefore, it is a prerequisite to select photo-stable dyes with high absorptivity and high quantum yield. The molecular brightness also depends on the excitation light intensity. Thus, the laser power can be successively increased to maximize the brightness parameter by concurrently keeping the triplet state formation within acceptable limits. On the other hand, for molecular count rates exceeding $\varepsilon \geq 1$, the noise is only determined by the stochastic nature of the diffusion process and further raising ε does not significantly improve the statistical accuracy of $G(t)$ [75]. In both the low and high counting-rate limit, the experimental error can be reduced by increasing the data acquisition time t_{acq}. Currently under development to combine the HTS-dictated high sample throughput rate with longer measurement times are multi-channel parallel detection technologies.

In order to estimate the minimal resolvable mass difference in HTS-compatible FCS we consider first an exemplary ligand-receptor binding assay involving a small labeled ligand of 6 kDa and a receptor protein of 42 kDa. The mass ratio between the bound and unbound reaction state is 8. For a typical diffusion time t_{diff1} of the unbound ligand of 150 µs (c.f. Table 17.1), the corresponding diffusion time of the 48 kDa ligand-receptor complex is twice as high, i.e., $t_{\text{diff2}} = 300$ µs. Using Eq. (17.18), the normalized standard deviations $\sigma(t_{\text{diff1}})/t_{\text{diff1}}$ and $\sigma(t_{\text{diff2}})/t_{\text{diff2}}$ as a function of the data acquisition time t_{acq} are calculated and depicted in Fig. 17.3. In addition, the influence of the brightness parameter ε^* is demonstrated by evaluating the experimental errors in the case

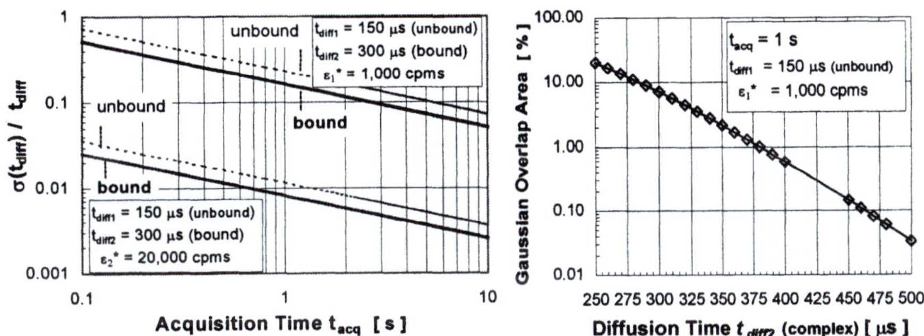

Fig. 17.3 A, B. Statistical accuracy of the autocorrelation functions as predicted by Eq. (17.18): A (left panel) calculated variances of the correlation times of a small 6 kDa ligand and an eight-times larger ligand-receptor complex with corresponding diffusion times of 150 µs and 300 µs, respectively. To demonstrate the brightness effect on the correlation time variance, the ligand is either tagged by a dim label ($\varepsilon^* = 1,000$ cpms) or by a moderately bright label ($\varepsilon^* = 20,000$ cpms). The resulting relative standard deviations $\sigma(t_{\text{diff}})/t_{\text{diff}}$ of the diffusion times of the free (unbound) and bound molecules are shown as a function of the acquisition time t_{acq}; At $t_{\text{acq}} = 1$s the dim label yields t_{diff}-errors ($\sigma(t_{\text{diff}})/t_{\text{diff}}$) of 23% (unbound) and 16% (bound). Correspondingly, the t_{diff}-errors obtained with the bright label are 1.2% (unbound) and 0.8% (bound). B (right panel) for the HTS-like 1 s acquisition time, the diffusion time of the bound complex is increased from 250 µs to 500 µs and the overlapping area between the Gaussian assumed density distributions has been numerically calculated. Only the dim label is depicted since the area integral for the bright label is always well below 0.01%

of a faint and moderately bright label. The corresponding brightnesses were assumed as 1,000 cpms and 20,000 cpms, respectively, and the counting interval Δt_s set to 10 µs. The relative error in the diffusion time decreases proportional to the square root of the acquisition time while the brightness increase linearly contributes to the error reduction. According to Eq. (17.18), the diffusion time of the ligand-receptor complex can be determined more accurately within a given acquisition time than that of the faster diffusing unbound ligand. However, the determination of t_{diff2} is experimentally more demanding (two-component mixture) such that the actual uncertainty of t_{diff1} is often smaller than that of t_{diff2}, i.e., $\sigma(t_{\mathrm{diff1}})/t_{\mathrm{diff1}} < \sigma(t_{\mathrm{diff2}})/t_{\mathrm{diff2}}$. An acquisition time of 1 s can be regarded as HTS typical and is therefore further evaluated. By using the 'dim' label ($\varepsilon^* = 1000$ cpms), the corresponding relative errors for the molecules' diffusion times in the unbound and bound reaction state are relatively high with 23 % and 16%, respectively. Since the uncertainty in the diffusion times scales linearly with the brightness of the used label, the accuracy in the determination of the diffusion times is improved by a factor of 20 with the bright label ($\varepsilon^* = 20,000$ cpms), i.e., $\sigma(t_{\mathrm{diff1}})/t_{\mathrm{diff1}} = 0.012$ % and $\sigma(t_{\mathrm{diff2}})/t_{\mathrm{diff2}} = 0.008$.

Note, the Koppel formula approximately predicts the correlation time uncertainties of a single-species system. The one-component approach can be extended to estimate the practical resolvable diffusion time difference of a two-component system by introducing an appropriate separation criterion between $\langle t_{\mathrm{diff1}} \rangle$ and $\langle t_{\mathrm{diff2}} \rangle$ in terms of their single-species variances. The basic concept is depicted in Fig. 17.4. Assuming Gaussian distributed t_{diff1} and t_{diff2} values, the distribution of the diffusion time values is described by two bell-shaped density functions centered at $\langle t_{\mathrm{diff1}} \rangle$ and $\langle t_{\mathrm{diff2}} \rangle$, respectively, and the variances $\sigma(t_{\mathrm{diff1}})$

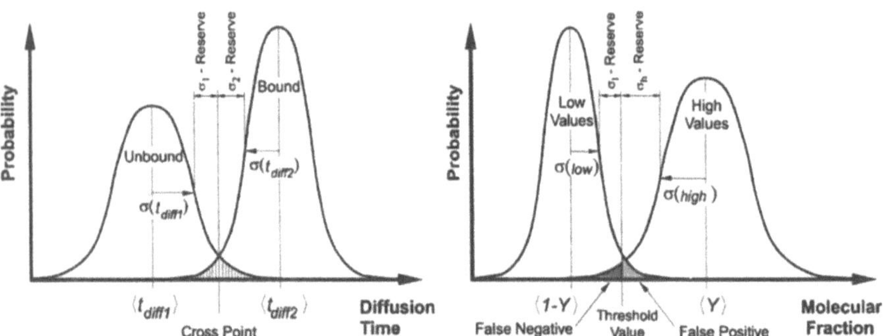

Fig. 17.4. A (left panel) Exemplary probability distribution of the diffusion time values t_{diff1} and t_{diff2} as obtained from a sample population of two kinds of molecules with average correlation times $\langle t_{\mathrm{diff1}} \rangle$ and $\langle t_{\mathrm{diff2}} \rangle$, respectively. The distance from the two (inward laying) half-maximum points to the crossing point of the density curves is referred to as σ-reserve. It is a measure for the ability to distinguish between the two diffusion times, and hence, the two molecular species. B (right panel) Theoretical probability distribution of the high and low controls. Within this context, the low and high values are the molecular fractions $1 - Y$ and Y of the unbound and bound reaction partners, respectively. Since the molecular fractions are fitted parameters (c.f. Eq. 17.16), the 'Y-statistics' crucially depends on the uncertainty of the remaining fitting parameters, in particular the variances in the fluorescence diffusion times (c.f. left panel in this figure)

and $\sigma(t_{\text{diff2}})$ can be approximately calculated by Eq. (17.18). Hence, an appropriate measure for the minimal resolvable correlation time difference is the common overlapping area of the two distribution curves; or correspondingly, the residual distances (σ-reserve) from the half maximum points of a density curves to their common crossing point. For our exemplary receptor-ligand binding reaction, the intersection point is at 216 µs which has been calculated numerically (Mathematica 3.0, Wolfram Research, Champaign, IL). By using the dim label, the σ-reserves obtained for the unbound and bound correlation times are 0.9 and 0.7, respectively. The overlapping area of the two Gaussian curves is 7% which corresponds to the uncertainty or false rate in the discrimination between the 150 µs and 300 µs diffusion times. In order to reduce the false rate to less than 1%, the acquisition time has to be increased to about 3.4 s, or alternatively, the diffusion time of the receptor-ligand complex should be increased by 27% to 380 µs. The corresponding bound-unbound mass ratio is on the order of 16. The dependence of the density curves' overlapping area as a function of their separation in correlation times is depicted in Fig. 17.3B. A significantly higher accuracy is achieved with the bright label. For the 1 s acquisition time the Gaussians' overlapping area is always well below 0.1% (data not shown). However, these false rate results are likely to be overestimated since Eq. (17.18) only holds for low molecular count rates. Again, Eq. (17.18) is used for a qualitative analysis and can only yield the correct order of magnitude.

In the context of HTS, the signals obtained from the unbound and bound reaction state are typically referred to as *low* and *high* values, respectively. Correspondingly, the low and high values in an FCS based screen are the molecular fraction parameters $1-Y$ and Y. In contrast to a conventional fluorescence intensity based assay, the statistics of the low and high values is not directly accessible because (1) the FCS data analysis is not performed on the raw signal (photon arrivals per time bin) but rather on its autocorrelation function, i.e., an already processed data set, and (2) the fractional parameters are only obtained in combination with other fitting parameters such as their associated correlation times which intrinsically are additional sources of noise. A possible approach in the determination of the $1-Y$ and Y variances is subsequently outlined.

In order to assess the statistical quality of an assay in primary screening it is common practice to perform a test screen. The test plates contain randomly distributed patterns of receptor-bound ('high controls') and receptor-unbound ('low controls') fluorescent ligands. First, all low controls are analyzed in terms of a one-component fitting function and the series of retrieved correlation times are averaged $\langle t_{\text{diff1}} \rangle$. The high controls are then analyzed by a two-component fitting function having the first correlation time fixed with the averaged low control value $\langle t_{\text{diff1}} \rangle$. The fitted second diffusion times are associated with that of the complex and again their mean value $\langle t_{\text{diff2}} \rangle$ is calculated. Using the two mean values as fixed parameters and leaving only the molecular fractions as fitting parameters, the low and high controls are again analyzed by the two-component model. The correspondingly retrieved molecular fraction $1-Y$ and Y are then plotted in a frequency histogram which allows one to assess the fidelity of the screen. An exemplary histogram is shown in Fig. 17.4B. In particular, the cross-

Table 17.2. To achieve a molecular misclassification rate of 0.5 % or better, the correspondingly required minimal variances $\sigma(\,)$ of the molecular fractions $1-Y$ and Y have been calculated for five different Y-threshold levels. For the numerical calculation the distribution of both Y-parameters is assumed to be Gaussian

Y-Threshold	$100 \times \sigma(1-Y)$ [%]	$100 \times \sigma(Y)$ [%]
0.1	3.88	34.9
0.2	6	31
0.3	11.5	27.1
0.4	15.5	23.2
0.5	30	19.4

talk region where high values are scattered into the low regime and vice versa provides valuable information on the robustness of an assay. In a primary HTscreen the discrimination between a bound (high Y) or unbound (low Y) component is typically achieved by setting a carefully selected threshold value for the Y-parameter. Defining for instance low values as negative (no effect) and high values as positive (binding event), the threshold can be calculated according to the acceptable rate of erroneously detected high values (false negatives) and low values (false positives). Table 17.2 summarizes the experimentally required accuracy in the molecular fraction parameters provided these values obey the Gaussian statistics. For a series of five increasing Y-threshold values, the maximal acceptable variances of the Y-parameters have been calculated such that both the false positive and the false negative rate are 0.5 % or less. As an illustrative numerical example, from 1000 tested samples which are classified via the 0.5 % threshold level, five 'hits' would be missed and five samples would be erroneously subjected to further validation cycles. For the tabulated 0.5 and 0.4 Y-threshold values, a correspondingly performed assay would be quite robust since variances of about 15 % are experimentally reasonable to achieve in HTS screening systems. Typical encountered variances of the $1-Y$ and Y parameter are in the range of 5 % and 15 %, respectively. However, real data can considerably deviate from the Gaussian distribution which might increase the misclassification rate. In order to regain the necessary experimentally accuracy, multi-dimensional detection schemes have to be applied such that the additionally acquired signal parameters (e.g., anisotropy) can be also taken into account by the data analysis (personal communication with D. Ullman, Evotec Biosystems, Germany).

As a general approach to the development of FCS-based screens it is recommended to address the following points:

1. Keep the fluorescently labeled substrate as small as possible.
2. Wherever possible, apply avidin or antibody technology to increase the molecular weight difference between bound and free component.
3. Do not waste time in working out specific labeling chemistries. Standard 6-carbon linkers between fluorophore and compound usually work.
4. Maximize the number of detected photons per molecule by using high quantum yield labels and via careful adjustment of the laser intensity and proper alignment of the optics.

5. Reduce the actual number of fitting parameters to measure up with the considerable statistical variance caused by the short data acquisition times. The variables eventually encompass only the molecular fractions of the species to be resolved while in particular the diffusion times, the background, and the structure parameter have to be determined in advance.
6. Take particular care to correct the background signal for dim labels.
7. Evaluate possible changes in the quantum yield upon binding which correspondingly have to be accounted for by appropriate fitting routines [75].

With the inherent problems of FCS analysis clearly recognized in multiple assays which were developed by the consortium partners (Evotec Biosystems, Novartis, and Smith-Kline Beecham) during the last three years it became clear that a molecular weight independent detection technology was of essential importance to make SMA an universal technique for drug screening. However, the need to distinguish single fluorescent molecules based on their specific molecular brightnesses is as important for basic science as it is for applied approaches. This probably explains that the intensity dependent single molecule analysis methods were developed nearly independently by two European groups (Evotec, MPI-Göttingen) and one US-American institute (LFD, University of Illinois at Urbaba-Champaign) since 1996. Only recently details were published [63–65, 72]. Within the Evotec-Novartis consortium collaboration FIDA-based assays were introduced in 1998 and further developed towards pharmaceutical screening. Based on this hands-on experience but also on multiple discussions with the developers of the related techniques, a summary over SMA methods will be given. Emphasis will be on the selection of the most appropriate technique for applications in drug screening. In particular an overview on the 'Photon Counting Histogram' (PCH) technique developed by Y. Chen, J. Mueller, and E. Gratton, the 'Fluorescence Intensity Distribution Analysis' (FIDA) developed by Evotec Biosystems (group Kask P and Gall K), and the 'Burst Integrated Fluorescence Lifetime' method (BIFL), developed by the MPI-Göttingen (group C. Seidel) will be shown.

17.2.4
Photon Counting Statistics: Poisson and Super-Poisson Analysis

Random processes are often described by means of probability distributions [62]. The correspondingly obtained readout parameters yield information about the amplitude distribution of the fluctuations. In a first approach referred to as Poisson analysis, the entrance of a particle into the sampling cavity is assumed to follow Poisson statistics [93]. The corresponding Poissonian model states that the distribution of intervals of entry events into the observation volume decreases exponentially with time:

$$N(\Delta t) = \alpha \exp(-\beta \Delta t) \tag{17.19}$$

where N is the number of times an event recurs after a time Δt, β is the event's characteristic recurrence frequency, and α is a proportionality constant. The recurrence time $\tau_R = 1/\beta$, i.e., the time for a particle to encounter the sampling cavity, is given by diffusion theory: $\tau_R = 1/(4\pi n R_0 D)$ where D is the diffusion coefficient, R_0 is the radius of the observation volume, and n is particle density in the solute. Experimentally, the entrance of a particle into the sampling volume is detectable via its accordingly generated photon burst. The distribution of dark intervals between photon burst has been measured and analyzed [94, 95]. For long time scales, the data can be reasonably well described by Eq. (17.18), but at short time scales, a positive deviation away from the Poisson curve was found, i.e., the probability to detect a molecule within a short time interval is much higher than predicted by Poisson statistics. The discrepancy has been partly attributed to optical trapping which results in a biased diffusion and correspondingly enhances the re-crossing number of the same molecule into the sampling cavity.

17.2.5
Photon Counting Histogram (PCH)

Additional sources of randomness and specific experimental parameters such as the inhomogeneous excitation profile have to be taken into account for a more accurate description of the probability distribution. A recent method, referred to as 'Photon Counting Histogram' or PCH (group Gratton), analyzes the distribution of the rate of occurrence of photon counts per selected time interval [65]. Such 'photon counting distribution curves' sensitively reflect the brightness distribution of the underlying sample population. The essential readout parameters of the PCH method are the molecular brightness, here referred to as ε, and the average particle number $\dfrac{\bar{N}}{V_0}$ in the observation volume V_0. The specific molecular brightness is expressed as the mean count rate per given particle, i.e., photon counts per molecule per time. It is a characteristic measure for a particular species since ε depends on the molecule's absorption cross section and quantum yield. In contrast to the autocorrelation method, PCH allows one to distinguish molecules with the same diffusion time as long as their individual fluorescence intensity differs. The concept is schematically shown in Fig. 17.5.

Fig. 17.5 A, B. A Concept of photon counting histogram method and fluorescence intensity distribution analysis (PCH/FIDA method-1). Molecules are analyzed based on their differences in specific molecular brightness and particle numbers. The brightness difference originates from a change in the quantum yield, from spectral shifts or from fluorescence resonance energy transfer. The top part of the figure schematically depicts four (❶ ❷ ❸ ❹) and two (❺ ❻) transit events of two kinds of molecules respectively, an unbound labeled oligonucleotide (ODN), and a transcription factor (TRF) bound ODN. Changes in the local environmental structure or polarity of the ODN fluorescence label (*gray dot*) upon binding to the TRF induce a change in the fluorescence intensity of the dye (shown by a *darker dot*). This is a prerequisite in PCH and FIDA to discriminate the two species. Provided that the photon sampling time (bin-time) is sufficiently shorter than the average diffusion time of the molecules, the stroboscopically sampled fluorescence emission (*blurred dots*) scales with the specific molecular brightness independent of the diffusion time and the number of transit events. For the two kind of molecules, free and complexed labeled ODN, the corresponding time series of sampled fluorescence photons are demonstrated by the two graphs in the *photon burst diagrams* of the

Experimentally, the statistics of photon count amplitudes, correspondingly referred to as the photon counting histogram or PCH, is generated from a sequence of photon counts and then analyzed in terms of the probability distribution $p(k)$ to observe k photon counts per selected sampling interval Δt_s. The bin width Δt_s is chosen short enough (e.g., 10 µs) to track the intensity fluctua-

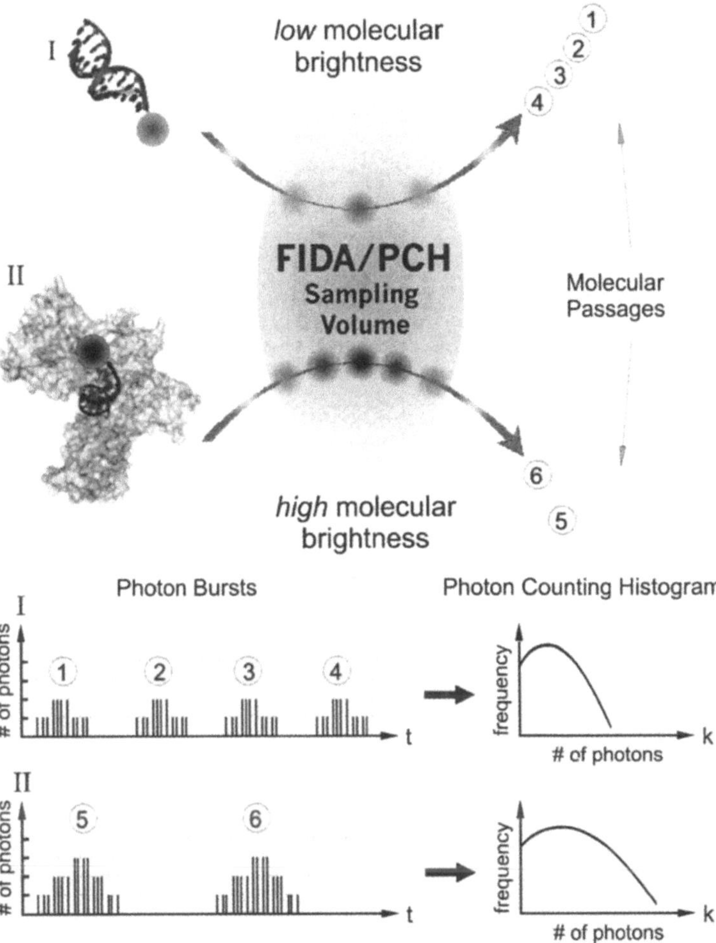

Fig. 17.5 A (continued)
figure. The specific molecular brightness is proportional to the maximal amplitude of the number of recorded photons (c.f. # of photon axis in the burst graphs). The poor photon statistics obtained from a single transit event is typically not sufficient to perform single-burst PCH or FIDA on the basis of the photon counting histogram, i.e., the distribution of the rate of occurrence of photon counts per bin-time interval. Therefore many transit events are included in the generation of the photon counting histogram. The increase in the molecular brightness between the 'dim' ODN and the 'bright' TRF-ODN complex is revealed by the calculated histograms as a corresponding 'stretch' of the photon distribution curve towards higher values on the histogram's photon count axis (k-axis). (For a detailed description of the terminology and symbols used in this figure please refer to Fig. 17.1)

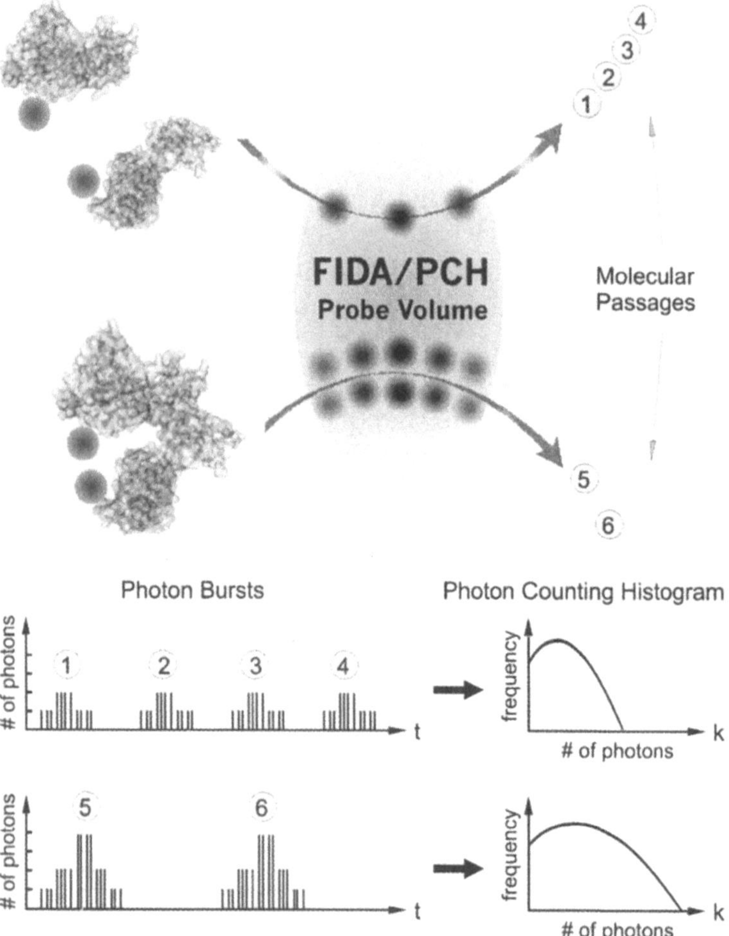

Fig. 17.5. B PCH/FIDA performed on a system with fluorescently labeled ligand AND receptor (PCH/FIDA method-2): The heterodimerization of a transcription factor is used as an example for an assay monitoring protein-protein interaction. Both binding partners are fluorescently tagged with either the same dye (1-D setup), or two different tracers (like a green and a red dye, 2-D setup). With two clearly separated positions of the two dyes on the complex, the intensity increase upon complexation of the two units can be as high as 100%. Very rarely fluorescence quenching reaches 100% or fluorescence enhancement effects (c.f. Fig. 17.5A) exceed 100%. The double labeling strategy is therefore superior to the single labeling method in absolute brightness change. With the availability of numerous new targets from genomics projects, efficient methods to setup assays for protein-protein interactions gain increased importance in pharmaceutical industry. The double labeling strategy applied in FIDA/PCH will probably set the standard for the robust screening for macromolecular complexation in a molecular independent way. (For a detailed description of the terminology and symbols used in this figure please refer to Fig. 17.1)

tions of the particular process of interest. Mathematically, PCH is based on Mandel's formula [96] which accounts for two sources of randomness, the statistical nature (shot noise) of the photon detection process and the random walk of molecules diffusing through an inhomogeneous excitation profile:

$$p(k) = \int_0^\infty Poi(k, \eta_\mathrm{I} I_\mathrm{D}) \, p(I_\mathrm{D}) \, dI_\mathrm{D} \tag{17.20}$$

Correspondingly, the first integrand term, $Poi(k, \eta_\mathrm{I} I_\mathrm{D})$, describes the Poissonian governed photon count statistic for light with constant intensity I_D impinging onto the photodetector. The mean number of detected photons $\langle k \rangle$ is proportional to I_D, $\langle k \rangle = \eta_\mathrm{I} I_\mathrm{D}$, where the proportionality factor η_I is the product of the detection efficiency η_det and Δt_S. The second integrand term, $p(I_\mathrm{D})$, is the intensity distribution function which accounts for the fluctuations in the light intensity originating from molecules randomly diffusing through an inhomogeneous excitation profile. Thus, the overall photon count distribution $p(k)$ is the weighted superposition of individual Poisson distributions for each of the intensity values I_D with the scaling amplitude p(I_D). Obviously, any fluctuation in the light intensity results in a broadening of the photon count distribution with respect to a pure Poisson distribution, and the PCH broadening becomes more pronounced as the strength of intensity fluctuations increases. The distribution of $p(k)$ is called super-Poissonian [66] since its actual variance $\langle \Delta k^2 \rangle$ is greater than its mean value, i.e., $\langle \Delta k^2 \rangle > \langle k \rangle$.

The theoretical expression for $p(k)$ can be obtained by starting from a single particle enclosed in a box of volume V_0 which is essentially illuminated by the point spread function of the excitation light. Based on Mandel's formula and using a scaled point spread function $\overline{\mathrm{PSF}}$ such that its volume V_PSF is equal to the volume defined for FCS [75], the PCH of a single diffusing particle can be written as [65]

$$p^{(1)}(k; V_0, \varepsilon) = \frac{1}{V_0} \int_{V_0} Poi(k, \varepsilon \overline{\mathrm{PSF}}(\vec{r})) \, d\vec{r} \tag{17.21}$$

where ε denotes the molecular brightness, i.e., the average number of detected photons per sampling time Δt_s for a particle diffusing through the observation volume V_PSF. As expected, the average number of photons $\langle k \rangle$ detected from the single V_0-box enclosed particle is essentially determined by the molecular brightness ε and the volume ratio V_PSF/V_0 which denotes the probability of finding the molecule within the volume of the scaled point spread function $\langle k \rangle = 1/V_0 \int_{V_0} \varepsilon \overline{\mathrm{PSF}}(\vec{r}) \, d\vec{r} = \varepsilon (V_\mathrm{PSF}/V_0)$. Correspondingly, $p^{(1)}(k; V_0, \varepsilon)$ depends on the shape of the $\overline{\mathrm{PSF}}$ and the parameter ε.

For N independent but still enclosed particles, the PCH is obtained by consecutively convoluting the corresponding single-particle PCH:

$$p^{(N)}(k; V_0, \varepsilon) = \underbrace{(p^{(1)} \otimes \cdots \otimes p^{(1)})}_{N\text{-times}}(k; V_0, \varepsilon) \tag{17.22}$$

where $\otimes$ denotes the numerical convolution operation. The next level of complexity opens the enclosed volume V_0. If the observation volume is much smaller than the reservoir, the number fluctuations of freely diffusing particles is described by Poissonian statistics $p_\#(N) = P(N,\bar{N})$ with $\bar{N}$ being the average number of molecules within V_0. $\bar{N}$ can be written in terms of the particle concentration C and the Avogadro's number N_A as $\bar{N} = C V_0 N_A$.

For an open system, it is intuitively clear that the particular choice of the reference volume is irrelevant [65], i.e., $p^{\bar{N}}(k; V_0) = p^{\bar{N}_1}(k; V_1)$ where $\bar{N}_1$ and the mean particle number found in another volume element V_1 of the reservoir. For practical purposes, the photon count probability is referenced to the standard volume used in FCS, i.e., the volume of the scaled point spread function V_{PSF}. As a result, the average number of detected photons $\langle k \rangle$ is simply the product of the molecular brightness ε and $\bar{N}_{PSF}$, the average number of particles within the observation volume, i.e., $\langle k \rangle = \varepsilon \bar{N}_{PSF}$.

The PCH for an open system containing N (identical) particles $\Pi(k, \bar{N}_{PSF}, \varepsilon)$, is the average of the individual N-particle probability functions $p^{(N)}(k, V_0, \varepsilon)$ but weighted by their probability to observe N particles $p_\#(N)$

$$\Pi(k; \bar{N}_{PSF}, \varepsilon) = \sum_{N=0}^{\infty} p^{(N)}(k; V_0, \varepsilon) p_\#(N) \tag{17.23}$$

where V_0 is identical with the volume of the point spread function V_{PSF}. The density function $\Pi(k, \bar{N}_{PSF}, \varepsilon)$ is the probability to observe k photons from a solution of identical particles with a concentration of $C = \bar{N}_{PSF}/V_0$.

The histogram of a single species is characterized by two parameters, the average number of particles $\bar{N}$ in the open reference volume and their molecular brightness ε. For a mixture of M independent species which all differ in their molecular brightnesses, the overall open system PCH function can be generated by successively convoluting the photon counting distribution of each individual species with one another.

For two species, the photon count distribution $\Pi(k; \bar{N}_1, \bar{N}_2, \varepsilon_1, \varepsilon_2)$ becomes

$$\Pi(k; \bar{N}_1, \bar{N}_2, \varepsilon_1, \varepsilon_2) = \Pi(k; \bar{N}_1, \varepsilon_1) \otimes \Pi(k; \bar{N}_2, \varepsilon_2) \tag{17.24}$$

where ε_1 and ε_2 respectively are the molecular brightnesses of species *1* and *2*, and $\bar{N}_1$ and $\bar{N}_2$ correspondingly denote the species' average particle numbers inside the reference volume. Thus, PCH analysis based on Eq. (17.24) can resolve a mixture of two molecular species only based on the difference in their molecular brightness regardless of their diffusion coefficients.

For illustrative purposes, the effect of the brightness parameter on the shape of the photon counting histogram for three distinct dyes are demonstrated in Fig. 17.6. (The figure has been courteously provided by J. Mueller, Y. Chen, and E. Gratton) At increasing molecular brightness the PCH curves are (left-)shifted towards photon count number k. The amplitude of a particular PCH depends on the molecular concentration. To allow for an easy comparison between the histograms the concentration of the three samples was kept similar. To demonstrate further the super-Poisson behavior of the PCH, the correspondingly predicted pure Poisson distribution has also been depicted. It can be clearly seen that the deviation between the PCH and the corresponding common Poisson curve becomes more evident at higher molecular brightnesses.

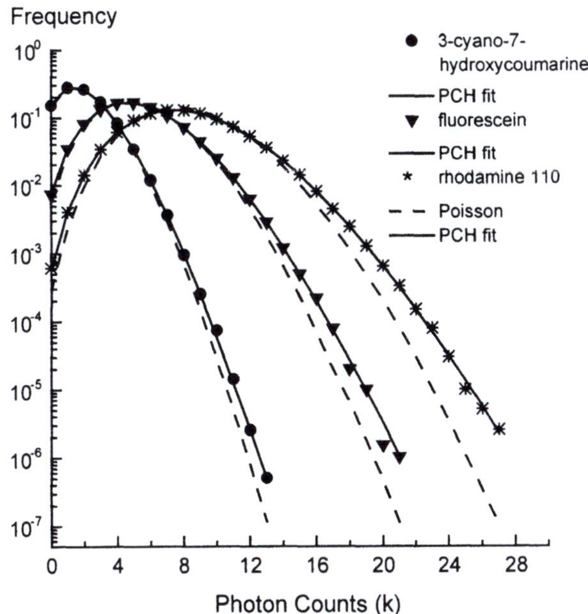

Fig. 17.6. Photon counting histograms for three dyes differing in their molecular brightness ε. The histograms of cyanohydroxycoumarin ($\bullet$), fluorescein ($\blacktriangledown$) and rhodamine 110 ($*$) taken with the same number of data points were fit to the theoretical PCH function $\Pi(k, \bar{N}, \varepsilon)$ shown as *solid lines*. The concentration of the three samples was kept similar to facilitate the comparison between the histograms. The fit recovered the average number of molecules $\bar{N}$ as 2.6, 3.3, and 3.0 for cyanohydroxycoumarin, fluorescein, and rhodamine 10, respectively. The molecular brightness ε of the dyes is rather different, however, with a value of 0.30 for cyanohydroxycoumarin, 0.65 for fluorescein and 1.10 for rhodamine 110. For each histogram a Poisson distribution with a mean equal to the average number of photon counts is plotted as a *dashed line*. The deviation between the Poisson distribution and the photon counting histogram increases markedly with increased molecular brightness ε. (Figure courteously obtained from Gratton E, Y. Chen, and J. Mueller, Laboratory for Fluorescence Dynamics, University of Illinois in Urbana-Champaign, IL)

17.2.6
Fluorescence Intensity Distribution Analysis (FIDA)

Analogues to PCH, the FIDA method (Kask/Gall) analyzes the distribution of the rate of occurrence of photon counts per selected time interval [63, 64]. FIDA allows one to distinguish quantitatively different species of molecules according to a difference in their brightnesses since the average fluorescence intensity emitted by a molecule depends on its absorption cross section and quantum yield. The specific brightness is expressed as a mean count rate per given particle, i.e., photon counts per molecule per time unit. The second essential readout parameters of the FIDA method are the molecular concentrations of different fluorescent species. Both parameters are determined from one individual 1D- or 2D-FIDA measurement. The species' concentrations used in a FIDA analysis are usually substantially lower than in FCS measurements, resulting in less than one particle within the confocal focus.

Experimentally, step (1) in FIDA is the detection of the time-binned number of photon counts in a repetitive way. The bin-time window used is typically between 10 µs and 50 µs. Step (2) is the determination of the frequency distribution of photon counts derived from the recorded time series of detected photons. The usually half logarithmic plot of the photon counting distribution can be referred to as the raw data of FIDA. Both theoretical and experimental raw data plots are depicted in Fig. 17.7 A,C, respectively. (Figure 17.7 has been kindly provided by Evotec Biosystems, Germany.) The 'photon counting distribution

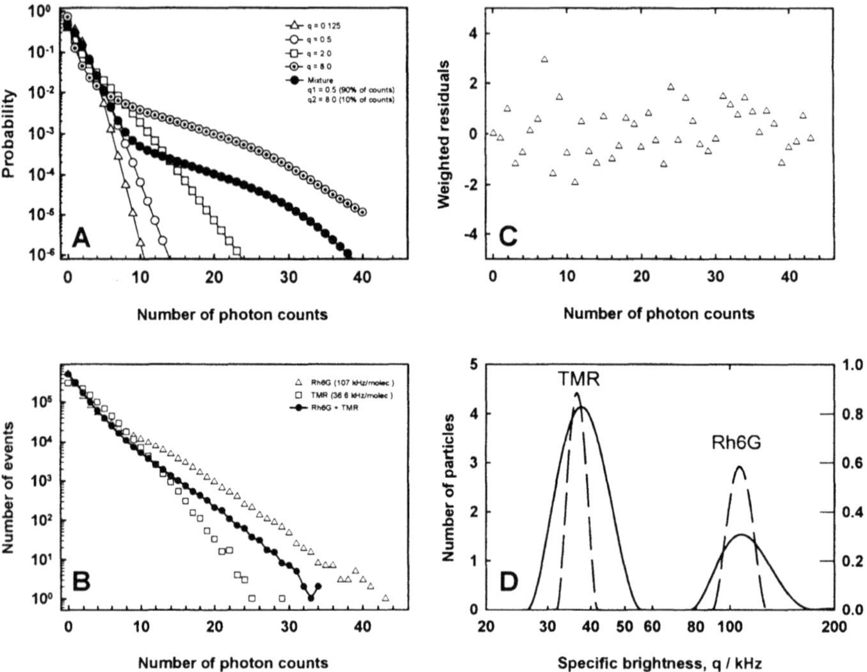

Fig. 17.7 A–D. Theoretical and experimental distributions of the number of photon counts: **A** expected probability distributions, P(n) for five cases of equal mean count rate, $\bar{n} = 1.0$. The *open symbols* correspond to solutions of single species but with different values of the mean count number per particle qT. The *solid line* is calculated for a mixture of two species, one with $qT = 0.5$ and the other with $qT = 0.8$; **B** measured distributions of the number of photon counts. The data for three samples are presented: the solutions of 0.5 nmol/l Rh6G, 1.5 nmol/l tetramethylrhodamine (TMR), and a mixture of them (0.8 nmol/l TMR, 0.1 nmol/l Rh6G). The time window of 40 µs was used, and the number of photon counts was determined 1,250,000 (N) times in each of the 50 s experiments; **C** weighted residuals from multicomponent fit analysis of the count number distribution measured for rhodamine 6G; **D** results of ITR analysis applied to the curves shown in Fig. 17.7 B. The *dashed lines* correspond to the solutions of single dyes (Rh6G and TMR) and the *solid line* correspond to their mixture. The ordinates give the mean number of particles within the confocal volume element (*left*, single dyes; *right*, mixture). (Figure courteously obtained from Gall K EVOTEC BioSystems AG, Germany)

curves' sensitively reflect the brightness distribution of the underlying sample population (see Fig. 17.7 A). The effect of the molecular brightness on the shape of the distribution curves can be best understood by considering a mixed sample population of two strongly diluted species which differ considerably in their brightnesses (but not necessarily in their diffusion times). Due to the low sample concentration no more than one molecule passes through the observation volume and correspondingly either a very bright or very dim photon burst is detected. Averaging over many transit events, the correspondingly obtained frequency distribution of photon counts would exhibit two (equally high) frequency peaks well separated on the photon count axis. Of course, such double-peaked 'raw data' histograms are hardly observed in real examples. Figure 17.7B demonstrates the difference between dim and bright species revealed in the distribution of photon counts. Already here one can see the steeper decay of the half logarithmic number of events vs number of photon counts plot for a darker dye (tetramethylrhodamine, TMR) relative to the brighter dye (Rhodamine6G) with the summary decay of the mixture placed between the pure dyes histogram curve.

Step (3) in FIDA is the calculation of the expected distribution of the number of photon counts. This is the basis for iterative adjustment of the fitting parameters, the concentrations and specific brightnesses, to the photon counting histogram by non-linear least squares fitting procedures. Central to fitting the photon count number distribution is the accurate description of the spatial sampling profile $B(\vec{r})$. It is a normalized and position dependent function. In fact, $B(\vec{r})$ is the product of excitation intensity and fluorescence collection efficiency, as a function of spatial coordinates of a molecule in the sample. Due to the (inevitable) non-perfect optical set-up, the experimental spatial brightness function deviates substantially from the theoretical form. The correspondingly necessary detailed measurement of the exact experimental spatial brightness function would be to time-consuming for the practical screening applications which FIDA is designed for. It was therefore key for FIDA developers to come up with an accurate empirical description of $B(\vec{r})$. The strategy introduced by Kask et al. [64] was to search for a simple but flexible mathematical description of $B(\vec{r})$, ignoring details of the complex 3 D-function. The theoretical distribution of the number of photon counts is calculated through compartmentalization of the sample volume into a series of spatial sections with sizes dV_i each having a nearly constant value of spatial brightness B_i. The contribution from a spatial section to the theoretical distribution of the number of photon counts does not depend on the shape of the section. Therefore, in calculation of the expected distribution of the number of photon counts it is possible to represent $B(\vec{r})$ by a relatively simple one-dimensional relationship between B_i and dV_i. This relationship was found to be best represented by an empirical function, a power series of a variable $x = \ln[B(0)/B(\vec{r})]$:

$$\frac{dV}{dx} = \sum_k a_k x^k \tag{17.25}$$

with only first, second and third power terms included. The coefficients a_k represent empirical characteristics of the optical equipment, the values of which are pre-determined from adjustment experiments on single species. For a single spatial section dV_i and single species, the expected distribution of the number of photon counts can be expressed as 'double Poissonian'

$$P_{dV_i}(n) = \sum_{m=0}^{\infty} P(m)\,P(n|m) \tag{17.26}$$

$$= \sum_{m=0}^{\infty} \frac{(CdV_i)^m}{m!}\,e^{-CdV_i}\,\frac{(mqTB(\vec{r}))^n}{n!}\,e^{-mqTB(\vec{r})}$$

The first Poisson distribution $P(m)$, with CdV_i as its mean value, describes the particles' stochastic occupation number in the spatial section dV_i. In consideration of the statistical nature of light, the second distribution $P(n|m)$ is the conditional Poisson probability of detecting n photons provided that there are m molecules in the sample volume V. If q denotes the molecular brightness (cpms), and T is the length of a sufficiently short sampling time interval, the conditional probability $P(n|m)$ has $mqTB_i$ as its mean value.

For an accurate analysis of two or more species which differ in less than 100 % of their specific brightness a very high data quality is necessary. Measurement times and calculation (= fitting) times of more than a few seconds are no-go arguments for uHTS. Although the calculation scheme using the convolutes of the Poisson distributions is straightforward, it can become unacceptably time-consuming for an HT screening process, when the contributions from different species as well as from background is included. The second 'trick' used in the FIDA algorithm is to replace the straightforward calculation scheme of the distribution of the number of photon counts (as a convolution integral over spatial sections) by a significantly faster calculation through the generating function

$$G(\xi) = \sum_n \xi^n P(n) \tag{17.27}$$

The Fourier-transform-like representation of the photon count distribution $P(n)$ by $G(\xi)$ including the use of ξ as complex argument $e^{i\varphi}$ results in a substantial simplification of the calculations scheme by replacing time-consuming convolution integrals by the species dependent sum of spatial integrals. The numerical calculation by means of the fast inverse Fourier transform

$$P(n) = FFT^{-1}\left(\exp\left[\sum_j C_j \int (e^{(e^{i\varphi}-1)q_j TB(r)} - 1)\,d^3r\right]\right) \tag{17.28}$$

is easily accessible on every PC. The FFT algorithm is fast enough to evaluate the experimental data by means of a 'classical' multicomponent fitting procedure for the estimation of the unknown concentrations C_j and the corresponding specific brightnesses q_j of the sample under study. Shown in Fig. 17.7 D is an alternative way of fitting FIDA data. Although nice for representation purposes, the distribution analysis-based ITR method (inverse transformation with linear regularization), with concentrations restricted to non-negative values, is much too slow for the evaluation of HTS assays.

17.2.7
Features and Issues of FIDA and PCH

Distribution analysis of photon counting statistics is a novel and exciting data analysis tool. Related to correlation analysis, methods like FIDA or PCH provide a complementary set of readout parameters – i.e., the respective concentrations and molecular brightnesses of different fluorescent species even with all species having the same diffusion time. The FIDA technique performed with one instrumental fluorescence excitation and emission channel (1D-FIDA) becomes the method of choice for HTS if even the 150 kDa MW shift induced by secondary antibody coupling cannot provide the necessary eight- to tenfold increase in MW for reliable and statistically significant increases in the translational diffusion time. Already 25% intensity changes of dye, in a mass independent way, leads to the calculation of the particle number and the specific brightness determined from one measurement on a ligand or substrate. The technology is being extended to two-color excitation and emission. With two different laser lines and either two filters or two polarizers in the detection path for measuring the signal intensities (e.g., by avalanche photodiodes), it is possible to measure fluorescence anisotropies and FRET or ground state quenching effects either intramolecularly or intermolecularly.

There are two general possibilities for the set up of a 1D-FIDA assay system as schematically depicted in Fig. 17.5A, B. (A) The fluorophore can be placed at a position at the macromolecule where the local increase or decrease in hydrophobicity or polarity occurring on protein complexation leads to changes in quantum yields and/or to spectral shifts through locally or globally induced alteration in the environment of the dye. For some protein protein interactions fluorescence intensity changes of 100% and more can be detected, providing 1D-FIDA measurements with good signal to noise ratios. In RNA protein interaction assays a normal strategy is to substitute flexible positions within the RNA, like bulged adenine nucleotides with (non-nucleosidic) dyes [97]. Approximately 20% increase in fluorescence brightness on binding of proteins or carbohydrate can easily be achieved with such modifications. This increase is not enough for a single dye 1D-FIDA approach. In such cases the second 1D-FIDA possibility has to be applied; two identical dyes attached to different molecules. (B) The two reaction partners can be labeled with the same dye at nearly equal stoichiometry. If the two individual parts of a protein-protein heterodimer contain, for example, one rhodamine dye molecule each, the complexation reaction leads to a molecular species with double brightness in relation to each monomer.

For resolving the thermodynamic/kinetic mechanism of the formation of higher order molecular species with stoichiometries ≥ 2, with co-operative binding characteristics, or with heterologous consumption, the FIDA approach can be extended to allow in-vitro assay experiments to be performed successfully which, only a short while ago, spectroscopists would not have dared to try.

In SMA experiments performed with one excitation wavelength but two detection channels, spectral shifts or FRET can be resolved with filters. The analogous experiments to steady state anisotropy can be performed with two polarizers in the emissions paths. Many ligand interactions with proteins and small molecules occur via a direct binding event followed by a conformational change leading to a thermodynamically stable compact structure. In addition to screening ligand and protein binding by changes in intensity of one detecting dye or FCS, the interaction can be monitored by following the conformational change of the (protein or nucleic acid) ligand. Site specific labeling, however, is a prerequisite for the experimental realization of such FRET based systems. With two positions which undergo the largest intramolecular distance change on complexation, for example labeled with a dye excitable at 488 nm (like rhodamine green) and one excitable at 543 nm (like tetramethylrhodamine), the FRET effect can be followed in 1D by donor (RG) quenching or acceptor enhancement (TMR) or in a two color detector set-up by monitoring each emission wavelength through interference filters. If the dyes used in 2D experiments are selected spectroscopically separate, i.e., the absorption spectrum of the red dye does not overlap with the emission spectrum of the green dye, very specific interaction experiments can be performed using burst coincidence analysis and cross correlation methodologies with considerably (approximately 90%) reduced detection times [60, 61, 98]. For such experiments, both the ligand and the target molecule must contain one dye. Sometimes the target protein is very sensitive to labeling. It can lose high affinity binding on complexation. Then the second label can also be applied indirectly via a non-neutralizing detection antibody.

FIDA has proven its potential by making cell surface receptor binding assays more precise by a more detailed kinetic and thermodynamic characterization. Monitoring the different degrees of brightness of a labeled molecule like EGF, Somatostatin or CRF (Corticotropin Releasing Factor) when bound to the cell surface receptor together with a two-dimensional beam scanner to screen the membrane, provides a much more detailed possibility to characterize receptor sub-populations and their quantitative interactions with the ligand than traditional methods. FCS features inherent drawbacks in using cells or cellular vesicles for developing receptor ligand interaction assays. The long translational diffusion time of these huge 'particles' destroys the correlation curve and makes fitting impossible. Already the 1D-FIDA approach improves the situation dramatically allowing one to distinguish free ligand and vesicles bearing multiple ligands. The low resolution of bound and free ligand with only a low number of occupied binding sites with 1D-FIDA is overcome by using the various 2D-FIDA detection possibilities. With the increased amount of data and the more complicated mathematical algorithms, data fitting times will increase substantially, making the applications in HTS a matter of high computing power. An especially attractive future concept is the application of combination SMA approaches to study coupled assay systems. The future of HTS will rely not only on miniaturization; it will be equally important to test more than one molecular interaction in a simple screening run. Screening "molecular machines" rather than individual interactions becomes feasible with multi-color timely coupled SMA [18].

17.2.8
Burst Integrated Lifetime (BIFL)

In the SMA detection techniques discussed so far, the signal analysis is performed over many molecular transit events. Therefore, the retrieved signal parameters (diffusion time, number densities, brightness) are average properties of the sample population under study. The recently invented 'Burst Integrated Fluorescence Lifetime (BIFL, group Seidel) method is a real-time spectroscopic tool which fully exploits the multi-dimensional fluorescence parameter space down to the single molecular level, i.e. intensity, lifetime, anisotropy (or rotational correlation time), and spectral resolution of individual fluorophores [2, 53, 72, 73]. Instrumentally, BIFL uses a pulsed laser source and a (confocal) microscope for small probe volume generation and efficient fluorescence photon collection. The laser pulses periodically excite the sample molecules in the 80 MHz range, but concurrently introduce time-tics along the experimental time axis. In a single-channel BIFL experiment dual-time base information is recorded simultaneously for each detected photon: (1) on a macroscopic time axis (~ms) the elapsed time, Δt, to the previously acquired photon, and (2) on a microscopic time scale (~10 ns) the photons' arrival times relative to the preceding laser pulse. The fast photon arrival events are time-registered by means of time-correlated single photon counting (TCSP) [99]. In a dual-channel setup for anisotropy measurements, the channel number is also acquired to identify the polarization of the actually recorded photon. Based on the macroscopic frequency of photon arrival events, the fluorescence signal can be reliably discriminated from the background which allows one to extract only well-defined photon bursts. For each selected burst the (microscopic) arrival time histogram is generated and subjected to intra-burst lifetime analysis. As a result, BIFL 'snapshots' each molecule crossing the sampling volume in terms of its characteristic fluorescence parameters e.g. lifetime and anisotropy. This correspondingly allows (lifetime) differentiation on a molecule by molecule basis. The snapshot concept is schematically shown in Fig. 17.8 and the corresponding experimental data can be seen from the three insets in Fig. 17.9. (Figure 17.9 has been kindly provided by C. Seidel, MPI-Goettingen, Germany). The arrival time histograms in insets I and II depict the distinct fluorescence decay pattern of two molecular species sequentially crossing the sampling volume. The sharp peak in inset III directly reveals the instrument response since the histogram is derived from the background BIFL trace. The instrument response accounts for the finite width of the laser pulses and the instrument's limited time-resolution. However, by accumulating all intra-burst determined lifetimes in a histogram, the heterogeneity of the studied population can be resolved in terms of both the separation between the species' associated τ-peaks and the distribution (variance) of the lifetime values around each τ-peak in the histogram. It has been demonstrated that BIFL is able to discriminate quantitatively common dyes on the basis of their individual fluorescence lifetime [29, 72], and to distinguish differently sized molecules by means of time-resolved single molecule anisotropy [73]. In principle, it is also possible to explore real-time dynamic molecular processes which evolve during a single molecular passage through the

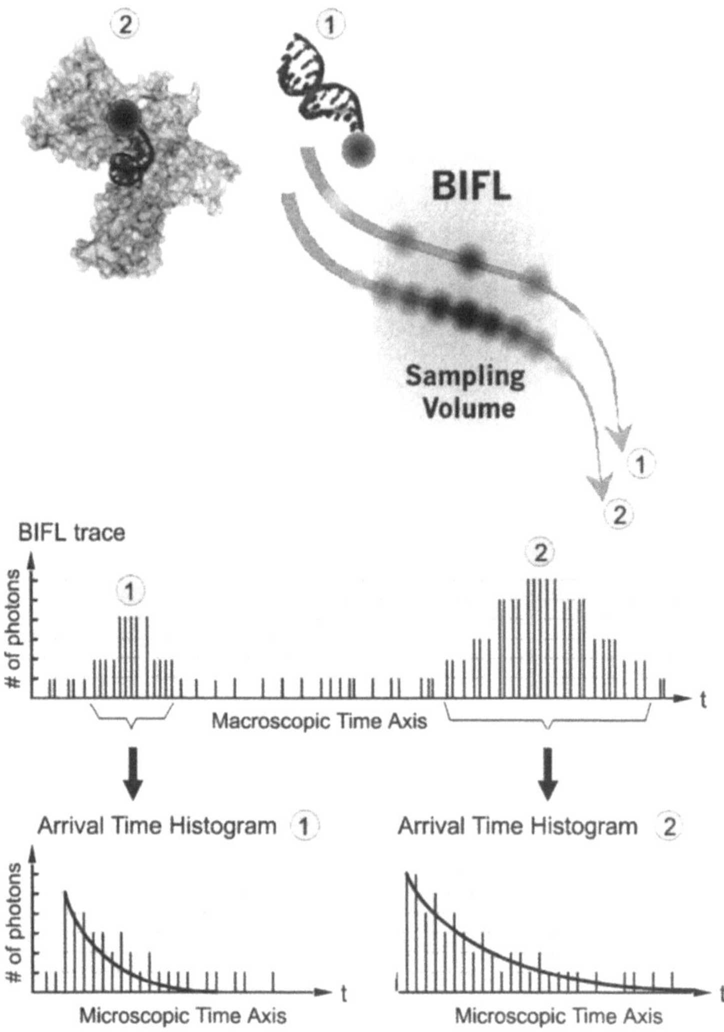

Fig. 17.8. Concept of the burst integrated fluorescence lifetime (BIFL) method: Different kinds of molecules are analyzed in terms of the difference in their characteristic fluorescence lifetimes and particle numbers. The fluorescence lifetime τ is an absolute signal parameter, and therefore ideally suited for molecular recognition analysis. While τ is largely independent on instrumental settings (e.g., laser power), sample concentrations, and translational/rotational motions of the molecules, the fluorescence lifetime can be very sensitive to intermolecular reactions including binding-induced quenching/enhancement. The τ-difference is a prerequisite in BIFL to discriminate the two species. The *top part* of the figure schematically depicts two temporally separated (❶ before ❷) transit events of an unbound labeled oligonucleotide (ODN) and a transcription factor (TRF) bound ODN. Within the simplified model chosen for explanation, the fluorescence lifetime of the ODN label is assumed to increase upon binding to the TRF (shown by a *darker dot*). The fast diffusing ODN is further assumed to be sampled three times (*3 blurred dots*) along its trajectory and the slower diffusing ODN-TRF complex is correspondingly sampled seven times (*7 blurred dots*) during its transit event. From the relative time lags between consecutively detected photons an absolute

Fig. 17.8 (continued)

(macroscopic) time axis can be constructed. This results in a so-called BIFL trace by plotting the number of detected photons per sampling interval vs time. The package of photons detected within the sampling time is represented by 3–5 adjacent *vertical lines* and their height (# of photons) corresponds to the fluorescence intensity (*gray-level*) sampled from the particle at a distinct spot (*blurred dot*) along its trajectory through the observation volume. The two transit events of the ODN (⬤) and the ODN-TRF (⬤) are evidenced by two distinct photon bursts. The bursts differ in their size since the total number of detected burst photons depends on the molecular diffusion time and the molecular brightness. The three and seven sampled points of ODN and ODN-TRF respectively correspond to two and four levels of photon count rate in the BIFL trace. By using an appropriate burst search algorithm, BIFL selects all well-defined photons burst from the BIFL trace (*black down-pointing arrows*). Within selected bursts a time-correlated single photons counting (TCSP) analysis is performed. This means that a histogram plot is constructed from the number of photon detection events vs their arrival time relative to the laser pulse (called microscopic time axis). Included in the arrival time histograms are the mono-exponential data fitting curves (*solid lines*) for lifetime evaluation. Hence, BIFL 'snapshots' each molecule in terms of its burst size and intra-burst fluorescence lifetime. The number densities of the species are obtained by classifying the selected photon bursts according to their intra-burst fluorescence lifetime

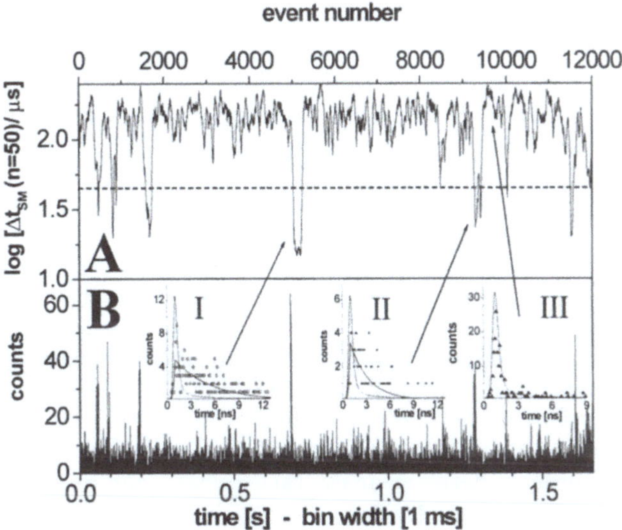

Fig. 17.9 A, B. Two equivalent BIFL-representations of a time-dependent signal trace of a mixture of rhodamine 6G (Rh6G) and rhodamine B (RhB): **A** time lag, Δt_{SM} (50), between consecutive photons of the smoothed data (averaged for n = 50 originally recorded lag times, Δt) vs the signal event number, and threshold value, Δt_{th}, for the subsequent burst selection ($\Delta t_{th} = 45$ μs, *dashed line*); **B** multi-channel scaler (MCS) trace with a bin width of 1 ms calculated from the Δt trace of A vs the macroscopic measurement time of the experiment. The different x-axes of A and B reveal slight distortions in the position of related fluorescence bursts. *Insets:* two typical fluorescence decays (I and II) and background signal (III) obtained from photons of the same measurement indicated by arrows in A: fluorescence decay (*open circles, full circles and triangles*), instrument response function (*dotted line*) and fit (*straight line*) by MLE: τ (I) = 3.9 ns, τ (II) = 2.0 ns. (Figure courteously obtained from Seidel CAM, Max-Plank-Institut fuer Biophysikalische Chemie, Goettingen, Germany)

probe volume. The corresponding evolution of the fluorescence lifetime can be analyzed either by a sliding scale or jump directed analysis [53].

Experimentally, step (1) in a BIFL analysis acquires the dual-time base information for each detected photon during the specified period of the experiment. The correspondingly generated data series, herein referred to as the BIFL trace, can be regarded as a record of photon events with the macroscopic and microscopic time data as event entries (c.f. Fig. 17.9 A, B for two possible representations of the BIFL trace).

In step (2), photon bursts are isolated via a burst search algorithm which separates signal carrying photons from background counts. This is accomplished by defining an appropriate threshold value for the macroscopic time variable Δt (c.f. Fig. 17.9 A). Since the count rate of background events is small (Poisson statistics), the time lags between consecutive background photons are extended and fluctuate largely. On the other hand, a single molecule transit event produces a distinct photon burst which is directly reflected by small time lags. Hence, a drop in Δt below the threshold value indicates the beginning of a signal burst while its end is notified by the next crossing of the time lag parameter above the threshold value. The (inherent) signal noise typically hampers the actual finding of the threshold value. Therefore, the 'raw' Δt-trace is smoothed by averaging the macroscopic time over an appropriate number of photon events (e.g. 50). Of course, both the smoothing factor and the threshold value have to be carefully determined in order to avoid artifacts in the subsequent analysis of the burst size distribution. Based on Poisson analysis, the threshold dependent probability of an erroneous burst identification can be calculated [72].

In step (3), the key point of the BIFL approach is to generate a burst size distribution (BSD) of a sample survey, i.e. the frequency histogram of the number of burst photons C_t of all step (2) selected photon bursts. The correspondingly obtained BSD is then extrapolated to the entire population (c.f. Step 6) to allow for a quantitative analysis of individual particles within a heterogeneous sample. The burst size C_t depends on individual experimental and molecular properties including the laser power, the instrument's detection efficiency, and the molecule's transit time through the probe volume. Further, a particular burst might be caused by more than one molecule (multi-molecule transit event); thus, C_t also depends on the average number of molecules within the probe volume N_{av}. In order to derive an analytical expression for the BSD, only single-transit events are considered first.

The probability $P_1(C_t)$ to detect C_t burst photons is calculated by integrating the product of the dwell-time dependent probability $P_1(C_t, t)$ to detect C_t burst photons and the probability distribution of dwell times Pt_1 over all dwell times t

$$P_1(C_t) = \int_0^\infty P_1(C_t, t)\, Pt_1\, dt \tag{17.29}$$

Equation (17.29) simply accounts for the fact that the number of detected burst photons depends on the time the corresponding molecule spent in the volume element. The diffusion of single molecules and the detection of single photons are both random in nature. As suggested by Qian [92], the photon statistic of a single molecule diffusing through the sampling volume can be described by a Poisson distribution

$$P_1(C_t, t) = \frac{C_{t,av}(t)^{C_t}}{C_t!} \exp\left(-C_{t,av}(t)\right) \tag{17.30}$$

with $C_{t,av}(t)$ denoting the average number of detected photons during the transit time t. $C_{t,av}(t)$ depends on the instrument's actual PSF (c.f. Eq. 17.8 or 9) and on the collection efficiency of the optical set-up. As a first approximation, the intensity profile of the excitation light $I(\vec{r})$ may be assumed as spatially constant throughout the detection volume element, i.e. $I(\vec{r}) = I$. Neglecting triplet kinetics and fluorescence saturation, $C_{t,av}$ is therefore linearly proportional to both the transit time t and the laser intensity I

$$C_{t,av}(t) = gIt, \quad \text{with} \quad g = \psi_F \sigma(_{ex})(hc_1/\lambda_{ex})^{-1} \tag{17.31}$$

The proportionality factor g depends on the instrument's optical detection efficiency Ψ, the quantum yield Φ_F of the fluorophore, the molecule's absorption cross section $\sigma(\lambda_{ex})$, and the photon energy $E_{ph} = hc_1/\lambda_{ex}$, where λ_{ex} is the excitation wavelength, and h and c_1 are the Planck constant and the speed of light, respectively. Note, the term gI corresponds to the brightness of the molecule in the experiment. Hence, Eq. (17.31) opens up a wide range of experimental possibilities to monitor molecular interactions based on brightness. However, the transit time dependent probability $P_1(C_t, t)$ accounting for the spatial inhomogeneous intensity distribution of the laser light has been only solved numerically [72].

Alternatively, an analytical term for $P_1(C_t, t)$ can be found by means of the irradiance shell approach proposed by Rigler and Mets [100]. Since randomly diffusing molecules explore each point in space with equal probability, the probability to measure C_t burst photons is proportional to the volume of a corresponding irradiance shell of constant light intensity I within the boundaries I and $I + \delta I$. For a 3-D Gaussian intensity profile, it has been shown that the dwell time dependent density function is [72]:

$$P_1(C_t, t) = K_{IS} C_t^{-1} \sqrt{\ln(gI_0 t/C_t)} \tag{17.32}$$

where I_0 is the light intensity at the center of the beam waist, K_{IS} is the probability normalization constant, and g denotes the instrument factor (cf. Eq. 17.31). It is important to note that both theoretical models for $P_1(C_t, t)$, i.e., Eqs. (17.30) and (17.32), are equal as shown in Fig. 2 of [72].

The remaining term to be solved in Eq. (17.29) is Pt_1, the density distribution of single-transit dwell times. Since BIFL only selects well-defined bursts for lifetime analysis, the short bursts caused by boundary re-crossing molecules are rejected while transit events through the entire probe volume correspondingly pass the burst search algorithm. For such central single-molecule transits Poisson statistics applies, and hence, the probability Pt_1 to measure a dwell time of length t exponentially decreases with time t [83]:

$$Pt_t(t) = 1/t_t \exp\left(-t/t_t\right) \tag{17.33}$$

where t_t is the average molecule dwell time in the probe volume.

Now, the next level of complexity in the description of the BSD includes the possibility of having more than one molecule concurrently diffusing through the open volume element. Hence, the experimentally observed mean burst duration t_B is longer than the average single-molecule transit time t_t. If $Pm_n(N_{av})$ denotes the probability of finding n molecules in the observation volume, the apparent transit time t_B is simply given by the weighted sum of the mean single-transit time t_t [83]:

$$t_B = \sum_{n=1}^{\infty} (nt_t) Pm_n(N_{av}) \tag{17.34}$$

and, the experimentally relevant dwell time density function Pt_m is given by

$$Pt_m(t) = 1/t_B \exp(-t/t_B) \tag{17.35}$$

The mean burst duration t_B can be easily obtained from the experimentally measured burst duration t by fitting Eq. (17.35) to the t-histogram.

Assuming well-separated molecular entry events into the sampling volume and a Poissonian governed entry statistics, the probability of finding n molecules in the probe volume is given by [72]

$$Pm_n(N_{av}) = \frac{\exp(-2N_{av})(1 - \exp(-N_{av}))^{n-1}}{\sum\limits_{n=1}^{\infty} \exp(-2N_{av})(1 - \exp(-N_{av}))^{n-1}} \tag{17.36}$$

where the denominator accounts for the normalization, i.e., $\sum\limits_{n} Pm_n(N_{av}) = 1$.

Analogous to PCH or FIDA (c.f. Eq. 17.22), the probability to detect C_t burst photons as a result of n molecules diffusing through the detection volume is given by iterative numerical convolution, i.e., $P_n(C_t) = P_1(C_t) \otimes ... \otimes P_1(C_t)$, n-times. Summation over all possible single- and multi-molecule transit events yields the theoretical burst size density function

$$P(C_t, N_{av}) = \sum_{n=1}^{\infty} P_{m_n}(N_{av}) P_n(C_t) \tag{17.37}$$

Since the burst search algorithms reject poorly defined bursts, the number of fluorescence burst actually taking place during the data acquisition time t_{acq} is unknown. Consequently, the theoretical density function $P(C_t, N_{av})$ has to be adapted in order to correctly describe the experimentally acquired burst size histogram $\beta(C_t, N_{av})$ [72]:

$$\beta(C_t, N_{av}, g) = P(C_t, N_{av}, g)(1 - \exp(-N_{av})(t_{mes}/t_B) \tag{17.38}$$

In analogy to FCS, FIDA, or PCH, the analysis of the burst size distribution by means of Eq. (17.38) as the fitting function, also yields the average number of molecules N_{av}, within the observation volume. The corresponding fitting parameters are N_{av}, and g while the mean burst duration t_B is simply calculated from the experimental data (c.f. text below Eq. 17.35). Note, the instrument's detection efficiency can be elegantly obtained from the second fitting parameter g provided that the quantum yield and absorption cross section of the used dye is known.

For the series of selected photon bursts, step (4) generates the corresponding series of arrival time histograms using the microscopic time information of the BIFL trace (c.f. insets I and II in Fig. 17.9 B).

Step (5) is fundamental for BIFL experiments since it identifies the detected single molecules by means of their characteristic fluorescence parameters, e.g., lifetimes. Correspondingly, the arrival time histograms are individually subjected to fluorescence lifetime analysis. Since a single burst typically contains no more than 200–300 detected photons, conventional chi-square minimization procedures as for instance used in TCSPC experiments are not optimal for the BIFL histograms [101]. An appropriate lifetime estimation procedure is the maximum likelihood estimator (MLE) [102], a pattern recognition technique which accounts for the multi-nomal (rather than the Gaussian) probability distribution of the small number of burst photons [53]. As in conventional least square fitting procedures, the instrument response function [86], the multi-scanning effect encountered for decay times exceeding to pulse-to-pulse time period [103], and the Raman scatter can be included in the basic MLE fitting function. For the sake of readability the corresponding MLE fitting function is only referenced here [29].

In step (6), the concentrations of the different molecular species are determined by classifying the selected photon bursts according to their intra-burst fluorescence decay rate. For the exemplary two-species mixture the burst series is lifetime-sorted with respect to a separation fluorescence lifetime τ_{th} (threshold). To equalize the misclassification rate for both species, the threshold value is chosen such that its threshold associated fluorescence decay is equally 'well' described by each of the two species' fluorescence decays. From the separated burst populations the corresponding burst size histograms $\beta(C_t, N_{av})$ are constructed and individually analyzed by Eq. (17.38) to obtain the molecular concentrations, or more specifically N_{av}, the average number of molecules within the sampling volume. Recent work showed that the above theoretical framework can be also used to analyze a sample mixture without lifetime classification of the molecules just by using a multi-component fit of individual brightnesses (personal communication with C. Seidel).

17.2.9
Features and Issues of BIFL

In order to achieve a reasonable standard deviation of the intra-burst fluorescence lifetime, only bursts exceeding a minimum number of photons (e.g., ≥ 50) are selected for lifetime analysis. Using a confocal microscope and reasonably high laser power, typical photon bursts comprise 100–200 photons. Of course, the average burst size can be further pushed by increasing the observation volume, the laser power, and the viscosity of the solution. However, the maximal burst size hardly exceeds 300 photons specifically due to photophysical and experimental limitations (photobleaching, collection efficiency). As a result, the statistics of the arrival time histogram makes it difficult to resolve two- and more intra-burst lifetime components [104]. In order to discriminate quantitatively different kinds of molecules on the basis of a mono-exponential lifetime function, no more than two molecules should cross the probe volume at the same time, hence, the sample solution has to be drastically diluted. Practical BIFL concentrations are in the picomolar range such that the mean occupation number of the confocal volume is less than about 0.2 particles. The low working concentration is considered to be the major difference to the previously described SMA methods. With respect to drug discovery, such low picomolar working concentrations drastically reduce the number of screenable drug targets to those having similar (picomolar) equilibrium constants. This is a serious shortcoming of the BIFL technique under the perspective of HTS. As we understand it, the only possibility to use the BIFL setup for standard HTS, monitoring molecular interactions with $K_d s$ from about 5 nmol/l to 500 nmol/l, is to analyze alternatively the BIFL trace in terms of the distribution of the macroscopic lag time Δt between successively recorded photons (Eggeling C, personal communication). But then the method corresponds to FIDA or PCH. In fact, the time lag distribution and the distribution of time-binned photons are closely related, and correspondingly the Δt-histogram yields the same signal parameters as PCH and FIDA.

While FIDA, or PCH describe the photon distribution over the full photon counting range, BIFL distinctively selects only well-defined (photon-exact) burst. The burst selection procedure is based on an appropriate threshold in terms of a minimum number of burst photons (e.g., more than 50 photons are required to become a BIFL validated burst). The resulting burst size distribution (C_t-histogram) is 'left-side' cut in comparison with a PCH or FIDA distribution. With respect to the molecular diffusion time obtained in FCS, the corresponding mean burst duration t_B measured in an FCS-equivalent BIFL set-up differs since BIFL only considers a well-defined subset of molecular transit events. Short bursts are typically due to background noise and boundary re-crossing molecules while photon bursts of extended length are mostly accidental coincidences (e.g., large impurities are diffusing through the observation volume). Hence, by subjecting detected bursts to a double-threshold selection criterion, i.e., the size of each burst has to pass an appropriately chosen upper and lower threshold value, BIFL particularly skips the background sections and high intensity artifacts of the BIFL trace. The focus of the data analysis on the actual (signal carrying) photons in the bursts certainly increases the statistical accuracy of the retrieved signal parameters.

Since BIFL discriminates different kinds of molecules via their characteristic lifetimes, the confidence level for single molecule identification crucially depends on the distribution of the measured intra-burst lifetimes, i.e., the variances of the peaks in the τ-histogram. The probability for misidentification of different fluorophores is proportional to the overlap area between the different lifetime distributions. The width of the variances is directly proportional to the corresponding number of burst photons used to calculate the fluorescence decay rate $\sigma \propto \sqrt{1/C_t}$; [102]). In contrast to FIDA or PCH, the accuracy in the molecular separation is therefore not affected by the concentration ratio between different molecular species because the lifetime is an absolute (molecular specific) parameter. Hence, the vital parameter in BIFL experiments is the total number of photons collected within a burst. Consequently, the instrument's collection and detection efficiency have to be optimized by using high NA objectives, high quantum yield photodetectors, and photon counting electronics with short dead times. A pinhole which is chosen slightly larger than the optimal confocal pinhole correspondingly enlarges the sampling volume and consequently the mean size of the bursts. However, increasing the sampling volume also increases the mean number of observed particles. Thus, the upper limit of the confocal volume is determined by the maximal tolerable sample concentration which guarantees that on average much less than one particle is occupying the sampling volume. Increasing the laser power might further increase the burst size until the correspondingly raised photobleaching rate starts to broaden the τ-variances. Such saturation effects have to be balanced carefully with the permanent experimental need to collect as many burst photons as possible.

Single molecule identification based on fluorophores with sufficiently different lifetimes has been demonstrated in various studies [6, 72, 105, 106]. In order to assess the performance of BIFL, the following paragraph briefly summarizes characteristic results in resolving two kinds of molecules. In context of the stringent statistical requirements for SMA in HTS as described in the FCS section, the simple example of the discrimination between the two dyes included below demonstrates the power of BIFL in the multidimensional dynamic characterization of single molecules but also its current restriction in HTS. For instance, a mixture of rhodamine 6G (R6G) and rhodamine B (RhB) has been discriminated at a confidence level of 5% [72]. The relative error in the fluorescence lifetime was 15% for both dyes. The burst-retrieved mean lifetime values of R6G and RhB were 2.4 ± 0.3 ns and 3.7 ± 0.6 ns, respectively. If the single intra-burst lifetime parameter is not sufficient for an accurate discrimination between individual members of a sample population, additional BIFL signal parameters such as the burst size C_t can be correlated in 2-D histograms. Analogous to a scatter plot in Fluorescence Activated Cell Sorting (FACS) [74], the individual species appear as distinct island (populations) in the 2-D plot and can be separated by an area-encompassing selection criterion. The correlated burst size and intra-burst lifetime analysis of the BIFL trace obtained from a 1:1 solution of tetramethylrhodamine isothiocyanate (TRITC) and rhodamine 6G (Rh6G) in a flowing sample stream achieved a classification accuracy of 93% and 99.8%, respectively [107]. The 7% misclassification of TRITC is due to photobleaching which primarily results in a corrupted burst size distribution. The bleaching rate of such photo-unstable dyes can be decreased by antifade reagent which correspondingly increases the accuracy of single molecule identification.

As in all SMA method, various saturation effects can erroneously affect the signal parameters in a BIFL experiment. Since the accuracy of the intra-burst lifetime crucially depends on the number of collected burst photons, BIFL typically uses considerably higher laser powers (~ 0.9 mW, [72]) than FCS, PCH, or FIDA experiments. However, the maximal burst counting rates reported in the literature (~ 250) are still below the critical value (c.f. introduction), such that triplet kinetic and fluorescence saturation are of minor importance in comparison with the more dominant effect of photobleaching.

The BIFL approach provides a general experimental platform to obtain the complete four-parameter fluorescence information of individual molecules during their passages through the observation volume. The open data structure allows one to combine traditional fluorescence spectroscopy such as TCSPC with recent techniques such as correlation spectroscopy and intensity distribution analysis [72, 108]. The BIFL-technique may become a valuable tool for assay-development and for a detailed characterization of ligand-target interactions in secondary screening.

17.3
Conclusion and Outlook

High throughput screening combines robotics, integrated and combined data management, miniaturized screening formats, and nanoliter handling of complex sample collections. The full power of such assay technologies, however, can only be exploited if homogeneous assay procedures result in high quality mechanistic inhibitor characterization. With the integration of translational diffusion (FCS), rotational diffusion (2D-FIDA-anisotropy), fluorescence brightness, spectral shifts, fluorescence resonance energy transfer (all 1D-FIDA or 2D-FIDA) and fluorescence lifetime as detection methods, SMA provides an attractive detection platform for future HTS. The major strategy in all SMA techniques has to aim for the maximization of the number of photons collected while concurrently minimizing the background noise. In particular, the molecular brightness has been recognized as the most vital signal parameter since it directly affects the achievable statistical accuracy of an assay. Bright fluorescence labels are therefore a prerequisite for achieving high quality HTS with SMA. Another parameter of concern is the photobleaching rate since already moderately high triplet rates significantly corrupt the observed signal parameter. Although currently used rhodamine and cyanine dyes are successfully applied in SMA the search and development of photochemically and photophysically more stable dyes is a future need for screening applications. A further bottleneck in assay development is the lack of generic labeling techniques to cope with the needs of SMA. Approaches to find new ways for generic labeling of proteins for fast assay development include fusion proteins, peptide tags for proteins and cell permeable dyes, enzymatic reactions, and small molecular dye adapters.

If the signal originating from a single fluorescence detection method does not provide the necessary accuracy to discriminate between individual species in a mixture, additional fluorescence property changes of molecular interaction such as anisotropy and fluorescence lifetime have to be acquired via a multichannel detection scheme. The concurrent global analysis [87] of at least two signal parameters (e.g., FCS and fluorescence anisotropy) significantly improves the probability for accurate molecular classification in compound screening and further allows the rejection of photobleaching caused artifacts. In the future it can be expected that only the correlated analysis of at least two signal parameters will measure up with the stringent needs of HTS. The nanomolar concentration range for the labeled species in the assay limits FCS- and FIDA-based screening systems to complexations with dissociation constants from nanomolar to low micromolar, with BIFL as lifetime analysis method (not run as FIDA method) falling out of the scope of applicable Kds for HTS. The disadvantage of the restricted concentration range is strongly outweighed by the numerous advantages of SMA including the insensitivity to miniaturization, the potential to integrate all fluorescence techniques, the improved well to well and assay to assay quality control, the concentrations determination free of artifacts, and the possibility for on-line mechanistic characterization. The current throughput rate of about 50,000 a day is anticipated to increase dramatically as soon as parallel confocal optics are implemented.

Acknowledgement. We thank C. Seidel and C. Eggeling, K. Gall and P. Kask, as well as E. Gratton, Y. Chen, and J. Mueller for providing us with key figures representing their SMA technologies, and for numerous intensive discussions which help us to evaluate the BIFL, FIDA, and the PCH detection method. We thank D. Ullmann and R. Guenther for critical discussions on the subject of assay statistics, and all NAT members, in particular K. Mueller, U. Hassiepen, and A. Schilb for the intensive discussions about the topic of SMA and editorial work. We thank P. Herrling, J.E. de Vries, R. Amstutz, and R. Naef for continuous support and encouragement.

References

1. Chen D, Dovichi NJ (1996) Single-molecule detection in capillary electrophoresis: molecular shot noise as a fundamental limit to chemical analysis. Anal Chem 68: 690–696
2. Keller RA, Ambrose WP, Goodwin PM, Jett JH, Martin JC, Wu M (1996) Single molecule fluorescence analysis in solution. Appl Spectrosc 50:12 A–32 A
3. Goodwin PM, Ambrose WP, Keller RA (1996) Single-molecule detection in liquids by laser-induced fluorescence. Acc Chem Res 29:607–613
4. Lyon WA, Nie S (1997) Confinement and detection of single molecules in submicrometer channels. Anal Chem 69:3400–3405
5. Lermer N, Barnes MD, Kung C-Y, Whitten WB, JM (1997) High-efficiency molecular counting in solution: single-molecule detection in electrodynamically focused microdroplet streams. Anal Chem 69:2115–2121
6. Zander C, Sauer M, Drexhage KH, Ko D-S, Schulz A, Wolfrum J, Brand L, Eggeling C, Seidel CAM (1996) Detection and characterization of single molecules in aqueous solution. Appl Phys B 63:517–523
7. Zander C, Drexhage K-H, Han K-T, Wolfrum J, Sauer M (1998) Single-molecule counting and identification in a microcapillary. Chem Phys Lett 286:457–465
8. Trautman JK, Macklin JJ, Brus LE, Betzig E (1994) Near-field spectroscopy of single molecules of room temperature. Nature 369:40
9. Xie XS, Dunn RC (1994) Probing single-molecule dynamics. Science 265:361–364
10. Bian RX, Dunn RC, Xie XS, Leung PT (1995) Single molecule emission characteristics in near-field microscopy. Phys Rev Lett 75:4772–4775
11. Mets U, Rigler RJ (1994) Submillisecond detection of single rhodamine molecules in water. Fluorescence 4:259–264
12. Nie S, Chiu DT, Zare RN (1995) Real-time detection of single molecules in solution by confocal fluorescence microscopy. Anal Chem 67:2849–2857
13. Macklin JJ, Trautman JK, Harris TD, Brus LE (1996) Imaging and time-resolved spectroscopy of single molecules at an interface. Science 272:255
14. Trautman JK, Macklin JJ (1996) Time-resolved spectroscopy of single molecules using near-field and far-field optic. Chem Phys 205:221–229

15. Dickson RM, Norris DJ, Tzeng YL, Moerner WE (1996) Three dimensional imaging of single molecules solvated in pores of poly(acrylamide) gels Science 274:966–969

16. Dickson RM, Cubitt AB, Tsien RY, Moerner WE (1997) On/off blinking and switching behavior of single molecules of green fluorescent protein. Nature 388:355–358

17. Xu X-H, Yeung ES (1997) Direct measurement of single-molecule diffusion and photo-decomposition in free solution. Science 275:1106–1109

18. Alberts B (1998) The cell as a collection of protein machines: preparing the next generation of molecular biologists. Cell 92:291–294

19. Auer M, Moore KJ, Meyer-Almes F-J, Guenther R, Pope AJ, Stoeckli KA (1998) Fluorescence correlation spectroscopy: lead discovery by miniaturized HTS. Drug Discovery Today 3:457–465

20. Loescher F, Boehme S, Martin J, Seeger S (1998) Counting of single protein molecules at interfaces and the application of this technique in early-stage-diagnosis. Anal Chem 70:3202–3205

21. Hirschfeld T (1976) Limits of analysis. Anal Chem 48:16A–31A

22. Affleck RL, Ambrose WP, Demas JN, Goodwin PM, Schecker JA, Wu M, Keller RA (1996) Reduction of luminescent background in ultrasensitive fluorescence detection by photo-bleaching. Anal Chem 68:2270–2276

23. Jett JH, Keller RA, Martin JC, Marrone BL, Moyzis RK, Ratliff RL, Seitzinger NK, Shera EB, Stewart CC (1989) High-speed DNA sequencing: an approach based upon fluorescence detection of single molecules. J Biomol Struct Dyn 7:301

24. Shera EB, Seitzinger NK, Davis LM, Keller RA, Soper SA (1990) Detection of Single Fluorescent Molecules. Chem Phys Lett 174:553–557

25. Goeppert-Mayer M (1931) Ueber Elementarakte mit zwei Quantenspruengen. Ann Phys (Leipzig) 9:273–294

26. Denk W, Strickler JH, Webb WW (1990) Two-photon laser scanning fluorescence micro-scopy. Science 248:73

27. Denk W, Piston DW, Webb WW (1995) Two-photon molecular excitation in laser-scan-ning microscopy. In: Pawley J (ed) Handbook of biological confocal microscopy. Plenum Press, New York, pp 445–458

28. Stelzer EHK, Hell S, Lindek S (1994) Nonlinear absorption extends confocal fluorescence microscopy into the ultra-violet regime and confines the illumination volume. Opt Commun 104:223–228

29. Brand L, Eggeling C, Zander C, Drexhage KH, Seidel CAM (1997) Single-molecule identification of Coumarin-120 by time-resolved detection: comparison of one and two-photon excitation in solution. J Phys Chem A 101:4313–4321

30. Nie S, Chiu DT, Zare RN (1994) Probing individual molecules with confocal fluorescence microscopy. Science 266:1018

31. Saleh B (1978) Photoelectron statistics, with applications to spectroscopy and optical communications, Springer, Berlin Heidelberg New York

32. Gardiner CW (1985) Handbook of stochastic methods, Springer, Berlin Heidelberg New York

33. Wiener N (1949) Extrapolation, intrapolation, smoothing of stationary time series. MIT Press, Cambridge, MA

34. Ehrenberg M, Rigler R (1974) Rotational Brownian motion and fluorescence intensity fluctuations. Chem Phys 4:390–401

35. Aragon SR, Pecora R (1975) Fluorescence correlation spectroscopy and Brownian rotational diffusion. Biopolymers 14:119–137

36. Koppel DE, Axelrod D, Schlessinger J, Elson EL, Webb WW (1976) Dynamics of fluores-cence marker concentration as a probe of mobility. Biophys J 16:1315–1329

37. Kask P, Piksarv P, Pooga M, Mets U (1989) Separation of the rotational contribution in fluorescence correlation experiments. Biophys J 55:213–220

38. Magde D, Webb WW, Elson EL (1978) Fluorescence correlation spectroscopy. III. Uniform translation and laminar flow. Biopolymers 17:361–376

39. Peterson NO, Johnson DC, Schlesinger MJ (1986) Scanning fluorescence spectroscopy. II. Application to virus glycoprotein aggregation. Biophys J 49:817–820

40. Palmer AG, Thompson NL (1989) High-order fluorescence correlation analysis of model protein clusters. Proc Natl Acad Sci USA 86:6148–6152

41. Magde D, Elson EL, Webb WW (1974) Fluorescence correlation spectroscopy. II. An experimental realization. Biopolymers 13:29–61

42. Rauer B, Neumann E, Widengren J, Rigler R (1996) Fluorescence correlation spectrometry of the interaction kinetics of tetramethylrhodamin alpha-bungarotoxin with torpedo californica acetylcholine receptor. Biophys Chem 58:3–12

43. Widengren J, Mets U, Rigler R (1995) fluorescence correlation spectroscopy of triplet states in solution – a theoretical and experimental study. J Phys Chem 99:13,368–13,379

44. Borejdo J (1979) Motion of myosin fragments during actin-activated ATPase: fluorescence correlation spectroscopy study. Biopolymers 18:2807–2820

45. Thompson NL, Axelrod D (1983) Immunoglobulin surface-binding kinetics studied by total internal reflection with fluorescence correlation spectroscopy. Biophys J 43:103–114

46. Kinjo M, Rigler R (1995) Ultrasensitive hybridization analysis using fluorescence correlation spectroscopy. Nucleic Acids Res 23:1795–1799

47. Schwille PF, Oehlenschlager WNG (1996) Quantitative hybridization kinetics of DNA probes to RNA in solution followed by diffusional fluorescence correlation analysis. Biochemistry 35:10,182–10,193

48. Berland KM, So PTC, Gratton E (1995) Two-photon fluorescence correlation spectroscopy: method and application to the intracellular environment. Biophys J 68:694–701

49. Schwille P, Haupts U, Maiti S, Webb WW (1999) Molecular dynamics in living cells observed by fluorescence correlation spectroscopy with one- and two-photon excitation. Biophys J 77:2251–2265

50. Brock R, Vàmosi G, Vereb G, Jovin TW (1999) Rapid characterization of green fluorescent protein fusion proteins on the molecular and cellular level by fluorescence correlation microscopy. Proc Natl Acad Sci USA 96:10,123–10,128

51. Weissman M, Schindler H, Feher G (1976) Determination of molecular weights by fluctuation spectroscopy: application to DNA. Proc Natl Acad Sci USA 73:2776–2780

52. Axelrod D, Koppel DE, Schlessinger J, Elson E, Webb WW (1976) Mobility measurement by analysis of fluorescence photobleaching recovery kinetics. Biophys J 16:1055–1069

53. Eggeling C, Fries JR, Brand L, Günther R, Seidel CAM (1998) Monitoring conformational dynamics of a single molecule by selective fluorescence spectroscopy. Proc Natl Acad Sci USA 95:1556–1561

54. Koppel DE (1974) Statistical accuracy in fluorescence correlation spectroscopy. Phys Rev A 10:1938–1945

55. Palmer AG, Thompson NL (1987) Molecular aggregation characterized by high order autocorrelation in fluorescence correlation spectroscopy. Biophys J 52:257–270

56. Qian H, Elson EL (1990) Distribution of molecular aggregation by analysis of fluctuation moments. Proc Natl Acad Sci USA 87:5479–5483

57. Qian H, Elson EL (1990) On the analysis of high order moments of fluorescence fluctuations. Biophys J 57:375–380

58. Eigen M, Rigler R (1994) Sorting single molecules: application to diagnostics and evolutionary biotechnology. Proc Natl Acad Sci USA 91:5740–5747

59. Schwille P, Meyer-Almes F-J, Rigler R (1997) Dual-color fluorescence cross-correlation spectroscopy for multicomponent diffusional analysis in solution. Biophys J 72:1878–1886

60. Koltermann A, Kettling U, Bieschke J, Winkler T, Eigen M (1998) Rapid assay processing by integration of dual-color fluorescence cross-correlation spectroscopy: high throughput screening for enzyme activity. Proc Natl Acad Sci USA 95:1421–1426

61. Winkler T, Kettling U, Koltermann A, Eigen M (1999) Confocal fluorescence coincidence analysis: an approach to ultra high-throughput screening. Biophysics 96:1375–1378

62. Bendat JS, Piersol AG (1971) Random data: analysis and measurement procedures. Wiley-Interscience, New York
63. Evotec Biosystems AG (1998) Int Patent Appl PCT/EP97/05619, Int Publ No WO98/16814
64. Kask P, Palo K, Ullmann D, Gall K (1999) Fluorescence-intensity distribution analysis and its application in biomolecular technology. Proc Natl Acad Sci USA 96:13,756–13,761
65. Chen Y, Müller JD, So PTC, Gratton E (1999) The photon counting histogram in fluorescence fluctuation spectroscopy. Biophys J 77:553–567
66. Teich MC, Saleh BEA (1988) Photon bunching and antibunching. In: Wolf E (ed) Progress in optics. North-Holland Publishing Company, Amsterdam, pp 1–104
67. Soper SA, Davis LM, Shera EB (1992) Detection and identification of single molecules in solution. J Opt Soc Am B 9:1761–1769
68. Birch DJS, Imhof RE (1991) Time-domain fluorescence spectroscopy using time-correlated single photon counting. In: Lakowicz JR (ed) Topics in fluorescence spectroscopy. 1. Techniques. Plenum Press, New York, pp 1–95
69. Efftink M (1991) Fluorescence techniques for studying protein structure. In: Suelter CH (ed) Methods of biochemical analysis. 35. Protein structure determination. Wiley, New York, pp 127–205
70. Millar DP (1996) Time-resolved fluorescence spectroscopy. Curr Opin Struct Biol 6:637–42
71. Prummer M, Hübner CG, Sick B, Hecht B, Renn A, Wild UP (1999) Multidimensional photon counting and sequential analysis for rapid identification of single molecules in ambient conditions. Poster and abstract, 5th International Workshop on Single Molecule Detection andUltrasensitive Analysis in Life Sciences, 29.9.–1.10.1999, Berlin-Adlershof
72. Fries JR, Brand L, Eggeling C, Köllner M, Seidel CAM (1998) Quantitative identification of different single molecules by selective time-resolved confocal fluorescence spectroscopy. J Phys Chem A 102:6601–6613
73. Schaffer J, Volkmer A, Eggeling C, Subramaniam V, Striker G, Seidel CAM (1999) Identification of single molecules in aqueous solution by time-resolved fluorescence anisotropy. J Phys Chem A 103:331–336
74. Shapiro HM (1995) Practical flow cytometry, 3rd edn. Wiley, New York, pp 327–330
75. Thompson NL (1991) Fluorescence correlation spectroscopy. In: Lakowicz JR (ed) Topics in fluorescence spectroscopy. 1. Techniques. Plenum Press, New York, pp 337–378
76. Born M, Wolf E (1980) Principles of optics. Pergamon, Oxford
77. Elson EL, Magde D (1974) Fluorescence correlation spectroscopy. I. Conceptual basis and theory. Biopolymers 13:1–27
78. Rigler R, Mets U, Widengren J, Kask P (1993) Fluorescence correlation spectroscopy with high count rate and low background: analysis of translational diffusion. Eur Biophys J 22:169–175
79. Lakowicz JR (1983) Principles of fluorescence spectroscopy, 3rd edn. Plenum Press, New York
80. Pawley JB (ed) (1995) Handbook of biological confocal microscopy, 2nd edn. Plenum Press, New York
81. Hiraoka Y, Sedat W, Agard DA (1990) Determination of three-dimensional image properties of a light microscope system. Biophys J 57:325–333
82. Edward JT (1970) Molecular volumes and the Stokes-Einstein Equation. J Chem Ed 47:261–270
83. Ko D-S, Sauer M, Nord S, Müller R, Wolfrum J (1997) Determination of the diffusion coefficient in dye solution at single molecule level. Chem Phys Lett 269:54–58
84. Widengren J, Rigler R, Mets U (1994) Triplet-state monitoring by fluorescence correlation spectroscopy. Fluorescence 4:255–258
85. Meseth U, Wohland T, Rigler R, Vogel H (1999) Resolution of fluorescence correlation measurements. Biophys J 76:1619–1631
86. Daly TJ, Doten RC, Rusche JR, Auer M (1995) The amino terminal domain of HIV-1 Rev is required for the discrimination of the RRE from nonspecific RNA. J Mol Biol 253:243–258

87. Beechem JM, Gratton E, Ameloot M, Knutson JR, Brand L (1991) The global analysis of fluorescence intensity and anisotropy decay data: second-generation theory and programs. In: Lakowicz JR (ed) Topics in fluorescence spectroscopy. 2. Principles. Plenum Press, New York, 241–301

88. Petersen NO, Hoddelius PL, Wiseman PW, Seger O, Magnusson KE (1993) Quantitation of membrane receptor distributions by image correlation spectroscopy: concept and application. Biophys J 65:1135–1146

89. Koppel DE, Morgan F, Cowan AE, Carson JH (1994) Scanning concentration correlation spectroscopy using the confocal laser microscope. Biophys J 66:502–507

90. Berland KM, So PTC, Chen Y, Mantulin WW, Gratton E (1996) Scanning two-photon fluctuation correlation spectroscopy: particle counting measurements for detection of molecular aggregation. Biophys J 71:410–420

91. Kask P, Gunther R, Axhausen P (1997) Statistical accuracy in fluorescence fluctuation experiments. Eur Biophys J 25:163–169

92. Qian H (1990) On the statistics of fluorescence correlation spectroscopy. Biophys Chem 38:49–57

93. Feller W (1957) An introduction to probability theory and its applications. Wiley, New York, 327–330

94. Chiu DT, Zare RN (1996) Biased diffusion, optical trapping, manipulation of single molecules in solution. J Am Chem Soc 118:6512–6513

95. Osborne MA, Balasubramania S, Furey WS, Klenerman D (1998) Optically biased diffusion of single molecules studied by confocal fluorescence microscopy. J Phys Chem B 102:3160–3167

96. Mandel L (1979) Sub-Poissonian photon statistics in resonance fluorescence. Optics Lett 4:205–207

97. Auer M, Seifert J-M, Wallace ST, Sleigh R (1999) In: Wallis M, Schroeder R (eds) Review book-chapter for RNA-binding antibiotics. Chapman & Hall

98. Kettling U, Koltermann A, Schwille P, Eigen M (1998) Real-time enzyme kinetics monitored by dual-color fluorescence cross-correlation spectroscopy. Proc Natl Acad Sci USA 95:1416–1420

99. O'Conner DV, Phillips D (1984) Time-correlated single photon counting. Academic Press, London

100. Rigler R, Mets Ü (1992) Diffusion of single molecules through a Gaussian laser beam. SPIE 1921:239–248

101. Bevington PR, Robinson DK (1992) Data reduction and error analysis for the physical sciences. McGraw-Hill, New York

102. Hall P, Selinger P (1981) Better estimates of exponential decay parameters. J Chem Phys 85:2941

103. Sakai Y, Hirayama S (1988) A fast deconvolution method to analyse fluorescence decays when the excitation pulse repetition period is less than the decay times. J Lumin 39:145–151

104. Sauer M, Arden-Jacob J, Drexhage KH, Göbel F, Lieberwirth U, Mühlegger K, Müller R, Wolfrum J, Zander C (1998) Time-resolved identification of individual mononucleotide molecules in aqueous solution with pulsed semiconductor lasers. Bioimaging 6:14–24

105. Mueller R, Zander C, Sauer M, Deimel M, Ko DS, Siebert S, Arden-Jacob J, Deltau G, Marx NJ, Drexhage KH, Wolfrum J (1996) J Chem Phys Lett 262:716–722

106. Enderlein J, Robbins DL, Ambrose WP, Goodwin PM, Keller RA (1997) Statistics of single-molecule detection. J Phys Chem B 101:3626–3632

107. Van Orden A, Machara NP, Goodwin PM, Keller RA (1998) Single-molecule identification in flowing sample streams by fluorescence burst size and intraburst fluorescence decay rate. Anal Chem 70:1444–1451

108. Eggeling C, Schaffer J, Seidel CAM, Korte J, Brehm G, Schneider S, Schrof W (1999) Multidimensional surface enhanced Raman spectroscopy of single dyes on single Ag-Particles. Phys Rev Lett (submitted)

CHAPTER 18

Picosecond Fluorescence Lifetime Imaging Spectroscopy as a New Tool for 3 D Structure Determination of Macromolecules in Living Cells

K. KEMNITZ

Picosecond fluorescence lifetime imaging microscopy (Picosecond FLIM), based on time-and space-correlated single photon counting (TSCSPC), is the method of choice to study living cells at minimal-invasive conditions that are required for preserving the living state. Time-resolved fluorescence imaging, at ultra-low excitation intensity and ultra-low level of labelling, i. e., *Minimal-Invasive Fluorescence Microscopy* (MIFM), became possible after the introduction of ultra-sensitive single photon counting imaging detectors, such as micro-channel plate (MCP) photomultipliers (PMT) with delay-line (DL) and quadrant anode (QA). The novel detectors have a time resolution of < 10 ps, 100 µm space resolution (250×250 channel), a dynamic range of $> 10^7$ and are capable of *Vehicle Micro-Spectroscopy* (VMS). Single-channel time-correlated single photon counting (TCSPC) microscopy was established 15 years ago and has since been applied by a growing community of cell biologists, due to superior performance and increasingly simple operation of pulsed picosecond laser systems. The new TSCSPC imaging detectors will further increase the attractiveness of the well-established method in ultra-sensitive studies of living cells.

18.1
Time- and Space-Correlated Single Photon Counting (TSCSPC) Spectroscopy and Microscopy

TSCSPC [1–9] spectroscopy and microscopy have emerged as superior imaging methods for the acquisition of fluorescence dynamics on the picosecond time scale of very weakly emitting sources, such that occur in living cells [10] under minimal-invasive conditions (Chap. 4). TSCSPC is based on the traditional, non-imaging time-correlated single photon counting (TCSPC) method [11], established 25 years ago. TCSPC is one of the most widely applied tools for high-quality determination of fluorescence lifetimes and is intensively used in basic spectroscopy as well as biological and medical research, due to its unique properties: it combines speed (picosecond and nanosecond time scale) with ultra-sensitivity, epitomised by single-molecule detection, and an ultra-wide dynamic range, enabling the researcher to study very intensely and very weakly emitting species simultaneously. Instrument response functions (IRF) below 20 ps FWHM [12] can be achieved with single-channel 6-µm MCP-PMT detectors, resulting in an effective time resolution of about 2 ps after deconvolution, thus approaching the limits of this technology. The present imaging detectors are based on 12-µm MCP two-stage chevron technology with an IRF under 40 ps FWHM [13], using a Ti-sapphire laser as excitation source. Recent advances in MCP-PMT technology, i. e. the introduction of 'coded' anodes, led to TSCSPC spectroscopy and microscopy [1–9], characterised by a drastic increase of space channels, simultaneous acquisition, and the further advantage of a continuous image. By replacing the disk anode of a standard MCP-PMT with a DL [14, 15] or QA [16, 17], we gain access to space information along x- and xy-directions, respectively.

18.1.1
DL-System

The delay-line splits the electric pulse, emanating from the 2nd MCP, and the difference in propagation time of both parts yields the space-coordinate [14, 15]. A 2D multi-channel analyser (MCA), equipped with transputer and local memory, calculates time- and space-coordinate from sum and difference (Fig. 18.1), respectively, of the outputs of both time-to-amplitude converters (TAC). An alternative MCP-timing scheme [1, 5] utilises an auxiliary timing signal from the 1st MCP (Fig. 18.2), resulting in independent TACs and sum/difference calculations are obsolete. The electronic read-out consists of standard NIM modules or, alternatively, of an integrated system, incorporating CFDs, TACs, and 2D-MCA on a single PC board, developed recently (Chap 2). Best field data so far: $FWHM(t) = 47$ ps, 200 space channels (100 µm), with a dynamic range of up to 3×10^6 of the non-cooled detector at only 1000 cps signal strength [1]. Excellent flatness of time resolution of individual space channels across the field-of-view, pre-requisite for high-quality lifetime imaging and lifetime separability (1.1 ns and 1.3 ns [6]), is demonstrated on picosecond (rose bengal/water: 89 ± 10 ps) and nanosecond (rhodamine B/H_2O: 1.67 ± 0.05 ns) time scale

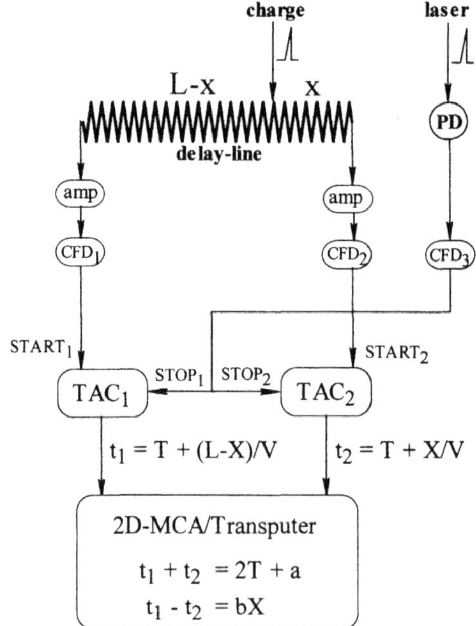

Fig. 18.1. Electronic scheme of DL-MCP-PMT

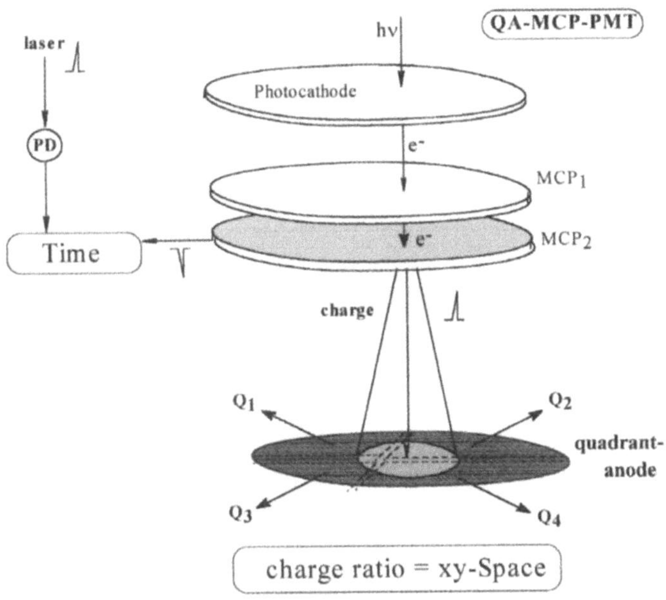

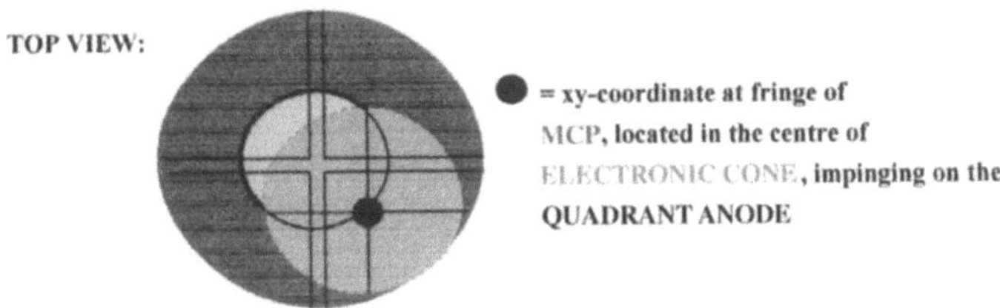

Fig. 18.2. QA-system with MCP-timing

and even extends to 3-exponential decays [6]. The fit usually includes 3 decades of rise [7], an indispensable feature for multi-exponential analysis. High-precision kinetics at subcellular resolution (fixed cells) was demonstrated several years ago [7], and probe-DNA and probe-protein complexes could be discriminated [6], as well as complexed and free NADH [2]. Novel applications in cell biology are currently explored in the frame of an EC Biotechnology Demonstration Project [9], with the aim of 3D structure determination of macromolecules in living cells (Chap. 2) by using Picosecond FLIM. Results of recent living cell studies include: (i) fluorescence anisotropy decays of EB-intercalated DNA in the cell nucleus [10], (ii) GFP-aggregation, studied by fluorescence anisotropy, (iii) protein-protein interaction, (iv) mitochondrial behaviour, (v) photosynthesis in plant cells, observing fluorescence dynamics of the reaction centre in individual chloroplasts, and (vi) research in photodynamic therapy (Chap. 5).

18.1.2
QA-System

The incident photon produces a cone-shaped cloud of electrons at the output face of the second MCP, as in the case of the DL-MCP-PMT above, whose centre hits the four segments of the quadrant-anode (Fig. 18.2) at a spatial position identical to that of the photon. The quadrant-anode splits the electric pulse into four portions, whose individual magnitudes depend on the xy-coordinate of the incident photon. The four charge packets are amplified, digitised, and the xy-coordinates calculated. The timing information is derived from an auxiliary pulse from the 1st MCP, as in the DL-system above, that is inverted, amplified, discriminated, and then used as start pulse into a TAC, together with a stop pulse from a photo-diode (or laser electronics) as in standard TCSPC. Best field data so far: $FWHM(t) = 80$ ps, 250×250 space channels (100 µm), 10^5 cps through-put in DC or 10^6 cps in burst-mode, and 0.5 µm (diffraction limited) space resolution, using a standard epifluorescence microscope [9]. An auxiliary electrode (patent pending) corrects the inherent distortions of simple 4-anode systems [16]. *Specifications of 5-anode version of QA-detector:* (1) 100,000 cps through-put, (2) dynamic range of $>10^7$, due to use of "hot", low-resistance MCPs that are not limited by saturation and allow high count rate within very small areas, resulting in a tremendously high S/N, (3) 100 µm space resolution, corresponding to 250×250 pixels on 25 mm diameter photocathode, (4) 20% quantum efficiency of photocathode, (5) photocathode can be selected from broad spectral range: 300–700 nm with 20 cps background in full cathode area of non-cooled system, 300–800 nm with 200 cps, 300–900 nm with 2,000 cps, (6) automatic HV shut-off, when too much light, (7) sophisticated pile-up rejection: acquisition with up to 10% of laser rep-rate without distortion is possible, (8)

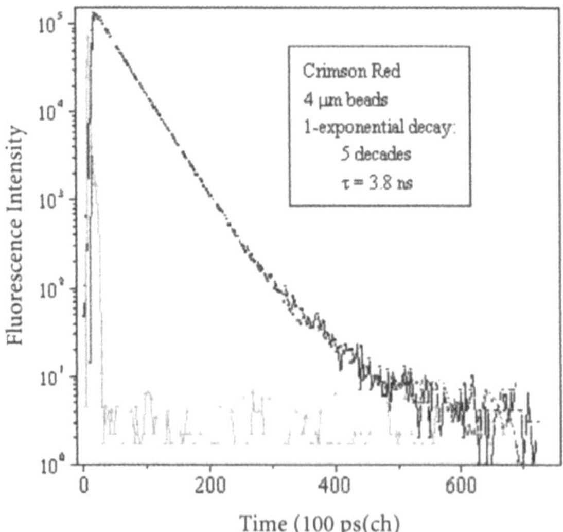

Fig. 18.3. Single-exponential time-space reference system

absolute arrival time with 10 MHz resolution, (9) internal TAC with 10, 25, 50 ns window at 4096 channels (external standard TAC can be used), (10) multi-parameter acquisition: system control by external electrical signals from electro-optic (or -mechanical) periphery, such as tuneable filters and polarisers. Additional coordinates of emission colour and polarisation will be added to standard coordinates of xy, t (TAC), and t(abs) (Chap. 3), (11) optional confocality by modular multi-lens array or Nipkov disk, (12) photocathode can be cooled by cold nitrogen gas, reducing background virtually to zero.

Figure 18.3 demonstrates the large dynamic range of the QA-system, displaying the fluorescence decay of 4 μm beads (Crimson Red), with single-exponential decay over 5 decades, representing an excellent time-space standard for TSCSPC microscopy.

18.2
EC Biotechnology Demonstration Project (BIO4-CT97–2177): Picosecond Fluorescence Lifetime Imaging as a New Tool for 3D Structure Determination of Macromolecules in Cells (http://www.europhoton.de)

18.2.1
Current State of Knowledge

Fluorescence lifetime imaging is the technique of choice to tackle molecular interactions in relation to their microenvironment. Energy transfer, collisional quenching, and physico-chemical changes by the molecular probe's microenvironment have a profound impact on fluorescence lifetime. Understanding biological functions within the cell requires a better understanding of the molecular interactions between macromolecules within their natural environment. By applying time-resolving techniques such as picosecond fluorescence lifetime imaging microscopy (Picosecond FLIM), using very fast and very sensitive detectors, it should be possible (i) to increase the spatial resolution by Förster resonance energy transfer (FRET) and other kinetic studies, (ii) to discriminate between coexisting states of complexation of the fluorescent probe, and (iii) to detect conformational changes in macromolecules such as DNA and proteins, with the ultimate goal, to follow the dynamics of biological processes in living cells. The novel technology has the potential of a profound impact both on cell biological and medical research.

The time-correlated single photon counting (TCSPC) method is one of the most widely applied tools for the determination of fluorescence lifetimes and is intensively used in basic spectroscopy as well as biological and medical research, due to its unique properties: it combines speed (picosecond and nanosecond time scale) with ultra-sensitivity, epitomised by single-molecule detection, and a wide dynamic range, enabling the researcher to study very intensely and very weakly emitting species simultaneously. A TCSPC system typically consists of a high-repetition rate laser, electronic modules, and a fast and sensitive detector, such as a photomultiplier tube (PMT), or more lately, an MCP-PMT. A recent advance in MCP-PMT technology, i.e. the introduction of 'coded' anodes, led to time- and space-correlated single photon counting (TSCSPC) spectroscopy and microscopy, characterised by a drastic increase of space channels and the further advantage of a continuous image. By making use of the superior characteristics of the novel 'coded-anode' detectors, i.e. the ultra-sensitive MCP-PMTs with conductive delay-line (DL) and quadrant anode (QA) (Figs. 18.1 and 18.2), respectively, simultaneous acquisition of fluorescence lifetime information in 200 or 250×250 space channels is possible, at a time resolution of 10 ps and a dynamic range of $> 10^6$. By applying bi-exponential global analysis in 80 wavelength channels of a binary mixture, for example, a lifetime separability of 20% could be achieved, i.e. 1.1 ns and 1.3 ns of POPOP and PPO, respectively, in cyclohexane could be recovered, at full resolution of both vibrational structures, and by applying Picosecond FLIM and 4-exponential global analysis, it was possible to discriminate 6 TOTO-DNA fluorescence lifetimes between 200 ps and 10 ns within a single cell and assign the lifetimes to individual subcellular domains of down to 1 µm size (Preparatory Award BIO4-CT95–9253 [6, 7]). By utilising the inherent potential to gain information usually obtained by fluorescence correlation spectroscopy (FCS), translational and rotational diffusion, flow, aggregation of macromolecules, as well as membrane dynamics will be studied in living cells, by applying the new concept of Vehicle Micro-Spectroscopy (VMS).

18.2.2
Demonstration Objectives

1. To construct a prototype microscopy imaging system of superior performance, based on single photon counting Picosecond FLIM, as a novel technique for 3D structure determination, and advance the novel prototype by improvements in electronics, software, and detector technology.
2. To set up FCS systems, two-photon excitation capabilities, and provide for confocal microscopy (3D imaging).
3. To establish novel time/space standards (fluorescent beads) for picosecond and micrometer scale, respectively.
4. To determine 3D macromolecular structures within subcellular compartments.
5. To study sub-structures of live cells.
6. Dissemination of technological and biological knowledge and know-how.

18.2.3
Work Content

General strategy for implementation is the introduction of novel TSCSPC spectroscopy to cell biology and construction of dual-detector Picosecond FLIM prototypes at three cell-biological end-users of differing biological interests, multi-disciplinary structure of project, combination of technology producers and users, and bringing together European leaders in cell biology, picosecond spectroscopy, and electronics development (SME).

18.2.4
Role of Partners

18.2.4.1
Technology Producers

EuroPhoton GmbH (K. Kemnitz). (i) Administrative and scientific coordination, (ii) development, optimisation, and set-up of QA/DL-microscope prototypes for 3D-picosecond imaging microscopy, (iii) development of novel detector prototype or improvement of standard version, (iv) improvement of time/space parameters of DL/QA systems, and development of novel time/space standards.

FAST ComTec GmbH (W. Wagner). Development of a high-throughput ($>2,000,000$ event $\times$ # of parameters/s), multi-parameter data acquisition and display system for the QA-detector, with (i) higher integration (time measurement and conversion within one instrument), (ii) modularity (number of parameters to be measured can be incremented in steps of four (up to 16)), (iii) shorter conversion time of the TAC/ADC.

TU Berlin (H.-J. Eckert). (i) Software development for scanning mechanism of DL-microscope systems, (ii) electronics development for scanning mechanism of DL-microscope systems, (iii) interface with motorised DL-microscope systems, (iv) time-resolved spectroscopy for partners, using existing DL-spectrometer, to study photophysics of novel probe molecules and probe-macromolecule entities.

Uni Paris-Sud (J.-C. Brochon). (i) Fast global kinetic analysis, using the quantified maximum entropy method, with 50×50 time-resolved pixels within an image and 5-exponential fluorescence decays, (ii) extension to 400×400 pixel and FRET analysis, (iii) introduction of user-defined kinetic models.

Imperial College (D. Phillips). (i) Set-up of QA-microscope system, (ii) development of QA-TSCSPC software, based on existing FAST ComTec MPA software, (iii) on-line kinetic analysis and display for measurement control, (iv) automated data interface for global analysis.

Becker & Hickl Gmb (W. Becker). Development of user-friendly hybrid electronics PC-board for the DL-detector system with high through-put and multi-scheme capability.

18.2.4.2
Technology Users

Goethe University Frankfurt (J. Bereiter-Hahn). (i) Set-up of DL/QA microscope prototype system for 3D imaging, including two-photon capability and VMS, (ii) biological demonstrations, such as cytoskeleton dynamics in living cells, nucleocytoplasmic transport, free ion dynamics, enzyme-cytoskeleton inter-action.

Curie Institute (M. Coppey-Moisan). (i) Set-up of DL/QA microscope prototype system, with two-photon capability and VMS, (ii) spatial and temporal calibrations of the prototype, (iii) biological demonstrations, such as DNA structure and dynamics at the level of individual DNA molecules and single fluorescent molecule, higher-order organisation of DNA in the chromatin of the living cell, protocols of experimental conditions which do not destroy the living cell during the acquisitions of fluorescence lifetime images, DNA-cationic lipids interaction, biological membrane fusion, viscosity measurement of nucleus in living cells, and conformational study of protein fusion.

CNR (T. Parasassi). (i) Kinetic measurements, using a frequency-domain time-resolved system (multi-frequency phase and modulation cross-correlation fluorometer), and comparison of data with that obtained by time-domain systems of project partners, (ii) biological studies, such as DNA-membrane interaction, modulation of virus-cell membrane interaction, cell membrane lipids architecture and dynamics.

18.3
Multi-Parameter TSCSPC

Recent development of novel single photon counting detectors made it possible to acquire and store the full set of system parameters, completely describing the physical properties of each individual photon:

$$h\nu = f\,[x, y, \Delta t\,(\text{TAC}), t\,(\text{abs}), \lambda\,(\text{em}), \|/\perp],$$

where x, y are the space coordinates of the impingent photon at the photo-cathode, $\Delta t\,(\text{TAC})$ the time difference correlating laser excitation pulse and fluorescence photon (yielding ps/ns fluorescence dynamics), $t\,(\text{abs})$ the absolute arrival time of each fluorescence photon at the detector (resulting in tracking capabilities and diffusional rates), $\lambda\,(\text{em})$ its emission wavelength (leading to time-resolved emission spectra), and $\|/\perp$ the direction of polarisation (providing anisotropy dynamics). All parameters can be acquired (pseudo)-simultaneously. Novel microscopes are under construction, combining DL and QA detectors for picosecond-resolved spectroscopy and imaging. A modular Nipkov disk [18] or pinhole-microlens disk [19] can be added for one-photon excitation and a rotating microlens array [20] for two-photon excitation, to achieve confocality (Fig. 18.4). Up to 20,000 confocal spots are used to scan simultaneously the sample, resulting in a dramatic reduction in peak power and photobleaching, as compared to standard single-spot galvanic scanning (Chap. 4).

These microscopes are employed in a novel spectroscopic technique, *Vehicle Micro-Spectroscopy* (Chap. 6), utilising the tracking capabilities of the system

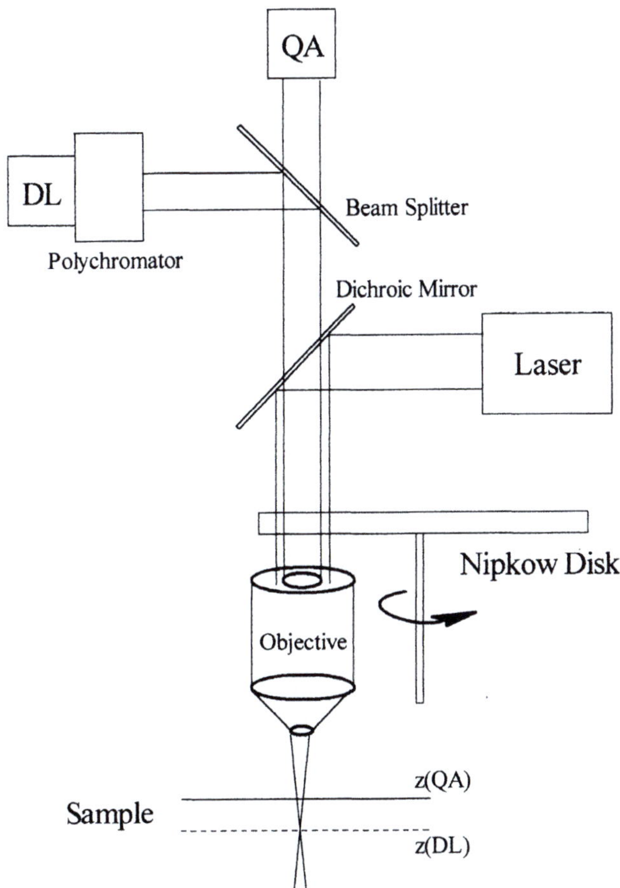

Fig. 18.4. New microscopes for Multi-Parameter TSCSPC: the DL/polychromator unit can be replaced by a QA/imaging-polychromator combination, resulting in 250 time-resolved spectra (at 200 wavelength channel each) across the image

to follow the path of emitters on the sub-second timescale, while acquiring its ps/ns fluorescence dynamics at each point of the trajectory. The resulting movie captures macromolecular diffusional dynamics and interactions in living cells, fluorescence anisotropy dynamics, and micro-environmental properties along the trajectory of the emitting vehicle. The vehicle is a suitably labelled carrier of probe molecules and can be a virus, a macromolecule, or a latex bead. The multi-dimensional set of photon coordinates is stored in list-mode, providing the option to correlate parameter sub-sets with each other.

18.4
Minimal-Invasive Fluorescence Microscopy (MIFM)

To study living cells, minimal-invasive conditions are required to preserve the living state. Ultra-sensitive time-resolved fluorescence imaging, at ultra-low excitation intensity and ultra-low level of labelling (MIFM), became possible after the introduction of QA- and DL-detector systems that are 10–100 times more sensitive than most advanced cooled CCD cameras. Time-correlation dramatically enhances the signal over the non-correlated noise, exemplified by the recent acquisition of fluorescence anisotropy dynamics of DNA labelled by ethidium bromide (EB) in the nucleus of living cells, detected at ultra-low labelling (one EB probe molecule in 10,000 base-pairs of DNA) and ultra-low excitation levels (20 mW/cm^2 range at sample) [10].

To achieve confocality, a Nipkov disk [18] or pinhole-microlens disk [19] can be added for one-photon and a rotating microlens array [20] for two-photon excitation (Fig. 18.4). In rotating disk systems, up to 20,000 excitational spots are used to scan simultaneously the sample, resulting in a dramatic reduction of peak power, at constant average excitation level, and consequently in a considerable reduction of photobleaching and photodynamic reactions, as compared to standard single-spot galvanic scanning.

Photodynamic reactions are easily induced, especially in live cell studies, even at ultra-low (≥ 1 mW/cm^2 on sample) excitation levels that are still far below excitation intensities of standard epifluorescence microscopy. Examples are: (i) DASPMI in mitochondria of living cells (Chap. 6), (ii) chlorophyll in living plant cells (Chap. 5), (iii) JC1 in mitochondrial live cell studies (personal communication: M. Coppey-Moisan), and (iv) intercalated Acridine Orange in living cells, where as little as about 10 mW/cm^2 on the sample had to be used in order to avoid photodynamic reactions [21]. A 10 mW/cm^2 steady-state illumination at 436 nm correspond to 2×10^{16} photons cm^{-2} s^{-1}. In pulsed excitation, the photon density within the short pulses is dramatically higher: at 10 mW/cm^2, it would reach 5×10^{21} photons cm^{-2} s^{-1}, at 1 ps pulse width and 4 MHz repetition rate. An even higher photon density is reached in single-point scan applications: if only 1 mW of the steady-state 488 nm argon-ion line is focused to 0.25 µm, the photon density in the centre will be 1.3×10^{24} photons cm^{-2} s^{-1} [22], and will still be higher with pulsed lasers (Ti-sapphire).

MIFM as described above, using an imaging QA-detector (in potential combination with a rotating lens-pinhole disk), is considered to be the only fluorescence microscope method in living cell studies, to completely avoid photodynamic reactions and bleaching. The build-up of triplet state population, which is among the major causes of photobleaching and photodynamic reactions [22], can be suppressed by sufficiently low photon density. Application of the QA-detector and TSCSPC microscopy has the additional advantage of VBMS (Chap. 6), detecting and interpreting potential photodynamic reactions, due to its ability to acquire movies of fluorescence dynamics (Figs. 18.18–18.20).

Decay of ethidium fluorescence in living cell

mitochondria

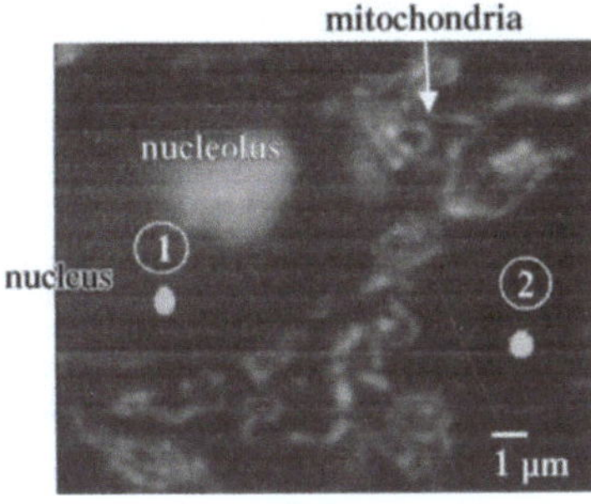

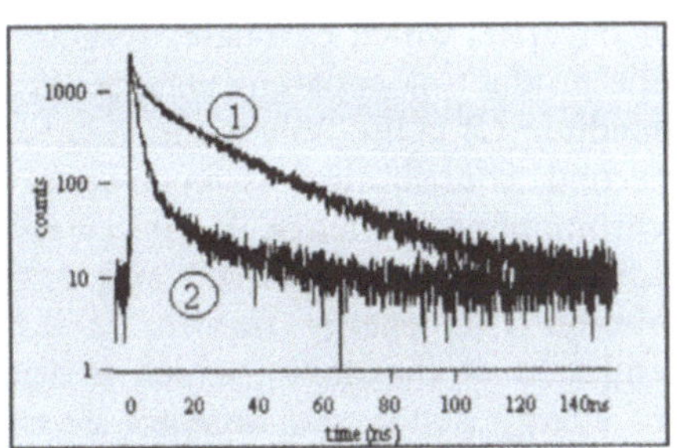

Steady state
Ethidium fluorescence
in living cell

Decay of ethidium fluorescence
1   in the nucleus of living cell
2   in the extra-cellular medium

Fig. 18.5. The presence of EB in nucleus and extra-cellular medium is detectable by time-correlated techniques. In intensified CCD systems, the EB signal is below noise level

18.5
Living Cells: Fluorescence Dynamics Imaging

18.5.1
Fluorescence and Fluorescence Anisotropy Decays of EB-Intercalated DNA in the Cell Nucleus: Collaboration with Maïté Coppey-Moisan (Institut Jacques Monod, Paris)

Physical parameters, describing the state of chromatinised DNA in living mammalian cells, were revealed by in situ fluorescence dynamic properties of ethidium bromide (EB) in its free (1.8 ns) and intercalated (22 ns) states [10] (Fig. 18.5). The lifetimes and anisotropy decays of this cationic chromophore were measured within the nuclear domain, by using the ultra-sensitive time-correlated single photon counting technique, confocal microscopy, and ultra-low probe concentrations. We found that, in living cells: (i) free EB molecules equilibrate between extracellular milieu and nucleus, demonstrating that the cation is naturally transported into the nucleus, (ii) the intercalation of EB into chromatinised DNA is strongly inhibited, with relaxation of the inhibition after mild (digitonin) cell treatment, (iii) intercalation sites are likely to be located in chromatin DNA, and (iv) the fluorescence anisotropy relaxation of intercalated molecules is very slow. The combination of fluorescence kinetic and fluorescence anisotropy dynamics indicates that the torsional dynamics of nuclear DNA is highly restrained in living cells.

18.5.2
GFP-Aggregation, Studied by Fluorescence and Fluorescence Anisotropy Dynamics: Collaboration with Maïté Coppey-Moisan (Institut Jacques Monod, Paris)

Time-Resolved Fluorescence Anisotropy Microscopy. Fluorescence depolarisation dynamics can principally contain contributions by (i) molecular rotation and by (ii) Förster resonance energy transfer (FRET) between chromophores. The first contribution is described by the Stokes-Einstein equation, $\Phi = \eta V/RT$, which correlates rotational time of a spherical molecule, Φ, with the product of apparent volume, V, of the molecule and viscosity, η, of the medium (R is the Boltzmann constant and T the absolute temperature). The second contribution is correlating energy transfer time, τ, with inter-chromophore distance, R, by $\tau = \tau_0(R/R_0)^6$, where R_0 is the Förster radius (distance of 50% transfer) and τ_0 the fluorescence lifetime when FRET is absent. This energy transfer is accompanied by a depolarisation of the emission of the donor, which is strongly dependent on mutual orientation of energy donor and acceptor dipoles. If donor and acceptor are identical (homo-transfer, as in case of homo-dimers), the observed fluorescence life-time of the donor is unchanged and the depolarisation is the only means to detect this transfer. Both aspects of anisotropy dynamics can be used to deduce the volume of tagged proteins and inter-molecular distance inside macromolecular structures, respectively, directly in living cells. GFP is a very good candidate for above anisotropy studies, since it decays mono-exponentially, when free or fused to other proteins (Fig. 18.6).

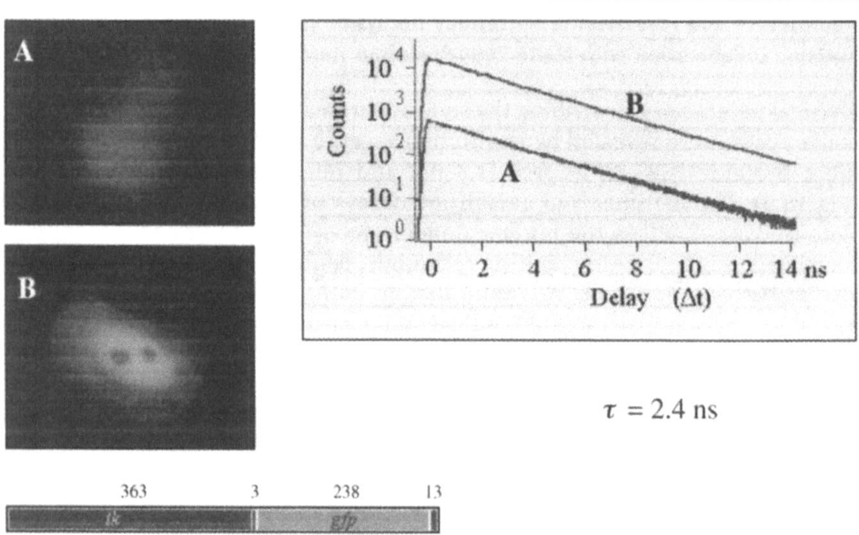

Fig. 18.6. Fluorescence decays of TK-GFP in living cells at low (**A**) and high (**B**) expression. The fluorescence dynamics is 1-exponential and independent of GFP concentration, indicating the presence of properly folded protein. The presence of mis-folded protein or free (cleaved) GFP would be revealed by deviations from 1-exponential behaviour

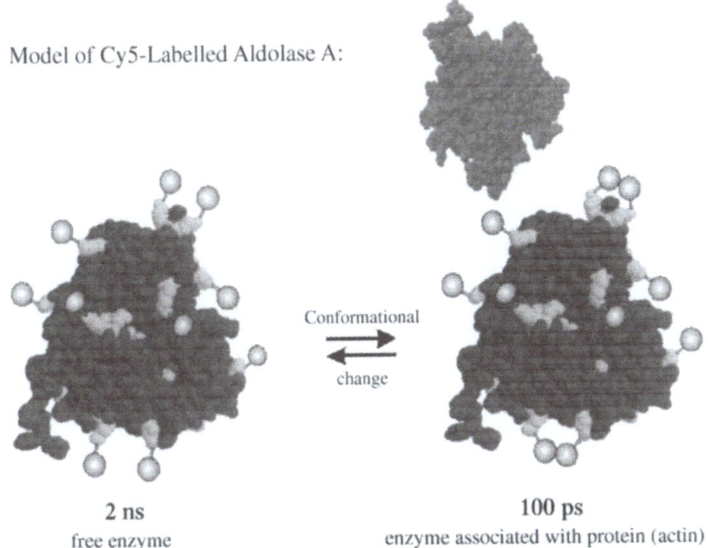

Fig. 18.7. Cartoon of protein-protein interaction (in vitro), interpreting the changes of cy5 fluorescence dynamics upon complex formation. Several dye molecules are bound to one enzyme molecule, allowing for the existence of multiple decay times, as the dye molecules may face a different environment. On binding of the enzyme to actin, the enzyme changes its conformation, thus altering the distance between the bound dye molecules. The large contribution of the short fluorescence lifetime of 200 ps (Fig. 18.8) might be due to enhanced dimer formation in the complex. Dimers of xanthene and cyanine dyes are known to have lifetimes in the 100 ps range [23]

18.5.3
Protein-Protein Interaction: Collaboration with Jürgen Bereiter-Hahn (Goethe University Frankfurt)

Glykolytic enzymes associate with cytoskeletal structures in living cells. This association is transient and depends on the presence of the appropriate substrate [24] and on the presence of growth factors as well. Only a fraction of about 20% might be associated with actin or other cytoskeletal elements. In the case of aldolase, binding to g-actin and to f-actin occurs (Fig. 18.7), inhibiting actin polymerisation. A strong activation of actin polymerisation is achieved by the addition of the natural substrate of aldolase, 1,6-fructose bisphosphate. In vitro studies of the binding of Cy5-labeled aldolase (derived from rabbit skeletal muscle) to alpha-actin (from skeletal muscle) revealed conformational changes of the molecule, detected by alteration of the relative contribution of the three decay times found for dye molecules which were covalently bound to the enzyme (Fig. 18.8).

These measurements are now the basis for investigations of aldolase actin interactions in living cells, which are done after injection of labelled aldolase into living endothelial cells. A comparison is made with fluorescence emission from adjacent cells, micro-injected with Cy5 only.

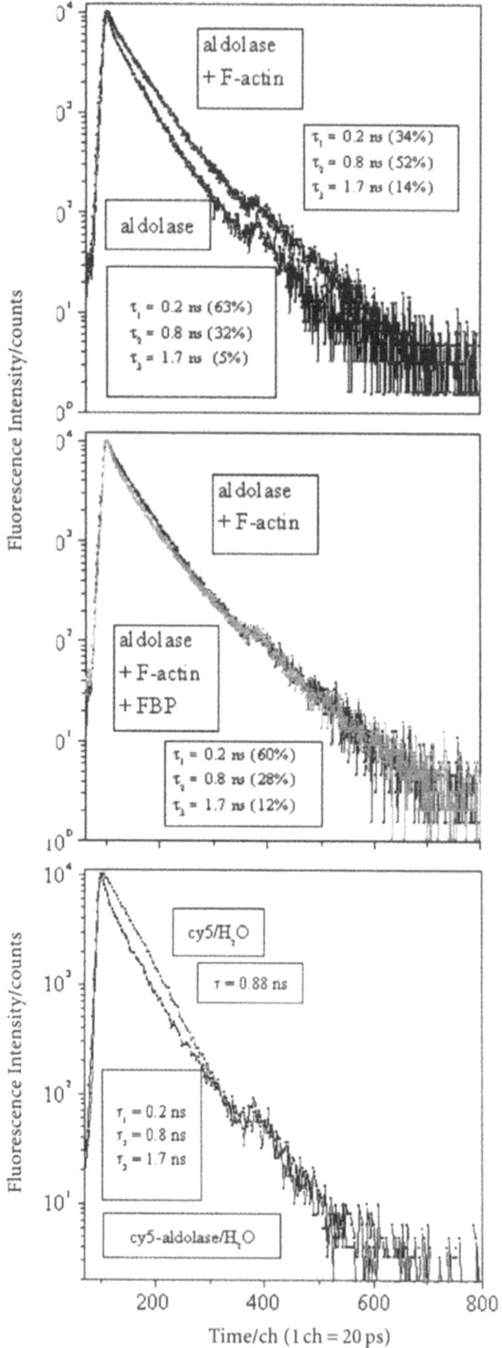

Fig. 18.8. Three-exponential fluorescence dynamics of cy5-labelled aldolase and complex formation with actin. $\lambda_{ex} = 635$ nm, $\lambda_{em} = 670$ nm

18.5.4
Mitochondria: Fluorescence Dynamics of DASPMI and Rhodamine 700: Collaboration with Jürgen Bereiter-Hahn (Goethe University Frankfurt)

DASPMI (dimethyl-aminostyryl-pyrmidinium methiodide). DASPMI is among the first dyes that were detected to stain selectively mitochondria in living cells [25, 26]. The fluorescence intensity in mitochondria is directly related to the mitochondrial membrane potential [25, 27–29]. Quantum yield and fluorescence lifetime of styryl dyes, such as DASPMI, depend on dielectric constant and viscosity of the micro-environment [30]. Therefore, a strong shift in fluorescence lifetime is to be expected on uptake into functioning mitochondria. In water $\tau = 60$ ps, while in mitochondria of living endothelial cells three decay times with 0.3 ns, 1.2 ns and 5 ns are found (Fig. 18.9), corresponding to a large increase in quantum efficiency in lipophilic media, as had been derived from classical intensity measurements before [28]. At low

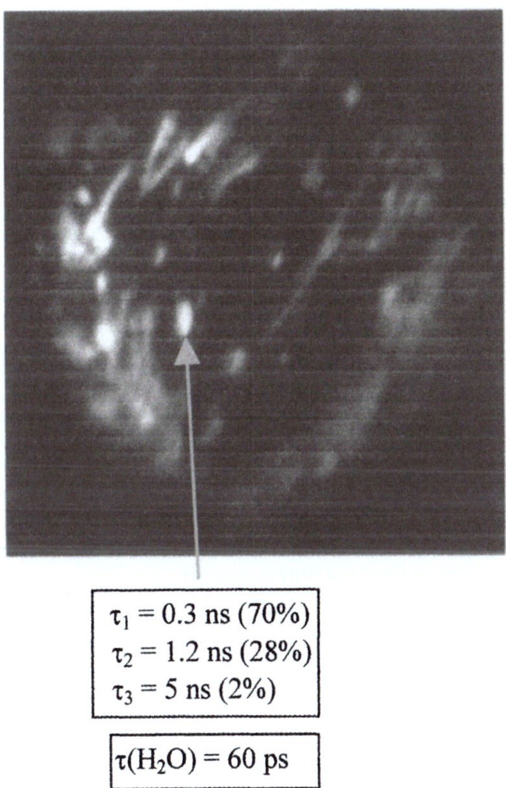

Fig. 18.9. QA-image and fluorescence dynamics of DASPMI-labelled mitochondria in living cells. Three-exponential fluorescence dynamics of single mitochondria/DASPMI: 50,000 cts/ mitochondrion are sufficient, when global analysis is applied with auto-fluorescence correction. $\lambda_{ex} = 476$ nm, $\lambda_{em} = 546$ nm

excitation intensities of about $1\,mW/cm^2$ at the sample (0.5 µm dye), the fluorescence lifetimes determines over a time interval of 20 min are extremely constant with a standard deviation of ± 5 ps only (Fig. 18.20), at constant pre-exponential factors. At about 5 times higher illumination, photodynamic changes are induced, as evidenced by irreversible morphological changes (Fig. 18.19) and a continuous increase of fluorescence lifetimes and change in pre-exponential factors (Fig. 18.20). A swelling of cristae and mitochondrial matrix might be responsible for observed fluorescence dynamics change.

Rhodamine 700. Most of the rhodamines selectively stain mitochondria in living cells [25–29], as does Rhodamine 700, when used in sub-micromolar concentration, accompanied by the appearance of two decay times: $\tau_2 = 0.3$–0.4 ns and $\tau_1 = 2.25$–2.4 ns (Figs. 18.10 and 18.11). In contrast, the rhodamine 700 fluorescence decay in solution is one-exponential, independent of pH, but very sensitive for changes in solvation environment (Tables 18.1 and 18.2), due to differences in hydrogen bonding [31].

Intracellular (mitochondrial) accumulation and fluorescence dynamics of rhodamine 700 (1 µm, $\lambda_{ex} = 635$ nm) is shown in Fig. 18.10. The total intensity in the focal plane increases with time and accumulation, accompanied by a transition from mono- to bi-exponentiality of the fluorescence dynamics. Because of the relatively high dye concentration, the mitochondria are facing phototoxicity, resulting in a decrease of the membrane potential, revealed by a decrease of the overall fluorescence intensity after about 15 min.

At concentrations below 1 µmol/l, the cells are less impaired, thus maintaining mitochondrial membrane potential and high fluorescence intensity (Fig. 18.11). The lifetime value of the major component seems to increase from 1.9 ns to 2.3 ns and remain constant (as in Fig. 18.10), while the ratio of the pre-exponential factors is fluctuating by almost 10%, much stronger than in comparable in vitro measurements, and possibly indicating cell dynamical processes. The long decay time corresponds to that in bovine serum albumin (regardless of the protein concentration) and thus indicates binding to matrix proteins, rather than accumulation in the inner membrane of the mitochondria.

Table 18.1. Rhodamine 700 lifetime as a function of pH

pH	5	6	7	8	9
Lifetime (ns)	1.57 ± 0.02	1.55 ± 0.02	1.55 ± 0.02	1.56 ± 0.02	1.6 ± 0.02

Table 18.2. Rhodamine 700 lifetime in different solvents

Solvent	Water	Ethanol	Chloroform
Lifetime (ns)	1.55 ± 0.02	2.90 ± 0.03	4.0 ± 0.02

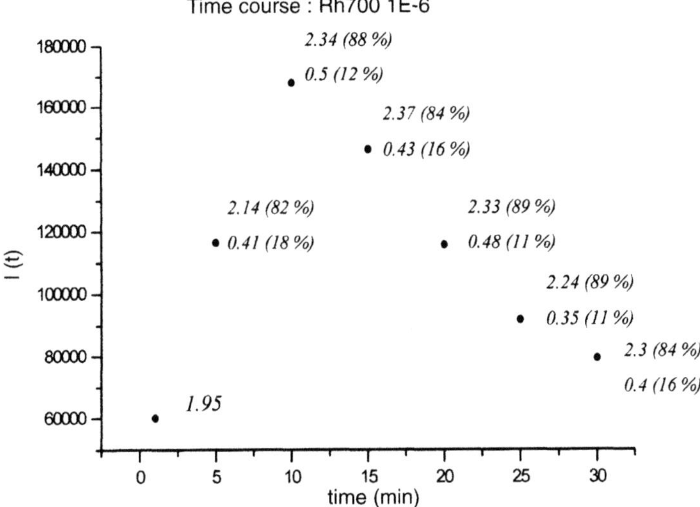

Fig. 18.10. Uptake of rhodamine 700 (1 µm in culture medium) in mitochondria of living XTH-2 cells, derived from *Xenopus* tadpole heart endothelia. $\lambda_{ex} = 635$ nm, $\lambda_{em} = 670$ nm

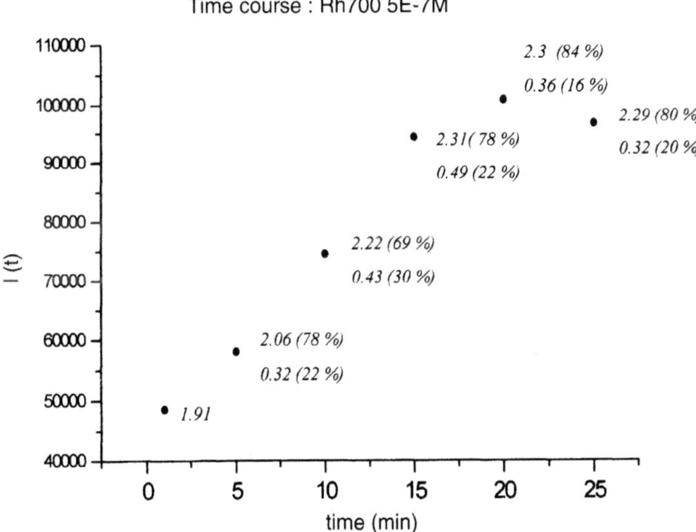

Fig. 18.11. Uptake of rhodamine 700 (0.5 µm in culture medium) and fluorescence dynamics in mitochondria of living XTH-2 cells. $\lambda_{ex} = 635$ nm, $\lambda_{em} = 670$ nm

18.5.5
Chloroplasts: Photosynthesis in Living Plant Cells by Observing Fluorescence Dynamics of the Reaction Centre in Individual Chloroplasts: Collaboration with Hann-Jörg Eckert (TU Berlin)

By applying picosecond fluorescence lifetime imaging microscopy (Picosecond FLIM) and spectroscopy, based on TSCSPC techniques, we were able to measure the 3-exponential chlorophyll *a* dynamics of individual chloroplasts in living cells of moss leafs and protonemata of *Ceratodon purpureus*. [32] Using the novel quadrant-anode QA-MCP-PMT detector, attached to a fluorescence microscope, we measured the spatial distribution (1 μm resolution) of intensity and lifetime components with picosecond time resolution. In addition, we investigated the fluorescence decay spectra of Photosystem II (PS II)-membrane fragments at various temperatures down to 10 K, by using a laser-fluorometer with delay-line (DL) anode MCP-PMT, which allows simultaneous acquisition of fluorescence dynamics in up to 200 wavelength channels [33].

Chl*a*-fluorescence is mainly emitted by the light harvesting complex of Photosystem II. After light absorption by the pigments of the antenna system of PS II (mainly Chl*a* and Chl*b* molecules), the electronic excitation migrates via singlet-singlet energy transfer to the reaction centre, where it induces a primary charge separation, forming the radical pair $P680^+Pheo^-$. The fluorescence lifetime not only depends on the radiative and non-radiative decay channels of the excited singlet state of Chl*a* (Chl*a**), but also on the trapping of electronic excitation by the reaction centre of PS II and subsequent charge separation. Measurements of fluorescence dynamics, therefore, provide a valuable minimal-invasive tool (Chap. 4) for analysing transfer of excitation energy and primary electron transfer in Photosystem II of plants.

Traditional investigations of chlorophyll fluorescence dynamics (fluorescence induction and lifetime) in chloroplast suspensions or thylakoid membrane fragments have significantly contributed to the current understanding of the primary processes in photosynthesis. The present in vivo measurements of individual chloroplasts (Fig. 18.12) at varying environmental conditions stand at the beginning, to our conviction, of a promising new avenue of exploration in plant cell biology.

Photoinhibition of individual chloroplasts in living plant cells was examined by irradiating, under the microscope, selected section of the sample by pulsed light of 480 nm (Figs. 18.13 and 18.14).

Upon irradiation, the observed fluorescence dynamics initially becomes faster (after 3 min exposure to 0.05 kW/m^2), but slows down after a second period of light exposure (8 min of 5 kW/m^2), developing a slow component of about 5 ns.

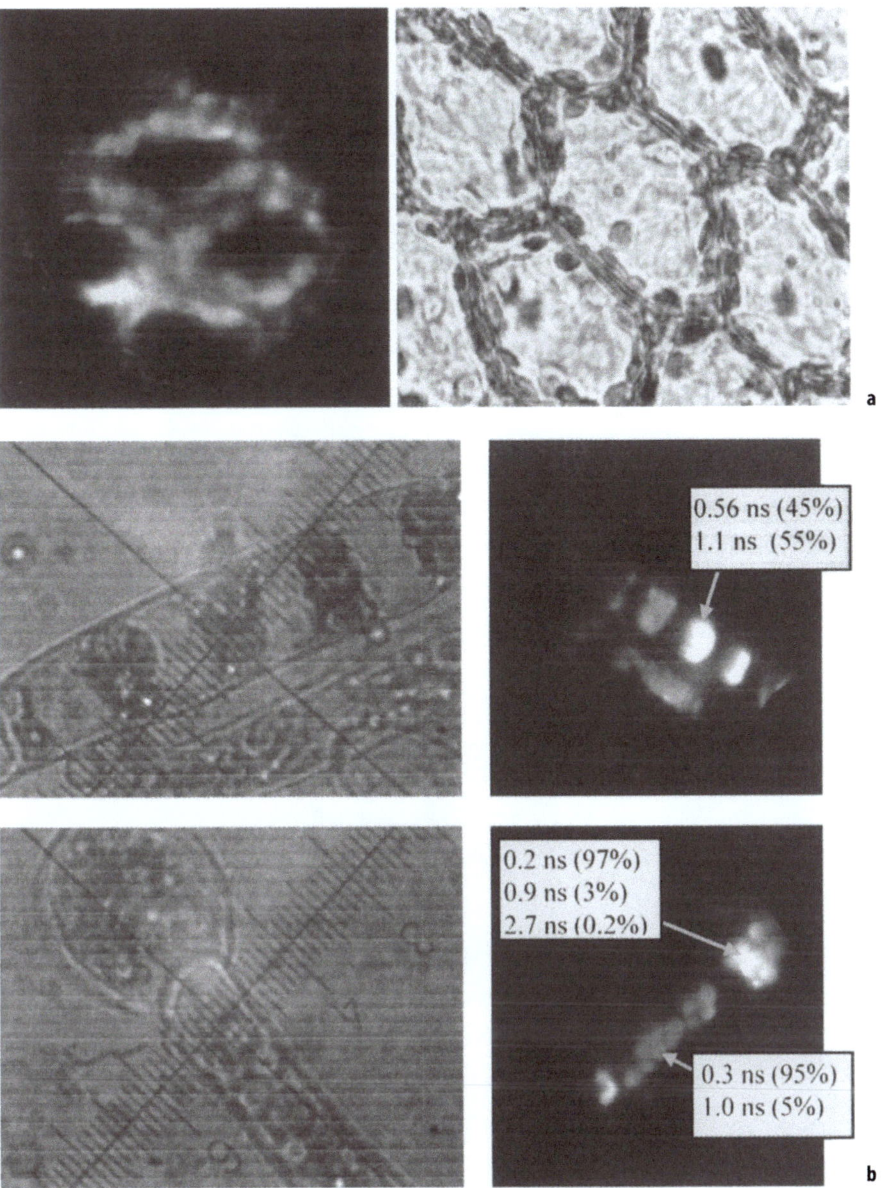

Fig. 18.12a, b. CCD and QA-image with fluorescence dynamics of Chla in: **a** moss leaves; **b** *Protonemata*. $\lambda_{ex} = 635$ nm, $\lambda_{em} = 670$ nm. Note the individual fluorescence decay behaviour of individual chloroplasts

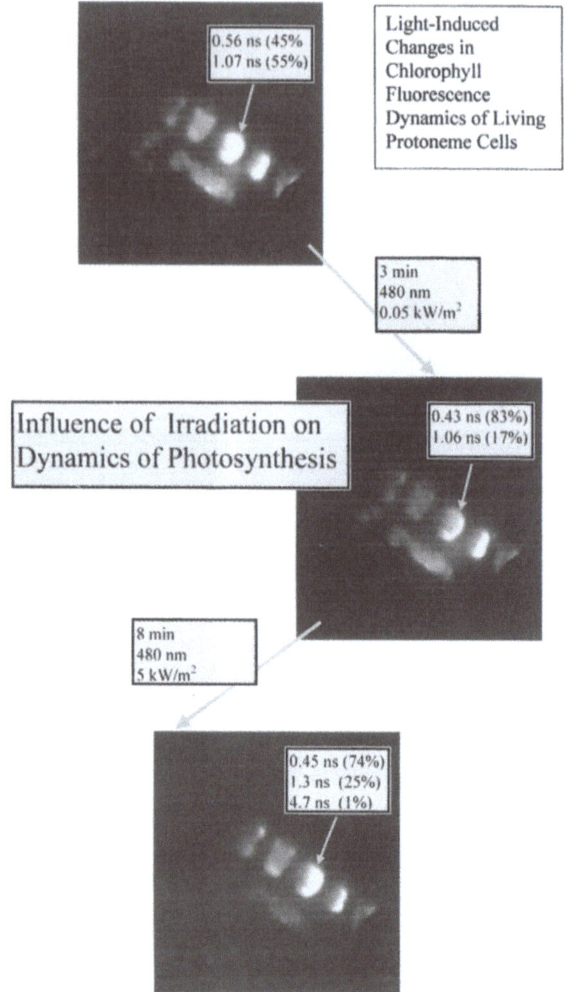

Fig. 18.13. QA-image and photoinhibition of Chla dynamics in *Protonemata* by increased light levels

18.5.6
The Acquisition of Fluorescence Lifetime Values from Intracellular Sulphonated Aluminium Phthalocyanines Using Confocal Point-Scan and Wide-Field QA Detection [32 b]. Collaboration with David Phillips (Imperial College, London)

The aims were threefold – to demonstrate the applicability of using the QA detector to obtain fluorescence lifetime values from intracellular sulphonated aluminium phthalocyanines, to compare the results obtained from wide-field detection using the QA detector with those obtained from a point-scan system, and to interpret the results obtained in terms of phthalocyanine localisation.

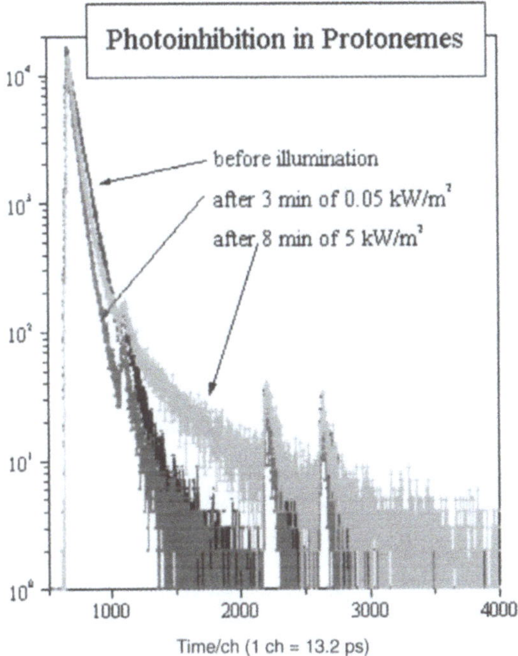

Fig. 18.14. Light-induced changes in fluorescence dynamics of Fig. 18.13. $\lambda_{ex} = 635$ nm, $\lambda_{em} = 670$ nm

18.5.6.1
Application of the QA Detector to Obtaining Fluorescence Lifetime Values from Intracellular Sulphonated Aluminium Phthalocyanines

With a time resolution of 10 ps and a dynamic range of $> 10^6$, the QA detector is ideal for investigating changes in the fluorescence lifetime of membrane-bound photosensitisers that result from photodynamic processes. Spatial resolution, in the transverse plane of $<1\,\mu$m, can be achieved from cell samples using a standard microscope set-up, allowing spatial resolution of sub-cellular structures. For flattened cells, such as epithelial cells and fibroblasts, which are of the order of 1 μm thick, inherent confocality is conferred on the sample. This means that there is no need to use a pin hole and fluorescence acquisition is enhanced.

In our investigations, CT3 murine fibroblasts were stained with di-sulphonated aluminium phthalocyanine. The spatial distribution of the AlPcS2 fluorescence, observed using the QA detector (Fig. 18.15), is consistent with membrane localisation. The nucleus exhibits no fluorescence.

The fluorescence lifetime values obtained by TSCSPC are shown in Table 18.3, and can be rationalised in terms of the local AlPcS2 environment.

The 'lifetime' value, $\tau_1 = 0.4$ ns, is in fact a laser artefact (fluorescence of 635 nm diode laser), the short lifetime value, $\tau_2 = 2.6$ ns, results from self-quenching of the membrane bound AlPcS2, whilst the longer lifetime value, $\tau_3 = 5.5$ ns, is from unquenched AlPcS2 present in the bathing solution.

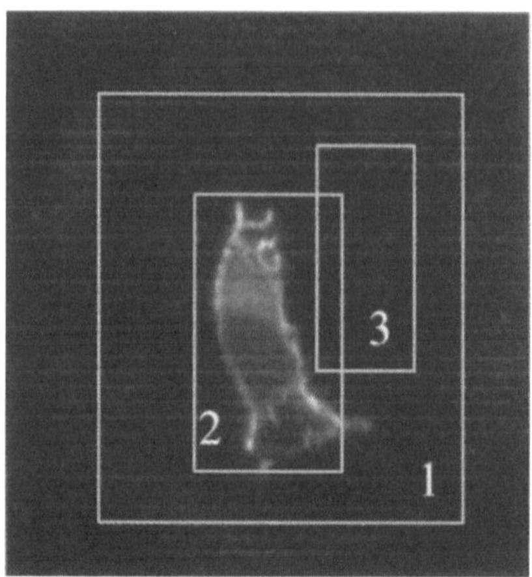

Fig. 18.15. QA image of CT3 murine fibroblast stained with AlPcS2. The observed fluorescence is from membrane-localised AlPcS2. λ_{ex} = 635 nm, λ_{em} = 670 nm

Table 18.3. Fluorescence lifetime values obtained by TSCSPC

	$A_1\,(\tau_1 = 0.4\ \text{ns})$	$A_2\,(\tau_2 = 2.6\ \text{ns})$	$A_3\,(\tau_3 = 5.5\ \text{ns})$
Region 1 (cell membranes and bathing solution)	27.1	22.0	50.9
Region 2 (cell membranes)	11.0	43.8	45.2
Region 3 (bathing solution)	45.3	7.0	47.7

18.5.6.2
The Application of Confocal Fluorescence Microscopy in Obtaining Fluorescence Lifetime Values from Intracellular Sulphonated Aluminium Phthalocyanines

Using a point-scan confocal system, a spatial resolution in the x-y plane of less than 1 μm and axial resolution of the order of 1 μm can be obtained. The point-scan system, which utilises a photomultiplier tube as a detector, has a time resolution similar to that of the QA system. Changes in the intracellular fluorescence intensity of AlPcS2 loaded 3T3-L1 are typically observed on irradiation, with a low initial fluorescence intensity increasing with irradiation time until a maximum is reached. Images showing the changes in intracellular fluorescence intensity over the period of photoirradiation are shown in Fig. 18.16.

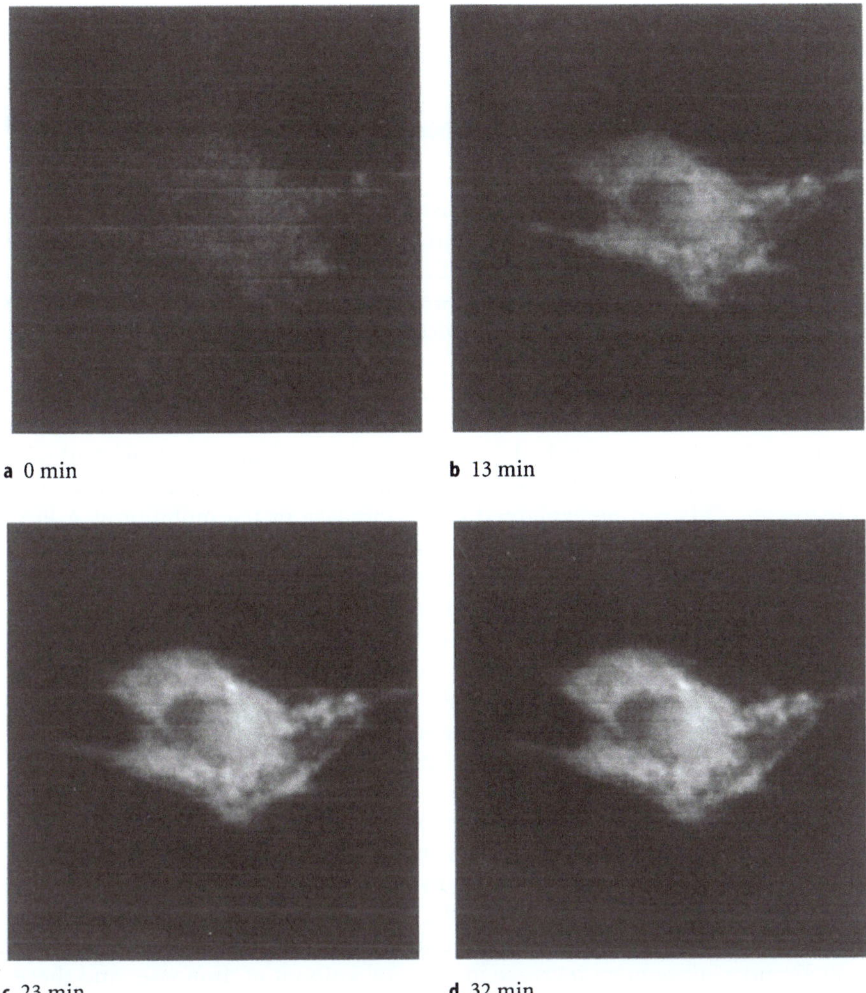

a 0 min

b 13 min

c 23 min

d 32 min

Fig. 18.16a–d. Increase of fluorescence intensity, caused by photoirradiation

18.5.6.3
Interpretation of the Results in Terms of Intracellular Phthalocyanine Localisation

The intracellular fluorescence decay data of AlPcS2 in leukaemic K562 cells acquired using TCSPC have previously been fitted to give a "short lifetime" ($\tau = 2.2$ ns) component and a "long lifetime" ($\tau = 6.1$ ns) component. This is comparable to initial results using the QA detector (2.6 ns and 5.5 ns).

Since AlPcS2 dimers do not fluoresce, it is possible that dynamic quenching of monomer fluorescence by dimers or by monomers during dimer formation could result in the appearance of the "short lifetime" component in the fluorescence decay when the SOE model (Eq. 18.1) is used to fit the kinetic data. Equation (18.2) describes the Smoluchowski-Collins-Kimball model of diffusion-controlled fluorescence quenching. Diffusion controlled fluorescence quenching in restricted geometries can be described by Eq. (18.3), describing self-quenching in heterogeneous media. Due to the inherent difficulty in recording an appropriate instrument response function (IRF), and the necessity of taking into account the convolution of the decay and the IRF, it is reasonable to omit the term containing exponent 2d in Eq. (18.4) (fractal model by Takami, Mataga, Duportail, and Lianos) and the data is fitted to Eq. (18.3) in which parameter b was fixed. The value of b was fixed to $1.82 \times 10^8 \, s^{-1}$ corresponding to a fluorescence lifetime value of 5.5 ns, the average lifetime of aqueous AlPcS2 (5 ns) and AlPcS2 in a non-polar medium (6 ns). Low parameter d values were found initially, indicating that the movement of AlPcS2 molecules is very restricted. The low initial fluorescence intensity and the highly quenched fluorescence decays found are consistent with fluorescence quenching. After longer irradiation times, the value of parameter d increases. This can be explained by an increase in the mobility of AlPcS2 molecules, probably resulting from changes in their local environment. If changes in the membrane structure result in an effective dilution of AlPcS2 an increase in aggregate dissociation and fluorescence intensity would be expected. That this happens is borne out by the steady-state data. Changes in the decay profiles, resulting from a decrease in the amount of fluorescence quenching and from an increase in the concentration of unquenched monomer, explain the increase in fluorescence intensity observed. It could be shown that fluorescence from AlPcS2 loaded 3T3-L1 murine fibroblasts is initially highly quenched and that on photosensitisation the degree of quenching is lowered. We have also demonstrated that upon photosensitisation 3T3-L1 murine fibroblasts exhibit an increase in fluorescence intensity. The decrease in quenching is explained by changes in the micro-environment which also results in a decrease in the restriction of molecular movement. Changes in the microenvironment are thought to result in a local dilution of AlPcS2 and a subsequent increase in the total amount of monomer and therefore of fluorescence intensity. The QA system will allow the construction of lifetime images, and therefore allow comparisons between sub-cellular compartments, allowing accurate assessment of photosensitiser induced changes at a sub-cellular level. This should assist in the design of photosensitisers to target particular cell regions, leading to the possibility of assessing "tissue specific photosensitisers".

$$I = \sum_{i=1}^{n} A_i \exp\left(\frac{-t}{\tau_i}\right) \tag{18.1}$$

$$I = a \exp(-bt - c\sqrt{t}) \tag{18.2}$$

$$I = a \exp(-bt - ct^d) \tag{18.3}$$

Eq. (18.3) being derived from Eq. (18.4)

$$I = a \exp\left(-bt - ct^{\,d} + et^{2d}\right) \qquad (18.4)$$

18.6
Vehicle Micro-Spectroscopy

Vehicle micro-spectroscopy (VMS) (Fig. 18.17) had been introduced for:

1. Tracking of moving vehicles, reporting on pH, Ca^{++}, viscosity, polarity, etc. (molecular probes) and inter-molecular distances (picosecond FRET), along the trajectory of the vehicle in living cells.
2. Acquisition of the full history (movie capability) of a prolonged experiment, including pico/nanosecond fluorescence dynamics of potential photobleaching and photodynamic reactions (Fig. 18–20).

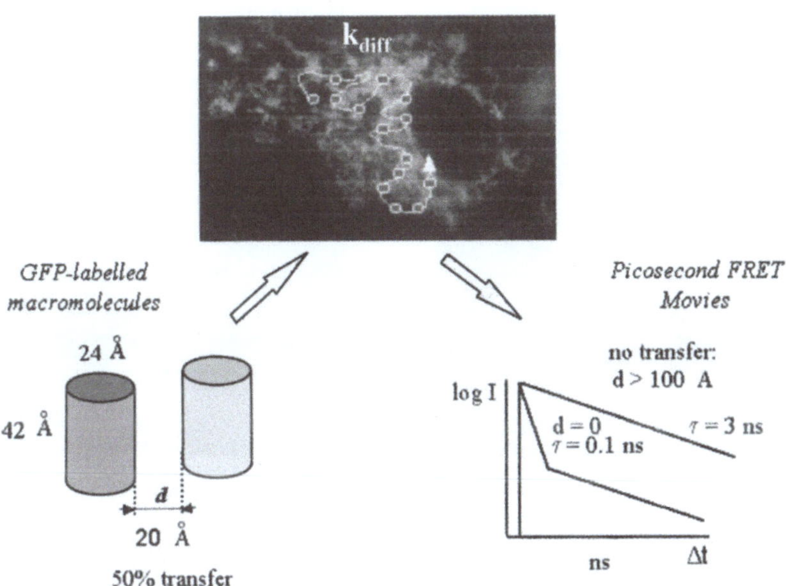

Fig. 18.17. Tracking capability on sub-second scale, combined with ps/ns fluorescence dynamics. Vehicle can be a labelled virus, macromolecule, latex bead, etc. and its movement by active transport or diffusion

0–5 min

5–10 min

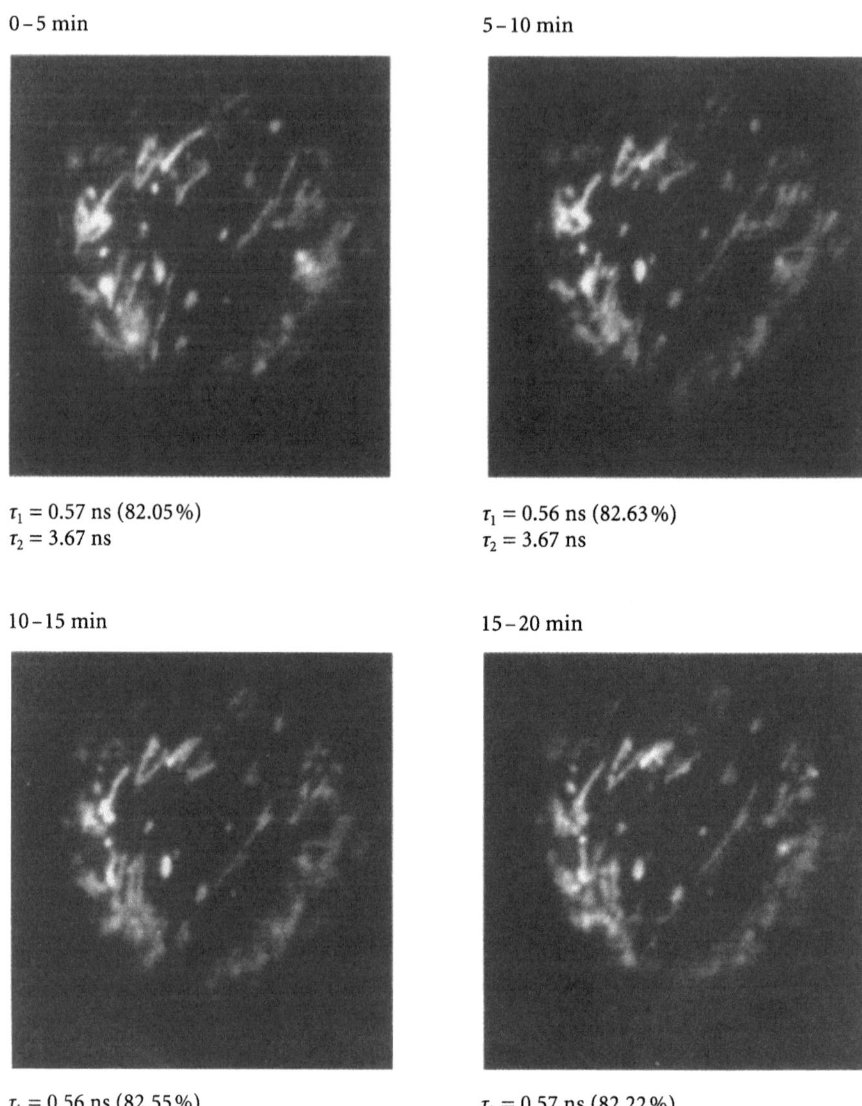

$\tau_1 = 0.57$ ns (82.05%)
$\tau_2 = 3.67$ ns

$\tau_1 = 0.56$ ns (82.63%)
$\tau_2 = 3.67$ ns

10–15 min

15–20 min

$\tau_1 = 0.56$ ns (82.55%)
$\tau_2 = 3.69$ ns

$\tau_1 = 0.57$ ns (82.22%)
$\tau_2 = 3.67$ ns

Fig. 18.18. DASPMI: living cell at low excitation level: QA-images and fluorescence dynamics at various stages of acquisition, at low-level of excitation (about 1 mW/cm² at sample). $\lambda_{ex} = 476$ nm, $\lambda_{em} = 546$ nm

0–5 min

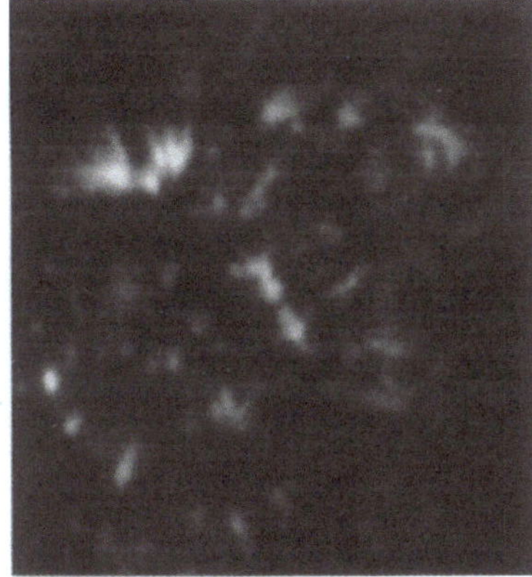

$\tau_1 = 0.49$ ns (83.6%)
$\tau_2 = 2.86$ ns

5–10 min

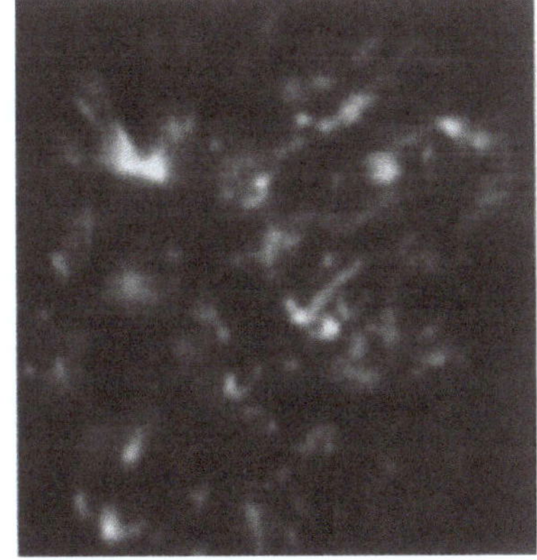

$\tau_1 = 0.51$ ns (82.0%)
$\tau_2 = 2.93$ ns

10–15 min

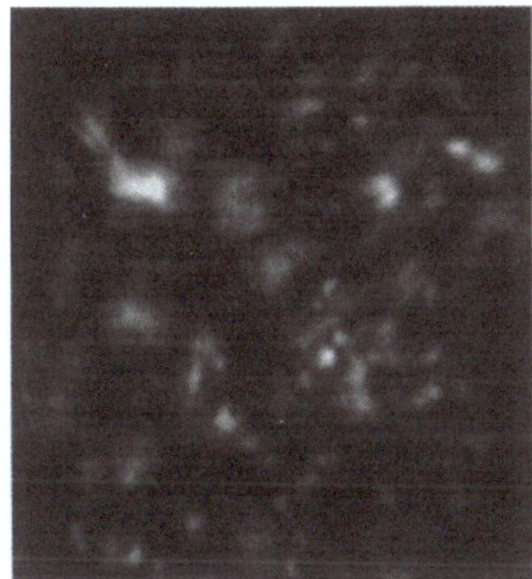

$\tau_1 = 0.53$ ns (81.3%)
$\tau_2 = 3.10$ ns

15–20 min

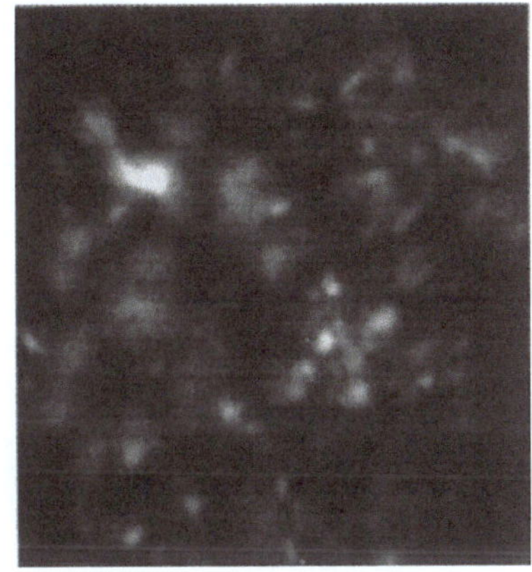

$\tau_1 = 0.53$ ns (81.1%)
$\tau_2 = 3.19$ ns

Fig. 18.19. DASPMI: living cell at "high" excitation level: QA-images and fluorescence dynamics at various stages of acquisition, at about five times the excitation level of Fig. 18.18. $\lambda_{ex} = 476$ nm, $\lambda_{em} = 546$ nm

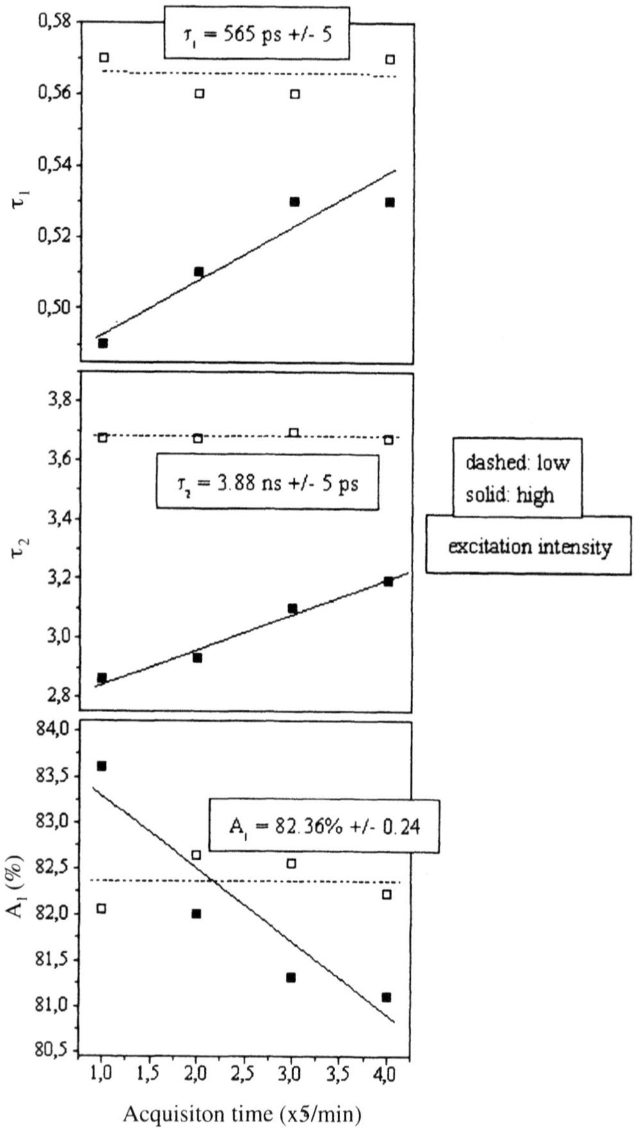

Fig. 18.20. Two-exponential analysis of fluorescence dynamics of DASPMI-labelled mitochondria in Figs. 18.18 and 18.19

References

1. Kemnitz K, Pfeifer L, Paul R, Fink F, Bergmann A (1995) Time- and space-correlated single photon counting spectroscopy. SPIE Proc 2628:2
2. Pfeifer L, Höhne W, Kemnitz K, Paul R, Fink F (1996) A novel simultaneous wavelength- and time-resolving fluorescence measurement system: application for the separation of biomolecular spectra. In: Corcoran VJ, Goldman TA (eds) Proceedings Lasers '95 Charleston, STS Press
3. Pfeifer L, Schmalzigaug K, Paul R, Lichey J, Kemnitz K, Fink F (1995) Time-resolved auto-fluorescence measurements for the differientiation of lung-tissue states. SPIE Proc 2627
4. Paul R, Pfeifer L, Kemnitz K, Fink F (1995) Ultrasensitive detection of cholecystokinin (CCK) by laserinduced time-resolved fluorescence diagnostics. SPIE Proc 2629
5. Kemnitz K, Pfeifer L, Ainbund MR (1997) Detector for multichannel spectroscopy and fluorescence lifetime imaging on the picosecond timescale. Nucl Instr Meth Sec A 387:86
6. Kemnitz K, Pfeifer L, Paul R, Coppey-Moisan M (1997) Novel detectors for fluorescence lifetime imaging on the picosecond timescale. J Fluorescence 7:93
7. Kemnitz K, Paul R, Coppey J, Coppey-Moisan M (1996) Fluorescence lifetime imaging of cells on the picosecond timescale. SPIE Proc 2926:177
8. Ainbund MR, Arzhantsev SY, Chikishev AY, Koroteev NI, Shkurinov AP, Toleutaev BN, Turbin EV, Lehmann A, Pfeifer L, Fink F, Kemnitz K (1998) Picosecond fluorescence lifetime imaging microscopy at 1 µm space- and 10 ps time-resolution: 50×50 ch MCP-PMT with quadrant-anode. In: Slavik J (ed) Fluorescent microscopy and fluorescent probes, vol 2. Plenum Press, New York
9. EC Biotechnology Demonstration Project (BIO4-CT97–2177): Picosecond fluorescence lifetime imaging microscopy as a new tool for 3D structure determination of macromolecules in cells, coordinator: Kemnitz K (EuroPhoton GmbH): 1.4 MECU, 9 partners, 3.5 years, start 1.11.97
10. Tramier M, Kemnitz K, Durieux C, Coppey J, Denjean P, Pansu RB, Coppey-Moisan M (2000) Restrained torsional dynamics of nuclear DNA in living proliferative mammalian cells. Biophys J 78:2614
11. O'Connor DV, Phillips D (1984) Time-correlated single photon counting. Academic Press
12. (a) 19.4 ps: Tamai N, Yamazaki T, Yamazaki I (1988) Chem Phys Lett 147:25; (b) 16 ps (Ti-Saph laser): Kemnitz K, unpublished data
13. 36 ps (Ti-Saph laser): Kemnitz K, unpublished data, obtained with disk anode MCP-PMT
14. Lampton M, Siegmund O, Raffanti R (1987) Delay line anodes for microchannel-plate spectrometers. Rev Sci Intr 58:2298
15. (a) Ainbund MR, Buevich OE, Kamalov VF, Menshikov GA, Toleutaev BN (1992) Simultaneous spectral and temporal resolution in single photon counting technique. Rev Sci Instr 63:3274. (b) US Pat 5,148,031
16. Lampton M, Malina RF (1976) Quadrant anode image sensor. Rev Sci Instr 47:1360
17. The present QA-MCP-PMT was developed within INTAS-94–4461 project, with Euro-Photon GmbH as coordinator; see also [8]
18. Zygo K3 Nipkov system (Syncotech GmbH)
19. Ultra View confocal system (EG&G, Wallac LSR)
20. Straub M, Hell SW (1998) Appl Phys Lett 73:1769
21. Delic J, Coppey J, Magdelenat H, Coppey-Moisan M (1991) Impossibility of Acridine Orange intercalation in nuclear DNA of the living cell. Exp Cell Res 194:147
22. Tsien RY, Waggoner A (1990) Fluorophores for confocal microscopy: photophysics and photochemistry. In: Pawley JB (ed) Handbook of biological confocal microscopy. Plenum Press, New York, p 169
23. Kemnitz K, Tamai N, Yamazaki I, Nakashima N, Yoshihara K (1986) Fluorescence decays and spectral properties of Rhodamine B in submono-, mono-, and multilayer systems. J Phys Chem 90:5094
24. Bereiter-Hahn J, Airas J, Blum S (1997) Supramolecular associations with the cytomatrix and their relevance in metabolic control: protein synthesis and glycolysis. Zoology 100:1

25. Bereiter-Hahn J (1976) Dimethylamino-styrylmethylpyridinium-iodine (DASPMI) as a fluorescent probe for mitochondria in situ. Biochim Biophys Acta
26. Bereiter-Hahn J (1990) Behavior of mitochondria in the living cell. Int Rev Cytol 122:1
27. Mewes HW, Rafael J (1981) The 2-(dimethylaminostyryl)-1-methylpyridinium cation as indicator for the mitochondrial membrane potential. FEBS Lett 131:7
28. Bereiter-Hahn J, Vöth M (1998) Do mitochondria contain zones with different membrane potential? Experimental Biology Online 3/12
29. Bereiter-Hahn J, Seipel KH, Vöth M, Ploem JS (1983) Fluorimetry of mitochondria in cells vitally stained with DASPMI or rhodamine 6G. Cell Biochem Function 1:147
30. Fromherz P, Ephardt H (1989) J Phys Chem 93:7717
31. Arden-Jacob J (1993) Neue langwellige Xanthen-Farbstoffe für Fluoreszenzsonden und Farbstofflaser, Dissertation, Universität Siegen (Drexhage KH)
32. (a) Eckert H-J, Bergmann A, Turbin EV, Strepetov AN, Kemnitz K, Picosecond (1999) Fluorescence lifetime imaging microscopy of individual chloroplasts in living cells; (b) Petrasek Z, Kemnitz K, Ostler R, Phillips D (1999) Second European Workshop on Picosecond Fluorescence Lifetime Imaging Microscopy as a new Tool for 3D Structure Determination of Macromolecules in Cells, July 1999, Rome
33. Bergmann A, Eichler H-J, Eckert H-J, Renger G (1998) Picosecond laser-fluorometer with simultaneous time and wavelength resolution for monitoring decay spectra of photo-inhibited Photosystem II particles at 277 K and 10 K. Photosyn Res 58:303

Part 5
Proteins and Their Interactions as Studied by Fluorescence

About the Prediction of Tryptophan Fluorescence Lifetimes and the Analysis of Fluorescence Changes in Multi-Tryptophan Proteins

A. SILLEN, Y. ENGELBORGHS

In the last five years we have studied the time resolved fluorescence of several proteins using multi-frequency phase fluorometry. Each protein has taught us new aspects of tryptophan fluorescence and this manuscript describes the accumulated experience. For a multi-tryptophan protein we consider the measured lifetimes to be amplitude-average lifetimes per family (short, medium, long). As a consequence the lifetimes of a multi-tryptophan protein can be calculated from the lifetimes of the individual tryptophan residues determined from the mutants. Additivity, defined in this way, indicates that the mutations have not altered the environment and the fluorescence properties of the remaining individual tryptophan residues, and that there is no energy exchange among their excited states. Non-additivity points to one or more of these complications. Interacting pairs can be identified by checking additivity pairwise. Alterations of protein fluorescence due to the effects of ligand binding or side chain modifications can be analyzed via the ratio of the quantum yield of the modified protein to that of the reference state. This ratio of quantum yields can be factorized in three factors. A first one describes the change in the apparent radiative rate constant reflecting a genuine change in the radiative properties or some static quenching. A second one describes the change in dynamic quenching and a third one identifies a change in the balance of the populations of the microstates and/or a change in local static quenching. In some cases the fluorescence rate constants of single tryptophan residues can be calculated from structural and spectral information. The radiative rate constant is largely determined by the electric field in the neighborhood, and is therefore linked to the wavelength of maximum emission. A pragmatic equation allows the calculation of k_r from this wavelength. In the absence of the well known side chain quenchers, the non-radiative rate constant seems to be determined by the distance between the CE3 atom of tryptophan and the carbonyl carbon of its peptide backbone. In this way a quantitative prediction of the lifetime was possible for a couple of proteins.

19.1
Interpreting Fluorescence Changes in Proteins

The analysis of the details of the fluorescence properties of multi-tryptophan proteins is complicated by the fact that most tryptophan residues in proteins display multi-exponential fluorescence decay [1]. The origin of this complex behavior is either dynamic, e.g., processes in the excited state or static, i.e., heterogeneity in the ground state. Dynamic processes in the excited state include energy transfer, electron transfer, dynamic quenching, dipolar relaxation processes, and other excited state reactions [2, 3]. Not all of them lead to multi-exponential behavior. Heterogeneity corresponds to the existence of different (local or micro-) conformations of the indole side chain itself due to different chi-angels [4, 5] while the overall fold of the protein does not alter. Explicit indications for dynamic processes in the excited state are the presence of negative amplitude fractions, and/or lifetimes that are emission wavelength dependent, and/or an abnormal initial anisotropy. Of course, many other experiments demonstrate the existence of dynamic properties of proteins on other time scales, e.g., H/D–exchange of the tryptophan, oxygen quenching etc.

The changes of the tryptophan fluorescence that occur in proteins upon a conformational change induced by a variety of phenomena, e.g., ligand binding, a change in pH, reduction of SS-bridges, oxidation of SH-groups, a mutation, etc., can be caused by a change in any of the mechanisms mentioned, including a reshuffling of the balance between the different local conformations. In order to analyze such a change in fluorescence we suggested calculation of the ratio of the quantum yields (Q/Q_0) of the modified protein relative to the reference situation (e.g., a mutation relative to the WT). We proposed division of this ratio into three factors [6, 7]. The first factor (f_{kr}) corresponds to the change in the average radiative rate constant $(\langle k_r \rangle)$, and reflects a change in environment and/or homogeneous static quenching. The second factor (f_{PR}) reflects population reshuffling and/or heterogeneous static quenching and the third factor (f_{DQ}) represents pure dynamic quenching. These factors are defined as follows [6, 8]:

$$\frac{Q}{Q_0} = \frac{\langle k_r \rangle \sum \alpha_i \tau_i}{\langle k_{r0} \rangle \sum \alpha_{0i} \tau_{0i}} = \frac{\langle k_r \rangle \sum \alpha_i \tau_{0i}}{\langle k_{r0} \rangle \sum \alpha_{0i} \tau_{0i}} \frac{\sum \alpha_i \tau_i}{\sum \alpha_i \tau_{0i}} = f_{kr} \times f_{PR} \times f_{DQ} \tag{19.1}$$

where τ is the fluorescence lifetime and α is the wavelength-independent amplitude fraction [8]. This splitting in factors is of course purely conceptual and none of the factors can be ascribed to uniquely defined phenomena, yet this analysis seems to give interesting insights in a number of situations, especially when one of the factors equals one. The future will show if it will become possible to find independent support for this analysis. In order to be able to calculate these factors different parameters have to be determined.

19.2
Determination of the Parameters

19.2.1
The Wavelength-Independent Amplitude Fraction α

This parameter can be calculated as described in Eq. (19.2) [8]:

$$\alpha_i = \frac{\int I_i(\lambda)/\tau_i \, d\lambda}{\sum_j \int I_i(\lambda)/\tau_j \, d\lambda} \tag{19.2}$$

where τ_i is the fluorescence lifetime of species i, and I_i is the fluorescence intensity of species i obtained from the decay-associated spectra (DAS). Decay-associated spectra are constructed by multiplying the intensity fraction with the intensity of the emission spectra at the respective wavelengths. A log-normal function [9] can be fitted to the associated intensities to obtain the decay-associated spectra at every wavelength:

$$I(\lambda) = I_m \exp\left[\frac{\ln 2}{\ln^2 \rho} \ln^2 \left(\frac{(a - 1/\lambda)}{(a - 1/\lambda_m)} \right) \right] \tag{19.3}$$

Here $I_m = I(\lambda_m)$ is the maximal fluorescence intensity; λ_m is the wavelength of the band maximum; $\rho = (1/\lambda_m - 1/\lambda_-)/(1/\lambda_+ - 1/\lambda_m)$ is the band asymmetry parameter; $a = 1/\lambda_m + (1/\lambda_+ - 1/\lambda_-)\rho/(\rho^2 - 1)$ is the function limiting point; $\lambda_+ = 10^7/(0.830 \times 10^7/\lambda_m + 7071)$ and $\lambda_- = 10^7/(1.1768 \times 10^7/\lambda_m - 7681)$ are the wavelength positions of half-maximal amplitudes. This allows integration of the fluorescence intensity of species i over all wavelengths.

19.2.2
The Radiative Rate Constant

The average radiative rate constant is normally determined by dividing the quantum yield by the wavelength-independent amplitude-average lifetime [8, 10]. This method, however, can give incorrect results if static quenching would occur. Unfortunately static quenching is difficult to identify (unless a decrease in the calculated k_r which is not accompanied by a change in λ_{max} is attributed to static quenching [7]). It has also been suggested that static quenching might be due to dynamic quenching of very high frequency that escapes detection by lifetime analysis [11].

Another approach is suggested by Strickler and Berg [12]. These authors relate the radiative rate constant with the absorption and the emission band of a fluorophore:

$$k_r = \left[2.88 \times 10^{-9} n^2 \frac{g_a}{g_b} \int \frac{\varepsilon(\bar{v})}{\bar{v}} d\bar{v} \right] \frac{\int I(\bar{v}) d\bar{v}}{\int \bar{v}^{-3} I(\bar{v}) d\bar{v}} \tag{19.4}$$

$$= 2.88 \times 10^{-9} n^2 \frac{g_a}{g_b} g(abs\, L_a) \cdot f(\lambda_{max})$$

where n is the refractive index of the medium, g_a and g_b are the degeneracies of the lower and upper state, respectively, ε is the molar absorption coefficient, $\bar{v}$ is the wavenumber, and I is the fluorescence intensity. The factor behind the square brackets $f(\lambda_{max})$ is responsible for the influence of the emission spectrum on the radiative rate constant. Using the log-normal function (Eq. 19.3) for the shape of the spectra, this factor is fully determined by its dependence on the wavelength of maximum-emission intensity. In this way we constructed a pragmatic equation that permits an easy calculation of this factor from the wavelength of maximum emission intensity [13]:

$$f(\lambda_{max}) = a\lambda_{max} + b\lambda_{max}^2 + c\lambda_{max}^3 + d \tag{19.5}$$

The following values for the parameters were obtained: $a = -0.122571$, $b = 1.34575 \times 10^{-4}$, $c = 2.52413 \times 10^{-8}$ and $d = 27.5089$.

Taking 1.5 for the refractive index of proteins [14] and adjusting the dimensions for k_r to ns^{-1}, Eq. (19.4) becomes for tryptophan:

$$k_r = 6.48 \times 10^{-9} g(abs\, L_a) f(\lambda_{max}) \tag{19.6}$$

where $g(abs\, L_a)$ is the factor related with the absorption band L_a of tryptophan as defined in Eq. (19.4). The L_a absorption band is very difficult to measure because it overlaps with the L_b absorption band [15]. Therefore an average factor can be calculated by fitting Eq. (19.6) to the $\langle k_r \rangle$ obtained from a series of different mutants or states of a protein, e.g., for one tryptophan in the protein NSCP a value of $(292 \pm 7) \times 10^4$ is obtained for $g(abs\, L_a)$ [13].

For a multi tryptophan protein the value of $\langle k_r \rangle$ is also sensitive to changes in the molar absorption coefficient of the individual tryptophan residues, since the individual radiative rate constants which are contributing are weighted by their absorption fraction.

19.3
Analysis of the Meaning of the Different Factors of Q/Q_0

19.3.1
Heterogeneous Static Quenching or Population Reshuffling (f_{PR})

The factor f_{PR} represents the change in amplitude fractions while keeping the lifetimes equal to those of the reference state. It therefore reflects a reequilibration of the balance of the microstates. Heterogeneous static quenching, however, also affects the amplitude fraction and therefore also this parameter. If both static quenching and an increase of the fluorescence due to population reshuffling occur, then f_{PR} is the minimum increase of the fluorescence intensity due to a change in the balance of the micro conformations. We have previously attempted to distinguish the two situations by assuming that a genuine change in k_r is accompanied by a change in λ_{max} [7].

The amplitude fraction can also change due to heterogeneous static quenching as well as due to the change in the microstates of tryptophan. However, if there is no change in the measured average radiative rate constant and the factor f_{PR} is different from 1, then this is an indication that there is possibly a population reshuffling.

19.3.1.1
Estimation of Microstates of Tryptophan Side Chains

There are two methods to determine the different possible conformations of tryptophan in a protein if the X-ray or NMR determined structure is known. The first is the calculation of an energy map. Minimum perturbation maps [16] are calculated by fixing the Trp side chain at a particular χ_1, χ_2 value and allowing the residues nearby Trp to relax conformationally so as to achieve an energy minimization, where all the other residues are fixed to their crystallographic or NMR coordinates. (The minimization procedure we used consisted of 100 steps by the steepest descent method, and 1000 steps with a tolerance of 0.1 by the Powell method.) This procedure is repeated in 10°-steps over the whole angular space of χ_1 and χ_2. Throughout this paper χ_1 is defined as the torsion angle around C_α-C_β by the bond connectivity N-C_α-C_β-C_γ and χ_2 is defined as the torsion angle around C_β-C_γ by the bond connectivity C_α-C_β-C_γ-$C_{\delta1}$. These maps display areas where Trp is in an energetically favored position. But these maps do not take into account that changes of conformations of Trp in proteins are dynamic processes that occur at room temperature rather than at 0 K. Also, the protein backbone is fixed in an energy map but small changes in the backbone conformation can allow other conformations to exist and thus are not revealed in energy maps. Therefore, instead of calculating a minimum perturbation map a different approach was taken to calculate the different conformations of tryptophan [13]. To determine the different micro conformations, the χ_1 and χ_2 angles of tryptophan were set at several different values in the computer model. From these starting positions a free molecular dynamics simulation was started to explore the conformational space. The relative populations of the different microstates relate to the amplitude fraction of the different lifetimes.

The molecular dynamics (MD) calculations were performed using an all-hydrogen-force field. The starting structure was first energy minimized. From this energy-minimized structure the χ_1 and χ_2 angles of Trp were set at the appropriate grid points. These structures were again energy minimized, warmed up to 300 K and then equilibrated. Consequently a normal MD simulation was done to monitor the χ_1 and χ_2 angles of Trp at every starting position. The starting positions which resulted in a distortion of the secondary structure due to an energetically and/or geometrical very bad starting structure were omitted. Different starting structures converged to the same energetically favored conformations. If there are enough starting values such an MD-map can explore enough conformational space to find the stable energy minima. The minimal potential energy U_0 [17] of the whole protein for the different micro conformations should be approximately the same. However, it is free energy and not potential energy that determines the relative populations. Therefore it was assumed that an estimate of the relative population of a certain conformation could be made by adding the number of initial positions from which the final conformation could be reached. Conformation A, for example, is reached 10 times out of 20 starting positions, hence conformation A is assumed to have a 50% chance of occurring and its relative population is therefore assumed to be also 50%.

19.3.2
The Factor of Pure Dynamic Quenching (f_{DQ})

The factor f_{DQ} represents that part of the dynamic quenching process that is due to the change in fluorescence lifetimes which occurs after the new balance between the microstates is established. It should be clear that a fluorescence decrease represented by f_{PR} could also have the nature of dynamic quenching, if it is due to the increase of the population of a pre-existing highly dynamically quenched state.

The most important dynamic quenchers in proteins are the disulfide bridges, the amino acid side chains of protonated histidine, cysteine, and possibly tyrosine, [2, 18] and the peptide bond [19]. All the other amino acid residues have only a minor ability of dynamic quenching [2]. Changes in dynamic quenching are thus due to a different accessibility of tryptophan for the mentioned groups or a different conformation of tryptophan relative to the peptide bond. In principal, if the structure of the protein and the conformations of tryptophan are known and there are no dynamic processes in the excited state, it should be possible to calculate the fluorescence lifetime of one conformation of tryptophan. If all the quenching constants of the amino acids are known [2] then with long molecular dynamics simulations it should be possible to obtain the collision frequency of tryptophan with its neighboring amino acid residues and calculate with this the non-radiative rate constant of dynamic quenching of the amino acids (if enough calculation time is available). The unknown is the non-radiative rate constant due to quenching of the peptide bond. The quenching mechanism of the peptide bond is probably electron transfer to the carbonyl carbon of the peptide bond from the CE3 atom of the indole group of tryptophan [13]. Thus the quenching can be described by the following equation derived for electron transfer by Marcus and Sutin [20]:

$$k_{ET}(R) = k_0 \exp\left(-\beta\left(R - R_0\right)\right) \tag{19.7}$$

where $k_{ET}(R)$ is the rate for electron transfer, k_0 is the rate of electron transfer at the van der Waals contact distance ($R_0 = 0.3$ nm), β is the range parameter which is depended on the medium, and R is the distance between the donor and acceptor of electron transfer. Thus if the distance of the carbonyl carbon of the peptide bond to the CE3 atom of tryptophan is known and the constants k_0 and β are known, the non-radiative rate constant of dynamic quenching by the peptide bond can be calculated. These two constants were determined for the proteins NSCP and p21 and yielded $k_0 = 25 \pm 3$ ns^{-1} and $\beta = 19 \pm 1$ nm^{-1}[13]. In those cases where none of the known quenching side chains is present, the knowledge of the non-radiative rate constant of dynamic quenching by the peptide bond, and of the radiative lifetime as calculated using Eq. (19.6), is enough to calculate the overall lifetime.

19.4
Examples

DsbA is a monomeric, periplasmic 21.1-kDa protein (189 aa) that is required for efficient disulfide bond formation in secretory proteins in the bacterial periplasm [21, 22]. The enzyme contains a single, catalytic disulfide with the active-site sequence C30-P31-H32-C33. The X-ray structure of oxidized DsbA [23] as well as the X-ray [24] and NMR-structure [25] of reduced DsbA has revealed that the enzyme possesses a thioredoxin-like domain (residues 1–62 and

139–189), a motif found in all known structures of disulfide oxidoreductases [26]. The sequence of the thioredoxin-like domain of DsbA is, however, only 10% identical with *E. coli* thioredoxin. DsbA possesses a second domain (residues 63–138) of unknown function, which is inserted into the thioredoxin motif and exclusively consists of α-helices. DsbA contains two tryptophans W76 and W126, which are not contained in the thioredoxin domain and are located in the α-helical domain. W76 is buried and about 1.2 nm separated from the disulfide, whereas W126 is even further away from the disulfide bridge (about 2 nm) and partially solvent-accessible.

The fluorescence properties of W76 were studied by making the mutant W126F. The fluorescence of W76 is strongly quenched upon oxidation of the protein [27]. This oxidation linked quenching is much less pronounced when Phe26, which makes the contact between W76 and the SS-bridge, is replaced by Leu. Analysis of the quenching factors in Q/Q_0 indicates that in the case of F26L the quenching effect upon oxidation is purely due to f_{pr} which indicates that it is due to a conformational effect [6].

The fluorescence properties of W126 were also investigated by replacing W76 by a phenylalanine, but also by replacing the possible quenchers Q74 and N127 by alanine, yielding the following set of variants: W76F, W76F/Q74A, W76F/N127A, and W76F/Q74 A/N127A29. Compared to Trp in solution, which has a quantum yield of 0.14, the fluorescence of W126 is highly quenched in both the oxidized ($Q = 0.013$) and reduced state ($Q = 0.012$) of DsbA. This seems to be largely due to an increase of dynamic quenching because the apparent radiative rate constant is the same as the radiative rate constant of tryptophan (0.053 ns^{-1} [27]), whereas the nonradiative rate constant is 3.8 ns^{-1} compared to 0.33 ns^{-1} for Trp in solution. The only two candidates for dynamic quenching within collisional distances were the amide groups of Q74 and N127. Replacing Q74 and N127 by alanine indeed reduced the nonradiative rate constant to lower values. The question can then be raised: how does N127 and Q74 quench the fluorescence of W126 and why is the remaining non-radiative rate constant still quite high (1.12 ns^{-1})? A detailed quenching analysis can reveal the origin of the increase in quantum yield. Upon removal of the amide of Q74 or N127 there is none or only a small decrease in quantum yield due to static quenching (f_{kr}) and also a decrease in the quantum yield due to dynamic quenching [7]. The only reason why the total quantum yield increases upon removal of the amide of Q74 or N127 is due to the factor f_{PR} which represents a fluorescence increase due to population reshuffling. This indicates that amide groups are not direct quenchers of tryptophan fluorescence. The cause of the strong quenching of W126 in WT is due to the high population of the shortest lifetime. Replacement of Q74 or N127 allows for micro conformations with higher lifetimes to become more populated and thus increases the fluorescence.

19.5
Comparison of a System with Multiple Fluorophores and Multiple Lifetimes with a System Containing One Fluorophore with Multiple Lifetimes

When comparing the lifetimes of a multi tryptophan protein with its single tryptophan-containing mutants, the situation can become very complicated. In some cases the single tryptophan-containing mutants display the same families of lifetimes as the wild type. This is the case for colicin [28] and Barnase [29–31]. The question that can be asked is whether additivity within the families of lifetimes can be used to explain the lifetime data of the WT protein. In this context we found it simplifying to use an average lifetime [28]. Salient features are more apparent in the average lifetime than in the full details of all the lifetimes and fractions. Here there are m fluorophores $(1,\ldots,j,\ldots,m)$ that all have n lifetimes $(1,\ldots,i,\ldots,n)$. The i-th observed lifetime is an average of the i-th lifetime of each fluorophore, appropriately weighted (for a full deviation of the equations see [8]).

The amplitude fraction in these averages is weighted by the molar absorption coefficient (ε) divided by the average intrinsic lifetime $\langle \tau_r \rangle$ (this is equal to the inverse of the average radiative rate constant). The equation of the amplitude average lifetime is

$$\langle \tau_i \rangle_a = \frac{\sum\limits_{j}^{m} \dfrac{\alpha_{ij}\varepsilon_j}{\langle \tau_r \rangle_j}}{\sum\limits_{j}^{m} \dfrac{\alpha_{ij}\varepsilon_j}{\langle \tau_r \rangle_j}} \tag{19.8}$$

The equation of the amplitude is:

$$\langle \alpha_i \rangle = \frac{\sum\limits_{k=1}^{k=m} \dfrac{\alpha_{ik}\varepsilon_k}{\langle \tau_r \rangle_k}}{\sum\limits_{i=1}^{i=n} \sum\limits_{k=1}^{k=m} \dfrac{\alpha_{ik}\varepsilon_k}{\langle \tau_r \rangle_k}} \tag{19.9}$$

19.5.1
Examples

By calculating these average lifetimes obtained from individual tryptophans and comparing with those of wild type we could clearly demonstrate the presence of energy transfer between a particular tryptophan pair. Barnase is a small monomeric enzyme that has been extensively used as a model for studying the principles that rule protein stability and protein folding [32, 33]. In barnase the tryptophans are found at positions 35, 71, and 94. The interaction of His18 with Trp94 causes a pH-dependent quenching of its fluorescence, following the ionization curve of the histidine residue ($pK_a = 7.75$) [29]. An excited-state lifetime study using multifrequency phase fluorometry on these mutants revealed the presence of energy transfer between Trp71 and Trp94 [30]. The fluorescence lifetimes of the proteins could be resolved and, in most cases, spectra and lifetimes could be attributed to single tryptophan residues using the method of subtraction. Trp35 displays a pH-independent lifetime of 4.3 to 4.8 ns. Trp71 and Trp94 behave as an energy transfer couple. In the absence of energy transfer, the lifetime of Trp71, as determined with the mutant W94F, was 4.7 ns and remains unaffected by pH. The mutant W71Y was used to resolve the lifetime of Trp94. Values of 0.82 ns and 1.57 ns were found at pH 5.8 and pH 8.9, respectively. Data on single tryptophan residues were obtained later [31] and can be used to check additivity or interactions in the WT. The best parameter to be used for this purpose in our experience turned out to be the amplitude average lifetime. This is again confirmed here. The data clearly show that the combination of the lifetimes of all three tryptophans do not reproduce the average lifetime of the WT. However, the combination where the single W-mutant W35F is combined with the lifetime data of W35 gives a very good approximation, indicating that energy transfer occurs between W71 and W94. The average lifetime of mutant W94F can neatly be obtained by combining the individual data of W71 and W35, indicating additivity and no interactions between W35 and W71. The data on the single tryptophan-containing mutants revealed that the method of subtraction has its limitations: only lifetimes that disappear upon the removal of a tryptophan residue will be attributed to this residue [31].

19.6
Conclusion

We propose a number of steps to help the analysis of the fluorescence properties of multi tryptophan proteins. Unfortunately the interpretation is not always unambiguous but when applied to the proteins we studied, very reasonable results were obtained. Future applications to a wider range of proteins will show if this analysis can be supported by independent evidence.

References

1. Ross JA, Schmid CJ, Brand L (1981) Time-resolved fluorescence of the two tryptophans in horse liver alcohol dehydrogenase. Biochemistry 20:4369–4377
2. Chen Y, Barkley MD (1998). Toward understanding tryprophan fluorescence in proteins. Biochemistry 37:9976–9982
3. Hudson BS (1999) An ionization/recombination mechanism for complexity of the fluorescence of tryptophan in proteins. Acc Chem Res 32:297–300
4. Dahms TES, Willis KJ, Szabo AG (1995) Conformational heterogeneity of tryptophan in a protein crystal. J Am Chem Soc 117:2321–2326
5. Mérola F, Rigler R, Holmgren A, Brochon J-C (1989) Picosecond tryptophan fluorescence of thioredoxin: evidence for discrete species in slow exchange. Biochemistry 28:3383–3398
6. Sillen A, Hennecke J, Roethlisberger D, Glockshuber R, Engelborghs Y (1999) Fluorescence quenching in the DsbA protein from *E. coli*: complete picture of the excited-state energy pathway and evidence for the reshuffling dynamics of the microstates of tryptophan. Proteins, Struct Funct Genet 37:253–263

7. Gastmans M, Volckaert G, Engelborghs Y (1999) Tryptophan microstate reshuffling upon the binding of cyclosporin A to human cyclophilin A. Proteins, Struct Funct Genet 35:464–474

8. Sillen A, Engelborghs Y (1998) The correct use of average fluorescence parameters. Photochem Photobiol 67:475–486

9. Burstein EA, Emelyanenko VI (1996) Log-normal description of fluorescence spectra of organic fluorophores. Photochem Photobiol 64:316–320

10. Szabo AG, Faerman C (1992) Dilemma of correlating fluorescence quantum yield and intensity decay times in single tryptophan mutant proteins. SPIE 1640:70–80

11. Webber SE (1997) The role of time-dependent measurements in elucidating static versus dynamic quenching processes. Photochem Photobiol 65:33–38

12. Strickler SJ, Berg RA (1962) Relationship between absorption intensity and fluorescence lifetime of molecules. J Chem Phys 37:814–822

13. Sillen A, Diaz F, Engelborghs Y (2000) A step towards the prediction of the fluorescence lifetimes of tryptophan residues in proteins based on structural and spectral data. Prot Sci 9:158–169

14. Desie G, Boens N, De Schryver FC (1986) Study of the time resolved tryptophan fluorescence of crystalline a-chymotrypsin. Biochemistry 25:8301–8308

15. Valeur B, Weber G (1977) Resolution of the fluorescence excitation spectrum of indole into the ^1L_a and ^1L_b excitation bands. Photochem Photobiol 25:441–444

16. Haydock C (1993) Protein side chain rotational isomerisation: a minimum perturbation mapping study. J Chem Phys 98:8199–8214

17. Schlitter J, Engels M, Krüger P (1994) Targetted molecular dynamics: a new approach for searching pathways of conformational transitions. J Mol Graphics 12:84–89

18. Vos R, Engelborghs Y (1994) A fluorescence study of tryptophan-histidine interaction in the peptide anantin and in solution. Photochem Photobiol 60:24–32

19. Chen Y, Liu B, Yu H-T, Barkley MD (1996) The peptide bond quenches indole fluorescence. J Am Chem Soc 11:9271–9278

20. Marcus RA, Sutin N (1985) Electron transfer in chemistry and biology. Biochim Biophys Acta 811:265–322

21. Bardwell JCA, McGovern K, Beckwith J (1991) Identification of a protein required for disulfide bond formation in vivo. Cell 67:581–589

22. Kamitani S, Akiyama Y, Ito K (1992) Identification of an *Escherichia coli* gene required for the formation of correctly folded alkaline phosphatase, a periplasmic enzyme. EMBO J 11:57–62

23. Guddat LW, Bardwell JC, Zander T, Martin JL (1997) The uncharged surface features surrounding the active site of *Escherichia coli* DsbA are conserved and are implicated in peptide binding. Protein Sci 6, 1148–1156

24. Guddat LW, Bardwell J, Martin JL (1998)Crystal structures of reduced and oxidized DsbA: investigation of domain motion and thiolate stabilization. Structure 6:757–767

25. Schirra HJ, Renner C, Czisch M, Huber-Wunderlich M, Holak TA, Glockshuber R (1998) Structure of reduced DsbA from *Escherichia coli* in solution. Biochemistry 37:6263–6276

26. Martin JL (1995) Thioredoxin – a fold for all reasons. Structure 3:245–250

27. Hennecke J, Sillen A, Huber-Wunderlich M, Engelborghs Y, Glockshuber R (1997) Quenching of tryptophan fluorescence by the active-site disulfide bridge in the DsbA Protein from *Escherichia coli*. Biochemistry 36:6391–6400

28. Vos R, Izard J, Baty D, Engelborghs Y (1995) Fluorescence study of the three tryptophan residues of the pore-forming domain of colicin A using multifrequency phase fluorometry. Biochemistry 34:1734–1743

29. Loewenthal R, Sancho J, Fersht AR (1991) Fluorescence spectrum of barnase: contribution of three tryptophan residues and a histidine-related pH dependence. Biochemistry 30:6775–6779

30. Willaert K, Loewenthal R, Sancho J, Froeyen M, Fersht AR (1992) Determination of the excited state lifetimes of the tryptophan residues in barnase, via multifrequency phase fluorometry of tryptophan mutants. Biochemistry 31:711–716

31. De Beuckeleer K, Volckaert G, Engelborghs Y (1999) Time resolved fluorescence and phosphorescence properties of the individual tryptophan residues of Barnase: evidence for protein-protein interactions. Proteins 36:42–53

32. Fersht AR (1993) Protein folding and stability: the pathway of folding of barnase. FEBS Lett 325:5–16

33. Corrales JF, Fersht AR (1995) The folding of GroEl-bound barnase as a model for chaperonin-mediated protein folding. Proc Natl Acad Sci USA 92:5326–5330

Application of Time-Resolved Fluorescence Spectroscopy to Studies of DNA-Protein Interactions and RNA Folding

D. P. MILLAR

20.1
Introduction

In the life sciences, fluorescence spectroscopy has been widely used as a tool to investigate the structure and dynamics of proteins [1, 2]. The widespread use of fluorescence spectroscopic techniques stems from the occurrence of the naturally fluorescent amino acids tryptophan and tyrosine in many proteins. A number of proteins have just a single tryptophan or tyrosine residue, considerably simplifying the interpretation of fluorescence data. Moreover, through the use of site-directed mutagenesis procedures, it is relatively straightforward to engineer proteins for fluorescence studies by incorporating single tryptophan residues at desired positions or by removing all but one tryptophan residue from a protein of interest.

In contrast, the application of fluorescence spectroscopy to nucleic acid systems has traditionally been hampered by the lack of suitable intrinsic fluorophores. DNA has no measurable intrinsic fluorescence in solution at room temperature, necessitating the use of extrinsic fluorophores as probes. Early efforts to study the internal motion dynamics of DNA by time-dependent fluorescence depolarization employed intercalating dyes such as ethidium bromide as probes [3, 4]. More recently, with the advent of efficient and rapid solid phase methods for oligonucleotide synthesis, and the ready availability of reagents to incorporate fluorophores during synthesis or for post-synthetic modification, it is a simple matter to label DNA or RNA oligomers at specific sites. Fluorescent probes can be attached to either the 3′ or 5′ ends of an oligonucleotide or conjugated to a modified base that can be placed anywhere in the sequence [5–7]. In addition, fluorescent base analogs can be substituted for normal bases in either DNA or RNA [8]. This flexibility in the choice of labeling scheme, together with a range of fluorescence spectroscopic techniques that can be employed, including time-resolved anisotropy and energy transfer, has greatly expanded the scope of fluorescence-based studies of nucleic acid systems.

These approaches have been used successfully to investigate rapid base motions in DNA [8], to elucidate the architecture of complex branched structures such as the Holliday junction [9, 10] and the hammerhead ribozyme [11, 12], and to monitor protein-induced conformational changes in DNA [13, 14]. Some of these applications have been reviewed previously [15]. This chapter will focus on recent studies from the author's laboratory in which time-resolved fluorescence spectroscopic techniques have been used to resolve different binding modes of a DNA polymerase interacting with a DNA substrate and to monitor a tertiary structure folding transition necessary for the catalytic activity of the hairpin ribozyme.

20.2
DNA Polymerase Proofreading

DNA polymerases are multifunctional enzymes involved in DNA replication and repair. In addition to the 5'-3' polymerase activity used to extend a nascent DNA chain, most DNA polymerases possess a separate 3'-5' exonuclease activity. The latter provides a proofreading function by selectively removing mismatched bases from the 3' end of the primer strand. These mismatches are incorporated into the DNA as a result of polymerase errors and must be removed to prevent mutations being introduced into the genome. The large proteolytic fragment of *Eschericia coli* DNA polymerase I, termed the Klenow fragment, is a relatively small and well characterized DNA polymerase that serves as a model system for more complex DNA replication enzymes [16]. This fragment retains the polymerase and 3'-5' exonuclease activities required for high fidelity synthesis of DNA. Crystallographic and mutational studies of Klenow fragment have revealed that the 3'-5' exonuclease active site is separated from the polymerase active center by 25–30 Å [16], raising the question of how the two sites work together during DNA synthesis. Although crystal structures exist for DNA-protein complexes with DNA bound at the 3'-5' exonuclease site [17], a solution method is needed to investigate the mechanisms that govern the competition between the two sites for binding the DNA substrate and that control movement of DNA within the enzyme.

20.2.1
Detecting the Two DNA Binding Modes of Klenow Fragment

The critical event in proofreading is the decision to shuttle a DNA substrate from the polymerase site (pol site) to the 3'-5' exonuclease site (exo site). To elucidate the molecular events governing this transition, it is advantageous to have a technique that can measure the distribution of a DNA substrate between the two active sites (Fig. 20.1). This is a challenging measurement because the DNA substrate shuttles rapidly between the two sites, making it difficult to resolve the two complexes on most experimental time scales. Time-resolved fluorescence spectroscopy has proven to be ideally suited to this task. By labeling the DNA substrate with a dansyl probe, it is possible to resolve separate populations bound to the pol and exo sites of the enzyme by fluorescence anisotropy decay measurements [18]. The short time scale of the anisotropy decay ensures that the two populations do not interconvert during the measurement. The key to the measurement is the proper placement of the fluorescent probe on the DNA substrate. Internal labeling of a base within the primer strand provides the best way of distinguishing the different DNA footprints associated with each mode

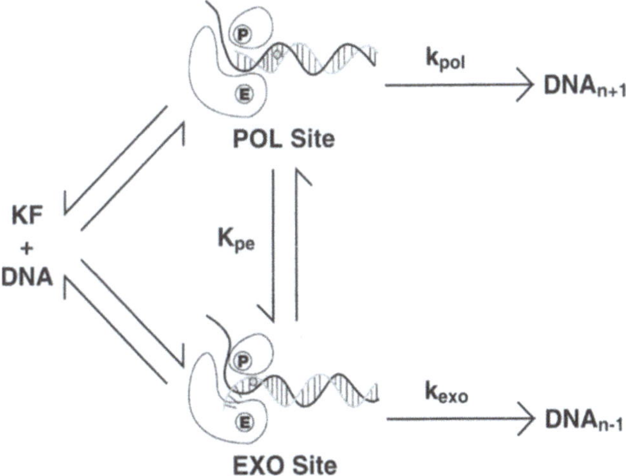

Fig. 20.1. Two modes of interaction between Klenow fragment (KF) and a DNA substrate. The primer 3′ terminus of the DNA substrate can occupy either the polymerase site (designated P) or the 3′-5′ exonuclease site (designated E). These two modes of binding correspond to polymerization and exonucleolysis reactions, respectively, which are inhibited during measurements of DNA binding and partitioning. The *circle located within the DNA helix* represents a dansyl probe covalently attached to a modified uridine base within the primer strand. The dansyl probe is exposed to solution when the primer terminus is bound at the polymerase site. Movement of DNA to the exonuclease site draws the probe into the interior of the protein. The primer-template duplex is partially single-stranded when bound at the exonuclease site. Reproduced from [18] with permission

of binding to the polymerase. In fact, the most sensitive position is seven bases upstream from the primer 3′ terminus [18, 19].

The time-resolved fluorescence anisotropy decay profile for a dansyl-labeled DNA substrate (labeled at position seven) bound to Klenow fragment exhibits an unusual "dip and rise" shape indicative of two different environments for the dansyl probe. Examples of such decays are shown in Fig. 20.2. The two environments can be assigned to DNA primer-templates bound at the pol site or the exo site (Fig. 20.1), on the basis of experiments with mismatched DNA substrates and with chemically-modified DNA substrates that bind tightly to the pol site [19]. The dansyl probe is largely exposed to solution when the primer 3′ terminus is bound at the pol site, resulting in a relatively short fluorescence lifetime. Shuttling of the DNA substrate to the exo site draws the dansyl probe into the interior of the enzyme, significantly lengthening the fluorescence lifetime. In addition, steric contacts between the probe and the enzyme severely restrict the free rotation of the probe. Accordingly, binding of DNA to the pol or exo sites can be distinguished by differences in both the fluorescence lifetime and anisotropy decay behavior of the dansyl probe.

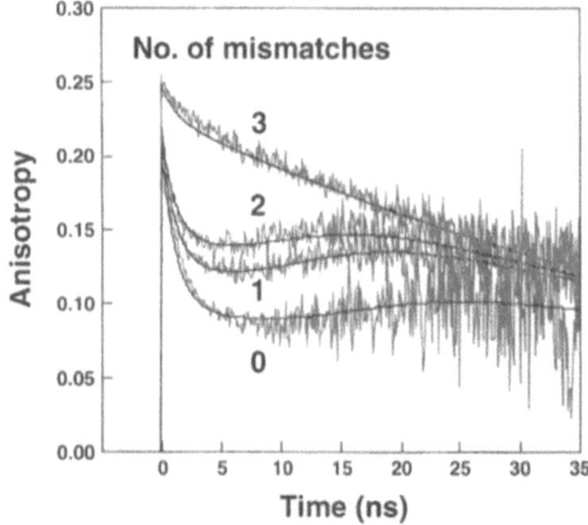

Fig. 20.2. Fluorescence anisotropy decay profiles for complexes of Klenow fragment and dansyl-labeled DNA substrates. DNA substrates contained 0–3 terminal mismatches, as indicated. Dansyl fluorophores were excited at 318 nm with the frequency-doubled output from a synchronously pumped DCM dye laser and the emission was monitored at 535 nm. The *solid lines* are from a global fit to a two-state model of exposed and buried probes (Eq. 20.1), corresponding to DNA substrates bound either to the polymerase site or 3′-5′ exonuclease site, respectively. Reproduced from [21] with permission

20.2.2
Time-Resolved Anisotropy for a Heterogeneous Mixture of Probe Environments

Owing to the differences in fluorescence lifetime and rotational behavior, the time-dependent anisotropy observed for a heterogeneous mixture of different probe environments can be quite complex. For labeled DNA substrates bound to Klenow fragment, the anisotropy can be represented in terms of contributions from two states of the dansyl probe, either exposed or buried (Eq. 20.1):

$$r(t) = f_e(t) \cdot r_e(t) + f_b(t) \cdot r_b(t) \tag{20.1}$$

where $r_e(t)$ and $r_b(t)$ are the anisotropy decay functions for the exposed and buried probes, respectively. These decay functions are quite different for the two populations because of the differences in probe mobility. Moreover, the contribution of each population to the observed anisotropy evolves over time because of the difference in fluorescence lifetimes. These contributions are expressed by the weighting factors $f_e(t)$ and $f_b(t)$, which in the case of the exposed probes can be represented as follows (Eq. 20.2):

$$f_e(t) = \frac{x_e \exp(-t/\tau_e)}{x_e \exp(-t/\tau_e) + x_b \exp(-t/\tau_b)} \tag{20.2}$$

where x_e and x_b are the equilibrium mole fractions of the exposed and buried dansyl probes, and τ_e and τ_b are the corresponding fluorescence lifetimes. An analogous expression applies for the contribution of the buried probes, $f_b(t)$. The expressions for $f_e(t)$ and $f_b(t)$ can be readily generalized to include more than one fluorescence lifetime for each probe population [18].

As a consequence of the different fluorescence lifetimes and rotational mobilities of the exposed and buried probes, the time-dependent fluorescence anisotropy described by Eq. (20.1) can exhibit a distinctive "dip and rise" pattern consisting of an initial rapid decline, a rising portion at intermediate times, followed by a slow decay at longer times. The precise shape of the anisotropy decay is strongly dependent upon the actual fractions of exposed and buried probes. The anisotropy decays recorded for a wide variety of matched and mismatched DNA substrates bound to Klenow fragment can be uniquely analyzed in terms of the two-state model (Eq. 20.1), using a common set of lifetime and rotational parameters to describe each probe population [18]. Examples of these global fits for a few representative DNA sequences are shown in Fig. 20.2. During the fitting, the fractions of exposed and buried probes are optimized for each data set. These correspond to the relative fractions of primer termini bound at either the pol site or the exo site, respectively. This information immediately yields the equilibrium constant K_{pe}, describing the partitioning of the DNA primer-template between the two active sites of the polymerase (Fig. 20.1).

20.2.3
Partitioning of Mismatched DNA Substrates Between pol and exo Sites

This method of examining partitioning of a DNA substrate between the pol and exo sites has the advantage that the reporter group is distant from the primer 3′ terminus. Thus, sequence changes can be introduced at the primer terminus without directly affecting the dansyl probe. Changes in fluorescence behavior can then be interpreted as a shift in the equilibrium distribution of the DNA substrate between pol and exo sites, rather than a direct effect on the dansyl probe itself. As a result, the time-resolved anisotropy method is ideal for characterizing the effects of mismatches, frameshifts, and other mutagenic phenomena that exert their effect at the 3′ end of the DNA primer. This approach has been used to characterize the interaction of Klenow fragment with a variety of different mispaired DNA substrates in order to assess the contribution of mismatched base pairs to the energetics of DNA proofreading [18]. Whereas matched sequences bind predominantly at the pol site of the enzyme, addition of an increasing number of mismatches causes the DNA to partition in favor of the exo site (Fig. 20.3). These observations support the idea that the occupancy of the exo site is correlated with the melting capacity of the DNA terminus, which increases with mismatching in the duplex sequence. This, in turn, is consistent with biochemical data indicating that localized melting and unwinding of the primer 3′ terminus is required for exonuclease activity on a duplex DNA substrate [20]. Crystallographic data also indicate that a duplex DNA substrate is bound to the exo site in a partially single-stranded form [17].

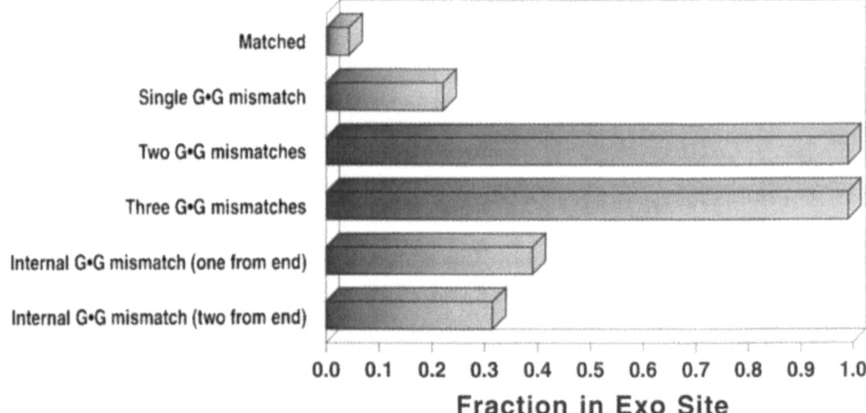

Fig. 20.3. Partitioning of DNA substrates between the polymerase and 3'-5' exonuclease sites of Klenow fragment. The fraction of bound DNA occupying the exo site is shown for DNA substrates containing various types and numbers of mismatched base pairs and for the corresponding perfectly matched sequence

20.2.4
Energetic Contributions of Protein Side Chains to DNA Partitioning

The ability of the polymerase to melt a region of the primer-template duplex raises interesting questions about the energetics of the DNA-protein interactions at the exo site. In particular, the free energy gained from the interactions of the single-stranded primer with the exo site must be sufficient to compensate for the loss of favorable base-base interactions in the duplex form of the substrate. To dissect the energetics of these interactions, the time-resolved fluorescence anisotropy technique was used to analyze the effects of protein mutations on the partitioning of DNA substrates between the pol and exo sites [21]. Mutations (alanine replacements) were introduced into amino acid side chains that are seen by X-ray crystallography to be in close proximity to the 3' terminus of a DNA substrate bound to the exo site. The residues examined are those involved in direct contacts with the terminal bases of the primer strand, that serve as ligands to two divalent metal ions at the active site, or that interact with the sugar-phosphate backbone upstream from the point of hydrolysis (Fig. 20.4).

Each mutation was observed to have a different effect on the partitioning of a DNA substrate between the two active sites of the polymerase, reflecting the loss of the binding energy contributed by the wild type side chain at the 3'-5' exonuclease site. The partitioning constants obtained for the mutant and wild type enzymes were combined to calculate $\Delta\Delta G$ values describing the contributions of individual amino acid side chains to the binding of a DNA substrate at the exo site. Some typical values are shown in Fig. 20.4. Leu[361] and Phe[473] are the most important for DNA binding, each contributing $3-4$ kJ mol^{-1}, consistent with structural data showing that these residues make intimate contacts with the

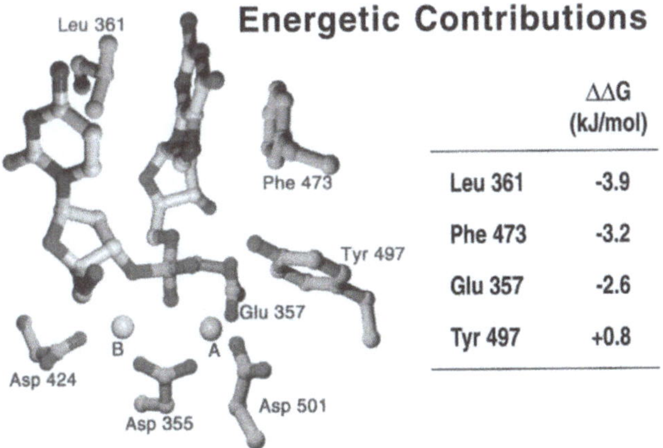

Fig. 20.4. Exonuclease active site of Klenow fragment and energetic contributions of individual amino acid residues to DNA substrate binding. A bound dinucleotide is shown together with the surrounding amino acid side chains (labeled). The two divalent metal ions at the active site are represented by spheres marked A and B. The contributions of individual amino acid side chains to the free energy of partitioning DNA into the exonuclease site are shown as $\Delta\Delta G$ values. These were calculated from measurements of the equilibrium distribution of a dansyl-labeled DNA substrate between the polymerase and 3′-5′ exonuclease site for mutant Klenow fragment enzymes containing single alanine replacements for the residues of interest. Adapted from [21] with permission

penultimate and terminal base (Fig. 20.4) [17]. Another interesting observation is that the contribution of Tyr[497] is unfavorable (Fig. 20.4), suggesting that the side chain of tyrosine actually interferes with DNA binding. This residue evidently plays a more subtle role in the exonuclease reaction than simply binding the DNA substrate, possibly acting to strain the substrate towards the transition state. Consistent with this, mutation of Tyr[497] significantly reduces the exonuclease activity of the enzyme [22].

These results demonstrate that the time-resolved anisotropy technique can be used to evaluate the energetic contributions associated with structurally-defined interactions. Moreover, information is obtained on the energetics of DNA-protein interactions at just one of the active sites of the polymerase, in contrast to standard binding measurements which simply yield an average of both sites.

20.3
Tertiary Structure Formation in the Hairpin Ribozyme

In addition to its role as an information carrier in gene expression, RNA can carry out a broad range of other functions, from chemical catalysis to the regulation of protein translation. To achieve these biological functions, RNA molecules must fold into specific three-dimensional structures that create active sites for catalysis or recognition motifs for the binding of particular proteins. Consequently, it is important to understand the nature of the molecular interactions that direct the folding of an RNA molecule and determine the details of the biologically active tertiary structure thus formed.

A useful model system for studying RNA folding is the hairpin ribozyme, a small catalytic RNA molecule that catalyzes cleavage and ligation reactions required for replication of a family of satellite RNAs associated with plant viruses [23]. A minimal hairpin ribozyme consists of two helix-loop-helix segments (denoted A and B) connected by a two-way helical junction. Previous evidence has shown that catalytic activity requires a sharp bend about the hinge of the junction, enabling a specific interdomain docking interaction [23]. The ribozyme-substrate complex can also form an alternative conformer in which the two helix-loop-helix domains adopt an extended structure. This extended conformer is a folding intermediate in the reaction pathway, preceding docking and cleavage [24]. No high resolution structural data are yet available on the hairpin ribozyme in either the extended or docked conformations. Moreover, little is known about the folding free energy landscape for this small catalytic RNA molecule.

Recent advances in X-ray crystallography and NMR spectroscopy have revealed atomic details of RNA structure [25], but are largely unable to report conformational changes essential to acquire biological function. Fluorescence spectroscopy is a very promising technique for the study of RNA folding because it can provide thermodynamic, kinetic, and structural information over a wide range of solution conditions. Fluorescence resonance energy transfer (FRET), a spectroscopic phenomenon involving long-range nonradiative coupling between two fluorophores, provides a powerful tool for the structural analysis of nucleic acids [7, 15, 26]. The technique can map distances over a range of 10–100 Å and produce moderate resolution structures. The most powerful feature of FRET, however, is the ability to detect the simultaneous existence of two or more conformations of a particular RNA molecule and to quantify free energy differences [27]. This capability, based on nanosecond time-resolved measurements of FRET (tr-FRET), provides a unique window into the rates and energetics of RNA conformational transitions.

20.3.1
tr-FRET Analysis of the Hairpin Ribozyme

Docked and extended tertiary structure conformers of the hairpin ribozyme are resolved by examining the effect of a tetramethylrhodamine acceptor on the fluorescence decay behavior of a fluorescein donor (Fig. 20.5). To quantify the equilibrium position between the two conformers and obtain long-range distance information, the nanosecond decay profile of the fluorescein donor is fitted with an expression incorporating one or more donor-acceptor (D-A) distance distributions:

$$I_{DA}(t) = \sum_k f_k \int_{R_{min}}^{R_{max}} \sum_i \alpha_i \exp\{(-t/\tau_i)[1+(R_0/R)^6]\} P_k(R)\, dR \qquad (20.3)$$

where the first summation refers to the number of distinct D-A species, each with fractional concentration f_k and distance distribution $P_k(R)$. In Eq. (20.3),

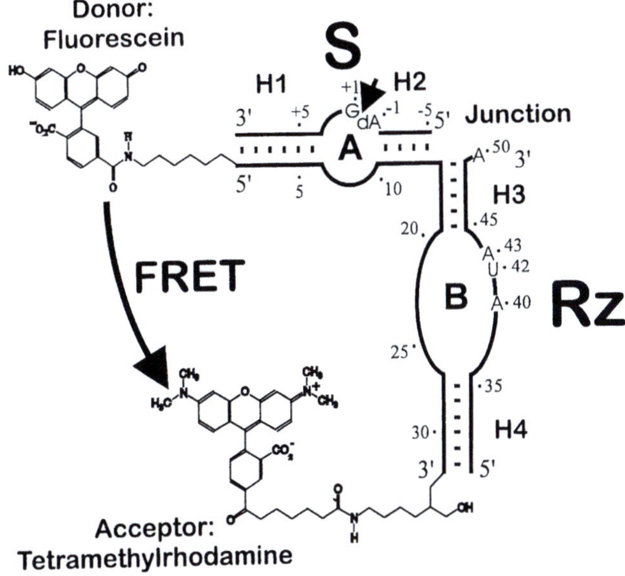

Fig. 20.5. Schematic of the doubly-labeled ribozyme-substrate complex used for tr-FRET measurements. The two-strand hairpin ribozyme (Rz) binds the 14 nucleotide substrate (S; the potential cleavage site is indicated by an *arrow*) to form domain A, containing helices H1 and H2 and symmetric internal loop A. This substrate-binding domain is joined by a flexible junction to domain B, containing helices H3 and H4 and asymmetric internal loop B. This design is referred to as a two-way junction (2WJ). Fluorescein and tetramethylrhodamine are coupled as a donor-acceptor pair to opposite ends of the complex. Cleavage is blocked by a 2′-deoxy modification of nucleotide A_{-1}. Reproduced from [27] with permission

R_{min} and R_{max} are the minimum and maximum D-A distances and R_0 is the critical transfer distance at which the energy transfer occurs with 50% efficiency. The second summation in Eq. (20.3) refers to the number of decay channels in the isolated donor, each with lifetime τ_i and amplitude α_i (determined from a sum of exponentials fit to a suitable donor-only molecule). Each distance distribution in Eq. (20.3) is represented by a weighted Gaussian distribution of the radial distance R:

$$P(R) = 4\pi R^2 c \exp\left[-a(R-b)^2\right] \tag{20.4}$$

Fitting the donor decay to Eqs. (20.3) and (20.4) yields the equilibrium distribution of conformers as well as the mean D-A distance and shape of each distribution.

Application of Eqs. (20.3) and (20.4) to the ribozyme-substrate complex reveals that two Gaussian distance distributions are required for best fit (Fig. 20.6). Under standard solution conditions (50 mmol/l tris-HCl, 12 mmol/l

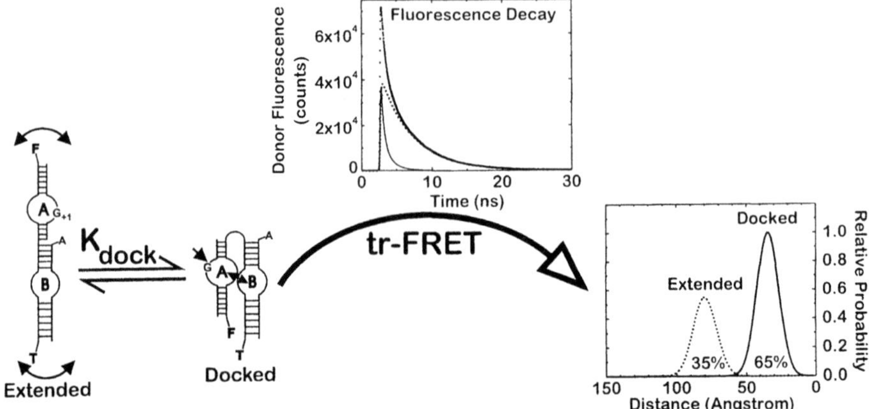

Fig. 20.6. Resolving extended and docked tertiary structure conformers by tr-FRET analysis. The emission decay of the fluorescein donor is deconvoluted into contributions from the docked and extended conformers according to Eq. (20.3). The calculated donor decays for the docked and extended conformers are shown below the experimental decay as *solid and dashed lines*, respectively. The corresponding D-A distance distributions are shown on the right. The heights of the distributions are scaled to reflect the relative fractions of the two conformers. Reproduced from [27] with permission

Mg^{2+}, 18 °C), 65% of the population is centered around a mean D-A distance of 34 Å (full width half maximum, 18 Å), and 35% is centered around 78 Å (fwhm, 18 Å). The longer distance is consistent with the overall length of an undocked, fully extended conformational isomer. The shorter distance species is interpreted as the docked, catalytically active conformer in which the bend at the domain hinge brings the fluorophores into closer proximity. The significant breadth of each distribution reflects fluorophore mobility as well as the intrinsic flexibility of the complex. When bending at the hinge is restricted by fusing the substrate and ribozyme strands, only extended conformers are observed [27].

The results of the tr-FRET analysis indicates that the hairpin ribozyme exists as an equilibrium between docked and extended tertiary structure conformers. The extended conformer is stabilized by helix-helix stacking interactions at the interdomain junction, while docking has to disrupt these stacks and leads to specific interdomain contacts of similar stability. To probe the nature and energetics of the interdomain contacts, mutations were introduced into residues in either loop A or loop B. Mutations that were previously shown to inhibit catalysis by the ribozyme are found to reduce the fraction of docked complexes, indicating that the inhibitory effect is due to less stable docking [27]. These experiments identify residues that are involved in important tertiary contacts between the loop A and B domains and can be used to test structural models of the docked complex.

20.3.2
Influence of the Interdomain Junction on Ribozyme Folding

The structure of the interdomain junction may be a key determinant in the proper folding of the hairpin ribozyme into the catalytically active docked conformation. In the tobacco ringspot virus satellite RNA, the hairpin ribozyme is embedded within a four-way helical junction. In contrast, the minimal hairpin

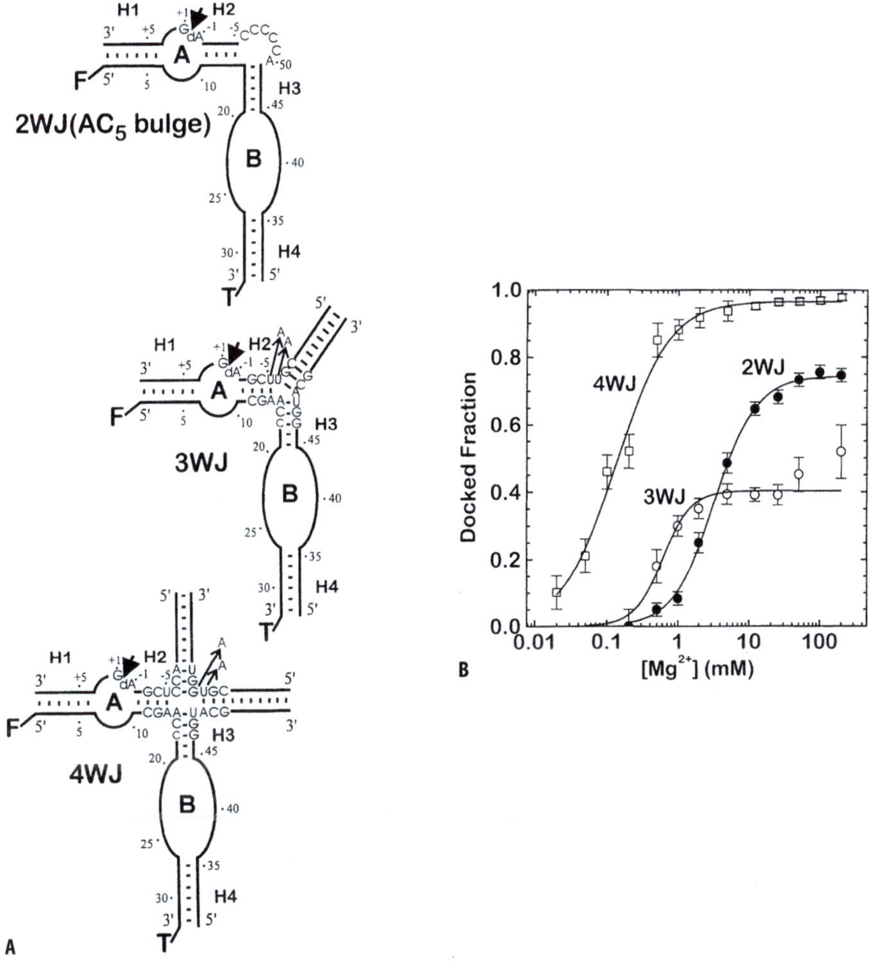

Fig. 20.7 A, B. Docking of ribozymes with modified junction designs: A ribozyme-substrate complexes with two- (2 WJ), three- (3 WJ), and four-way junctions (4 WJ). Only the sequence around the junction is shown. In the bulged version of the two-way junction, the substrate is covalently joined to the ribozyme by a flexible A (C)$_5$ linker. Fluorescein and tetramethylrhodamine are denoted by F and T, respectively. In all cases, the potential cleavage site (*short arrow*) is blocked by a 2′-deoxy modification of nucleotide A$_{-1}$; B stability of tertiary structure folding of different ribozyme-substrate complexes, as reflected in the dependence of the docked fraction on Mg^{2+} concentration (at pH 7.5, 18 °C). Reproduced from [27] with permission

ribozyme design described above contains just a two-way helical junction between the loop A and loop B domains.

tr-FRET analysis of ribozyme-substrate complexes containing two-, three-, and four-way junctions (Fig. 20.7a) reveals that the junction geometry has a profound effect on formation of the active tertiary structure. In particular, the four-way junction strongly favors formation of the docked complex, leading to an equilibrium in which 95% of molecules are docked, whereas the three-way junction leads to only 39% docked complex under standard conditions. Under physiological conditions (1–2 mmol/l Mg^{2+}), the relative enhancement of hairpin ribozyme folding by the four-way junction is even more dramatic. Here, the two-way junction yields the smallest fraction of docked complex, whereas the four-way junction still strongly favors docking (Fig. 20.7b).

These results demonstrate that a four-way helical junction, as found in the naturally occurring form of the hairpin ribozyme, dramatically stabilizes the docked and catalytically active conformer. In comparison to two- and three-way junctions, it appears to be nature's choice as a scaffold to facilitate docking of distant domains of an RNA molecule.

20.4
Conclusions and Outlook

The studies reviewed in this article illustrate how time-resolved fluorescence spectroscopic techniques can be used to resolve heterogeneous mixtures of different DNA-protein complexes or multiple conformational states of an RNA molecule. In addition to polymerases, other DNA binding proteins and DNA-processing enzymes can also interact with DNA in more than one mode of binding. The time-resolved fluorescence anisotropy technique should be useful for dissecting the energetics of multi-state protein-DNA binding equilibria in a range of systems. In addition, many RNA molecules can fold into alternative tertiary structures that interact differently with metal ion or protein cofactors. Transitions among the various conformational states are required for many of the biological functions of RNA. It is anticipated that the tr-FRET approach described here will find widespread use in studying the energetics and dynamics of RNA conformational transitions.

Acknowledgement. A number of coworkers and collaborators have contributed to the work described in this article. In the author's laboratory, I wish to thank Christopher Guest, Remo Hochstrasser, Ted Carver, Richard Fee, Wai-Chung Lam, Elizabeth Thompson, Michael Bailey, and Edwin Van der Schans for their contributions. The studies of hairpin ribozyme folding were carried out in collaboration with John Burke and Nils Walter at the University of Vermont. The author gratefully acknowledges the financial support of the Institute of General Medical Sciences (grant GM44060) and the National Science Foundation (grant MCB 9604568).

References

1. Eftink MR (1991) Fluorescence techniques for studying protein structure. In: Suelter CH (ed) Methods of biochemical analysis, vol 35: protein structure determination. Wiley, New York, pp 127–205
2. Millar DP (1996) Time-resolved fluorescence spectroscopy. Curr Opin Struct Biol 6:637–642
3. Wahl P, Paoletti J, Le Pecq JB (1970) Decay of fluorescence emission anisotropy of the ethidium bromide-DNA complex: evidence for an internal motion in DNA. Proc Natl Acad Sci USA 65:417–421
4. Millar DP, Robbins RJ, Zewail AH (1980) Direct observation of the torsional dynamics of DNA and RNA by picosecond spectroscopy. Proc Natl Acad Sci USA 77:5593–5597
5. Waggoner A (1995) Covalent labeling of proteins and nucleic acids with fluorophores. Meth Enzymol 246:362–373
6. Jameson DJ, Sawyer WH (1995) Fluorescence anisotropy applied to biomolecular interactions. Meth Enzymol 246:283–300
7. Yang M, Millar DP (1997) Fluorescence resonance energy transfer as a probe of DNA structure and function. Meth Enzymol 278:417–444
8. Guest CR, Hochstrasser RA, Sowers LC, Millar DP (1991) Dynamics of mismatched base pairs in DNA. Biochemistry 30:3271–3279
9. Clegg RM, Murchie AIH, Zechel A, Carlberg C, Diekmann S, Lilley DMJ (1992) Fluorescence resonance energy transfer analysis of the structure of the four-way DNA junction. Biochemistry 31:4846–4856
10. Eis PE, Millar DP (1993) Conformational distributions of a four-way DNA junction revealed by time-resolved fluorescence resonance energy tranfser. Biochemistry 32:13,852–13,860
11. Tuschl T, Gohlke C, Jovin TM, Westof E, Eckstein F (1994) A three-dimensional model for the hammerhead ribozyme based on fluorescence measurements. Science 266:785–789
12. Bassi GS, Murchie AIH, Walter F, Clegg RM, Lilley DMJ (1997) Ion-induced folding of the hammerhead ribozyme: a fluorescence resonance energy transfer study. EMBO J 16:7481–7489
13. Allan BW, Beechem JM, Lindstrom WM, Reich NO (1998) Direct real time observation of base flipping by the *Eco*RI DNA methyltransferase. J Biol Chem 273:2368–2373
14. Hochstrasser RA, Carver TE, Sowers LC, Millar DP (1994) Melting of a DNA helix terminus within the active site of a DNA polymerase. Biochemistry 33:11,971–11,979
15. Millar DP (1996) Fluorescence studies of DNA and RNA structure and dynamics. Curr Opin Struct Biol 6:322–326
16. Joyce CM, Steitz TA (1994) Function and structure relationships in DNA polymerases. Annu Rev Biochem 63:777–822
17. Beese LS, Derbyshire V, Steitz TA (1993) Structure of DNA polymerase I Klenow fragment bound to duplex DNA. Science 260:352–355
18. Carver TE, Hochstrasser RA, Millar DP (1994) Proofreading DNA: recognition of aberrant DNA termini by the Klenow fragment of DNA polymerase I. Proc Natl Acad Sci USA 91:10,670–10,674
19. Guest CR, Hochstrasser RA, Dupuy CG, Allen DJ, Benkovic SJ, Millar DP (1991) Interaction of DNA with the Klenow fragment of DNA polymerase I studied by time-resolved fluorescence spectroscopy. Biochemistry 30:8759–8770
20. Cowart M, Gibson KJ, Allen DJ, Benkovic SJ (1989) DNA substrate structural requirements for the exonuclease and polymerase activities of procaryotic and phage DNA polymerases. Biochemistry 28:1975–1983
21. Lam WC, Van der Schans EJC, Joyce CM, Millar DP (1998) Effects of mutations on the partitioning of DNA substrates between the polymerase and 3′-5′ exonuclease sites of DNA polymerase I (Klenow fragment). Biochemistry 37:1513–1522
22. Derbyshire V, Grindley ND, Joyce CM (1991) The 3′-5′ exonuclease of DNA polymerase I of *Escherichia coli*: contribution of each amino acid at the active site to the reaction. EMBO J 10:17–24

23. Walter NG, Burke JM (1998) The hairpin ribozyme: structure, assembly and catalysis. Curr Opin Chem Biol 2:24–30
24. Walter NG, Hample KJ, Brown KM, Burke JM (1998) Tertiary structure formation in the hairpin ribozyme monitored by fluorescence resonance energy transfer. EMBO J 17:2378–2391
25. Conn GL, Draper DE (1998) RNA structure. Curr Opin Struct Biol 8:278–285
26. Clegg RM (1992) Fluorescence resonance energy transfer and nucleic acids. Meth Enzymol 211:353–388
27. Walter NG, Burke JM, Millar DP (1999) Stability of hairpin ribozyme tertiary structure is governed by the interdomain junction. Nature Struct Biol 6:544–549

Rare Earth Cryptates and TRACE Technology as Tools for Probing Molecular Interactions in Biology

B. ALPHA-BAZIN, H. BAZIN, M. PRÉAUDAT, E. TRINQUET, G. MATHIS

21.1
Introduction

Molecular interactions are of tremendous importance in many processes involved in life. For example, in cell function and regulation, key mechanisms like receptor activation, transcription, and gene regulation are based on molecular interactions [1].

Many of these interactions are nowadays the object of thorough investigations from scientists discovering the implication of such mechanisms in diseases like cancer or viral infection.

The discovery of radioimmunoanalysis in the 1960s allowed the specific quantification of proteins and haptens (antigens) through the use of specific ligands (the antibodies) and sensitive radioactive labels like ^{125}I, ^3H [2]. However, this technique and the following non-isotopic equivalent are based on the physical separation between the antigen/antibody biomolecular complex and the excess of label (heterogeneous techniques). This was considered very early as a handicap and the researchers focused their efforts towards methods based on non-isotopic labels and physical principles related to size or proximity effects for devising homogeneous (non-separation) assays [3–7].

Homogeneous assays are based on the modulation of the label signal during the biological interaction under investigation. The ideal homogeneous method should possess a general methodology to probe all types of analytes or biological interaction with a high sensitivity and a strong resistance to media interactions.

Indeed, a homogeneous analytical method eliminates the need for cumbersome washing in view of assay automation. Moreover, avoiding bound to free label separation steps are of great interest for the understanding of interaction and dynamic studies of the molecular interaction.

For years, a number of analytical techniques, labels, and physicochemical principles have been explored in an attempt to devise such assays.

21.2
Fluorescence and Homogeneous Assays

Fluorescence which is, in theory, the most sensitive analytical technique and enables several types of physicochemical processes to modulate the label emission was therefore considered very early on as a technique of choice for designing homogeneous assays. Polarization, quenching, time correlation, lifetime variation, as well as fluorescence resonance energy transfer (FRET) have been explored [3, 6–9].

However, severe limitations still remain for most of these techniques. They can be classified into two types:

1. Those originating from the modulation which ideally must provide a high sensitivity and no restriction to assay design.
2. Those from the assay medium like optical properties variation (serum, biological fluid, etc.), non specific interaction of the label with media constituents, highly variable fluorescence background, and light diffusion from proteins, molecules, and aggregates in the media.

The very stringent set of properties required for a general and sensitive homogeneous fluorescent assay technique partly explains why the majority of such assays have been limited to academic research or are restricted to a particular type of analyte or present only low to medium range sensitivity [3, 9].

Among these techniques, FRET is of particular interest. After the establishment of the theory on FRET by Förster [10], who postulated the dependence on the inverse sixth power of the distance between an excited fluorescent donor and a nearby acceptor molecule for the rate of transfer, this technique has been used extensively in biology.

For commonly known donor-acceptor pairs, the distance R_0 (for which transfer is 50% efficient) is in the 1–7 nm range. Therefore, FRET has been used as a spectroscopic ruler to reveal proximity relationship in biological macromolecules, as suggested by pioneering work of Stryer [11], and others [12, 13].

There being very few techniques that allow such distances and interactions to be determined in solution, many domains of biology were explored with FRET, like enzymatic activity [14], protein-DNA interactions [15], and cell surface lectins receptor-ligand interactions [16]. Only recently, DNA hybridization has attracted much attention and become a field of investigation for FRET experiments, either for structural investigations, complementary oligonucleotide hybridization, or detection of gene translocation [17–19]. Because the processes involved in FRET occur within distances characteristic of antigen-antibody interactions, homogeneous fluoroimmunoassays were also developed [20, 21].

This impressive amount of work should not mask the difficulties encountered when using FRET. For example, the measured fluorescence intensities must be corrected for autofluorescence, light scattering, contribution from unbound fluorophores, and cell debris.

An interesting approach based on the use of a long-lived donor in a FRET experiment, allowing measurement of the acceptor by time-resolved fluorescence, was demonstrated in an immunoassay with pyrene butyrate and B phycoerythrin by Morrison [21]. However, the efficiency of the long-lived acceptor's signal gated detection is greatly outweighed by optical filtering of the donor/ acceptor emission spectra.

21.2.1
Time Resolved Fluorescence and Rare Earth Complexes

Owing to the natural fluorescence of compounds and proteins present in biological fluids or serum, the use of conventional fluorophores leads to serious limitations of sensitivity. Most of these background signals being short-lived, the use of long-lived labels and of time resolved fluorescence techniques allows one to reduce interference.

Due to their very specific photophysical and spectral properties, complexes of rare earth ions are of central interest for applications of fluorescence in biology [12, 22–25]. In fact, the outstanding electronic configuration of rare earth trivalent ions partly shields the optically active electrons. Thus, characteristics line type emission between excited state and ground state are obtained. Moreover, these electronic transitions being forbidden by quantum mechanistic rules, the emission lifetime of such ions are usually long (from microseconds to milliseconds). Unfortunately, for the same reason, their light collection efficiency is very poor. Therefore, rare earth chelates have been designed using the organic chelator as a light-harvesting device. The collected energy is then transferred by intramolecular non-radiative processes from the singlet to the triplet state of the ligand, then from the triplet of the ligand to the emissive level of the rare earth ion which emits its characteristic long lived luminescence [26].

However, to be considered as suitable luminescent probes in biological assays, rare earth complexes must fulfill a very stringent set of properties including thermodynamic and kinetic stability, complexation selectivity, high luminescence quantum yield, and the ability to be fixed on biomolecules. Moreover, working directly in biological fluids, immunity against quenching is of crucial importance.

21.2.2
Rare Earth Chelates

In practice, these requirements are not entirely fulfilled by the known rare earth chelates, and the lack of an "ideal" chelate leads to different solutions in devising analytical immunoassay. In a first approach an Eu^{3+} ion is used as the label and is carried by the antibody through the immunological reaction by the use of coupled EDTA or diethylenetriaminepentaacetic acid (DTPA) derivatives [24]. The formation of the highly fluorescent complex is achieved after the immunological reaction, by the addition of an excess of a suitable ligand and the addition of a synergistic agent in a detergent solution. This approach has been used successfully but presents two main limitations: vulnerability to Eu^{3+} contamination and, in view of a homogeneous assay, the non-availability of a fluorescent molecule bound directly to the antibody during the biological interaction.

The second approach is the opposite and uses the labeling of the immunoreactants by the chelator, the fluorescent complex being formed through the use of an excess of Eu^{3+} [27]. For the commonly known labels using this second approach, the main limitation is poor sensitivity [28] and the absence of the long-lived tracer during the biological interaction.

The combination of energy absorption, efficient energy transfer properties, and chelate stability in the same molecule is still the objective of research groups in this field [23, 29–31].

This lack of an ideal molecule may explain why rare earth chelate labels have been of little use in homogeneous assays.

21.3
TRACE Technology

21.3.1
Rare Earth Cryptates as a New Type of Fluorescent Label

The above-mentioned approach for designing rare earth chelates was driven mainly by thermodynamic considerations focusing research towards complexes possessing a high formation constant.

In contrast, considering the specific constraints of homogenous assays, we identified the dissociation kinetic as a key parameter for the selection of appropriate complexes. Therefore, the cryptates, which are formed by the inclusion of a cation into the three-dimensional cavity of the ligand, were investigated [32]. Considering kinetic stability, the main difference between cryptates and chelates is related to the fact that, in the dissociation process, the transition state for the cryptate involves a conformation where one of the bridgehead nitrogens points its electronic lone pair outward, leading to a high activation energy. This characteristic leads to the formation of exceptionally inert rare earth cryptate [33]. An antenna effect for the collection of the excitation energy is obtained, for example, by the incorporation of three bipyridine units in the cryptand structure [34]. The incorporation of fluoride ions in the Eu^{3+} coordination sphere allows a total shielding through the formation of ion pairs [35].

Heterobifunctional reagents are used to conjugate the Eu trisbipyridine cryptate (TBP Eu^{3+}) diamine derivative to antigens, antibodies, proteins, peptides, and oligonucleotides (Fig. 21.1). The photophysical properties of the photo-

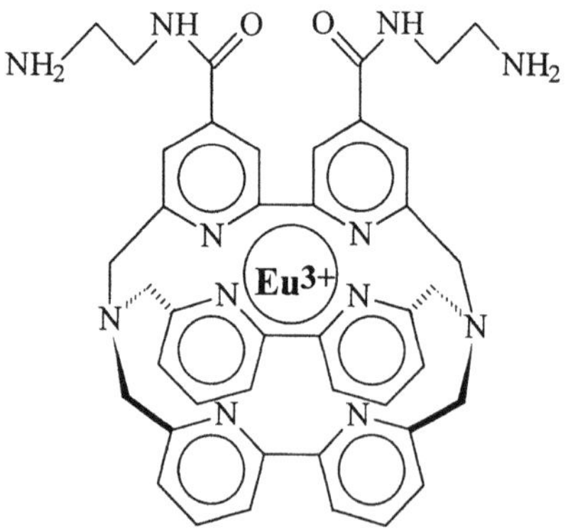

Fig. 21.1. Europium trisbipyridine cryptate structure with amine groups available for labeling various biomolecules

active cryptate are conserved in the conjugates [36, 37]. Unlike most other rare earth chelates, TPB Eu^{3+} possesses within the same molecule the properties required for use in homogeneous bioassays, namely light harvesting unit, defined stoichiometry, and high kinetic stability in biological media.

21.3.2
Modulation Processes and Homogeneous Assays

As reviewed above, homogenous assay techniques based on FRET have been widely used for academic research in probing interaction processes in biology but still suffer from serious handicaps.

Therefore, we researched methods to solve the drawbacks encountered when using FRET techniques in the "real world".

The homogeneous technique we have developed is based on FRET between TBP Eu^{3+} donor and a second fluorescent label (acceptor), Allophycocyanin, a 105-kDa phycobiliprotein [38]. Allophycocyanin was crosslinked to yield XL665 for increasing stability. It possesses a set of photophysical properties tuned for TBP Eu^{3+}: a high molar absorptivity at the cryptate emission wavelength which allows a high transfer efficiency (Ro ≈ 9.5 nm), emission in a spectral range where the cryptate signal is insignificant, and high quantum yield (≈ 70%) not quenched in biological fluids (Fig. 21.2).

In FRET the lifetime of the acceptor's emission includes a contribution equal to the donor's lifetime in presence of energy transfer [19, 39]. Therefore, using

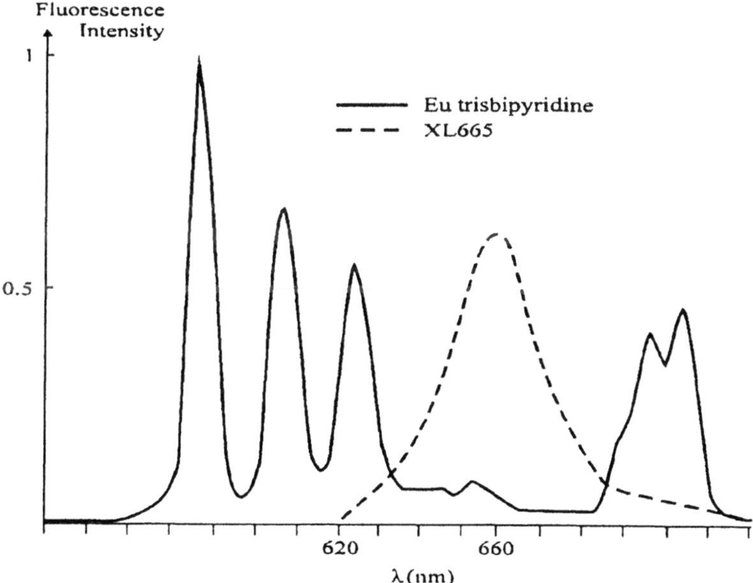

Fig. 21.2. TBP Eu^{3+} and XL665 emission spectra showing the spectral selectivity obtained with the chosen FRET pair

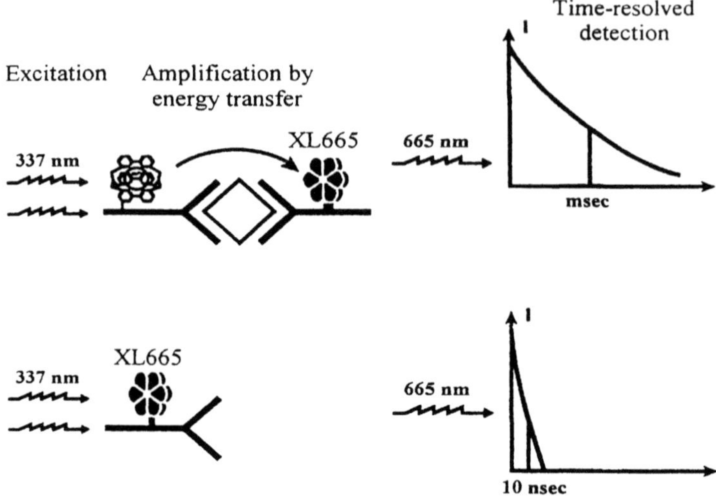

Fig. 21.3. Temporal selectivity on the XL665 signal in a TRACE immunoassay: a 50 µs gated detection allows one to isolate the FRET generated signal characteristic of the biological interaction under investigation

the long-lived TBP Eu^{3+} as donor leads to a long-lived emission of the XL665 that can be measured by time-resolved fluorescence. At the XL665 emission wavelength, we can then make a clear distinction between the long-lived signal from the acceptor engaged in the FRET process with the TBP Eu^{3+} and the freely diffusing XL665 which emits a short-lived signal. Therefore a temporal selectivity is obtained by time resolved fluorescence at the XL665 emission wavelength which allows one to isolate a FRET signal free from short-lived fluorescent background (Fig. 21.3) [40]. The TBP Eu^{3+}/XL665 pair leads to a high transfer efficiency (50–95% for distances in the $15 \cdot 100$ nm range) and the quantum yield of XL665 is high ($\approx 70\%$). Therefore the Eu^{3+} deactivation pathway using the FRET process and XL665 emission is more efficient than the radiative Eu^{3+} one ($\approx 30\%$). Moreover, as the TBP Eu^{3+} spectra is spread over 100 nm, the FRET process tend to "concentrate" the energy into the XL665 emission spectra. These highlighted properties are at the origin of the name of the technology we have developed: "TRACE" for Time Resolved Amplified Cryptate Emission [40].

21.3.3
Dual Wavelength Detection

A major problem when using homogeneous fluorescence techniques based on intensity measurements is the inner filter effect from the absorbing media. Sample dilution and rate measurement have been used to correct this drawback but always at the expense of analytical sensitivity [3]. By measuring the variation in apparent decay time at the donor emission wavelength by phase modulation in FRET, assays may overcome the problem [41], but not in the presence of a large excess of labeled ligands.

The exceptionally large spectral selectivity obtained with the TBP Eu^{3+}/ XL665 pair allows the measurement of the cryptate emission signal without contribution of the acceptor. We verified that in most cases the TBP Eu^{3+} signal reflected the media absorption at the excitation wavelength.

Therefore, the acceptor signal at 665 nm, which is also proportional to the optical properties of the media, can be divided by the cryptate signal yielding a measurement which is then independent of the optical characteristics of the media at the excitation wavelength. The XL665 to TBP Eu^{3+} signal ratio depends solely on the specific biological interactions under study [42].

21.3.4
TRACE Application in Immunoanalysis

A prolactin immunoassay was first used as a model to demonstrate the application of these principles. Two monoclonal antibodies raised against prolactin were labeled with TBP Eu^{3+} cryptate and acceptor XL665. After the mixing of the labeled antibodies and the sample, and excitation by a 337 nm nitrogen laser, typical signals of TBP Eu^{3+} cryptate and acceptor XL665 are emitted. An interference filter centered at 665 nm allows a first selection to be performed, which isolates the signal of the acceptor XL665-labeled antibody from that of the TBP Eu^{3+} cryptate labeled antibody. This signal has two contributors: a long-lived fluorescence from the acceptor XL665 label in the immune complex, and a short-lived signal from the free acceptor XL665 label conjugate, excited by a laser. The signal at 665 nm being measured after a delay of 10–100 µs, the contribution of the free acceptor XL665 label fluorescence is suppressed. In others words, a spectral and a temporal selection isolate the amplified XL665 signal from those of the TBP Eu^{3+} cryptate and free XL665 labels. The measured signal amplified over that of the cryptate label in the immune complex is therefore only proportional to the concentration of the immune complex. The detection limit, determined as the zero standard value +2 SD, is 0.3 µg/l which compares very well with heterogeneous assays based on radioactive labels [43].

21.3.5
Kinetic Measurements

The method being homogeneous and the measurement non-destructive, data acquisition can be performed continuously and the kinetic behavior of each sample be acquired and analyzed. Figure 21.4 shows such data monitored for 60 s and for different characteristic AFP (Alpha Foeto Protein) sample concentrations.

The shape of the kinetic curves demonstrates that, before 10 s, the behavior is quite different and depends highly on the concentration of the sample. The signal level measured within the first minute allows us to forecast an automatic dilution taking into account the concentration estimation [44]. Further investigations using the TRACE tools for the study of the mechanism involved in immunometric assays showed their capacity to be used for gaining information on the reaction kinetics. The data acquired can be useful both for assay optimization and theoretical and modeling studies [45, 46].

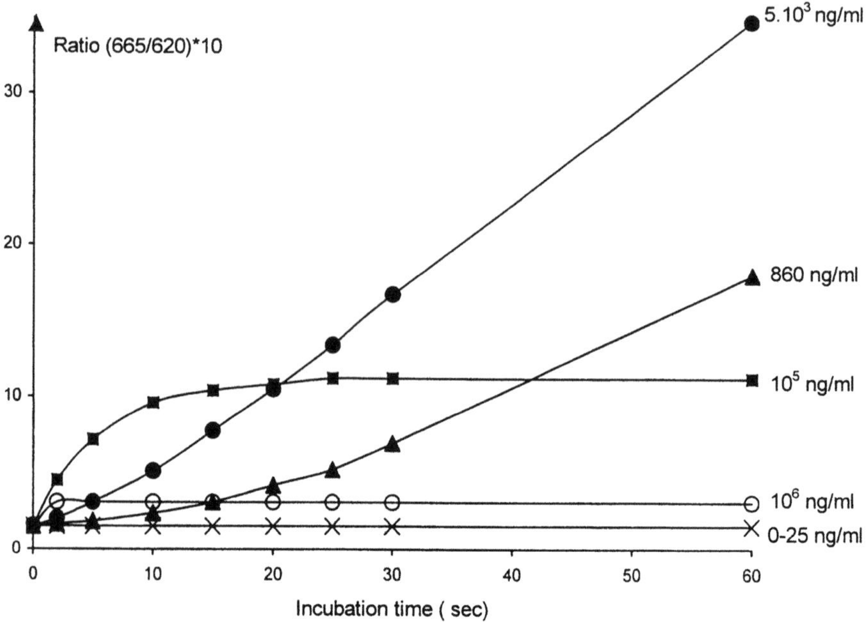

Fig. 21.4. The fluorescence ratio 665/620 is plotted vs time, for different AFP sample concentrations. The different shapes allow easily a quick estimation of the sample concentration range

21.4
TRACE for Probing Molecular Interactions in Life Science

In many domains of biology, the studies or events based on short-range interactions involving biomolecules is crucial for the understanding of molecular mechanism or for analytical purposes. Of increasing interest are the various mechanisms involved in the transfer of information to and inside the cell [47]. In particular, groups involved in drug discovery are seeking new tools to probe these interactions. Accordingly, we have expanded to other domains of biology the TRACE principles and reagents. Several assays, chosen from the signaling pathways involved in cellular communication and expression, have been published and demonstrate the versatility of this method for studying such interactions. TRACE and related reagents and technologies are now marketed under the HTRF name (HTRF for Homogeneous Time-Resolved Fluorescence).[1]

---

[1] HTRF is a trade name of Packard.

21.4.1
Cell Surface Receptor Studies

The binding of a ligand to its receptor on a cell surface is a key event in cellular communication. Therefore, several assay formats were developed for probing this very general type of interaction. In the case of the Epidermal Growth Factor Receptor (EGFR), a TBP Eu^{3+} cryptate-labeled Epidermal Growth Factor peptide (EGF) and an XL665 labeled monoclonal antibody to the EGFR are mixed with a receptor-containing preparation. After incubation and excitation at the cryptate excitation wavelength, the amplified specific signal is measured by time-resolved fluorescence and allows measuring receptor concentration or probing EGFR inhibitors. Moreover, the technique being homogeneous, it allows kinetic measurement for association/dissociation studies.

Radioligand assays (RLAs) are the most common procedures for receptor-ligand studies. Because they involve physical separation of bound and free ligand, disturbance of the equilibrium conditions cannot be avoided, and automation is difficult. In scintillation proximity assays (SPAs), binding of ligand activates fluorophores integrated in microbeads coated with the receptor [40]. Using SPAs solves the RLA problems related to equilibrium conditions and also enables one to obtain kinetic data. However, this technique does not overcome the need to use radioactivity (with the attendant problems of waste management and high counting time) and a solid phase, even if the solid phase is dispersed (problems of homogenization and non-specific binding). In contrast, the results we have obtained for the assay of EGF/EGFR by using TBP Eu^{3+} cryptate are promising on both analytical and methodological considerations [48].

The technology was also used to determine the stoichiometry of subunits within a hetero oligomeric γ-aminobutyric acid ($GABA_A$) receptor in intact cells. By using α_1, β_2, γ_2, $GABA_A$ receptor subunits tagged with c-Myc by site directed mutagenesis and anti c-Myc monoclonal antibodies labeled with TBP Eu^{3+} and XL665, McKernan and coworkers were able to determine the receptor stoichiometry [49].

21.4.2
Receptor Tyrosine Kinase Assay

The regulation of enzyme activities by reversible phosphorylation is crucial for the control of cellular processes after extracellular activation. After the development of molecular cloning, discovery of protein tyrosine, serine, and threonine kinases increased dramatically. The TRACE technology was adapted to investigate this field. A synthetic tyrosine containing substrate Poly (GAT) [50], which is recognized by tyrosine specific kinase, was labeled with biotin. The substrate was phosphorylated in the presence of ATP and A431 cell extracts containing EGFR preactivated by EGF. XL665 labeled streptavidin and cryptate-labeled anti-phosphotyrosine antibody were added to generate the specific signal, which was proportional to the phosphorylated tyrosine concentration. A K_m of 3.6 µmol/l for ATP and an IC_{50} of 5 µmol/l for Tyrphostin 47 were obtained, results that compare very well with published values [51, 52].

To our knowledge, this was the first non-isotopic homogeneous kinase activity assay described. Most of the corresponding heterogeneous assays incorporate ^{32}P in the substrate and separation is achieved by precipitation of the polymer or adsorption onto specific supports. Other kinase assays were also studied with the TRACE technology showing the universality of the method [50–52].

21.4.3
Protein-Protein Interactions

Protein-protein dimerization is a key mechanism for information transfer in biological systems. For example, transcription factors and growth factor-type receptors are activated through protein interactions [47]. The protein products of the Jun/Fos oncogenes require a functional protein-protein interaction domain, the "leucine zipper domain", to exert their transcriptional regulatory activity on the binding to the AP-1 site on DNA. To study the mechanism of specificity in Jun/Fos heterodimer formation, investigators synthesized the 40 amino acid peptides corresponding to the leucine zipper domains of the two proteins and studied the physicochemical properties of the interaction by circular dichroism or SPA [53]. We then used this model to evaluate the adaptation of our technique for monitoring protein heterodimerization and studying weak peptide-peptide interactions [54].

A biotinylated Jun peptide and an XL665 labeled Fos peptide were mixed with TBP Eu^{3+} cryptate-labeled streptavidin. The cryptate fluorescence is amplified by energy transfer only when both XL665 and cryptate are brought into proximity by the formation of the peptide heterodimer. For free donor or acceptor in solution, the mean distance is too large for transfer (Fig. 21.5).

Electrostatic interactions being involved in the dimer formation, we explored the effect of ionic strength on the binding by changing the KF concentrations in the assay from 50 mM to 500 mM. The results, showing a decrease of the specific signal for an increase in the ionic strength, are comparable with those already observed with SPA or circular dichroism [53].

This small peptide dimerization assay also highlights the fact that TBP Eu^{3+} and XL665 labels can be used without impairing electrostatic or hydrophobic binding [54].

An interesting indirect approach of the technology was published by Mellor and coworkers for the study of CD28/CD86 protein-protein binding assay [55]. CD28 provides the major costimulation signal for CD4 positive T cells. Ligation with its natural ligands CD80 and CD86 leads to signals that are required for the production of interleukin-2, a process implicated in programmed cell death. An indirect approach was taken whereby XL665 is linked to an antihuman domain fused to CD28. The CD86, a fusion protein with a rat immunoglobulin domain, is bound to a biotinylated sheep anti-rat antibody, which is complexed with a TBP Eu^{3+} labeled streptavidin. The assay format named "cassette format" by the author allows an easy adaptation of the generic reagents to the replacement of, for example, CD28 with CTLA-4 (fused with human immunoglobulin domain) and/or CD80 (fused with rat immunoglobulin domain) in place of CD86.

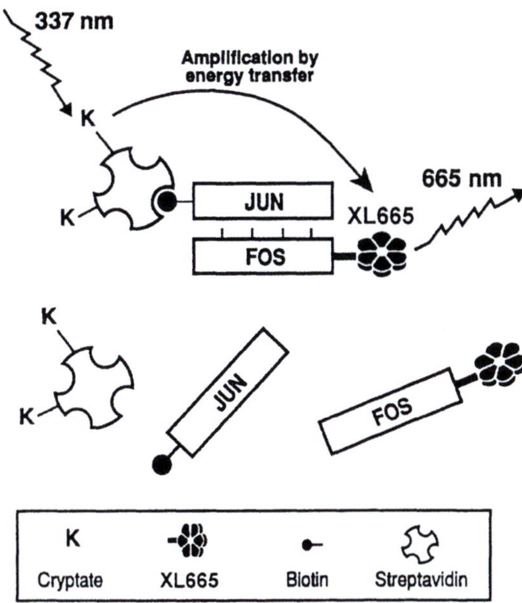

Fig. 21.5. Jun/Fos heterodimerization homogeneous assay scheme. The signal emitted at 665 nm after excitation at 337 nm was proportional to the concentration of heterodimers formed

21.4.4
Protease Assays

Protease activity is of the utmost importance in the studies related to viral infection and propagation. FRET constructs have already been set up for the discovery of protease inhibitors [14] but applications to the "real world" to screen libraries of compounds particularly of natural compounds are difficult due to the highly variable optical quality of the media.

Manly and coworkers have used our technology to screen Herpes Simplex Virus protease inhibitor [56]. A peptide containing a sequence recognized by the protease was labeled on one side with XL665 and biotin on the other. Upon the addition of streptavidin TBP Eu^{3+} conjugate, a specific long-lived signal was measured at 665 nm. In the presence of the protease which cleaves the peptide, the signal decreases. The level of the decrease then depends on the efficiency and/or concentration of inhibitors.

Cummings and coworkers built up an HIV protease format by tagging both end of a peptide respectively with biotin and a sequence containing phosphotyrosine. The specific signal is obtained by adding Streptavidin-XL665 and an anti phosphotyrosine antibody labeled with TBP Eu^{3+} [57] This is an example of a totally indirect protease assay.

In the same vein, H. Takemoto and coworkers performed a Caspase 3 indirect assay. A 7-amino acid peptide was labeled at both ends with biotin and dinitrophenyl (DNP) groups. Signal generation was obtained after addition of an TBP Eu^{3+} labeled anti-DNP antibody and XL665 labeled streptavidin [58].

21.4.5
Applications in Molecular Biochemistry

21.4.5.1
Nucleic Acid Hybridization

Nucleic acid hybridization has become a powerful technique in many fields of life science. For diagnostic purposes, many different types of assays have been used [59], particularly those involving rare earth chelates. Use of the polymerase chain reaction has circumvented some of the problems related to sensitivity [36, 37, 60, 61]. Different homogeneous formats based on FRET have also been developed [18, 19]. To investigate the features of the technique and labels in this situation, we labeled two complementary 21-base oligonucleotides at the 5' ends, one with ^{32}P or with TPB Eu^{3+} and the other with biotin. The duplexes formed were captured for a heterogeneous radioactive assay with streptavidin coated wells or XL665 labeled streptavidin for a homogeneous time-resolved fluorescence assay. Results showed that the two systems had equivalent analytical sensitivity, but the homogeneous method presents better capturing efficiency and speed.

21.4.5.2
Incorporation of TBP Eu^{3+} Labeled Nucleotides in DNA and RNA

Modified nucleoside 5'-triphosphate are widely used in biochemistry and molecular biology. In most applications the radioactively labeled DNA and RNA probes were progressively replaced by non-radioactive ones. As fluorescent nucleotides, europium chelates nucleotides were mostly used for the labeling of DNA probes associated with a heterogeneous detection based on time-resolved fluorimetry [62]. Analogs of uridine triphosphate (UTP) and deoxynucleotide triphosphate (dUTP) conjugated to a europium cryptate moiety have been synthesized. They were successfully tested as substrate for the most commonly used polymerase. Their direct enzymatic incorporation into the corresponding RNA or DNA entities was studied by using the TRACE technology [63, 64].

21.4.5.2.1
Terminal Deoxynucleotidyl Transferase (TdT) Reaction

TdT possesses a polymerase activity which does not need any template adding nucleotides at the 3' end of an ODN primer. Therefore 5' biotinylated primer was elongated in presence of TBP Eu^{3+} dUTP conjugate alone or in admixture with thymidine 5' triphosphate (dTTP). After the addition of XL665 labeled streptavidin the efficiency of the incorporation was related to the specific emission at 665 nm. This polymerase activity can also be used in a preparative purpose for the 3'-end labeling of DNA strands with cryptate [63, 64].

21.4.5.2.2
DNA Polymerase Reactions

Efficient incorporation of TBP Eu^{3+} labeled dUTP has been obtained with Klenow and *Taq* DNA polymerase. The extension reactions were tested by incubation in presence of Klenow (37 °C) or *Taq* DNA polymerase, a 5′-biotin oligonucleotide, a single stranded DNA template, TBP Eu^{3+} labeled dUTP and the four natural deoxynucleoside triphosphates. After the addition of XL665 labeled streptavidin, a signal proportional to polymerase activity is measured at 665 nm. This format can be used to design homogeneous assay for polymerase activity or to label DNA probes with long lived fluorophores [63].

Furthermore, a new PCR detection format can be designed through FRET between TBP Eu^{3+} moieties incorporated along the amplimer and a nested XL665 labeled probe.

Incorporation of TBP Eu^{3+} labeled dUTP into DNA was also demonstrated using reverse transcriptase, polyA as template, and biotinylated primer [65]. Another strategy using biotinylated dUTP and a cryptate labeled primer has also been used [64, 66, 67]

21.4.5.2.3
RNA Polymerase Reactions

Bacteriophage RNA polymerase, T^7 RNA polymerase being the one most commonly used, initiates the synthesis of RNA on double-stranded DNA template carrying the appropriate bacteriophage-specific promoter. TBP Eu^{3+} and biotin labeled CTP were incorporated together during in vitro transcription promoted by T^7 RNA polymerase. The RNA strands formed, containing both TBP Eu^{3+} and biotin, were detected after the addition of XL665 labeled streptavidin. Studies show an efficient incorporation of TBP Eu^{3+} UTP in the RNA construct [68].

21.4.6
"Cassettes" Formats as a Generic Tool

The 95 Å R_0 value obtained with the TBP Eu^{3+}/XL665 pair is the largest ever published for a donor-acceptor pair. Thus, it allows probing large distances through space, particularly in biomolecular constructs and aggregates [40, 55]

Researchers have therefore devised formats that allow the use of universal detection reagents for an easier and more rapid development of assays. Basically, each member of the biological interaction of interest is labeled with a genetically engineered tag like, for example, GST, as illustrated by Zhou and coworkers to probe nuclear receptor with their coactivator [69]. A chemically fixed hapten like biotin, DNP, or a phosphorylated peptide could also be used [57, 58].

By addition of TBP Eu^{3+} or XL665 anti-tag conjugates, the interacting tagged biomolecules are labeled indirectly. This versatile strategy is now widely used in the pharmaceutical industry induced by the development of generic anti-tag antibodies labeled with XL665 or TBP Eu^{3+} [70–73].

21.5
Conclusion

This new homogeneous fluoroassay technique using FRET principle with Eu^{3+} trisbipyridine cryptate as donor and amplification by non-radiative energy transfer allows a double discrimination of the emitted signals. Spectral selectivity allows separation of TPB Eu^{3+} cryptate and XL665 acceptor signals, and the fluorescence lifetime selection provides the separation of bound from free XL665 signals. Due to TBP Eu^{3+} cryptate stability at low concentration, and double wavelength detection which allows real time correction for optical transmission, the assay is free from media interactions. The use of TRACE technology to explore characteristics of typical interactions encountered in the real world of biology and more particularly in cell communications highlights the versatility and specific advantages of the technique. Moreover, the homogeneous nature of the assays combined with the non-destructive mode of measurement allows one to perform kinetic measurements to characterize the interaction under investigation [46].

The demonstrated ease in labeling different type of molecules, peptides, and oligonucleotides with TBP Eu^{3+} and XL665, as well as the straightforward mixing and measuring processes that are available only in homogeneous methods, allow the development of assays that involve only a minimal perturbation of equilibrium or steric environment. Also, the high efficiency of energy transfer allows one to probe large biological structures or assemblages. The millisecond range signal lifetime allows structural investigations by easy measurement of transfer efficiency and thus makes possible the determination of donor-acceptor distances.

Further clinical applications may be envisaged for TRACE technology in molecular biology, flow cytometry for cell surface mapping, and fluorescence microscopy. Basic studies in biology may also profit by this new tool for exploring processes where distance relationship or supramolecular structures are involved. Also the field of drug discovery is a challenging domain where new analytical techniques are required for random screening of products. The huge increase in the size of libraries from automated peptide and combinatorial chemical synthesis highlighted the need of new tools for the rapid adaptation of new targets for automated high throughput screening. The specific features of TBP Eu^{3+} cryptate and amplification by energy transfer described here will be decisive for this purpose, particularly in providing new tools for rapid investigation of molecular interactions.

Acknowledgements. Fruitful discussions and constant support from JM Lehn were highly appreciated. We thank J Becquart and C Pernelle (Rhône Poulenc Rorer, Centre de Recherche de Vitry, France) for efficient collaboration in the validation of the technology. Fruitful contacts with pharmaceutical research groups at Merck (J. Hermes group), Zeneca (J. Major group), Bristol Myers Squibb (S. Manly group), Glaxo (J. Mellor group), and Shionogi (H. Takemoto group) were also appreciated.

References

1. Austin DJ, Schreiber SI (1994) Chem Biol 1:131
2. Yallow RS, Berson SA (1960) J Clin Invest 39:1157
3. Ullman EF, Bellet NF, Brinkley JM, Zuk RF (1980) Homogeneous fluorescence immuno-assays. In: Nakamura RM, Dito WR, Tucker ES (eds) Immunoassays: clinical laboratory techniques for the 1980s. AR. Liss, New York, pp 13–43 and references cited therein
4. Patel A, Davies CJ, Campbell AK, McCapra F (1983) Anal Biochem 129:162
5. Henderson DR, Friedman SB, Harris JD, Manning WB, Zoccoli M (1986) Clin Chem 32:1637
6. Briggs J, Elings VB, Nicoli DF (1981) Science 212:1266
7. Bright FV, McGown LB (1985) Talanta 32:15
8. Kronick MN, Little WA (1975) J Immunol Meth 8:235
9. Dandliker WB, Feigen G (1961) Biochem Biophys Res Commun 5:299
10. Förster T (1948) Ann Physik 2:55
11. Stryer L (1978) Ann Rev Biochem 47:819
12. Selvin PR (1995) Fluorescence resonance energy transfer. Methods Enzymol 246:300
13. Clegg RM (1992) Fluorescence resonance energy transfer and nucleic acids. Methods Enzymol 211:353
14. Matayoshi ED, Wang GT, Krafft GA, Erickson J (1990) Science 247:954
15. Wingender E (1982) Anal Biochem 121:146
16. Schreiber AB, Hoebeke J, Vray J, Strosberg D (1980) FEBS Lett 11:303
17. Merguy JL, Boutorine AS, Garestier T, Belloc F, Rougée M, Bulychev NV, Koshkin AA, Bourson J, Lebedev AV, Valeur B (1994) Nucleic Acids Res 22:920
18. Clegg RM, Murchie AIH, Zechel A, Lilley DM (1993) Proc Natl Acad Sci USA 90:2994
19. Morrison LE, Hadler TC, Stols ML (1989) Anal Biochem 183:231
20. Ullman EF, Schwartzberg M, Rubenstein KE (1976) J Biol Chem 251:4172
21. Morrison LE (1988) Anal Biochem 174:101
22. Wieder I (1978) Background rejection in fluorescent immunoassay. In: Knapp W, Hobular K, Wick G (eds) Proceeding of the VI International Conference on Immunofluorescence. Amsterdam-New York, Elsevier Biomedical Press, pp 67–80
23. Marshall NJ, Dakubu S, Jackson T, Ekins RP (1981) Pulsed light, time resolved fluoro-immunoassay. In: Albertini A, Ekins R (eds) Monoclonal antibodies and development in immunoassay. Elsevier North Holland Biomedical Press, pp 101–108
24. Soini E, Lovgren T (1987) Crit Rev Anal Chem 18:105
25. Soini E, Hemmilä I (1979) Clin Chem 25:353
26. Bunzli JCG (1987) Complexes with synthetic ionophores. In: Gschneider KA, Eyring L (eds) Handbook on the physics and chemistry of rare earths, vol 9. North Holland, Amsterdam, pp 321–394
27. Evangelista RA, Pollak A, Allore B, Templeton E, Morton RC, Diamandis EP (1988) Clin Biochem 21:73
28. Morton RC, Diamandis EP (1990) Anal Chem 62:1841
29. Toner JL (1990) Lanthanide chelates as luminescent probes. In: Atwood J) (ed) Inclusion phenomena and molecular recognition. Plenum Press, New York, pp 185–197
30. Mukkala VM, Sund C, Kwiatkowski M, Pasanen P, Högberg M, Kankare J, Takalo H (1992) Helv Chim Acta 75:1621
31. Latua M, Takalo H, Mukkala VM, Matachescu C, Rodriguez-Ubis JC, Kankare J (1997) Luminescence J 75:149
32. Alpha B, Lehn JM, Mathis G (1987) Angew Chem Int Ed Engl 26:266
33. Gansow OA, Kausar AR, Triplett KM, Weaver MJ, Yee EL (1977) J Am Chem Soc 99:7087
34. Alpha B, Ballardini R, Balzani V, Lehn JM, Perathoner S, Sabbatini N (1990) Photochem Photobiol 52:299
35. Mathis G, Dumont C, Jolu EJP (1990) Pat WO 92:01224
36. Prat O, Lopez E, Mathis G (1991) Anal Biochem 195:283
37. Lopez E, Chypre C, Alpha B, Mathis G (1993) Clin Chem 39:196

38. Oi VT, Glazer A, Stryer L (1982) J Cell Biol 93:981
39. Schiller PW (1976) Measurement of intramolecular distances by energy transfer. In: Chem FC, Edelhoch H (eds) Biochemical fluorescence concepts. Dekker, New York, vol 1, pp 285–303
40. Mathis G (1993) Clin Chem 39:1953
41. Ozinskas AJ, Malak H, Joshi J, Szmacinski H, Britz J, Thomson RB (1993) Anal Biochem 213:264
42. Mabile M, Mathis G, Jolu EJP, Pouyat D, Dumont C (1991) Pat WO 92:13264
43. Mathis G, Socquet F, Viguier M, Darbouret B, Jolu EJP (1993) Clin Chem 39:1251
44. Mathis G, Socquet F, Viguier M, Darbouret B (1997) Anticancer Res 17:3011
45. Zuber E, Rosso L, Darbouret B, Socquet F, Mathis G, Flandrois JP (1997) J Immunoassay 18:21
46. Zuber E, Mathis G, Flandrois JP (1997) Anal Biochem 251:79
47. Kienhuis CBM, Geurts-Moespot A, Ross HA, Foekens JA, Swinkels LMJW, Koenders PG, Ireson JC, Benraad TJ (1992) J Recept Res 12:389
48. Mathis G, Préaudat M, Trinquet E (1994) Homogeneous EGF receptor binding assay using rare earth cryptates, amplification by non radiative energy transfer and time resolved fluorescence [abstract]. High throughput screening for drug development. CHI Conference, Philadelphia, USA
49. Farrar SJ, Whiting PJ, Bonnert TP, McKernan RM (1999) J Biol Chem 274:10,100
50. Braun S, Raymond WE, Racher E (1984) J Biol Chem 259:2051
51. Kolb AJ, Kaplita PV, Hayes DJ, Park YW, Pernelle C, Major JS, Mathis G (1998) DDT 3:1359
52. Park YW, Cummings RT, Wu L, Zheng S, Cameron PM, Woods A, Zaller D, Marcy A, Hermes JD (1999) Anal Biochem 269:94
53. Pernelle C, Clerc FF, Dureuil C, Bracco L, Tocque B (1993) Biochemistry 32:11,682
54. Mathis G, Préaudat M, Trinquet E, Pernelle C, Trouillas M (1994) A new homogeneous method using rare earth cryptates and amplification by energy transfer for characterization of the Fos and Jun leucine zipper peptides dimerization [abstract]. High throughput screening for drug development. CHI Conference, Philadelphia, USA, Sept 26–28
55. Mellor GW, Burden NM, Préaudat M, Joseph Y, Cooksley SB, Ellis JH, Banks MN (1998) J Biomolecular Screening 3:91
56. Kolb JM, Yamnaka G, Manly S (1996) J Biomolecular Screening 1:203
57. Cummings RT, McGovern H, Zheng S, Park YW, Hermes JD (1999) Anal Biochem 269:79
58. Préaudat M, Ouled-Diaf J, Alpha-Bazin B, Mathis G, Mitsugi T, Aono Y, Takahashi K, Takemoto H (1999). A homogeneous Caspase 3 activity assay using HTRF technology [poster]. SBS Conference, Edinburgh, Scotland, Sept 13–17
59. Matthews JA, Kricka LJ (1988) Anal Biochem 169:1
60. Hurskainen P, Dahlen P, Ylikoski J, Kwiatkowski M, Sütari H, Lövgren T (1991) Nucleic Acids Res 19:1057
61. Chan A, Diamandis E, Kradjen M (1993) Anal Chem 39:196
62. Lovgren T, Litia A, Hurskainen P, Dahlen P (1995) Detection of lanthanide chelate by time resolved fluorescence. In: Kricka LJ (ed) Non isotopic probing blotting and sequencing. Academic Press, pp 348–350
63. Alpha-Bazin B, Bazin H, Guillemer S, Sauvaigo S, Mathis G (2000) Europium cryptate labeled deoxyuridine-triphosphate analog: synthesis and enzymatic incorporation. Nucleosides Nucleotides and Nucleic Acids, 19:1463
64. Alpha-Bazin B, Bazin H, Guillemer S, Mathis G (1999). Europium cryptate labeled oligonucleotides in homogeneous time-resolved FRET DNA based assays. J Fluoresc (submitted)
65. Alpha-Bazin B, Bazin H, Guillemer S, Boissy L, Tanchou V, Mathis G (1999) Europium cryptate tethered nucleosides triphosphate for non radioactive labeling and detection of DNA and RNA [poster]. SBS Conference, Edinburgh, Scotland, Sept 13–17
66. Alpha-Bazin B, Mathis G (1999) Nucleosides Nucleotides 18(6/7):1277
67. Zhan JH, Chen T, Oldenburg KR (1999) A HTRF⟩ high throughput assay for reverse transcriptase using generic labeled reagents [poster]. SBS Conference, Edinburgh, Scotland, Sept 13–17

68. Alpha-Bazin B, Bazin H, Boissy L, Mathis G (2000) Europium cryptate-tethered ribo-nuclotide for the labeling of RNA and its detection by time-resolved amplification of cryptate emission. Anal Biochem 286:17
69. Zhou G, Cummings R, Li Y, Mitra S, Wilkinson HA, Elbrecht A, Hermes JA, Schaeffer JM, Smith RG, Moller DE (1998) Mol Endocrinol 12:1594
70. Degorce F, Achard S, Cœur B, Cougouluegne F, Ouled-Diaf J, Alpha-Bazin B, Préaudat M, Amoravain M, Carter-Allen K, Seguin P, Mathis G (1999) HTRF check kits for GST, 6-HIS, C-MYC and FLAG tag accessibility. SBS Conference, Edinburgh, Scotland, Sept 13–17
71. Degorce F, Achard S, Ouled-Diaf J, Alpha-Bazin B, Carter-Allen K, Seguin P, Mathis G (1999) Comparison of HTRF indirect systems for mouse and rat monoclonal antibody detection. SBS Conference, Edinburgh, Scotland, Sept 13–17
72. Amoravain M, Cœur B, Degorce F, Seguin P, Carter-Allen K, Mathis G (1999) New anti-FLAG conjugates for HTRF epitope tag recognition to be used in assay development and high throughput screening. SBS Conference, Edinburgh, Scotland, Sept 13–17
73. Degorce F, Beaumont F, Cougouluegne F, Alpha-Bazin B, Bazin H, Carter-Allen K, Seguin P, Mathis G (1999) Use of 2,4-dinitrophenyl-*N*-hydroxysuccinimide ester (DNP-NHS) for protein labeling and its detection by anti-DNP HTRF generic reagents. SBS Conference, Edinburgh, Scotland, Sept 13–17

Tracking Molecular Dynamics of Flavoproteins with Time-Resolved Fluorescence Spectroscopy

P. A. W. VAN DEN BERG, A. J. W. G. VISSER

22.1
Intrinsic Protein Fluorescence

The intrinsic fluorescence of tryptophan containing proteins is nowadays commonly used to study the physical and dynamic properties of this prominent class of biomacromolecules [1–5]. Although tryptophan fluorescence experiments have provided detailed insight in many protein systems, the applicability of the technique is strongly dependent on the number and position of the tryptophan residues, as well as on the specific research objective. Reason for this is the well-known complexity of the photophysical properties of the indole ring. Due to the degeneracy of energy levels in the excited state (^1L_A and ^1L_B) the observed nonexponential fluorescence decays are rather rule than exception, even in proteins that contain only one tryptophan in a rigid protein environment. In addition, fast depolarization of the fluorescence originating from interconversion between these two states complicates the interpretation of time-resolved fluorescence anisotropy measurements [6]. Particularly for the investigation of dynamic events such as protein motions, these fluorescence characteristics of the indole moiety prove a serious obstacle.

A possible way to overcome these problems is to attach an external fluorescence label such as an artificial dye to the protein, or to make a fusion with the natural fluorescent Green Fluorescent Protein (GFP) [7]. Fluorescence resonance energy transfer measurements on biomacromolecules which contain both an (artificial) energy donating and an energy accepting moiety are now routinely performed to gain information on the geometric and dynamic properties of proteins, protein complexes, nucleic acids or mixtures of them [8]. For the investigation of large overall protein motions fluorescence labeling has proven to be particularly suitable. However, for the study of more subtle processes such as protein dynamics in the active center of enzymes, where protein packing obstructs the use of external labels, the natural fluorescence of the protein is still the most appropriate tool.

22.2
Flavins and Flavoproteins

A particularly interesting class of enzymes for investigating conformational dynamics in the active site of proteins are the so-called flavoenzymes.

Flavoenzymes are involved in numerous redox processes in metabolic oxidation-reduction, photobiology and biological electron transport (for an overview see [9]). Members of this widespread class of enzymes contain as redox-active prosthetic group a naturally fluorescent flavin cofactor that emits light in the green spectral region. The fluorescence characteristics of this flavin cofactor strongly depend on the physical characteristics of the molecular environment of the fluorophore. Therefore, flavoenzymes offer the unique possibility to probe the conformational dynamics of the active site directly by time-resolved fluorescence and fluorescence anisotropy detection of this specific prosthetic group.

Flavin cofactors are derivatives of riboflavin, a compound better known as vitamin B2. While bacteria and plants are able to synthesize vitamin B2 themselves, higher organisms are dependent on the uptake of riboflavin via nourishment. The enzymes flavokinase and FAD synthase can subsequently convert riboflavin into flavin mononucleotide (FMN) and flavin adenine dinucleotide (FAD), which are the cofactors commonly found in flavoproteins (Fig. 22.1).

Both from the biochemical and the biophysical point of view, the essential part of the flavin cofactor is the isoalloxazine ring. The cofactor is able to act as an redox-mediator through this three-membered ring system, which can exist in the oxidized flavoquinone, one-electron reduced flavo-semiquinone and two-electron reduced flavo-1,5-dihydroquinone states. In general, the atom N5 – and to a lesser extent C4α – of the isoalloxazine ring acts as electrophilic site during catalysis. The redox-properties of the flavin are subtly modulated by the protein-environment, which makes it a very versatile molecule that is involved in a wide range of enzymatic reactions.

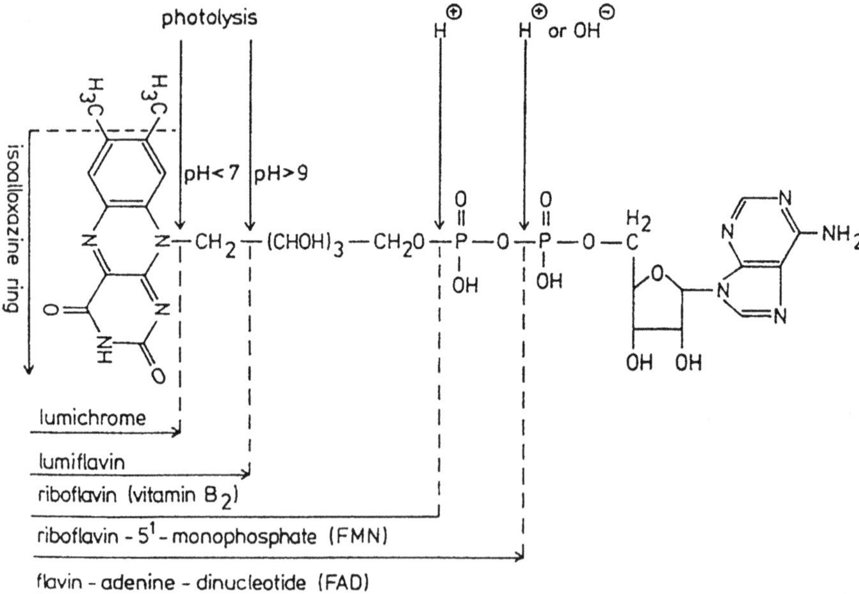

Fig. 22.1. Structure of FAD, FMN and riboflavin

The unique photophysical properties of the flavin ring have made this chromophore a rewarding tool for the whole spectrum of sophisticated optical spectroscopic techniques that are currently available. Owing to their relatively high extinction coefficient in the blue spectral region, oxidized flavins are easily recognized as yellow pigments. As the absorption of unpolarized as well as polarized light is influenced by the protein environment, optical techniques are commonly used to study flavoproteins and their interactions (e.g., free flavins do not have a pronounced circular dichroism spectrum in contrast to the bound species in many flavoenzymes). In general, however, fluorescence techniques provide the most sensitive tool for studying the physical and dynamic properties of the direct environment of the chromophore. Indeed this axiom applies to a whole range of flavoenzymes.

Flavins and flavoproteins fluoresce in the green spectral region which is a rarity for natural probes that allows highly selective detection. The fluorescence spectral characteristics as well as the fluorescence quantum yield of flavins strongly depend on the environmental factors such as dielectric constant, refractive index, and solvent polarity. In contrast to oxidized flavin, the flavin in the reduced state is hardly fluorescent. As a consequence of this low intrinsic fluorescence and the complex photophysical and photochemical behavior of the reduced state, fluorescence studies are normally restricted to oxidized flavins and flavoenzymes. In view of this, the content of this chapter will be restricted to oxidized flavins only. In aqueous solution, riboflavin and FMN possess a rather high fluorescence quantum yield ($Q = 0.26$; [10]). FAD, however, is much less fluorescent ($Q = 0.03$; [10]) because of the formation of an intramolecular complex between the flavin and adenine moieties. This reduction in quantum yield was shown to result from both static and dynamic quenching of the flavin fluorescence [11, 12]. In general, the flavin cofactor is bound to the protein in an extended conformation. Consequently, the intramolecular quenching of the FAD molecule is removed, implying that the fluorescence characteristics are determined by the protein environment in the very same way as those of bound FMN.

The electronic, photophysical, and structural features of the flavin molecule have been studied extensively [13–16]. The optical transition moments from ground state to both the first and the second singlet excited states have been determined by Johansson et al. [17]. The angle between the absorption and emission dipole moment of the first electronic transition is about 16°, as ascertained from the fundamental anisotropy of the flavin. The direction of the emission dipole moment was deduced from homo-energy transfer studies [18].

22.3
Flavin as a Fluorescent Probe for Flavoprotein Dynamics

As mentioned above, in nearly all flavoenzymes the FAD cofactor is bound in an extended conformation. The fluorescence characteristics of protein-bound flavins are therefore dependent on the exact amino-acid composition of the direct environment of the isoalloxazine ring as well as its solvent accessibility.

The most fluorescent flavoprotein is undoubtedly the yellow fluorescent protein from bioluminescent bacteria with the exceptionally high fluorescence quantum yield of about 0.6 [19]. However, contrary to flavoenzymes whose primary function is redox catalysis, this flavoprotein was specially designed with the purpose of emitting light. It is therefore not surprising that the fluorescence quantum yield of regular flavoenzymes is considerably lower. With a quantum yield of about 0.1, lipoamide dehydrogenase and thioredoxin reductase belong to the most fluorescent flavoenzymes known thus far. In many flavoproteins however, the flavin fluorescence is so heavily quenched ($Q \ll 0.01$) that they were traditionally regarded as 'non-fluorescent'. However, recent studies in which modern fluorescence methodologies with an ultra-high time-resolution were applied, have shown that the idea of 'non-fluorescent flavoproteins' should be abandoned (see below).

An important aspect in studying flavoproteins by fluorescence methods is the binding of the prosthetic group. In most flavoproteins, the flavin cofactor is non-covalently bound. In a small number of flavoenzymes however, the flavin ring is covalently linked to a histidine, cysteine or tyrosine of the polypeptide chain [20, 21]. Well-known examples of these covalent flavoproteins are for instance succinate dehydrogenase (Krebbs cycle) and monoamine oxidase (inactivation of neurotransmitters). The dissociability of the flavin cofactor can, in principle, seriously hamper fluorescence studies. Fortunately, in quite a lot of flavoproteins – among which are most of the disulfide-oxidoreductases from *E. coli* – the affinity for the prosthetic group appears to be quite high with equilibrium dissociation constants in the (sub)nanomolar range. However, other flavoproteins such as yellow fluorescence protein loose their flavin cofactor relatively easily ($40 \, nmol/l < K_d < 1 \, \mu mol/l$). Although the presence of free flavin should be avoided in all fluorescence experiments on flavoproteins, extreme care should be exercised in time-resolved fluorescence investigations on conformational dynamics of proteins. For a proper interpretation of fluorescence lifetime and anisotropy data in terms of different species or enzyme conformations, the absence of free flavin is essential, especially for enzymes with a low flavin fluorescence quantum yield.

In addition to suitable characteristics of the protein chromophore, other prerequisites exist that are equally important for a successful investigation of active-site conformational dynamics by fluorescence methods. One of these prerequisites obviously is the availability of a high-resolution three-dimensional structure of the protein. When studying the conformational dynamics in relation to the catalytic function, an additional requirement is the availability of detailed information concerning the catalytic mechanism of the enzyme. Both these conditions are now well met by a whole pool of flavoenzymes. Owing to their general occurrence and the wide range of biologically important reactions that they catalyze, flavoenzymes have since long time been subject of investigation.

Apart from detailed knowledge about the catalytic mechanism and structural information at the atomic level, a third criterion for studying functionally important motions of enzymes is the indication that these motions could actually exist in the enzyme of choice. These indications often arise from information on the three-dimensional structure of the enzyme, e.g., when crystal structure data show more than one conformation, or when the resolved structure does not provide a clear pathway for the catalytic reaction. Kinetic and spectroscopic studies, however, can yield such signs as well. Various flavoenzymes from different subclasses have been shown to satisfy this criterion as well. Taking into account the unique fluorescence properties of the flavin cofactor, these flavoenzymes are particularly interesting candidates for studying catalysis-related conformational dynamics with fluorescence methods. In the next sections, the results of a few of these flavoenzymes which have been recently studied will be discussed. However, prior to this discussion, a brief overview will be provided on the current insights in protein dynamics itself.

22.4
The Intrinsic Flexibility of Proteins

Since the mid-1970s, many experimental studies have suggested the existence of a whole landscape of conformational (sub)states within the protein native state, among which transitions can occur through activated dynamics of the polypeptide chain [22–26]. In this concept, conformational states are defined as regions in the configurational space surrounding local potential energy minima, each separated by potential energy barriers. In principle, these activation energies can vary from practically zero to values in the order of 100 kJ/mol. An interesting point of this model is that protein substates which are nearly isoenergetic may significantly vary in other properties such as binding constants or catalytic rate constants. Conformational transitions of enzymes from (sub)-states less efficient in catalysis (at that particular point in the reaction pathway) to 'more efficient' substates can thus be regarded as 'functionally important motions'.

Dynamic processes in proteins occur on a wide range of time scales. Distinction has to be made, between vibrational motions (within a single conformational (sub)state) and motions that are actually related to a conformational transition. Vibrational dynamics are characterized by a spectrum of frequencies ranging from 10^{14} s^{-1} [27] to 10^{10} s^{-1} [28, 29]. Modes in the high-frequency range comprise mainly the stretching and bending of bonds, whereas those in the low-frequency range encompass mainly the collective vibrational torsional motions in dihedral angles about the bonds. Conformational transition dynamics are characterized by a spectrum of relaxation times. Under physiological conditions, this spectrum of relaxation times ranges from 10^{-11} s (local side chain rotations or hydrogen bond rearrangements on the protein exterior) to hours or even years.

Contrary to the vibrational modes, the general mechanisms of motion for conformational transitions are still not very clear. One of the few 'general rules' known at this moment is that conformational transitions within the protein native state do not take place in the entire body of a protein domain, but are predominantly found in liquid-like regions on the protein exterior that surround more solid-like secondary structure elements. Based on crystallographic data from free proteins existing in more than one conformation, and on comparisons between free proteins and protein/substrate complexes, Gerstein and coworkers [30] have classified the structural changes involved in conformational transitions into basically two types: hinge motions and shear motions. Hinge motions occur in strands as well as in β-sheets and α-helices that are not constrained by tertiary packing interactions. Shear motions are found in more closely packed segments of the polypeptide chain. Whereas hinge motions can generate rather large changes in the three-dimensional structure of the protein (especially in the case of strands and β-sheets), shear motions generally result in small structural changes only. By combination of the hinge and shear mechanisms, a multitude of different kinds of motions can be described. The principle of hinge and shear motions was found to apply to both large inter-domain motions of proteins, as well as to motions of small protein fragments such as individual loops or α-helices. Gerstein et al. [30] noted that proteins that have a predominantly hinged domain motion usually possess two domains connected by linking regions that are relatively unconstrained by packing. Enzymes with two crystal conformations which show such a hinge motion are, e. g., lactoferrin [31, 32], adenylate kinase [33, 34] and glutamate dehydrogenase [35]. As with hinge motions, proteins with shear motions also tend to have specific architectural features. One example is a layered structure in which one layer can slide over another. Although shear motions have been found for many different interfaces, a characteristic feature are helix-helix interfaces in which the helices are crossed (interhelical angle between 60° and 90°). Well known examples, for which two different conformations were found that can interconvert mainly via shear motions, are the enzymes citrate synthase [36, 37], and alcohol dehydrogenase [38, 39].

Another general feature of conformational transitions is the impact of substrate binding: in many enzymes the substrate-binding site is located at the interface of different domains. When the binding site is easily accessible for the substrate in solution – and in many cases it is – the enzyme can be regarded as having an 'open conformation'. Often, the protein domains can close around such a binding site. In general, substrate binding stabilizes a 'closed' conformation of the enzyme. The opposite is true for enzymes in which the substrate-binding site is shielded from the solvent by part of the protein, for instance via a loop region. The occurrence of thermal fluctuations, protein-protein interactions, or the interaction with a second substrate are then required to 'open up' this conformation so that the binding site becomes accessible to the substrate. Contrary to the closed state, which usually yields a single crystal structure conformation, structural data indicate that the open state can consist of a range of different open conformations. For instance, X-ray crystallographic studies revealed more than one open conformation for mutants of T4 lysozyme [40, 41] and the leucine/isoleucine/valine-binding protein [42]. However, care should be exercised not to cling to the idea of 'open' and 'closed' conformations as single, discrete, rigid conformations; many other studies confirm the continuously dynamic character of the enzyme in both closed and open state, which is invisible with crystallography due to the crystal packing forces. The evidence presently available suggests that the open and closed states of enzymes only differ slightly in energy; at room temperature they will be in dynamic equilibrium [30 and references therein].

The current models for describing conformational dynamics can be divided into two classes: the 'protein glass' and the 'protein machine' (for a review see [43]). Both classes have in common that the spectrum of relaxation times describing conformational transitions seems to be quasi-continuous in the range from 10^{-11} s to 10^{-7} s. In the 'protein glass' models, the spectral density of relaxation times is assumed to vary according to a power law, which causes the dynamics to be alike in all time scales. This kind of time-scaling can arise from a hierarchy of potential barriers ('fractal times'), such as in the famous model of Frauenfelder et al. [23–26]. In their 'tiers of substates' model, each particular conformational (sub)state surrounded by high energy barriers comprises a multitude of conformational substates with lower energy barriers. According to this model, the height of the activation energy for a conformational transition in each tier is directly related to the magnitude of the protein motion involved. However, recent evidence suggests that such time-scaling can often equally well originate from a hierarchy of bottlenecks (the entropy barrier heights) in the network connecting conformations that can be directly interconverted into one another (the 'fractal space') [43–45]. The assumption of infinite time-scaling, however, makes the protein glass model rather unrealistic, and limits its applicability to only a few levels of the hierarchy [24, 26]. Free of this disadvantage is the 'protein machine' class of models. In these models, the activated dynamics of conformational transitions are represented by a quasi-continuous motion along a few 'mechanical coordinates' [43]. These mechanical coordinates are, e.g., angles describing the mutual orientation of rigid fragments of secondary structures or domains. Conformational transitions can only take place directly between two adjacent 'conformational coordinates'. Variation of the angle between two structural elements of a protein goes through a series of well-defined successive conformational transitions involving interactions at the atomic level such as rearrangements of hydrogen bonds within the protein, and those between the protein and the solvent. A hierarchy of 'mechanical elements' can exist in the form of side chains, secondary structure elements, and domains. According to the protein machine model, a biochemical reaction can be gated by specific conformational transitions. The concept of the protein machine has, although perhaps implicitly, been used by various authors to describe biochemical reaction kinetics. An example of this is the use of Kramers theory of reaction rates in the spatial diffusion limit [46] for the interpretation of specific enzymatic reactions in solvents of various viscosities [47, 48]. At this point of time, a statistical theory is needed that adequately describes the involvement of conformational dynamics in biochemical processes. An initial attempt hereto was made by Kurzyñsky [43, 44], who has used the concept of the protein machine to describe an enzymatic reaction involving a single covalent transformation being gated by the enzyme intramolecular dynamics.

22.5
Functionally Important Motions in Flavoenzymes; an Introduction to Glutathione Reductase, Thioredoxin Reductase and *p*-Hydroxybenzoate Hydroxylase

With this general picture of conformational dynamics and protein motions in mind, let us now return to the flavoenzymes. Based on the chemical reactivity, and the concomitant differences in the protein structure, flavoenzymes have been grouped in a variety of different flavoenzyme families. Due to this richness of structures and reactivity, one can expect virtually all types of protein motions to occur in flavoenzymes. As with other proteins, the first indications for the existence of functionally important motions mainly arise from information on the three-dimensional structure. In this section we would like to introduce three flavoenzymes for which crystal structure data have suggested different types of functionally important motions, and which were recently studied by time-resolved fluorescence methods. These enzymes are glutathione reductase, thioredoxin reductase, and *p*-hydroxybenzoate hydrolase, and a description of them will help to guide us through the current insights in fluorescence dynamics.

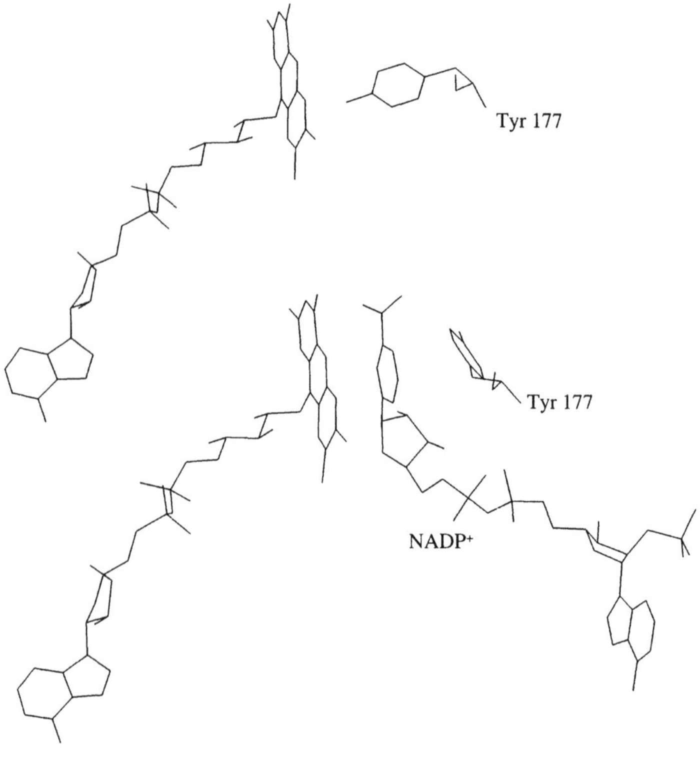

Fig. 22.2. The relative positions of the isoalloxazine ring of FAD and Tyr 177 in the structure of *E. coli* glutathione reductase (GR) (*top panel*) and GR complexed with NADP+ (*lower panel*). Note that in the latter case Tyr 177 is moved away from the isoalloxazine

Glutathione reductase (GR) is an example of a flavoenzyme in which a small motion of an amino acid side chain plays an evident role in catalysis. The enzyme is one of the best-studied flavoproteins arising from its crucial function in a variety of cellular processes; by catalyzing the NADPH-dependent reduction of oxidized glutathione (GSSG), the enzyme is responsible for maintaining a high GSH over GSSG ratio in cells (for a review see [49]). Glutathione reductase belongs to the pyridine nucleotide disulfide-oxidoreductase family, which includes, among others, thioredoxin reductase (see below) and lipoamide dehydrogenase. The enzyme is a homodimer and contains one redox-active disulfide bridge and one molecule of FAD per ~50 kDa subunit. High-resolution crystal structures have provided detailed structural information on GR from various sources including human erythrocytes and *E. coli* [50–54]. One of the most compelling structural characteristics is the position of the tyrosine adjacent to the flavin (Tyr177 in *E. coli* GR and Tyr197 in erythrocyte GR): this tyrosine residue blocks the active site thereby preventing the binding of NADPH. The catalytic mechanism, which was based on three-dimensional structures of free and substrate-bound forms of the human enzyme [55], therefore includes a movement of this tyrosine away from the flavin (Fig. 22.2).

Rotation of a complete protein domain is involved in the catalytic mechanism of *E. coli* thioredoxin reductase (TrxR). This enzyme catalyzes the NADPH-dependent reduction of the protein substrate thioredoxin, which is involved in cellular processes such as ribonucleotide reduction and protein folding (for a review see [49, 56]). Whereas the human enzyme resembles glutathione reductase, the *E. coli* enzyme shows a remarkably different crystal structure [57, 58], in which no obvious path for the flow of electrons from NADPH to thioredoxin is found. The monomers (35 kDa) of the homodimeric enzyme consist of an NADPH-binding domain and an FAD-binding domain, connected by a double-stranded β-sheet. In the crystal structure there is no bindingsite present for thioredoxin. Moreover, NADPH is bound far away from the flavin ring (17 Å), and its access to the isoalloxazine ring is blocked. By graphically rotating the NADPH domain over 66°, Waksman et al. [57] showed that the above-mentioned problems could be overcome (Fig. 22.3). In the rotated structure the nicotinamide and isoalloxazine rings are in close contact with each other, and the redox-active disulfides move from the inside (near the flavin) to the surface of the protein where they become accessible for the protein substrate. It was therefore proposed that *E. coli* thioredoxin reductase has two conformational states: a conformation in which the rings of NADPH and FAD are juxtaposed (FR), and the form corresponding to the crystal structure (FO) (Fig. 22.3). Fluorescence studies on wild-type TrxR and mutant enzyme TrxR C138S have now provided strong evidence for the existence of the FO and FR conformations in solution [59] (see below).

A completely different type of motion, namely that of the isoalloxazine ring of the flavin cofactor itself, is found in the enzyme *p*-hydroxybenzoate hydroxylase (PHBH) [60, 61]. This aromatic hydroxylase catalyzes the conversion of *p*-hydroxybenzoate into 3,4-dihydroxybenzoate using NADPH and molecular oxygen (for a review see [62]). Crystallographic studies of binary enzyme-substrate complexes showed the isoalloxazine part of the FAD cofactor in two

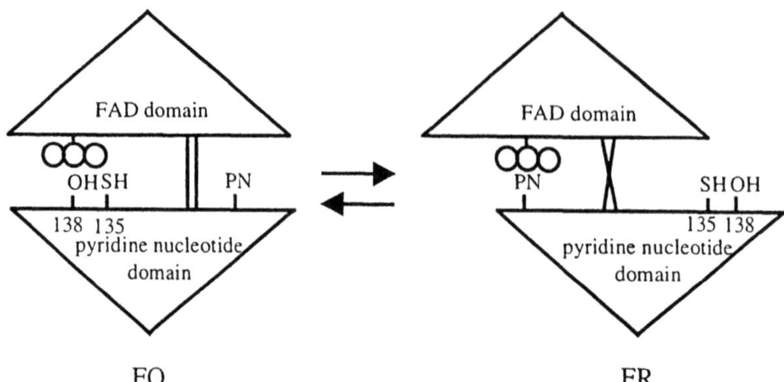

Fig. 22.3. Diagram of the mutant C138S of *E. coli* thioredoxin reductase (TrxR) in the FO and FR conformers. *The lines* connecting the FAD domain and the pyridine nucleotide domain depict the double-stranded β-sheet. FAD is represented by *three circles*, PN indicates bound pyridine nucleotide

distinct conformations [60, 63]. The 'in' conformation, in which the flavin ring is located in the active site, was found for the enzyme saturated with *p*-hydroxy-benzoate. Binding of substrates bearing a hydroxyl group at the 2-position (2,4-dihydroxybenzoate, 2-hydroxy-4-aminobenzoate) led to the 'out' conformation, in which the isoalloxazine ring has moved towards the surface of the protein (Fig. 22.4). Absorption difference spectra of the binary enzyme-substrate complexes were found to reflect these conformational differences. Crystallographic

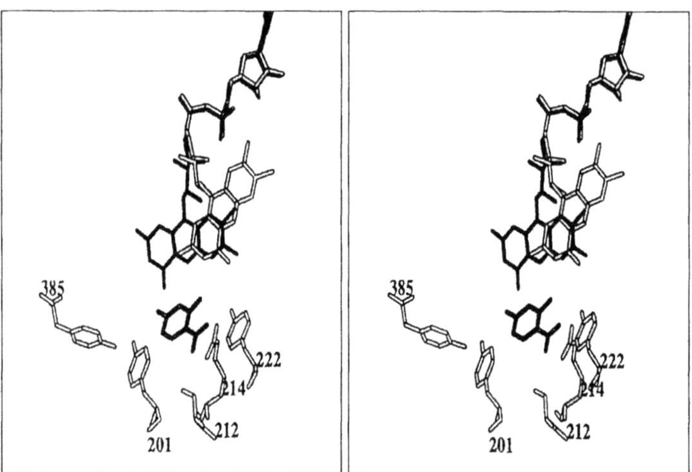

Fig. 22.4. Structure of wild-type *p*-hydroxybenzoate hydroxylase (PHBH) in the complex with 2,4-dihydroxybenzoate. The flavin occupies the 'out' conformation (*open bonds*). The 'in' conformation of the flavin in wild-type PHBH complexed with *p*-hydroxybenzoate is also shown (*solid bonds*)

data [61] and spectral data [64] on mutant enzymes, in which Tyr 222 – one of the three active-site tyrosines in PHBH – was replaced, demonstrated that this particular residue is involved in flavin motion. In PHBHn, mobility of the flavin itself is essential for catalysis: while the flavin in the 'in' position provides a suitable environment for efficient hydroxylation of the substrate (shielded from the solvent), the swinging 'out' of the flavin creates a pathway for substrate binding and product release. Recent crystallographic data of the substrate-free enzyme showed that the isoalloxazine ring is indeed flexible, and on average located at a position intermediate between the 'in' and 'out' conformation [65].

With a global picture of these three enzymes in mind, we shall now turn our attention to the insights in protein dynamics and mobility that modern fluorescence techniques can offer.

22.6
Current Insights in Flavoprotein Active-Site Dynamics from Fluorescence: the Drive to Higher Time-Resolution, the Revised Interpretation of Heterogeneous Fluorescence Decays, and the Introduction of a New Mechanism for Fluorescence Depolarization

In the last decade, fluorescence spectroscopy has yielded insight in the active-site dynamics of a range of different flavoproteins [18, 59, 66–73]. This current information on the active-site dynamics of flavoenzymes has mainly been derived from fluorescence lifetime spectra. At first sight this may seem somewhat odd as, contrary to lifetime data, time-resolved fluorescence anisotropy data yield direct information on the mobility of the fluorophore. Correlation times recovered from time-resolved fluorescence anisotropy have indeed yielded valuable information on the overall protein tumbling of flavoenzymes, the possible dissociation of the flavin cofactor and the existence of local mobility of the flavin ring on the nanosecond time-scale. However, the first two types of correlation times, although quite relevant to check the system, do not give information on the active-site dynamics itself. As many flavoenzymes have a considerable mass and only a short fluorescence lifetime, overall protein tumbling is in fact often hard to resolve at all. As mentioned earlier, dissociation of the flavin cofactor seriously hampers the interpretation of fluorescence data and should therefore be avoided in studying protein dynamics. This requirement is best met by choosing flavoenzymes in which the prosthetic group is tightly bound ($K_d < 1–10$ nmol/l). In these flavoenzymes, however, local mobility of the flavin ring is often absent and the flavin appears to be rigidly bound on the (sub)nanosecond time-scale. Furthermore, the interpretation of fluorescence anisotropy data can be complicated by the presence of other sources of rapid fluorescence depolarization, which can be easily misinterpreted as local mobility.

A good example of this is the enzyme lipoamide dehydrogenase, a homodimeric flavoenzyme, which is closely related to glutathione reductase. Due to its relatively high fluorescence quantum yield, lipoamide dehydrogenase was one of the first flavoenzymes studied by time-resolved fluorescence and fluorescence anisotropy measurements [74, 75]. For this enzyme, a small, rapidly depolarizing process was first assigned to local mobility of the isoalloxazine ring around its longitudinal axis. Later measurements performed with improved equipment and analysis techniques showed that the process was temperature invariant, and that not mobility but homo-energy transfer between the flavin cofactors from the two different subunits is the source of depolarization [18]. Many flavo-enzymes are in fact dimers, tetramers or even octamers, so that homo-energy transfer is often a process to be taken into account. However, fluorescence depolarization arising from homo-energy transfer can be recognized rather easily by calculating the expected time constant from the inter-flavin distance and orientation of the flavin rings – which can be determined from a high-resolution crystal structure –, and by testing the temperature invariance of the process [18, 68, 70]. Recent studies on flavoenzymes have shown that fluores-cence depolarization can also arise from other processes which are not related to the mobility of the fluorophore. Further on we will discuss this newly dis-covered mechanism of fluorescence depolarization that evidently appears in the enzymes glutathione reductase and NADH peroxidase (see below).

The misinterpretation of time-resolved fluorescence data is, however, not restricted to anisotropy data only: not everything that shows up as dynamic fluorescence quenching in lifetime patterns should be explained in terms of dynamics and conformational differences as well. For some time, the hetero-geneous fluorescence decays that are found in many flavoenzymes and in vir-tually all tryptophan-containing enzymes, were thought to originate solely from dynamics (relaxation processes) on the time-scale of fluorescence and/or the existence of different protein conformational states. While analysis of the fluorescence decay through a quasi-continuous distribution of time constants often resulted in interpretations in terms of relaxational processes, analysis by a sum of exponentials frequently led to the suggestion of the existence of just as many protein conformational states. For many protein systems, either one or both of these explanations may indeed contribute to the heterogeneous fluores-cence decays found. However, an interesting paper by Bajzer and Prendergast [76] opened up a completely new way to explain heterogeneous fluorescence decays. The basic assumption in explaining fluorescence lifetime data in terms of conformational substates is that in each protein conformation the inter-actions between the chromophore and its direct environment will result in a single fluorescence lifetime. Bajzer and Prendergast [76] demonstrated that this assumption is not necessarily correct by showing that the non-exponential fluorescence decay of several tryptophan-containing proteins can be explained by energy transfer to different acceptor sites in the protein, which all contribute with a certain probability to de-excitation of the donor. In this so-called 'multiple quenching sites' model, the heterogeneity in fluorescence decay can arise from a multiplicity of competing interactions that involve transfer of energy in a broad sense between the light-excited chromophore and different sites in the molecule. In contrast to the conformationally determined models of quenching, which all assume energy transfer via collisional quenching, the model of 'multiple quenching sites' is not dependent on collisions, but includes other de-excitation processes such as fluorescence resonance energy transfer and electron transfer. A multi-exponential description of the fluorescence decay can then be explained without invoking multiple protein conformations.

The first flavoprotein that was evaluated in the light of this new concept was *E. coli* glutathione reductase [69]. An earlier time-resolved fluorescence study on this hardly fluorescent enzyme from human erythrocytes [68] had yielded a fluorescence lifetime spectrum with five components, in which an ultra-short lifetime component of about 30 ps clearly dominated. This ultra-short component, which was near the detection limit of the set-up used at that time, was interpreted as an excited-state interaction between the flavin and the juxtaposed tyrosine residue that blocks the active site (see above). The five lifetime components were explained by the existence of (at least) five conformational substates of the enzyme, with different positions and/or orientations of this tyrosine residue [68]. Moreover, it was proposed that interconversion between the assumed substates corresponded to gross structural changes in the protein.

A more detailed study on *E. coli* glutathione reductase, and on mutants in which this flavin-shielding tyrosine was replaced by either a phenylalanine (GR Y177F) or a glycine (GR Y177G), has shed light on the role of this particular tyrosine and on the existence of different protein conformations [69]. For wild-type glutathione reductase, this study confirmed the lifetime spectrum with five components (Fig. 22.5 B). Due to the enhanced time-resolution of the equipment, the predominant ultra-short lifetime component was found to be signifi-

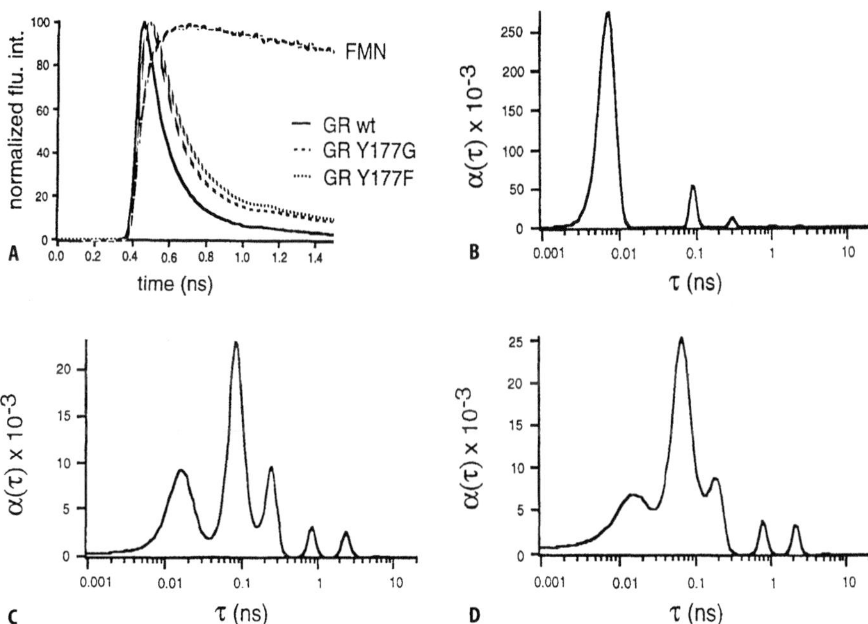

Fig. 22.5. **A** Experimental total decay of wild-type GR and mutants Y177F and Y177G in 50 mmol/l potassium phosphate buffer (pH 7.6) and FMN in 50 mmol/l potassium phosphate buffer (pH 7.0) at 293 K. Only the initial part of the 16-ns experimental time window is shown. **B–D** Fluorescence lifetime distribution of wild-type GR (**B**) and mutants Y177F (**C**) and Y177G (**D**) in 50 mmol/l potassium phosphate buffer (pH 7.6) at 293 K. The figure has been reproduced from [69] with permission of the Biophysical Society

cantly shorter (7 ps), and again close to the detection limit. The mutant enzymes, however, yielded striking new results: these enzymes, which lacked the flavin-shielding tyrosine, still showed a rapid fluorescence decay with respect to a non-quenched flavin compound such as FMN in aqueous solution (Fig. 22.5 A). Data analysis again resulted in five lifetime components, four of which were practically identical to that of wild-type enzyme. However, the ultrafast 7 ps lifetime component had disappeared, and quenching was now largely determined by a longer component of about 90 ps (Fig. 22.5 C, D). From the lifetime data of the *E. coli* glutathione reductase enzymes, three interesting issues could be derived. First, the data provided strong evidence for 'the multiple quenching sites' model. Second, the molecular mechanism of ultra-rapid fluorescence quenching in glutathione reductase was disclosed. Third, the study showed that ultra-fast processes in the picosecond domain are quite relevant to analyze dynamic processes in biological systems. In the next sections we will discuss the first two topics in a more general context. We will return to the third point of consideration in discussing recent studies on the enzyme *E. coli* thioredoxin reductase (see below).

For 'the multiple quenching sites' model, the *E. coli* glutathione reductase enzymes yielded convincing results. The fact that the mutant enzymes, which lack the tyrosine residue causing the ultra-short fluorescence lifetime, still exhibited a rapid heterogeneous fluorescence decay could only be explained by the existence of quenching sites in the protein other than the juxtaposed tyrosine. The similarity between the four longer fluorescence lifetimes found for both wild-type GR and the mutant enzymes strongly indicated that these quenching sites also contribute, albeit to a minor extent, to the flavin fluorescence decay of the wild-type enzyme. The position of the specific site and its distance to the flavin, together with the nature of the quenching interaction, then determines the probability by which each site contributes to fluorescence quenching. However, conformational fluctuations on the (sub)nanosecond time-scale as well as on longer time-scales will create a heterogeneous ensemble in which the rate and probability of fluorescence quenching by a specific site is controlled by the precise micro-structure and dynamics of the light-excited enzyme. Fluorescence lifetime data of other flavoenzymes such as NADH peroxidase [70] and *p*-hydroxybenzoate hydroxylase (P.A.W. van den Berg, unpublished results) have also confirmed the validity of 'the multiple quenching sites' model. In both these enzymes, the replacement of a tyrosine residue juxtaposed to the flavin by a non-aromatic residue (Y159A in *E. coli* NADH peroxidase and Y222V in *P. fluorescens p*-hydroxybenzoate hydroxylase) resulted in a simplified fluorescence lifetime pattern as well. The study on NADH peroxidase also revealed the nature of a second quenching site, namely the active-site cysteine, Cys42. In general, the active-site cysteines of disulfide oxidoreductases that are responsible for oxidation of the flavin after reduction by a nicotinamide cofactor are likely quenching sites [69].

The data of the tyrosine mutants of glutathione reductase, corroborated by those of NADH peroxidase [70], riboflavin binding protein and glucose oxidase [72], put a new perspective to the concept of 'non-fluorescent flavoproteins'. As mentioned before, a considerable number of flavoenzymes exhibit such a low quantum yield of flavin fluorescence that they were traditionally regarded as 'non-fluorescent'. When high-resolution structures of various flavoproteins became available, it was noted that these enzymes generally contained one or more aromatic residues, particularly tryptophans or tyrosines, in close proximity to the flavin. The strongly quenched flavin fluorescence was then explained by the formation of a ground-state complex between the isoalloxazine ring and the aromatic amino acid leading to static quenching ('dark complex'). In the late 1980s, when time-resolved fluorescence techniques were developed to a sub-nanosecond time-resolution, the first picosecond lifetime components were reported for flavoenzymes with a tryptophan juxtaposed to the flavin, such as flavodoxin [77]. The detection of such a short flavin fluorescence lifetime, which is by definition reflecting a dynamic process of quenching, led to the proposal of a new mechanism of quenching. For flavodoxin from *Desulfovibrio*, it was suggested that photo-induced electron transfer from the electron rich aromatic side chain of the tryptophan to the light-excited flavin was responsible for this ultra-rapid process (30 ps) [77]. Picosecond-absorption measurements on this enzyme proved the existence of a transient exciplex absorption band with an identical lifetime that could indeed be attributed to electron transfer from tryptophan to the excited flavin [78]. The ultra-short flavin fluorescence lifetime components recently found for glutathione reductase (Fig. 22.5 B), NADH peroxidase [70], and p-hydroxybenzoate hydroxylase (Fig. 22.8) demonstrated that this mechanism of quenching is generally applicable for flavoproteins. The temperature-invariant 7 ps lifetime component of *E. coli* glutathione reductase has been compared with the time-constant expected for photo-induced electron transfer which was calculated from the Rehm-Weller equation [79, 80]. This calculation confirmed the existence of a highly efficient quenching route via electron transfer from the tyrosine to the light-excited flavin [69]. The recent experiments strongly suggest that, in principle, there are no 'non-fluorescent' flavoproteins, and that the expression should be interpreted as a relative one. An illustrative example in this respect is the enzyme ferredoxin NADP$^+$ reductase (FNR) from the cyanobacterium *Anabaena*. In this enzyme, the isoalloxazine ring of the FAD cofactor is sandwiched between two tyrosine residues (one ring is parallel to the isoalloxazine, the other tyrosine is under a certain angle) [81]. Based on the mechanism of quenching described above, it is not surprising that flavin fluorescence in this enzyme is extremely quenched. Fluorescence lifetime spectra of FNR obtained with the very same set-up as used for *E. coli* glutathione reductase yielded just one single decay component at the limit of detection (about 3 ps, unpublished results of A. Arakaki). Obviously, this enzyme is one bridge too far for measurements with 'standard' time-correlated single photon counting techniques. However, by applying modern fluorescence methodologies with an ultra-high time-resolution – preferably on the subpicosecond time-scale – investigation of these 'non-fluorescent' systems is now becoming feasible (see below).

Let us now return to fluorescence depolarization as a source of information on protein dynamics. It was while studying the *E. coli* glutathione reductase enzymes that we first encountered a quite unexpected phenomenon which has led us to propose a new mechanism of fluorescence depolarization. Analysis of the time-resolved anisotropy decay of glutathione reductase revealed two processes of flavin fluorescence depolarization. The first one, with a time constant of about 6 ns, could be assigned to intramolecular energy transfer between the two flavins [68, 69]. The second process – a rapid anisotropy decay with a time constant of about 2 ns at 293 K and a large amplitude – was in human erythrocyte glutathione reductase interpreted as local mobility of the isoalloxazine ring [68]. However, the anisotropy data of the *E. coli* mutants rendered this explanation highly unlikely. In contrast to wild-type enzyme in which this rapid depolarization process completely dominated the anisotropy decay, the *E. coli* mutants Y177F and Y177G showed hardly any depolarization at all; only a minor amplitude process corresponding to homo-energy transfer between the flavins was resolved [69]. Judging from the fluorescence lifetime data, there was no reason to assume that the flavin would be mobile in wild-type enzyme and not in the tyrosine mutants; in fact, fluorescence lifetime data over a broad range of temperatures indicated a more flexible structure for the mutant enzymes. Based on these results, the idea arose that not internal mobility of the flavin itself, but a transient charge-transfer complex between tyrosine 177 and the light-excited flavin, is the source of rapid depolarization. Excitation of the flavin induces a relaxation process in the protein environment that leads to the formation of a charge-transfer complex between the flavin and tyrosine in which the direction of the emission dipole moment of the flavin is changed. Studies on wild-type enzyme in which the effect of temperature and substrate analogues binding was investigated have yielded evidence for the direct involvement of tyrosine 177 and the relaxational nature of the process (P. A. W. van den Berg et al., manuscript in preparation).

Supporting evidence for this new mechanism of fluorescence depolarization was found in studies on the enzyme NADH peroxidase from *E. coli* [70]. The active-site structure of this enzyme strongly resembles that of the closely related glutathione reductase; in the crystal structure, the NADH binding site is blocked by a tyrosine residue (Tyr159) that is juxtaposed to the flavin [82]. As with glutathione reductase, structural information revealed that binding of the nicotinamide cofactor is coupled with a movement of this tyrosine away from the flavin [83]. The fluorescence anisotropy decay of wild-type NADH peroxidase showed a rapidly depolarizing process (Fig. 22.6). In the mutant enzyme Y159A, this rapid depolarization no longer appeared (Fig. 22.6), and the remaining small amount of depolarization was only determined by homo-energy transfer between the four flavins in the tetramer [70]. Besides a temperature dependence, the rapid depolarization in wild-type enzyme also showed a clear wavelength dependence; on the red side of the emission spectrum (567 nm), depolarization occurred much faster than on main-band detection (526 nm). This effect confirms the relaxational character of the observed phenomenon: the formation of a transient complex between the light-excited flavin and the tyrosine results in a more stabilized charge-transfer excited state, that is hence red-shifted compared to the first-excited singlet state. Similar behavior in fluorescence depolarization was found for the enzyme *p*-hydroxybenzoate hydroxylase (unpublished results). Binding of the aromatic substrate to the enzyme induced a rapid fluorescence depolarization. These findings indicate that the formation of a

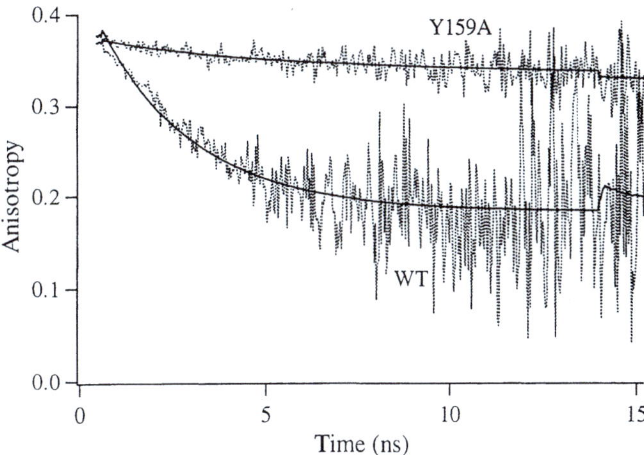

Fig. 22.6. Experimental (*dotted curves*) and fitted (*solid line*) fluorescence anisotropy decays for wild-type NADH peroxidase and Y159A mutant in 50 mmol/l potassium phosphate buffer (pH 7.0) at 277 K. For the wild-type enzyme the fit was according to a three-exponential decay model with the following amplitudes and correlation times: $\beta_1 = 0.19$, $\phi_1 = 2.9$ ns (relaxation), $\beta_2 = 0.01$, $\phi_2 = 12.3$ ns (energy transfer, fixed), $\beta_3 = 0.17$, $\phi_3 = 132$ ns (rotation of tetramer, fixed). For the Y159A mutant there was no rapid relaxation and only the latter two correlation times were sufficient to approximate the anisotropy: $\beta_2 = 0.01$, $\phi_2 = 12.3$ ns (energy transfer, fixed), $\beta_3 = 0.36$, $\phi_3 = 132$ ns (rotation of tetramer, fixed). The figure has been reproduced from [70] with permission of the American Chemical Society

charge-transfer complex, in which the emission dipole moment moves out of the plane of the isoalloxazine ring (Fig. 22.7), should be taken into account as a possible mechanism of flavin fluorescence depolarization in all systems where aromatic amino acids or substrates are at Van der Waals distance of the isoalloxazine ring.

Earlier in this paragraph we have discussed that not every visible fluorescence lifetime corresponds to a single protein conformation. Recently, however, we were confronted with the fact that the opposite can be true as well. Even in systems where clear evidence is available for the existence of different protein conformations, it can be quite difficult – if not impossible – to visualize and identify these various conformations by means of fluorescence lifetime data. Here we will discuss this complication taking the enzymes *E. coli* thioredoxin reductase and *p*-hydroxybenzoate hydroxylase as an example. For *E. coli* thioredoxin reductase (TrxR), strong indications were found for the involvement of a large domain rotation in catalysis yielding two highly different protein conformations (see above). Support for this concept was provided by structural information [56, 57] as well as by kinetic data on cross-linked complexes of the enzyme and substrate [84, 85] and spectroscopic studies. Firm evidence was acquired from the investigation of a specific mutant enzyme, TrxR C138S [59]. Contrary to wild-type enzyme, TrxR C138S was found to be little fluorescent (7% of wild-type TrxR). Absorption spectra also revealed significant differences

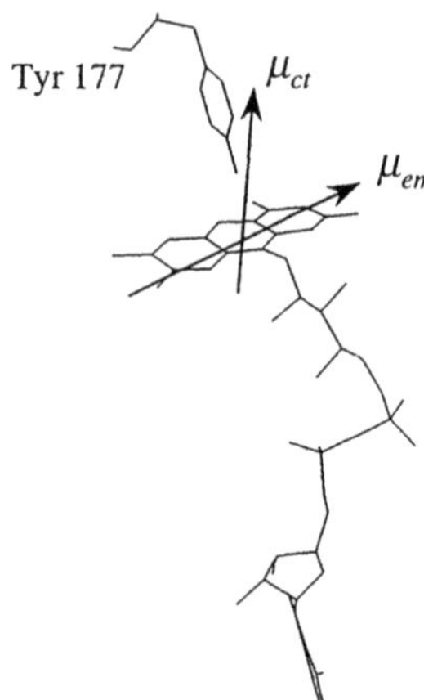

Fig. 22.7. Schematic representation of the depolarization caused by a change in direction of the emission transition moment from the initial 'in-plane' (of the isoalloxazine ring) direction (μ_{em}) to an 'out-of-plane' direction corresponding to the emission transition moment of the charge-transfer excited state (μ_{ct}). The example is taken from the structure of *E. coli* GR (see Fig. 22.2)

between wild-type and mutant enzymes. It was then proposed that in TrxR C138S, the crystal structure conformation (FO) is highly stabilized by a hydrogen bond between the serine residue and the N5 of the flavin, which causes strong fluorescence quenching. Wild-type TrxR was postulated to be predominantly in the FR conformation, in which no quenching residues are in close contact with the flavin. Steady-state fluorescence titrations of both enzymes with a thiol-specific reagent (phenylmercuric acetate (PMA)) and the NADPH analogue 3-aminopyridine adenine dinucleotide (AADP$^+$) corroborated this model [59].

Although the fluorescence intensity of the different enzyme/substrate (analogue) complexes varied highly, time-resolved fluorescence investigations on these complexes revealed hardly any differences in the fluorescence lifetime spectra. Only after applying highly sophisticated fluorescence methods, that yield a time-resolution up to several hundreds of femtoseconds in combination with time-resolved spectral data, were we able to visualize and identify both the FO and FR conformation (Van den Berg et al., manuscript in preparation). These results show that fluorescence lifetimes of just a few picoseconds, or even hundreds of femtoseconds, can disclose important information on the conformational dynamics of proteins. Recently, Mataga et al. [72] have been the first to report femtosecond fluorescence quenching in flavoproteins. By means of fluorescence up-conversion techniques they were able to detect such ultra-rapid processes in the formerly 'non-fluorescent' flavoenzymes riboflavin binding protein and glucose oxidase. The recent results on thioredoxin reductase prove that an ultra-high time-resolution is not only a necessity to study 'non-fluorescent' flavoenzymes, but also an important tool to investigate properly every protein system irrespective of its quantum yield. The studies presented in this chapter reflect the current drive to fluorescence methods with a better time-resolution. Time has now come to no longer ignore indications for the existence of rapid events, but to fully investigate them.

However, even fluorescence techniques with an ultra-high time-resolution and studies of complex formation with substrate analogues are not always sufficient to reveal the presence of different protein conformers. An example of this is the enzyme *p*-hydroxybenzoate hydroxylase (PHBH). High-resolution crystal structures of wild-type PHBH and mutant enzymes, either in the free state or complexed with aromatic substrates, proved the existence of different conformations (see above). However, time-resolved fluorescence measurements of these enzyme-substrate complexes did not reveal specific lifetime patterns for the 'in' and 'out' conformations of the flavin. The reason for this is found in the active-site structure of PHBH and the geometric differences between the two conformations. Close to the flavin (< 10 Å) various quenching sites are located, among which three tyrosines, of which two of these are involved in substrate activation (Tyr201 and Tyr385), and the third one (Tyr222) in flavin motion. As a consequence, the fluorescence decay of wild-type PHBH is rather complex and is best described by five lifetime components (Fig. 22.8). Unfortunately, in the 'in' and 'out' conformation, the shortest distance between these tyrosines and the isoalloxazine ring differs only little on average, so that they are all likely to contribute to fluorescence quenching in both conformations. An additional complication is the fact that binding of the aromatic substrates itself causes efficient

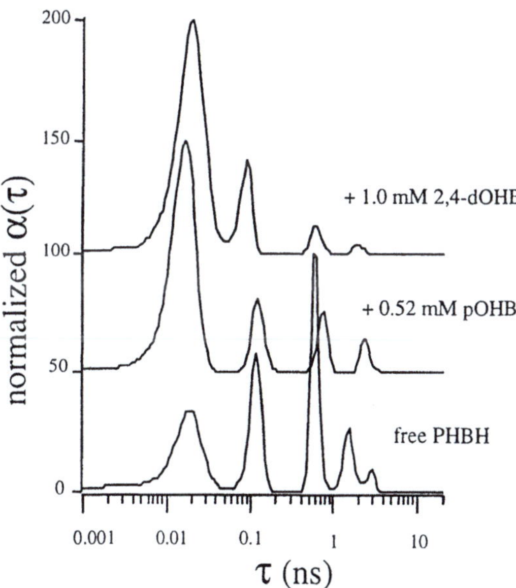

Fig. 22.8. Normalized fluorescence lifetime distributions of PHBH, uncomplexed as well as saturated with *p*-hydroxybenzoate (pOHB) and 2,4-dihydroxybenzoate (2,4-dOHB) in 50 mmol/l potassium phosphate buffer (pH 7.5) at 293 K. For clarity, the lifetime distributions were scaled to 100, and vertical offset of 50 was used

fluorescence quenching (Fig. 22.8), which is most probably due to a similar electron transfer mechanism as described above for the tyrosine-containing reductases. As a result, the presence of the 'in' or 'out' conformation is only reflected in the relative contributions of the longer lifetimes, which are small compared to the amplitude of the lifetime originating from quenching by the substrate. In the uncomplexed state, the different conformations of PHBH cannot be visualized either: in substrate-free PHBH, the flavin was shown to be flexible [65]. As time-resolved fluorescence data are obtained from a large number of molecules, the fluorescence lifetime pattern of uncomplexed PHBH is an ensemble average in which the patterns corresponding to pure 'in' and pure 'out' are mixed. New methods, however, have now been developed to overcome this limitation.

22.7
From Ensembles to Single Molecules

Until recently it was only possible to perform fluorescence measurements on ensembles of molecules. Consequently, only data were obtained which show the average properties of the investigated systems. This imposes an intrinsic barrier to studying protein conformational dynamics, where the deviation from average itself is the subject of investigation. Recent progress in the field of opto-electronics has made it possible to detect individual molecules. Single-molecule detection (SMD) allows one to examine the features of individual members of heterogeneous populations of molecule. In principle, by registering the physical and chemical properties of a large number of single molecules, and by analyzing the time trajectories of the observed properties, a full picture of the population including its subpopulations and the distributions of properties can be acquired. Static disorder, which is the intrinsic heterogeneity of a specific property of molecules of the same (genetic) population irrespective of time, and dynamic disorder, which shows the time-dependent fluctuations of this property of an individual molecule, can be detected and distinguished. SMD is especially advantageous for investigating fluctuating systems under equilibrium conditions. However, SMD can also provide information on, for instance, dynamic properties and reaction pathways of molecules in non-equilibrated systems. Transient molecular structures such as protein folding intermediates, and transient molecular interactions such as the binding and release of substrates and products during enzymatic catalysis, can be directly visualized.

Single-molecule experiments make use of various detection methods. The patch clamp technique (for a review see [86]), atomic force microscopy [87, 88], scanning tunneling microscopy [89, 90], and enzymatic assays of highly diluted systems [91] have been successfully applied to gain information on individual molecules. Most of the current SMD methods, however, are based on the fluorescence detection of single molecules tagged with fluorescent labels (for excellent reviews see [92–94]). Although a wide range of techniques has been developed for answering specific research questions, these techniques are based on similar principles. Upon selecting an extremely small volume element (femtoliter) with a focussed laser beam that repetitively excites the molecule within the confocal spot, a burst of fluorescence photons from a single molecule can be generated. For each single molecule, this burst can be analyzed for properties such as its brightness, duration, repetition frequency ('on' and 'off' times), anisotropy, and spectrum. The different analyses can provide information on a variety of molecular properties including size, fluorescence lifetime, constants of diffusion and rotation, and concentration. In addition, molecular interactions such as binding, co-localization, and even enzymatic turnover can be visualized.

Fluorescence SMD techniques have been developed for freely diffusing molecules as well as for surface-bound systems. The main advantage of the latter is the increased time-period for which one and the same molecule can be investigated. Molecules, which are either attached to a surface via a non-invasive linker [95–98] or entrapped in a gel system [99], can be spotted with a fluorescence microscope and continuously observed until photobleaching irreversibly destroys the molecules under study. Besides single-molecule burst analyses and single-pair fluorescence resonance energy transfer observations (see below), which can be applied to both surface-bound and freely diffusing molecules, fluorescence SMD methods have been developed that specifically employ the surface attachment. For instance, surface plasmon resonance fluorescence microscopy was used to study motions of the fluorescently labeled motor protein coupled to the ATPase reaction, which was bound to metal surfaces [100]. An elegant technique to study conformational dynamics of macromolecules such as polymers and DNA forms single-particle tracking, in which one end of the polymer is tethered to an immobile surface, and a label attached to the other end is then tracked by video-enhanced optical spectroscopy [101].

Fluorescence correlation spectroscopy (FCS) is probably the most famous fluorescence SMD technique for freely diffusing molecules. In FCS, of which the theoretical concept dates back to the 1970s [102, 103], the fluctuations in fluorescence intensity of single fluorophores in time are analyzed via the autocorrelation function [104, 105]. This technique is particularly suitable for retrieving information on diffusion properties and concentrations. By the recent development of two-color FCS, this technique is rapidly expanding as a tool for molecular interaction and co-localization studies [106, 107]. As a considerable amount of nanosecond information is disregarded in the analysis via the autocorrelation function, FCS seems less suited for studying conformational dynamics on that time-scale.

In 1996, Edman et al. [108] were the first to report on a conformational transition at the single-molecule level. In the latter study, the fluorescence decay of a single molecule was determined from integrating data over the entire time-window of observation, thereby losing essential kinetic information. Since then, sensitive techniques have been developed to study conformational fluctuations of biomolecules free in solution, including kinetics. A very powerful approach in this respect is single-pair fluorescence resonance energy transfer (spFRET; for a beautiful overview see [92]). In spFRET, the resonance energy transfer efficiency from the donor label to the acceptor label of a doubly labeled system is used to distinguish subpopulations with different intramolecular distances or orientations [109]. Another highly promising technique for single-molecule dynamics is BIFL (Burst-Integrated Fluorescence Lifetime) spectroscopy, which allows simultaneous registration of fluorescence intensity, lifetime and anisotropy for freely diffusing molecules in a multi-dimensional way [110–112]. This method combines the advantages of time-resolved and fluorescence correlation spectroscopy. Molecular conformations can then be monitored directly through the fluorescence lifetimes, and concomitantly a statistical analysis of the kinetic and physical properties of the fluorophore can be performed.

The progress in SMD has currently resulted in the first reports on single-molecule enzymatics (for a minireview see [113]). Turnover numbers of individual enzyme molecules have now been determined for a few motor proteins [114–118] and a nuclease [119]. In principle, it is possible to determine catalytic rates from the changes in fluorescence accompanying catalytic conversion and departing from the reactant as well as from the enzyme itself. The latter approach has the advantage that any conformational fluctuations of the individual enzyme molecule itself may be directly linked to its catalytic action.

In principle, single-molecule flavin fluorescence detection could be the perfect method to relate the conformational dynamics of flavoenzymes to the biochemical function. In practice, however, SMD of flavins is limited by the molecule's physical and photochemical characteristics. Flavin compounds are known to participate in a variety of photochemical reactions (for a review see [13]). Well-known examples are the photolysis of the ribityl side chain and the photoreduction by electron-donating compounds such as amines and EDTA. Besides the redox-properties of the flavin, this complication originates from the high triplet quantum yield of flavins (> 0.5 for FMN, 0.15 for FAD) and the long lifetime of the triplet state, which strongly promotes photoreactions and irreversible photobleaching. These long-lived triplet states are easily involved in photo-oxidation processes, in which the triplet molecule reacts with ground-state molecular oxygen to yield the highly reactive singlet oxygen. Another route for photobleaching is caused by triplet-triplet absorption, which results in very reactive higher excited states. In addition to this photochemical complication in SMD experiments, intersystem crossing to the triplet state renders the molecule invisible until it relaxes back to the ground-state. This leads to a serious deterioration of the signal-to-noise ratio. The above-mentioned complications were indeed observed in FCS experiments on free FMN and – albeit to a lesser extent due to its less unfavorable triplet properties – FAD [120]. In flavoenzymes, the protein environment generally reduces the quantum yield and the lifetime of the triplet state, thereby protecting the flavin from undesirable photochemical and photophysical reactions. Preliminary FCS studies on a few flavoenzymes (riboflavin-bound blue fluorescence protein, lipoamide dehydrogenase) have nevertheless shown that, upon applying higher intensities, these problems may appear in flavoenzymes as well, and that they may even lead to photodissociation of the flavin chromophore [120] (P. A. W. van den Berg, unpublished results).

However, a spectacular breakthrough in SMD of flavoenzymes was recently reported for cholesterol oxidase [99]. Enzymatic turnovers of single cholesterol oxidase molecules were observed in real time by monitoring flavin fluorescence. The cholesterol oxidase was confined in an agarose gel, which prevented translational diffusion of the enzyme, but allowed free exchange of substrates and products. Flavin fluorescence images of these single immobilized enzymes were used to monitor the enzyme's redox state. As in the reduced state, the flavin is 'non-fluorescent' compared with the oxidized flavin, the successive 'on-times' and 'off-times' reflect the enzyme's redox cycle and thus enzymatic turnover. Statistical analysis of multiple single-molecule trajectories revealed a detailed picture of the fluctuations in terms of reaction rate. Enzymatic turnover was found to be somewhat dependent on its previous turnovers. This memory effect was attributed to slow conformational changes of the enzyme; a hypothesis supported by the observation of spontaneous spectral fluctuations of the enzyme-bound FAD on the same time-scale [99].

22.8
Prospects for Studying Conformational Dynamics by Flavin Fluorescence Detection

The spectacular developments in single-molecule detection undoubtedly predict a bright future for a ubiquitous application of this approach in exploring conformational dynamics. Maturation of the various techniques and the possible development of new methods will result in a versatile toolbox for studying dynamics in different biomolecular systems. Although the emphasis may be put on single-molecule studies, a substantial role will remain for fluorescence techniques based on ensemble detection. In addition to being a reliable source of information on average properties of biomolecules, these methods continue to be indispensable in time domains that are not (yet) accessible to single-molecule detection. Particularly in the ultra-fast time domain ((sub)picosecond), ensemble detected time-resolved fluorescence techniques are just beginning to shed light on dynamic, electronic, and physical events. In addition, it may prove difficult – if not impossible – to extend the usage of SMD to all biomolecular systems. Systems which are subject to rapid photochemical processes and photodegradation are still difficult to tackle with fluorescence single-molecule detection. The same holds true for molecules having a low fluorescence quantum yield or extinction coefficient, and systems with a high triplet quantum yield. Retrieving information on dynamic events from an ensemble of molecules that underwent only few excitations may then be preferable to interpreting highly noisy SMD data or to registering merely the photochemical death of a single molecule.

In this respect, flavoenzymes are not the easiest systems. Notwithstanding the impressive results of Lu et al. [99] described above, the photochemical and photophysical features of the flavin cofactor will complicate the general usage of single-molecule flavin fluorescence detection in this class of enzymes. Besides the intrinsic high triplet quantum yield of the flavin, and the extremely low flavin fluorescence quantum yield of many flavoproteins, the dissociability of the flavin cofactor can be a major obstacle for single-molecule detection (see above). The further development of SMD techniques based on two-photon excitation will provide an important tool to overcome the problem of rapid photodestruction of the flavin. Optimization of the signal to noise ratio by selectively focussing on a single surface-bound flavoenzyme may prove essential for investigating this class of enzymes by flavin fluorescence SMD. The method of sample preparation used by Lu et al. [99], in which the protein was rapidly confined in an agarose gel, allows registration of the single-molecule trajectory over a maximum period of time without losing the properties of the system in solution such as enzymatic activity. Scanning nearfield optical microscopy on such spin-coated systems might generate interesting results on flavoenzymes as well. Dissociation of the chromophore may be circumvented by using flavoenzymes with a covalently attached flavin cofactor.

Another approach to overcome some of the above-mentioned problems is to use a second, external fluorescent label. While direct excitation of the label yields information on global dynamic processes of single flavoproteins whose flavin fluorescence is severely quenched, spFRET between the flavin and the label may provide more detailed insight in the protein dynamics. An interesting development in this respect would be the construction of flavoenzyme-GFP adducts. Due to the overlap in the spectral cross-sections of flavin and GFP, the latter fluorescence emission can be used as a sensitive antenna for the redox state, and thus the enzymatic activity, of single flavoenzymes. Considering GFP as the donor, spFRET to the flavin will only affect the donor fluorescence when the flavin is in the oxidized state; in the reduced state of the flavin, the spectral cross-sections no longer overlap. Single-molecule detection and particularly single-pair fluorescence resonance energy transfer can thus provide a detailed picture of the conformational dynamics of flavoenzymes and its relation to the catalytic function.

Acknowledgement. We greatly appreciate the assistance of the following persons during the preparation of this chapter: Willem van Berkel for helpful suggestions, Scott Mulrooney and Charles Williams Jr for collaboration on thioredoxin reductase, Adrian Arakaki for communicating unpublished results of ferredoxin NADP$^+$ reductase, Nina Visser and Adrie Westphal for their help with preparation of the figures.

References

1. Beechem JM, Brand L (1985) Time-resolved fluorescence of proteins. Annu Rev Biochem 54:43–71
2. Eftink MR (1991) Fluorescence techniques for studying protein structure. Meth Biochem Anal 35:127–205
3. Demchenko AP (1992) Fluorescence and dynamics in proteins. In: Lakowicz JR (ed) Topics in fluorescence spectroscopy, vol 3. Plenum Press, New York, pp 65–111
4. Millar DP (1996) Time-resolved fluorescence spectroscopy. Curr Opin Struct Biol 6:637–642
5. Lakowicz JR (1999) Principles of fluorescence spectroscopy, 2nd edn. Plenum Press, New York
6. Ruggiero AJ, Todd DC, Fleming GR (1990) Subpicosecond anisotropy studies of tryptophan in water. J Am Chem Soc 112:1003–1014
7. Tsien RY (1998) The green fluorescent protein. Ann Rev Biochem 67:509–544
8. Clegg RM (1996) Fluorescence resonance energy transfer. In: Wang XF, Herman B (eds) Fluorescence imaging spectroscopy and microscopy. Wiley, New York, pp 179–252
9. Müller F (1991) Chemistry and biochemistry of flavoenzymes, vols I, II, III. CRC Press, Boca Raton
10. Weber G (1950) Fluorescence of riboflavin and flavin-adenine dinucleotide. Biochem J 47:114–121
11. Spencer RD, Weber G (1972) Thermodynamics and kinetics of the intramolecular complex in flavin-adenine dinucleotide. In: Åkeson Å, Ehrenberg A (eds) Structure and function of oxidation reduction enzymes. Pergamon, Oxford, pp 393–399

12. Visser AJWG (1984) Kinetics of stacking interactions in flavin adenine dinucleotide from time-resolved flavin fluorescence. Photochem Photobiol 40:703–706

13. Heelis PF (1991) The photochemistry of flavins. In: Müller F (ed) Chemistry and biochemistry of flavoenzymes, vol I. CRC Press, Boca Raton, pp 171–200

14. Zheng YJ, Ornstein RL (1996) A theoretical study of the structures of flavin in different oxidation and protonation states. J Am Chem Soc 118:9402–9408

15. Platenkamp RJ, Palmer MH, Visser AJWG (1980) Ab initio molecular orbital studies of flavin radicals and the lowest triplet state of isoalloxazine. J Mol Struct 67:45–64

16. Platenkamp RJ, Palmer MH, Visser AJWG (1987) Ab initio molecular orbital studies of closed shell flavins. Eur Biophys J 14:393–402

17. Johansson LB-Å, Davidsson Å, Lindblom G, Razi Naqvi K (1979) Electronic transitions in the isoalloxazine ring and orientation of flavins in mode membranes studied by polarized light spectroscopy. Biochemistry 18:4249–4253

18. Bastiaens PIH, Van Hoek A, Benen JAE, Brochon JC, Visser AJWG (1992) Conformational dynamics and intersubunit energy transfer in wild-type and mutant lipoamide dehydrogenase from *Azotobacter vinelandii*. Biophys J 63:839–853

19. Visser AJWG, Van Hoek A, Visser NV, Lee Y, Ghisla S (1997) Time-resolved fluorescence study of the dissociation of FMN from the yellow fluorescence protein from *Vibrio fischeri*. Photochem Photobiol 65:570–575

20. Singer TP, McIntire WS (1984) Covalent attachment of flavin to flavoproteins: occurrence, assay and synthesis. Meth Enzymol 106:369–378

21. Decker KF (1991) Covalent flavoproteins. In: Müller F (ed) Chemistry and biochemistry of flavoenzymes, vol II. CRC Press, Boca Raton, pp 343–375

22. Karplus M, McCammon JA (1983) Dynamics of proteins: elements and function. Ann Rev Biochem 53:263–300

23. Frauenfelder H, Gratton E (1986) Protein dynamics and hydration. Meth Enzym 127:207–216

24. Frauenfelder H, Parak F, Young RD (1988) Conformational substates in proteins. Annu Rev Biophys Biophys Chem 17:451–479

25. Frauenfelder H, Alberding NA, Ansari A, Braunstein D, Cowen BR, Hong MK, Iben IET, Johnson JB, Luck S, Marden MC, Mourant JR, Ormos P, Reinisch L, Scholl R, Schulte A, Shyamsunder E, Sorensen LB, Steinbach PJ, Xie A, Young RD, Yue KT (1990) Proteins and pressure. J Phys Chem 94:1024–1037

26. Frauenfelder H, Sligar SG, Wolynes PG (1991) The energy landscapes and motions of proteins. Science 254:1598–1603

27. Krimm S, Bandekar J (1986) Vibrational spectroscopy and conformation of peptides, polypeptides, and proteins. Adv Protein Chem 38:181–360

28. Gō N, Noguti T, Nishikawa T (1983) Dynamics of a small globular protein in terms of low-frequency vibrational modes. Proc Natl Acad Sci USA 80:3696–3700

29. Brooks B, Karplus M (1983) Harmonic dynamics of proteins: normal modes and fluctuations in bovine pancreatic trypsin inhibitor. Proc Natl Acad Sci USA 80:6571–6575

30. Gerstein M, Lesk AM, Chothia C (1994) Structural mechanisms for domain movements in proteins. Biochemistry 33:6739–6749

31. Anderson BF, Baker HM, Norris GE, Rumball SV, Baker EN (1990) Apolactoferrin structure demonstrates ligand-induced conformational change in transferrins. Nature 344:784–787

32. Gerstein M, Anderson BF, Norris GE, Baker EN, Lesk AM, Chothia C (1993) Domain closure in lactoferrin. Two hinges produce a see-saw motion between alternative close-packed interfaces. J Mol Biol 234:357–372

33. Schulz GE, Muller CW, Diederichs K (1990) Induced-fit movements in adenylate kinases. J Mol Biol 213:627–630

34. Gerstein M, Schulz G, Chothia C (1993) Domain closure in adenylate kinase. Joints on either side of two helices close like neighboring fingers. J Mol Biol 229:494–501

35. Stillman TJ, Baker PJ, Britton KL, Rice DW (1993) Conformational flexibility in glutamate dehydrogenase. Role of water in substrate recognition and catalysis. J Mol Biol 234:1131–1139

36. Remington S, Wiegand G, Huber R (1982) Crystallographic refinement and atomic models of two different forms of citrate synthase at 2.7 and 1.7 Å resolution. J Mol Biol 158:111–152

37. Lesk AM, Chothia C (1984) Mechanisms of domain closure in proteins. J Mol Biol 174:175–191

38. Eklund H, Samama JP, Wallén L, Brändén CI, Åkeson Å, Jones TA (1981) Structure of a triclinic ternary complex of horse liver alcohol dehydrogenase at 2.9 Å resolution. J Mol Biol 146:561–587

39. Colonna-Cesari F, Perahia D, Karplus M, Eklund H, Brāndén CI, Tapia O (1986) Interdomain motion in liver alcohol dehydrogenase. Structural and energetic analysis of the hinge bending mode. J Biol Chem 261:15,273–15,280

40. Faber HR, Matthews BW (1990) A mutant T4 lysozyme displays five different crystal conformations. Nature 348:263–266

41. Dixon MM, Nicholson H, Shewchuk L, Baase WA, Matthews BW (1992) Structure of a hinge-bending bacteriophage T4 lysozyme mutant, Ile3 → Pro. J Mol Biol 227:917–933

42. Sharff AJ, Rodseth LE, Spurlino JC, Quiocho FA (1992) Crystallographic evidence of a large ligand-induced hinge-twist motion between the two domains of the maltodextrin binding protein involved in active transport. Biochemistry 31:10,657–10,663

43. Kurzyński M (1998) A synthetic picture of intramolecular dynamics of proteins. Towards a contemporary statistical theory of biochemical processes. Prog Biophys Mol Biol 69:23–82

44. Kurzyński M (1997) Protein machine model of enzymatic reactions gated by enzyme internal dynamics. Biophys Chem 65:1–28

45. Kurzyński M (1998) Time course of reactions controlled and gated by intramolecular dynamics of proteins: predictions of the model of random walk on fractal lattices. Proc Natl Acad Sci USA 95:11,685–11,690

46. Frauenfelder H, Wolynes PG (1985) Rate theories and puzzles of hemeprotein kinetics. Science 229:337–345

47. Gavish B, Werber MM (1979) Viscosity-dependent structural fluctuations in enzyme catalysis. Biochemistry 18:1269–1275

48. Beece D, Eisenstein L, Frauenfelder H, Good D, Marden MC, Reinisch L, Reynolds AH, Sorensen LB, Yue KT (1980) Solvent viscosity and protein dynamics. Biochemistry 19:5147–5157

49. Williams CH Jr (1992) Lipoamide dehydrogenase, glutathione reductase, thioredoxin reductase, and mercuric ion reductase – a family of flavoenzyme transhydrogenases. In: Müller F (ed) Chemistry and biochemistry of flavoenzymes, vol III. CRC Press, Boca Raton, pp 121–211

50. Thieme R, Pai EF, Schirmer RH, Schulz GE (1981) Three-dimensional structure of glutathione reductase at 2 Å resolution. J Mol Biol 152:763–782

51. Karplus PA, Schulz GE (1987) Refined structure of glutathione reductase at 1.54 Å resolution. J Mol Biol 195:701–729

52. Karplus PA, Schulz GE (1989) Substrate binding and catalysis by glutathione reductase as derived from refined enzyme: substrate crystal structures at 2 Å resolution. J Mol Biol 210:163–180

53. Ermler U, Schulz GE (1991) The three-dimensional structure of glutathione reductase from Escherichia coli at 3.0 Å resolution. Proteins Struct Funct Genet 9:174–179

54. Mittl PRE, Schulz GE (1994) Structure of glutathione reductase from Escherichia coli at 1.86 Å resolution: comparison with the enzyme from human erythrocytes. Protein Sci 3:799–809

55. Pai EF, Schulz GE (1983) The catalytic mechanism of glutathione reductase as derived from X-ray diffraction analyses of reaction intermediates. J Biol Chem 258:1752–1757

56. Williams CH Jr (1995) Mechanism and structure of thioredoxin reductase from Escherichia coli. FASEB J 9:1267–1276

57. Waksman G, Krishna TSR, Williams CH Jr, Kuriyan J (1994) Crystal structure of *Escherichia coli* thioredoxin reductase refined at 2 Å resolution. Implications for a large conformational change during catalysis. J Mol Biol 236:800–816

58. Kuriyan J, Krishna TSR, Wong L, Guenther B, Pahler A, Williams CH Jr, Model P (1991) Convergent evolution of similar function in two structurally divergent enzymes. Nature 352:172–174

59. Mulrooney SB, Williams CH Jr (1997) Evidence for two conformational states of thioredoxin reductase from *Escherichia coli*: use of intrinsic and extrinsic quenchers of flavin fluorescence as probes to observe domain rotation. Protein Sci 6:2188–2195

60. Schreuder HA, Mattevi A, Obmolova G, Kalk KH, Hol WGJ, Van der Bolt FJT, Van Berkel WJH (1994) Crystal structures of wild-type *p*-hydroxybenzoate hydroxylase complexed with 4-aminobenzoate, 2,4-dihydroxybenzoate and 2-hydroxy-4-aminobenzoate and of the Tyr222Ala mutant, complexed with 2-hydroxy-4-aminobenzoate. Evidence for a proton channel and a new binding mode of the flavin ring. Biochemistry 33:10,161–10,170

61. Gatti DL, Palfey BA, Lah MS, Entsch B, Massey V, Ballou DP, Ludwig ML (1994) The mobile flavin of 4-OH benzoate hydroxylase. Science 266:110–114

62. Entsch B, Van Berkel WJH (1995) Structure and mechanism of para-hydroxybenzoate hydroxylase. FASEB J 9:476–483

63. Schreuder HA, Prick PAJ, Wieringa RK, Vriend G, Wilson KS, Hol WGJ, Drenth J (1989) Crystal structure of the *p*-hydroxybenzoate hydroxylase-substrate complex refined at 1.9 Å resolution. J Mol Biol 208:679–696

64. Van der Bolt FJT, Vervoort J, Van Berkel WJH (1996) Flavin motion in *p*-hydroxybenzoate hydroxylase. Substrate and effector specificity of the Tyr222Ala mutant. Eur J Biochem 237:592–600

65. Eppink MHM, Van Berkel WJH, Tepliakov A, Schreuder HA (1999) Crystal structures of unactivated *p*-hydroxybenzoate hydroxylase. In: Ghisla S, Kroneck P, Macheroux, P, Sund H (eds) Flavins and flavoproteins 1999. Agency for Scientific Publications, Berlin (in press)

66. Tanaka F, Tamai N, Yamazaki I (1989) Picosecond-resolved fluorescence spectra of D-amino-acid oxidase. A new fluorescent species of the coenzyme. Biochemistry 28:4259–4262

67. Tanaka F, Tamai N, Yamazaki I, Nakashima N, Yoshihara K (1989) Temperature-induced changes in the coenzyme environment of D-amino acid oxidase revealed by the multiple decays of FAD fluorescence. Biophys J 28:901–909

68. Bastiaens PIH, Van Hoek A, Wolkers WF, Brochon JC, Visser AJWG (1992) Comparison of the dynamical structures of lipoamide dehydrogenase and glutathione reductase by time-resolved polarized flavin fluorescence. Biochemistry 31:7050–7060

69. Van den Berg PAW, Van Hoek A, Walentas CD, Perham RN, Visser AJWG (1998) Flavin fluorescence dynamics and photoinduced electron transfer in *Escherichia coli* glutathione reductase. Biophys J 74:2046–2058

70. Visser AJWG, Van den Berg PAW, Visser NV, Van Hoek A, Van den Burg HA, Parsonage D, Claiborne A (1998) Time-resolved fluorescence of flavin adenine dinucleotide in wild-type and mutant NADH peroxidase. Elucidation of quenching sites and discovery of a new fluorescence depolarization mechanism. J Phys Chem B 102:10,431–10,439

71. Brunner K, Tortschanoff A, Hemmens B, Andrew PJ, Mayer B, Kungl AJ (1998) Sensitivity of flavin fluorescence dynamics in neuronal nitric oxide to cofactor-induced conformational changes and dimerization. Biochemistry 37:17,545–17,553

72. Mataga N, Chrosrowjan H, Shibata Y, Tanaka F (1998) Ultrafast fluorescence quenching dynamics of flavin chromophores in protein nanospace. J Phys Chem B 102:7081–7084

73. Boteva R, Visser AJWG, Filippi B, Vriend G, Veenhuis M, Van der Klei IJ (1999) Conformational transitions accompanying oligomerization of yeast alcohol oxidase, a peroxisomal flavoenzyme. Biochemistry 38:5034–5044

74. Visser AJWG, Grande HJ, Veeger C (1980) Rapid relaxation processes in pig heart lipoamide dehydrogenase revealed by subnanosecond resolved fluorometry. Biophys Chem 12:35–49

75. De Kok A, Visser AJWG (1987) Flavin binding site differences between lipoamide dehydrogenase and glutathione dehydrogenase as revealed by static and time-resolved flavin fluorescence. FEBS Lett 218:135–138

76. Bajzer Z, Prendergast FG (1993) A model for multiexponential tryptophan fluorescence intensity decay in proteins. Biophys J 65:2313–2323

77. Visser AJWG, Van Hoek A, Kulinski T, Le Gall J (1987) Time-resolved fluorescence studies of flavodoxin. Demonstration of picosecond fluorescence lifetimes of FMN in *Desulfovibrio* flavodoxins. FEBS Lett 224:406–410

78. Karen A, Sawada MT, Tanaka F, Mataga N (1987) Dynamics of excited flavoproteins-picosecond laser photolysis studies. Photochem Photobiol 45:49–53

79. Rehm D, Weller A (1970) Kinetics of fluorescence quenching by electron and H-atom transfer. Israel J Chem, 21st Farkas Memorial Symp 8:259–271

80. Marcus RA, Sutin N (1985) Electron transfers in chemistry and biology. Biochim Biophys Acta 811:265–322

81. Serre L, Vellieux FMD, Medina M, Gomez-Moreno C, Fontecilla-Camps JC, Frey M (1996) X-ray structure of the ferredoxin: $NADP^+$ reductase from the cyanobacterium *Anabaena* PCC 7119 at 1.8 Å resolution, and crystallographic studies of $NADP^+$ binding at 2.25 Å resolution. J Mol Biol 263:20–39

82. Stehle T, Ahmed SA, Claiborne A, Schulz GE (1991) Structure of NADH peroxidase from *Streptococcus faecalis* 10C1 refined at 2.16 Å resolution. J Mol Biol 221:1325–1344

83. Stehle T, Claiborne A, Schulz GE (1993) NADH binding site and catalysis of NADH peroxidase. Eur J Biochem 211:221–226

84. Wang PF, Veine DM, Ahn SH, Williams CH Jr (1996) A stable mixed disulfide between thioredoxin reductase and its substrate, thioredoxin: preparation and characterization. Biochemistry 35:4812–4819

85. Veine DM, Mulrooney SB, Wang PF, Williams CH Jr (1998) Formation and properties of mixed disulfides between thioredoxin reductase from *Escherichia coli* and thioredoxin: evidence that cysteine-138 functions to initiate dithiol-disulfide interchange and to accept the reducing equivalent from reduced flavin. Protein Sci 7:1441–1450

86. Sakmann B, Neher E (eds) (1995) Single-channel recording, 2nd edn. Plenum, New York

87. Radmacher M, Fritz M, Hansma HG, Hansma PK (1994) Direct observation of enzyme activity with the atomic force microscope. Science 265:1577–1579

88. Rees WA, Keller RW, Vesenka JP, Yang G, Bustamante C (1993) Evidence of DNA bending in transcription complexes imaged by scanning force microscopy. Science 260:1646–1649

89. Baro AM, Miranda R, Alaman J, Garcia N, Binnig G, Röhrer H, Gerber C, Carrascosa JL (1985) Determination of surface topography of biological specimens at high resolution by scanning tunneling microscopy. Nature: 315:253–254

90. Binnig G, Quate CF, Gerber C (1986) Atomic force microscope. Phys Rev Lett 56:930–933

91. Xue QF, Yeung ES (1995). Differences in the chemical reactivity of individual molecules of an enzyme. Nature 373:681–683

92. Weiss S (1999) Fluorescence spectroscopy of single biomolecules. Science 283:1676–1683

93. Xie XS, Trautman JK (1999) Optical studies of single molecules at room temperature. Annu Rev Phys Chem 49:441–480

94. Nie SM, Zare RN (1997) Optical detection of single molecules. Annu Rev Biophys Biomol Struct 26:567–596

95. Macklin JJ, Trautman JK, Harris TD, Brus LE (1996) Imaging and time-resolved spectroscopy of single molecules at an interface. Science 272:255–258

96. Ha T, Enderle Th, Chemla DS, Selvin PR, Weiss S (1996) Single molecule dynamics studied by polarization modulation. Phys Rev Lett 77:3979–3982

97. Ha T, Glass J, Enderle T, Chemla DS, Weiss S (1998) Hindered rotational diffusion and rotational jumps of single molecules. Phys Rev Lett 80:2093–2096

98. Wennmalm S, Edman L, Rigler R (1997) Conformational fluctuations in single DNA molecules. Proc Natl Acad Sci USA 94:10,641–10,646

99. Lu HP, Xun L, Xie XS (1998) Single-molecule enzymatic dynamics. Science 282:1877–1882

100. Yokota H, Saito K, Yanagida T (1998) Single molecule imaging of fluorescently labeled proteins on metal by surface plasmons in aqueous solution. Phys Rev Lett 80:4606–4609
101. Qian H, Elson EL (1999) Quantitative study of polymer conformation and dynamics by single-particle tracking. Biophys J 76:1598–1605
102. Elson EL, Magde D (1974) Fluorescence correlation spectroscopy. I. Conceptual basis and theory. Biopolymers 13:1–27
103. Ehrenberg M, Rigler R (1974) Rotational Brownian motion and fluorescence intensity fluctuations. Chem Phys 4:390–401
104. Rigler R, Mets Ü, Widengren J, Kask P (1993) Fluorescence correlation spectroscopy with high count rate and low background: analysis of translational diffusion. Eur Biophys J 22:169–175
105. Eigen M, Rigler R (1994) Sorting single molecules: application to diagnostics and evolutionary biotechnology. Proc Natl Acad Sci USA 91:5740–5747
106. Schwille P, Meyer-Almes FJ, Rigler R (1997) Dual-color fluorescence cross-correlation spectroscopy for multicomponent diffusional analysis in solution. Biophys J 72:1878–1886
107. Kettling U, Koltermann A, Schwille P, Eigen M (1998) Real-time enzyme kinetics monitored by dual-color fluorescence cross-correlation spectroscopy. Proc Natl Acad Sci USA 95:1416–1420
108. Edman L, Mets U, Rigler R (1996) Conformational transitions monitored for single molecules in solution. Proc Natl Acad Sci USA 93:6710–6715
109. Deniz AA, Dahan M, Grunwell JR, Ha T, Faulhaber AE, Chemla DS, Weiss S, Schultz PG (1999) Single-pair fluorescence energy transfer on freely diffusing molecules: observation of Förster distance dependence and subpopulations. Proc Natl Acad Sci USA 96:3670–3675
110. Eggeling C, Fries JR, Brand L, Günther R, Seidel CAM (1998) Monitoring conformational dynamics of a single molecule by selective fluorescence spectroscopy. Proc Natl Acad Sci USA 95:1556–1561
111. Fries R, Brand L, Eggeling C, Köllner M, Seidel CAM (1998) Quantitative identification of different single molecules by selective time-resolved confocal fluorescence spectroscopy. J Phys Chem A 102:6601–6613
112. Schaffer J, Volkmer A, Eggeling C, Subramaniam V, Striker G, Seidel CAM (1999) Identification of single molecules in aqueous solution by time-resolved fluorescence anisotropy. J Phys Chem A 103:331–336
113. Xie XS, Lu HP (1998) Single-molecule enzymology. J Biol Chem 274:15,967–15,970
114. Funatsu T, Harada Y, Tokunaga M, Saito K, Yanagida T (1995) Imaging of single fluorescent molecules and individual ATP turnovers by single myosin molecules in aqueous solution. Nature 374:555–559
115. Vale RD, Funatsu T, Pierce DW, Romberg L, Harada Y, Yanagida T (1996) Direct observation of single kinesin molecules moving along microtubules. Nature 380:451–453
116. Noji H, Yasuda R, Yoshida M, Kinosita K Jr (1997) Direct observation of the rotation of F_1-ATPase. Nature 386:299–302
117. Yasuda R, Moji H, Kinoshita K, Yoshida M (1998) F_1-ATPase is a highly efficient motor that rotates with discrete 120° steps. Cell 93:1117–1124
118. Ishijima A, Kojima H, Funatsu T, Tokunaga M, Higuchi H, Tanaka H, Yanagida T (1998) Simultaneous observation of individual ATPase and mechanical events by a single myosin molecule during interaction with actin. Cell 92:161–171
119. Ha T, Ting AY, Liang J, Caldwell WB, Deniz AA, Chemla DS, Schultz PG, Weiss S (1999) Single-molecule fluorescence spectroscopy of enzyme conformational dynamics and cleavage mechanism. Proc Natl Acad Sci USA 96:893–898
120. Visser AJWG, Van den Berg PAW, Hink MA, Petushkov VN (2000) Fluorescence correlation spectroscopy of flavins and flavoproteins. In: Elson EL, Rigler R (eds) Fluorescence correlation spectroscopy. Theory and application. Springer, Berlin Heidelberg New York (in press)

Subject Index

If you have any concerns about our products,
you can contact us on
ProductSafety@springernature.com

In case Publisher is established outside the EU,
the EU authorized representative is:
**Springer Nature Customer Service Center GmbH
Europaplatz 3, 69115 Heidelberg, Germany**

Printed by Libri Plureos GmbH
in Hamburg, Germany